Introduction to
High-Energy Heavy-Ion Collisions

Introduction to
High-Energy Heavy-Ion Collisions

Cheuk-Yin Wong

Oak Ridge National Laboratory
USA

World Scientific
Singapore • New Jersey • London • Hong Kong

Published by

World Scientific Publishing Co. Pte. Ltd.

5 Toh Tuck Link, Singapore 596224

USA office: 27 Warren Street, Suite 401-402, Hackensack, NJ 07601

UK office: 57 Shelton Street, Covent Garden, London WC2H 9HE

Library of Congress Cataloging-in-Publication Data
Wong, Cheuk-Yin.
    Introduction to high-energy heavy-ion collisions / Cheuk-Yin Wong.
    p. cm.
    Includes index.
    ISBN-13 978-981-02-0263-7 -- ISBN-10 981-02-0263-6
    ISBN-13 978-981-02-0264-4 (pbk) -- ISBN-10 981-02-0264-4 (pbk)
    1. Heavy ion collisions. I. Title.
    QC794.8.H8W65   1193
    539.7'57--dc20   93-28743
                                                              CIP

British Library Cataloguing-in-Publication Data
A catalogue record for this book is available from the British Library.

To my parents, my teachers, and my wife.

Preface

The study of high-energy heavy-ion collisions is an emerging field of research. Being at the interface between particle physics and nuclear physics, it carries the burden that many topics which are familiar to workers in one branch may not be familiar to workers in the other branch. Furthermore, students and researchers entering into this field need to learn a vast set of material which are scattered in many different areas and textbooks. The task is made all the more difficult because many of the contemporary problems in this field fall within the realm of nonperturbative quantum chromodynamics. The lack of nonperturbative solutions of quantum chromodynamics for these problems makes it inherently hard to discuss many topics with theoretical lucidity. The absence of definitive solutions also brings with it a proliferation of models. The same physical phenomenon is often described in many different ways with different concepts and modes of operation.

This book is written for students and new-comers who are entering the fascinating field of high-energy heavy-ion collisions. The objective is to present an elementary introduction to the major topics which one encounters in the literature. The emphasis is on the acquisition of basic tools and concepts. Accordingly, the pace of description is deliberate and unhurried, and the mathematical derivations are given as fully as possible, to show the origins of how the different concepts and models arise.

For the purpose of further assisting the readers, "supplements" and "exercises" have been included throughout the book, to give detailed descriptions of new concepts and detailed derivations of mathematical results. They are described as fully and as simply as possible so that the readers can follow the logical and mathematical steps without much difficulty. The supplements and exercises may appear too detailed to the experts and may be by-passed by those who are familiar with their contents. They are nevertheless included here in the hope that a thorough knowledge of the origin of the concepts presented in this book will assist the readers for further explorations on their own, using for example the excellent and extensive literature in the series of Quark Matter meetings through the years:

1. *Quark Matter '80*, Proceedings of the Workshop on Future Relativistic Heavy Ion Experiments, GSI, Darmstadt, Germany, 1980.
2. *Quark Matter '82, Quark Matter Formation and Heavy Ion Collisions*, Proceedings of the Bielefeld Workshop at Bielefeld, Germany, edited by M. Jacob and H. Satz, World Scientific Publishing Company, (1982).
3. *Quark Matter '83*, Proceedings of the Third International Conference on Ultra-Relativistic Nucleus-Nucleus Collisions at Brookhaven, New York, U. S. A.,

edited by T. W. Ludlam, and H. E. Wegener, published in Nucl. Phys. A418, (1984).

4. *Quark Matter '84*, Proceedings of the Fourth International Conference on Ultra-Relativistic Nucleus-Nucleus Collisions at Helsinki, Finland, edited by K. Kajantie, published by Springer-Verlag, Berlin, (1984).

5. *Quark Matter '86*, Proceedings of the Fifth International Conference on Ultra-Relativistic Nucleus-Nucleus Collisions at Asilomar, California, U. S. A., edited by L. S. Schroeder, and M. Gyulassy, published in Nucl. Phys. A461, (1987).

6. *Quark Matter '87*, Proceedings of the Sixth International Conference on Ultra-Relativistic Nucleus-Nucleus Collisions at Asilomar, California, U. S. A., edited by H. Satz, H. J. Specht, and R. Stock, published in Zeit. Physik C38, (1988).

7. *Quark Matter '88*, Proceedings of the Seventh International Conference on Ultra-Relativistic Nucleus-Nucleus Collisions at Lenox, Massachusetts, U. S. A., edited by G. Baym, P. Braun-Munzinger and S. Nagamiya, published in Nucl. Phys. A498, (1991).

8. *Quark Matter '90*, Proceedings of the Eighth International Conference on Ultra-Relativistic Nucleus-Nucleus Collisions at Menton, France, edited by J. P. Blaizot, C. Gerschel, B. Pire, and A. Romana, published in Nucl. Phys. A525, (1988).

9. *Quark Matter '91*, Proceedings of the Ninth International Conference on Ultra-Relativistic Nucleus-Nucleus Collisions at Gatlinburg, Tennessee, U. S. A., edited by T. C. Awes, F. E. Obenshain, F. Plasil, M. R. Strayer, and C. Y. Wong, published in Nucl. Phys. A544 (1992).

10. *Quark Matter '93*, Proceedings of the Tenth International Conference on Ultra-Relativistic Nucleus-Nucleus Collisions at Borlänge, Sweden, edited by E. Stenlund, H.-Å. Gustafsson, A. Oskarsson, and I. Otterlund, published in Nucl. Phys. A566 (1994).

This book has been written on such a level that the readers are assumed to have some familiarity with quantum mechanics and statistical mechanics. Because of the diverse nature of the subject matter, a few chapters may require some simple knowledge of quantum field theory. For those readers with a more focussed interest, the material can be subdivided so that many chapters can be independently studied. The table of contents in the beginning and the summary chapter at the end provide a guide to the contents of the book.

The author would like to express his heartfelt appreciation to Dr. T. C. Awes who has read carefully the entire manuscript in minute detail, including all mathematical derivations and formulas, and has provided the author with constructive comments and numerous insightful corrections. Dr. T. C. Awes' helpfulness has brought forth much of the improvement of the book. The author wishes to thank Dr. E. C. Halbert for reading the manuscript and for her valuable suggestions. The author is indebted to Prof. J. Kapusta, Prof. Che-Ming Ko, Dr. F. Plasil, Prof. Y. Sharon, Prof. T. Udagawa, Dr. R. Vogt, Prof. Ren-Chuen Wang, Prof. Jian-Shi Wu and Mr. Stephan Haas for valuable discussions and very helpful suggestions. The author would like to thank Dr. H. Gutbrod and the WA80 Collaboration for their permission to use the color pictures of their detectors on the book cover, to show an artistic perspective of high-energy heavy-ion collisions. The author is grateful to Ms. Karen Barry for providing many TEX-typesetting macro programs. The author also wishes to thank Ms. A. Tate and Ms. E. H. Chionh for correcting many typographical errors.

While much effort has been made to eliminate misprints and errors, it is unavoidable that they may still be present. The author would appreciate receiving communications concerning corrections and improvements so that should future printing be contemplated, they may be included in later editions. The author's Internet address is wong@orph01.phy.ornl.gov and fax number is (615)-574-4745.

This research was supported by the Division of Nuclear Physics, U.S. Department of Energy under Contract No. DE-AC05-84OR21400 with Martin Marietta Energy Systems, Inc.

Cheuk-Yin Wong
Oak Ridge, 1994

Contents

Contents

18.   Signatures for the Quark-Gluon Plasma (V)

1. Introduction

What are the interesting physical phenomena which can be explored by high-energy nucleus-nucleus collisions? To answer this question, it is useful to note that one of the most important characteristics of high-energy nucleus-nucleus collisions is the large amount of energy involved. When a large fraction of this energy is deposited in a small region of space in a short duration of time, the energy density can be very high [1]. As a concrete example, we may consider the Relativistic Heavy-Ion Collider (RHIC), which is being constructed at Brookhaven National Laboratory. The Collider is designed to accelerate nuclei to an energy of about 100 GeV per nucleon. In the collision of a gold nucleus with another gold nucleus in such a collider, the energy carried by each nucleus is about 100×197 GeV, or 19.7 TeV, and the center-of-mass energy $\sqrt{s}$ is about 2×19.7 TeV, or 39.4 TeV. The magnitude of energy involved in nucleus-nucleus collisions is very large indeed. To attain even greater center-of-mass energies, there is a recent proposal to accelerate heavy-ions in the Large Hadron Collider (LHC) being planned at CERN to a center-of-mass energy $\sqrt{s}$ of about 3 TeV per nucleon, which will lead to a center-of-mass energy of about 1262 TeV for the collision of Pb on Pb [2,3].

How is the energy distributed after such a high-energy nucleus-nucleus collision? We can consider the head-on collision case in the center-of-mass system. If the collision of the two nuclei were very 'transparent' so that the colliding nuclei lose only small fractions of their energies after the head-on collision, then the two nuclei would proceed forward retaining much of their initial energies, and the energy deposited in the central region of the colliding system would be small. In that case, nothing of great consequences would be expected from high-energy nucleus-nucleus collisions. This is however not the case! The experimental data which have been accumulated in the past two decades suggest that inelastic nuclear collisions have large cross sections and are highly inelastic. Consider first the case of a high-energy nucleon-nucleon inelastic collision. The inelastic nucleon-nucleon cross section is the dominant component of the total nucleon-nucleon cross section, and the two colliding nucleons, on the average, lose a large fraction (about half) of their energies. The energy lost by the nucleons is deposited in the vicinity of the center of mass and is subsequently carried away by pions and other mesons. In a central nucleus-nucleus collision, there are many inelastic nucleon-nucleon collisions. Qualitatively speaking, as far as the energy deposition in the vicinity of the center of mass is concerned, the effect of the many inelastic nucleon-nucleon collisions in a nucleus-nucleus reaction is roughly additive in nature. The more the number of the inelastic nucleon-nucleon collisions, the greater is the amount of energy deposited in the vicinity of the center of mass. Furthermore, because of Lorentz contraction, the two colliding nuclei appear initially as two thin disks. The elementary nucleon-nucleon collisions between the two nuclei occur nearly at the same time and in close spatial proximity. In consequence, as the colliding baryon matter recedes from each other after the collision (if they are not stopped as a result of the collision), *a large amount of energy is deposited in a small region of space in a short duration of time.* In this region, the energy density is therefore very large [1]. An energy density of the order of a few GeV/fm^3 may be achieved. This energy density, an order of magnitude greater than the energy density of nuclear matter in equilibrium, may favor the formation of new forms of matter such as the quark-gluon plasma. The search for new forms of matter under extreme conditions of high energy densities and high temperatures is an important objective of high-energy heavy-ion collisions.

On the theoretical side, the status of our understanding of modern physics is dominated by the puzzles of symmetry breaking and unseen isolated quarks [4,5]. As emphasized not the least by T. D. Lee, high-energy heavy-ion collisions may provide a valuable tool to examine the puzzles of symmetry breaking and quark confinement in strongly interacting quark-gluon systems [4].

Symmetry breaking phenomena occur in many physical systems, such as in ferromagnets, electro-weak interactions and in strong interaction physics. We shall illustrate the concept of symmetry breaking with the example of ferromagnetism. In a ferromagnetic system, the Hamiltonian which describes the interaction between the spins of the atoms at various sites is rotational invariant. For example, in the simple Heisenberg model of ferromagnetism, we describe the system as a collection of spins $\{s_i\}$ at lattice sites $\{i\}$. The Hamiltonian of the system is taken to be the sum of $-J\mathbf{s}_i \cdot \mathbf{s}_j$, where $J(>0)$ is the strength of the interaction and the sum is carried over the set of nearest neighbors i and j. The scalar product $\mathbf{s}_i \cdot \mathbf{s}_j$ is independent of the orientation of the coordinate system and thus the Hamiltonian possesses rotational symmetry. Because of the scalar product $\mathbf{s}_i \cdot \mathbf{s}_j$, the energy of the system is the lowest when the spins of two nearest neighbors line up in the same direction, but there is a complete symmetry with respect to the choice of that spin direction. Clearly, when a large fraction of the spins at various distant sites of the system line up in the same direction, the energy of the system will be low and the configuration will be favorable. This alignment of the spins of the system occurs at low temperatures to produce a macroscopic ferromagnetic system. Inside the ferromagnet, because there is a particular direction along which the spins of the atoms at different distant sites tend to align, there is no rotational symmetry. The rotational symmetry is said to be *spontaneously broken* for the ferromagnetic system in that state. Many different states, with the spin vector pointing in other directions will have the same energy. We have here the case that the Hamiltonian is symmetric with respect to rotational symmetry, but the state of the system does not possess rotational symmetry anymore. Because the spins have the tendency to point in a particular direction inside the macroscopic system of the ferromagnet, the rotational symmetry of the Hamiltonian is not an apparent symmetry but is a *hidden symmetry*. For an observer in the ferromagnetic system, the (hidden) rotational symmetry of the Hamiltonian can be deduced only after a careful analysis and a thorough understanding of the origin of the preferred spin direction in the ferromagnetic system. As the temperature increases, there is a tendency to disrupt the alignment of the spins at different sites of the system due to the thermal motion. When the temperature is greater than the critical Curie temperature, the tendency to disrupt the spin alignment overwhelms the tendency to align, and the ground state of the system restores itself to a state with a complete symmetry with respect to the different orientations. There is no preferred direction along which the spins at various sites of the system will align and the system resides in a state with rotational symmetry. The broken symmetry of the ferromagnetic system at low temperatures is said to be *restored* at high temperatures. There is a phase transition from the ferromagnetic phase at low temperatures with spin alignment to the non-ferromagnetic phase at high temperatures without spin alignment.

It is generally held that the phenomenon of symmetry breaking occurs also in strongly interacting systems and that it is connected with the puzzle of the lack of the observation of isolated quarks. We can make an analogy with the ferromagnetic system by describing a strongly interacting system in the arena of a space-time lattice which has lattice sites and lattice links joining the nearest neighboring lattice sites [6]. On the scaffold of lattice sites and links, one can form elementary squares by joining any four nearest neighboring links. The collection of any four nearest neighboring links in the form of an elementary closed square is called a *plaquette*. There are many elementary squares (plaquettes) in the scaffold of the lattice, each of which is formed by joining four nearest neighboring links.

In a strongly interacting system, the quanta of the fields are quarks and gluons and a configuration of the system is described by specifying the quark field and the gluon field at all space-time points. In the arena of the lattice with its lattice sites and lattice links, a configuration of the strongly interacting quark-gluon system is described with a set of quark field at the lattice sites and gluon field residing on the links between lattice sites. Instead of using the gluon field directly, it is more convenient to use a function of the gluon field, the *link variable*, to describe the gluon field at a link of the lattice. There is an abstract color space associated with the gluon field and the quark fields. The link variable is a matrix in color space, analogous to the Pauli spin matrices in spin space. The Lagrangian of the gluon field in the system is obtained by taking the trace of the product of the four link variables along the links in an elementary square (a plaquette), and adding up the traces of the products from all elementary squares in the lattice (see Chapters 10 and 11). As determined by the Lagrangian, the energy of the system is lowest when the four link variables in a plaquette are correlated in color space.

For the sake of intuitive understanding, we can consider a link variable in color space roughly as the analogue of a spin variable in spin space in a ferromagnetic system. Accordingly, the correlation of the four link variables residing on the four nearest neighboring links of an elementary square can be considered as the analogue of the alignment for the spins residing on two nearest neighboring sites in ferromagnetism. There is thus a tendency for the four link variables residing on the four links of a plaquette to correlate in color space, and by extension, for the link variables on different spatially-distant links of the system to correlate in color space, so as to lower the energy of the system. When a large fraction of the link variables on various spatially-distant links of the system are correlated in color space, the energy of the system will be low and the configuration will be favorable. On the other hand, the effect of entropy due to thermal motion is to disrupt the correlation of the link variables residing on the four nearest neighboring links of an elementary square (plaquette). By extension, the effect of thermal motion is to disrupt the correlation of the link variables on different spatially-distant links of the system. At low temperatures, the tendency to correlate overwhelms the tendency to disrupt the correlation, and the system acquires a non-zero average value for the link variables. With this correlation of the link variables in color space at various links of the lattice, the symmetry of the system with respect to a random orientation of the link variable in color space is spontaneously broken. If one calculates the interaction between a quark and an antiquark in this phase, one finds that the interaction is attractive and is proportional to the distance between the quark and the antiquark. The energy needed to isolate a quark from an antiquark is infinite and thus the quarks are confined. The system is a condensate of gluons, quarks and antiquarks. Hadrons are examples of strongly interacting particles in which quarks and gluons are confined.

As the temperature increases, the tendency to disrupt the correlation overwhelms the tendency to correlate and the ground state of the system restores itself to a state with symmetry. Quarks and gluons are then deconfined to form a quark-gluon plasma. There is a phase transition from the confined phase at low temperatures to the deconfined phase at high temperatures [6]. Various estimates of the transition temperature place it at about 200 MeV and the energy density of the quark-gluon plasma at a few GeV/fm^3. This range of temperature and energy density fall within the realm achievable by high-energy heavy-ion collisions. It is therefore of interest to utilize such collisions to explore possible existence of the quark-gluon plasma. Experimental study and identification of the quark-gluon plasma may provide new insight into the question of quark confinement. Furthermore, the creation of the domain of high energy density, albeit within a small region of space and time, may allow one to study matter under unusual conditions such as those thought to exist in the early history of the universe.

We expect the occurrence of high energy density regions in two different situations: in the 'stopping' or 'baryon-rich quark-gluon plasma' region of collision energies with $\sqrt{s} \sim 5-10$ GeV per nucleon, and in the 'baryon-free quark-gluon plasma' region of collision energy with $\sqrt{s} \gtrsim 100$ GeV per nucleon. In the first situation, we envisage the slowing-down of the baryons in the center-of-mass frame so that the nuclear matter is nearly stopped in that frame. The type of quark-gluon plasma which may be formed in this region is a baryon-rich quark-gluon plasma. At higher energies, the baryons cannot be completely stopped. They are slowed down but still recede away from the center of inelastic collisions. When the baryons are well separated, the energy which is deposited by the colliding nucleons may become liberated in the 'central rapidity' region between receding baryons. The additive effect of many such nucleon-nucleon collisions may produce a quark-gluon plasma with only a small baryon content. As the net baryon content of the early universe is very small, this type of quark-gluon plasma is of special astrophysical interest [7].

A quantitative understanding of the detailed dynamics of nucleus-nucleus collisions is useful for an assessment of the possibility of quark-gluon plasma formation. The knowledge of the dynamics will also help one separate out the signals which are expected in the hadronic phase and the signals which arise in the quark-gluon plasma phase. We need to know how a baryon may be slowed down in its passage through a nucleus and how particles are produced in a nuclear environment. How the heavy-ion collisions proceed is the subject of current investigations.

High-energy heavy-ion physics is an emerging field, both experimentally and theoretically. The development is still in a state of flux. Many models have been proposed and more will come as many problems are not completely resolved. It is appropriate to review the status of present research and to prepare here a set of tools so that newcomers can make use of these tools to make contributions in this field.

§References for Chapter 1

1. J. D. Bjorken, Phys. Rev. D27, 140 (1983).
2. C. Bubbia, in Proceedings of the XXVI International Conference on High Energy Physics, Dallas, Texas, August, 1992, edited by R. Sanford, AIP Conference Proceedings No. 272, Vol. I, p. 321.
3. H. Satz, Nucl. Phys. A544, 371c (1992).
4. T. D. Lee, Nucl. Phys. A538, 3c (1992).
5. An excellent introduction on the phenomena of symmetry breaking and quark confinement can be found in K. Gottfried and V. F. Weisskopf, *Concepts of Particle Physics*, Vols. I and II, Oxford University Press, Oxford, 1984, and T. D. Lee, *Particle Physics and Introduction to Field Theory*, Harwood Academic Publisher, Chur, Switzerland, 1981.
6. For a review of the status of lattice results in finite temperature QCD, see J. B. Kogut, Nucl. Phys. A461, 327c (1987), A. Ukawa, Nucl. Phys. A498, 227c (1989), B. Petersson, Nucl. Phys. A525, 237c (1991) and T. Hatsuda, Nucl. Phys. A544, 27c (1992), and references cited in these references.
7. S. Weinberg, *The First Three Minutes*, Basic Books, N.Y., 1977.

2. Kinematic Variables

In relativistic heavy-ion collisions and in many other high-energy reaction processes, it is convenient to use kinematic variables which have simple properties under a change of the frame of reference. The light-cone variables x_+ and x_-, the rapidity variable y, and the pseudorapidity variable η are kinematic variables which have simple properties under a Lorentz transformation. They are commonly used. The Feynman scaling variable, x_F, which is related to the light-cone variables x_+ and x_-, is also often used. It is worthwhile to discuss these variables in detail to establish the proper language for relativistic reactions.

§2.1 Notation and Conventions

In this book, we shall follow the notation of Bjorken and Drell, *Relativistic Quantum Mechanics*, (McGraw-Hill Book Company, N.Y. 1964). We use the *natural units* $c = \hbar = 1$. The space-time co-ordinates of a point x are denoted by a *contravariant vector* with components x^μ:

$$x^\mu = (x^0, x^1, x^2, x^3) = (t, \boldsymbol{x}) = (t, x, y, z). \tag{2.1}$$

The momentum vector p is similarly defined by a contravariant vector with components p^μ:

$$p^\mu = (p^0, p^1, p^2, p^3) = (E, \boldsymbol{p}) = (E, \boldsymbol{p}_T, p_z) = (E, p_x, p_y, p_z). \tag{2.2}$$

We shall adopt the space-time *metric tensor* $g_{\mu\nu}$ in the form

$$g_{\mu\nu} = \begin{pmatrix} 1 & 0 & 0 & 0 \\ 0 & -1 & 0 & 0 \\ 0 & 0 & -1 & 0 \\ 0 & 0 & 0 & -1 \end{pmatrix}. \tag{2.3}$$

The *covariant vector* x_μ is related to the contravariant vector x^μ through the metric tensor $g_{\mu\nu}$ by

$$x_\mu \equiv (x_0, x_1, x_2, x_3) \equiv g_{\mu\nu} x^\nu = (t, -x, -y, -z), \tag{2.4}$$

where we use the notation that a repeated index implies a summation with respect to that index, unless indicated otherwise. Conversely, the contravariant vector x^μ is related to the corresponding covariant vector x_ν by

$$x^\mu \equiv g^{\mu\nu} x_\nu , \qquad (2.5)$$

where the metric tensor $g^{\mu\nu}$ is

$$g^{\mu\nu} = \begin{pmatrix} 1 & 0 & 0 & 0 \\ 0 & -1 & 0 & 0 \\ 0 & 0 & -1 & 0 \\ 0 & 0 & 0 & -1 \end{pmatrix}. \qquad (2.6)$$

The *scalar product of two vectors* a and b is defined as

$$a \cdot b \equiv a^\mu b_\mu = g_{\mu\nu} a^\mu b^\nu = a^0 b^0 - \boldsymbol{a} \cdot \boldsymbol{b}.$$

The *four-momentum operator* p^μ in coordinate representation is

$$p^\mu = i\partial^\mu = (i\partial^0, i\partial^1, i\partial^2, i\partial^3) \qquad (2.7a)$$

$$= (i\partial_0, -i\partial_1, -i\partial_2, -i\partial_3) = (i\frac{\partial}{\partial x^0}, -i\frac{\partial}{\partial x^1}, -i\frac{\partial}{\partial x^2}, -i\frac{\partial}{\partial x^3}).(2.7b)$$

The covariant operator ∂_i is the usual *gradient operator* ∇ defined by

$$\nabla = (\nabla_x, \nabla_y, \nabla_z) = (\partial_1, \partial_2, \partial_3) = (\partial/\partial x^1, \partial/\partial x^2, \partial/\partial x^3). \qquad (2.8)$$

⊕[Supplement 2.1
　　The placement of the indices in the tensor notation of contravariant and co-variant vectors is confusing for beginning students when they are dealing with relativistic kinematics. In the "common" notation for vectors, only subscript indices are used and contravariant and covariant vectors are not distinguished. This is permissible in the Euclidean space, where $g_{\mu\nu} = \delta_{\mu\nu}$ and there is no significant distinction between a contravariant vector and a covariant vector. However, in Minkowski space with the metric tensor (2.3), they are different types of vectors as they have different properties under a coordinate transformation. In the tensor notation, the contravariant vectors have superscript indices but the covariant vectors have subscript indices. The tensor notation makes a clear distinction between these two types of vectors. The easiest way to understand the distinction between them is to remember that the suffixes 'contra-' and 'co-' refer to a comparison with the gradient vector. 'Contravariant' corresponds to the nomenclature 'contragradient', and 'covariant' corresponds to the term 'cogradient' [1]. A quantity which trans-forms like a gradient vector is 'cogradient' and is therefore a covariant vector with a subscript index. A quantity which transforms like the coordinates is 'contragradient' and is a contravariant vector with a superscript index.
　　Accordingly, the coordinate x^μ and the momentum p^μ are contravariant four-vectors and they have superscript indices. The covariant vectors x_μ and p_μ, with subscript indices derived from these vectors, are different vectors because the signs of their space components are changed as in Eq. (2.4).

On the other hand, because superscript indices are clumsy to use, the common notation for a vector uses subscript indices to refer to the components of any vector. It does not distinguish a contravariant vector from a covariant vector. Unfortunately, the most commonly used vectors, such as the coordinate vector and the momentum vector, are contravariant vectors. They have superscript indices in the tensor notation but subscript indices in the common notation. It is unavoidable that one uses both notations for these quantities. Therefore, when these quantities are written out, it is necessary to make a mental note as to which notation is adopted.

It is worth recommending that one adheres to the tensor notation as much as possible, to insure that one gets the correct signs in an algebraic manipulation with vector quantities. As an example, we can follow the line of reasoning which gives the coordinate representation of the momentum operator (2.7). There, the signs in front of the ∂ operators in Eqs. (2.7a) and (2.7b) are often a source of confusion for beginning students. We begin by recalling that the coordinate vector x^μ and the momentum vector p^μ are contravariant vectors. The gradient operator ∂_μ is a covariant vector, and it is the derivative with respect to the μ component of x. As x is a contravariant vector, ∂_μ is the derivative with respect to x^μ:

$$\partial_\mu = \frac{\partial}{\partial x^\mu} \, .$$

All other vectors such as p_μ, x_μ, and ∂^μ are derived in terms of the basic vectors x^μ, p^μ, and ∂_μ by using the metric tensors $g_{\mu\nu}$ and $g^{\mu\nu}$ as in Eqs. (2.4) and (2.5).

Being a contravariant vector, the momentum vector has a coordinate representation which is naturally another contravariant vector $i\partial^\mu$. If we want to write out the operator $i\partial^\mu$ explicitly in terms of the derivatives of the coordinates, it is necessary to express ∂^μ in terms of the covariant operator ∂_μ by using the metric tensor $g^{\mu\nu}$. Thus, we have

$$p^\mu = i\partial^\mu$$
$$= ig^{\mu\nu}\partial_\nu \, .$$

Writing the covariant operator ∂_ν explicitly as the derivative with respect to the contravariant vector x^μ, we have

$$p^\mu = ig^{\mu\nu}\frac{\partial}{\partial x^\nu}$$
$$= (i\frac{\partial}{\partial x^0}, -i\frac{\partial}{\partial x^1}, -i\frac{\partial}{\partial x^2}, -i\frac{\partial}{\partial x^3})$$

which is the expression in Eq. (2.7b). $\mathbf{1} \oplus$

§2.2 Light-Cone Variables

In many high-energy reaction processes, a detected particle can be identified as originating from one of the colliding particles. For example, in the reaction

$$b + a \rightarrow c + X \, ,$$

where c is a detected particle, c may sometimes be considered as fragmenting from the incident beam particle b or from the target particle a.

A reaction in which the detected particle c is described as originating from the beam particle b is called a *projectile fragmentation* reaction. The region of the momentum of c in which this type of reaction is dominant is referred to as the projectile fragmentation region. It lies in the forward direction with respect to the beam axis. Similarly, a reaction in which the detected particle c can be described as originating from the target particle a is called a *target fragmentation* reaction. The kinematic region in which this type of reaction is dominant is the target fragmentation region, which is near the region of momentum where the target particle is initially at rest.

In a reaction, kinematic quantities along the direction of the incident beam, which we shall identify as the longitudinal axis, have properties quite different from those along the transverse directions perpendicular to the beam axis. We shall designate the longitudinal axis as the z-axis. For convenience, we use the same symbol to represent a particle and its four-momentum. For example, $c = (c_0, \mathbf{c}_T, c_z)$, where c_0 is the energy of the particle c, c_z is its longitudinal momentum and $\mathbf{c}_T$ is its two-dimensional transverse momentum in the plane perpendicular to the longitudinal axis.

Two linear combinations of c_0 and c_z have special properties under a Lorentz transformation in the z-direction. The quantity

$$c_+ = c_0 + c_z \,, \tag{2.9a}$$

is called the *forward light-cone momentum* of c, while the quantity

$$c_- = c_0 - c_z \,, \tag{2.9b}$$

is called the *backward light-cone momentum* of c. For an energetic particle traveling in the forward direction (along the beam direction), its forward light-cone momentum c_+ is large, while its backward light-cone momentum c_- is small. Conversely, for a particle traveling in the backward direction opposite to that of the beam, its backward light-cone momentum is large, while its forward light-cone momentum is small.

The forward light-cone momentum of any particle in one frame is related to the forward light-cone momentum of the same particle in another boosted Lorentz frame by a constant factor. Therefore, if one considers a daughter particle c as fragmenting from a parent particle b, then the ratio of the forward light-cone momentum of c relative to that of b is independent of the Lorentz frame (see Exercise 2.1). It is convenient to introduce the *forward light-cone variable* x_+ of c relative to b, defined as the ratio of the forward light-cone momentum of the daughter particle c relative to the forward light-cone momentum of the parent particle b:

$$x_+ = \frac{c_0 + c_z}{b_0 + b_z} . \tag{2.10}$$

The forward light-cone variable x_+ is always positive. Because a daughter particle cannot possess a forward light-cone momentum greater than the forward light-cone momentum of its parent particle, the upper limit of the forward light-cone variable x_+ is 1. It is a Lorentz-invariant quantity, independent of the Lorentz frame (see Exercise 2.1).

We have been motivated to introduce the light-cone variable x_+ to specify the relationship between the forward light-cone momentum of a daughter particle and the forward light-cone momentum of its parent particle. However, because of its Lorentz invariant property, x_+ is sometimes used to specify the relationship between the momentum of a particle c relative to another reference particle b, whether the particle b is the parent particle of c or not. In this instance, the forward light-cone momentum of the reference particle b provides the scale by which the momentum of particle c is measured.

•⟦ Exercise 2.1
Show that x_+ is a Lorentz-invariant quantity.

◇Solution:
Under a Lorentz transformation, one goes from a frame F to a moving frame F' by boosting the frame F with a velocity β in the z-direction. The four-momentum $c = (c_0, \mathbf{c}_T, c_z)$ in the frame F is transformed into the four-momentum $c' = (c'_0, \mathbf{c}'_T, c'_z)$ in the frame F'. The momenta in the two different frames F and F' are related by the Lorentz transformation

$$c'_0 = \gamma(c_0 - \beta c_z) , \tag{1}$$

$$c'_z = \gamma(c_z - \beta c_0) , \tag{2}$$

and

$$\mathbf{c}'_T = \mathbf{c}_T ,$$

where

$$\gamma = \frac{1}{\sqrt{1 - \beta^2}} .$$

Therefore, from Eqs. (1) and (2), the forward light-cone momentum $c_0 + c_z$ in the frame F is related to that in the frame F' by the factor $\gamma(1 - \beta)$:

$$c'_0 + c'_z = \gamma(1 - \beta)(c_0 + c_z) . \tag{3}$$

Similarly, for another particle b, the forward light-cone momentum $b_0 + b_z$ in frame F is related to its forward light-cone momentum in frame F' by

$$b'_0 + b'_z = \gamma(1 - \beta)(b_0 + b_z). \tag{4}$$

If one considers particle c to be the daughter particle and particle b to be the parent particle (or reference particle), it is useful to introduce the forward light-cone variable x_+, defined as the ratio of the forward light-cone momentum of c relative to the forward light-cone momentum of b. In the frame F, the light-cone variable x_+ is

$$x_+ = \frac{c_0 + c_z}{b_0 + b_z}. \tag{5}$$

In the frame F', the light-cone variable x'_+ is

$$x'_+ = \frac{c'_0 + c'_z}{b'_0 + b'_z}. \tag{6}$$

From Eqs. (3), (4), (5) and (6), we see that

$$x_+ = x'_+.$$

Therefore, the light-cone variable x_+ is independent of the frame of reference. ∎•

In exactly the same way, one finds that under the change of Lorentz frame, the backward light-cone momentum of one particle in one frame is related to its backward light-cone momentum in another frame by a constant factor. The factor is $\gamma(1 + \beta)$ for the backward light-cone momenta, whereas for the forward light-cone momenta, the factor is $\gamma(1 - \beta)$ (see Eqs. (3) and (4) of Exercise 2.1). Therefore, if one considers a daughter particle c as fragmenting from a parent target particle a, it is convenient to introduce the *backward light-cone variable* x_- defined as

$$x_- = \frac{c_0 - c_z}{a_0 - a_z}, \tag{2.11}$$

which is a Lorentz-invariant quantity, independent of the Lorentz frame of reference. It is a positive quantity and its upper limit is 1.

In many problems, one deals exclusively with the forward region and the backward light-cone momentum is not considered. In these cases, the forward light-cone variable x_+ is often simply referred to as the *light-cone variable* and is denoted by x. There is no ambiguity that it would be confused with the backward light-cone variable x_-. At very high energies when the energy and the longitudinal momentum are approximately the same, the light-cone variable x is just the longitudinal momentum fraction of the daughter particle c relative to the parent particle b. For this reason, the variable x is sometimes called the *longitudinal momentum fraction*, or simply the *momentum fraction* of c relative to b.

The daughter particle c may be a particle detected as a free particle in a detector. In this case, the particle is not subject to any interaction, and the components of its four-momentum obey the relation appropriate for a free particle:

$$c^2 = c_0^2 - \mathbf{c}^2 = m_c^2, \qquad (2.12)$$

where m_c is the rest mass of c.

In the space of the momentum components of c, Eq. (2.12) can be considered as expressing m_c^2 as a function of the variables c_0 and $\mathbf{c}$, which is represented graphically by a hyperboloid of the variables c_0 and $\mathbf{c}$ characterized by the mass of the particle. Equation (2.12) is called the *mass-shell condition*. A free particle is said to be *on the mass shell* because its energy and momentum obey a simple relation with reference to the rest mass of the particle. The four-momentum c now has only three degrees of freedom, and it can be represented, for example, by $(x_+, \mathbf{c}_T)$ when the forward momentum of the parent particle b is known.

In some problems, the beam particle b is considered to be a composite system consisting of a constituent particle c and the other parts X. Sometimes, a reaction between the beam particle and the target particle is described as the interaction between the beam constituent c with the target particle. The particle c is not a free particle, and it is still subject to interactions with the other parts X. The four-momentum of c will not obey the mass-shell relation (2.12). Its four degrees of freedom can be specified by the Lorentz-invariant quantities $(x_+, c^2, \mathbf{c}_T)$, when the forward light-cone momentum of b is known. The particle c is said to be *off the mass shell*, or simply off shell. There is a simple transformation which gives $(c_0, \mathbf{c}_T, c_z)$ in terms of $(x_+, c^2, \mathbf{c}_T)$ (see Exercise 2.2). Similarly, when the particle c under consideration is a constituent of the target particle a, its four degrees of freedom can be specified by the Lorentz-invariant quantities $(x_-, c^2, \mathbf{c}_T)$ when the backward light-cone momentum of a is known.

In a typical high-energy collision process, the distribution of the transverse momentum $dN/d\mathbf{c}_T$ of the produced particles has a peak at zero transverse momentum and has a width of the order of a few hundred MeV/c. This width is small compared to the longitudinal momentum of the reaction. The momentum of the produced particles is therefore 'limited' in the transverse direction. It is often convenient to separate out the transverse degree of freedom from the longitudinal degree of freedom. For these problems, it is useful to write Eq. (2.12) in the form

$$c_0^2 - c_z^2 = m_c^2 + c_T^2 = m_{cT}^2 , \qquad (2.13)$$

where m_{cT} is called the *transverse mass* of particle c. It contains contributions from both the rest mass and the transverse momentum of the particle.

•⟦ Exercise 2.2
The variables (c_0, c_T, c_z) depend on the frame of reference. In many problems, one makes a change of variables from (c_0, c_T, c_z) to the Lorentz-invariant variables (x_+, c^2, c_T). Express the momentum components c_0 and c_z in terms of the forward light-cone variable x_+ and the Lorentz-invariant quantity c^2 when the particle c is produced from a reaction in which the beam particle is b. Also, find the relation between the differential elements $dc_0 dc_z dc_T$ and $dx_+ dc^2 dc_T$.

◇Solution:
From Eq. (2.10), we have

$$c_0 + c_z = x_+(b_0 + b_z) .$$

We also have

$$c_0 - c_z = (c_0^2 - c_z^2)/(c_0 + c_z) = (c^2 + c_T^2)/(c_0 + c_z) .$$

By adding and subtracting these two equations, we obtain

$$c_0 = \frac{1}{2}\left[x_+(b_0 + b_z) + \frac{c^2 + c_T^2}{x_+(b_0 + b_z)} \right], \qquad (1)$$

and

$$c_z = \frac{1}{2}\left[x_+(b_0 + b_z) - \frac{c^2 + c_T^2}{x_+(b_0 + b_z)} \right], \qquad (2)$$

which expresses c_0 and c_z in terms of x_+ and c^2. From these equations, we find

$$dc_0 = \frac{1}{2}\left[dx_+ \left\{ (b_0 + b_z) - \frac{c^2 + c_T^2}{x_+^2(b_0 + b_z)} \right\} + \frac{dc^2}{x_+(b_0 + b_z)} \right], \qquad (3)$$

and

$$dc_z = \frac{1}{2}\left[dx_+ \left\{ (b_0 + b_z) + \frac{c^2 + c_T^2}{x_+^2(b_0 + b_z)} \right\} - \frac{dc^2}{x_+(b_0 + b_z)} \right]. \qquad (4)$$

The relation between the differential elements $dc_0 dc_z$ and $dc^2 dx_+$ can be found by using the well-known Jacobian determinant formula. An intuitively simple way to get this result is by viewing the differential element $dc_0 dc_z$ as a surface area element. It has the meaning of the magnitude of the cross product $dc_0 \times dc_z$. Similarly, $dx_+ dc^2$ is a surface element in the space with coordinates x_+ and c^2, and it has the meaning of the magnitude of the cross product $dx_+ \times dc^2$ in that space. Hence, using Eqs. (3-4), we carry out the cross product and we have

$$dc_0 dc_z = \mid dc_0 \times dc_z \mid = \frac{1}{2} \frac{\mid dx_+ \times dc^2 \mid}{x_+} = \frac{1}{2} \frac{dx_+}{x_+} dc^2 . \tag{5}$$

Therefore, the invariant volume element d^4c can be expressed in terms of dx_+, dc^2 and dc_T by

$$d^4c = dc_0 dc_z d\mathbf{c}_T = \frac{dx_+}{2x_+} dc^2 d\mathbf{c}_T . \tag{6}$$

The relations obtained here are useful when one makes a change of variables from $(c_0, \mathbf{c}_T, c_z)$ to $(x_+, c^2, \mathbf{c}_T)$, whether the particle c is on the mass shell or not. If the particle c is on the mass-shell, the mass-shell condition can be represented by a delta function $\delta(c^2 - m^2)$ constraining the variable c^2 to the value m^2. ▮•

•⟦Exercise 2.3
The Feynman scaling variable x_F for a detected particle c is defined as

$$x_F = \frac{c_z^*}{c_z^*(max)} , \tag{1}$$

where the asterisks stand for quantities in the center-of-mass system. Obtain the relationship between the Feynman scaling variable x_F and the forward light-cone momentum fraction x_+.

◇Solution:
Consider the reaction $b + a \rightarrow c + X$. We shall work in the center-of-mass system, and start by expressing the denominator of Eq. (1) in terms of rest masses and the center-of-mass collision energy. This denominator is the maximum value of c_z^*. Clearly, c_z^* attains its maximum value when the undetected collection of particles 'X' in the reaction $b + a \rightarrow c + X$ consists of a single particle with a rest mass m_X, which corresponds to the minimum value of the rest mass of X allowed by the conservation laws (of baryon numbers, charge, etc). In this situation, the magnitude of the three-momentum of c is equal to the magnitude of the three-momentum of X. We have, therefore, from the energy conservation condition

$$\sqrt{(c_z^*(max))^2 + m_c^2} + \sqrt{(c_z^*(max))^2 + m_X^2} = \sqrt{s} ,$$

where $\sqrt{s}$ is the center-of-mass energy of the collision process. This equation can be solved for $c_z^*(max)$ in terms of s, and we obtain

$$c_z^*(max) = \frac{\lambda(s, m_c^2, m_X^2)}{2\sqrt{s}} , \tag{2}$$

where the function λ is defined by

$$\lambda^2(s, m_1^2, m_2^2) = s^2 + m_1^4 + m_2^4 - 2(sm_1^2 + sm_2^2 + m_1^2 m_2^2) . \tag{3}$$

From the definition of the Feynman scaling variable, the longitudinal momentum of c in the center-of-mass system is

$$c_z^* = x_F \frac{\lambda(s, m_c^2, m_x^2)}{2\sqrt{s}}. \tag{4}$$

We shall use (4) in the definition of x_+, so as to relate x_+ with x_F.

The energy of the particle c in the center-of-mass system is (see Eqs. (2.13) and (4))

$$c_0^* = \sqrt{(c_z^*)^2 + m_{cT}^2} = \sqrt{\frac{x_F^2 \lambda^2(s, m_c^2, m_x^2)}{4s} + c_T^2 + m_c^2}. \tag{5}$$

Furthermore, we need to know the energy and the longitudinal momentum of the parent beam particle in the center-of-mass system to obtain the forward light-cone variable. In this system, the magnitude of the three-momentum of b is equal to the magnitude of the three-momentum of a. Therefore, from the definition of the center-of-mass energy $\sqrt{s}$, we have

$$\sqrt{(b_z^*)^2 + m_b^2} + \sqrt{(b_z^*)^2 + m_a^2} = \sqrt{s}.$$

We then have

$$b_z^* = \frac{\lambda(s, m_a^2, m_b^2)}{2\sqrt{s}}, \tag{6}$$

and

$$b_0^* = \frac{s + m_b^2 - m_a^2}{2\sqrt{s}}. \tag{7}$$

The forward light-cone momentum x_+ is defined by (2.10) as

$$x_+ = \frac{c_0^* + c_z^*}{b_0^* + b_z^*}, \tag{8}$$

where c_0^*, c_z^*, b_0^*, and b_z^* are available from Eqs. (4)-(7) to express x_+ as a function of the Feynman scaling variable x_F.

To obtain the inverse relation, we use Eqs. (1), and (2) and Eq. (2) of Exercise 2.2, to get a relation that expresses x_F as a function of the light-cone variable,

$$x_F = \frac{1}{2}\left[x_+(b_0^* + b_z^*) - \frac{m_{cT}^2}{x_+(b_0^* + b_z^*)}\right]\frac{2\sqrt{s}}{\lambda(s, m_c^2, m_x^2)}, \tag{9}$$

where b_0^* and b_z^* are given by Eqs. (6) and (7) to express x_F as functions of x_+.

It is interesting to examine the case when the rest masses are small compared to the center-of-mass energy $\sqrt{s}$. Then, because $c_z^*(max) \approx \sqrt{s}/2$, and

$$b_0^* + b_z^* \approx \sqrt{s}(1 - m_a^2/s),$$

we have

$$x_+ \approx \frac{1}{2}\left\{x_F\left(1 - \frac{m_c^2 + m_x^2}{s}\right) + \sqrt{x_F^2\left(1 - \frac{m_c^2 + m_x^2}{s}\right)^2 + \frac{4m_{cT}^2}{s}}\right\}\frac{1}{1 - m_a^2/s},$$

and

$$x_F \approx \left\{ x_+\left(1 - \frac{m_a^2}{s}\right) - \frac{m_{cT}^2}{x_+ s(1 - m_a^2/s)} \right\} \frac{1}{1 - (m_c^2 + m_x^2)/s}.$$

One notices that while the forward light-cone variable is always positive, the Feynman scaling variable can be zero and negative. In the case of very high energies and $x_+ \gg 0$ or $x_F \gg 0$, the light-cone variable x_+ coincides with the Feynman scaling variable x_F. On the other hand, when x_+ or x_F are small, or when x_F is negative, the light-cone variable x_+ differs substantially from the Feynman scaling variable x_F.   ⟧•

§2.3 Rapidity Variable

Another useful variable used commonly to describe the kinematic condition of a particle is the *rapidity variable* y. The rapidity of a particle is defined in terms of its energy-momentum components p_0 and p_z by

$$y = \frac{1}{2} \ln\left(\frac{p_0 + p_z}{p_0 - p_z}\right). \tag{2.14}$$

It is a dimensionless quantity related to the ratio of the forward light-cone momentum to the backward light-cone momentum. It can be either positive or negative. In the nonrelativistic limit, the rapidity of a particle travelling in the longitudinal direction is equal to the velocity of the particle in units of the speed of light (see Exercise 2.4). The rapidity variable depends on the frame of reference, but the dependence is very simple. The rapidity of the particle in one frame of reference is related to the rapidity in another Lorentz frame of reference by an additive constant (see Exercise 2.5).

If the particle c is a free particle, it is then on the mass shell. Its four-momentum has only three degrees of freedom and can be represented by $(y, \boldsymbol{p}_T)$. We obtain below a simple transformation which gives $(p_0, \boldsymbol{p})$ in terms of $(y, \boldsymbol{p}_T)$.

From the definition (2.14), we have

$$e^y = \sqrt{\frac{p_0 + p_z}{p_0 - p_z}}, \tag{2.15a}$$

and

$$e^{-y} = \sqrt{\frac{p_0 - p_z}{p_0 + p_z}}. \tag{2.15b}$$

Adding Eqs. (2.15a) and (2.15b), we get the relation between the energy p_0 and the rapidity y of the particle c:

$$p_0 = m_T \cosh y, \tag{2.16}$$

where m_T is the transverse mass of the particle:

$$m_T^2 = m^2 + p_T^2 \,.$$

Subtracting Eq. (2.15b) from Eq. (2.15a), we obtain the relation between the longitudinal momentum p_z and the rapidity y of the particle

$$p_z = m_T \sinh y \,. \tag{2.17}$$

Equations (2.16) and (2.17) are useful relations linking the components of the momentum with the rapidity variable.

•⟦ Exercise 2.4.
What is the rapidity of a particle traveling in the positive z direction with a velocity β ?

◇ Solution:
The energy of the particle is

$$p_0 = \gamma m \,,$$

where m is the rest mass of the particle. The momentum of the particle in the longitudinal direction is

$$p_z = \gamma \beta m \,.$$

From the definition of the rapidity variable, Eq. (2.14), the rapidity of the particle traveling with a velocity β in the positive z direction is then

$$y_\beta = \frac{1}{2} \ln \left(\frac{1+\beta}{1-\beta} \right) \,. \tag{1}$$

Note that this is independent of the mass of the particle. When β is small, an expansion of y_β in terms of β leads to

$$y_\beta = \beta + O(\beta^3) \,.$$

Thus, in the nonrelativistic case, the rapidity of a particle is equal approximately to the longitudinal velocity of the particle.   ⟧•

•⟦ Exercise 2.5.
Find the relationship between the rapidity y of a particle in a laboratory frame F and the rapidity y' of the particle in a boosted Lorentz frame F' which moves with a velocity β in the z-direction.

◇ Solution:
The rapidity y' of the particle c in the new frame F' is defined by

$$y' = \frac{1}{2} \ln \left(\frac{p_0' + p_z'}{p_0' - p_z'} \right) \,.$$

Under the Lorentz transformation, the energy p'_0 and the longitudinal momentum p'_z in the frame F' are related to the energy p_0 and the longitudinal momentum p_z in the frame F' by

$$p'_0 = \gamma(p_0 - \beta p_z),$$
$$p'_z = \gamma(p_z - \beta p_0),$$

where β is the velocity of F' relative to F. Therefore, the rapidity y' in the frame F' is

$$y' = \frac{1}{2}\ln\left[\frac{\gamma(1-\beta)(p_0+p_z)}{\gamma(1+\beta)(p_0-p_z)}\right]$$
$$= y + \frac{1}{2}\ln\left(\frac{1-\beta}{1+\beta}\right)$$
$$= y - \frac{1}{2}\ln\left(\frac{1+\beta}{1-\beta}\right),$$

where y is the rapidity of the particle c in the laboratory frame F. (Note that here the symbol β is the velocity of the frame F' relative to the frame F, whereas in Exercise 2.4, β was the velocity of the particle in the frame F.)]•

From the above results in Exercise 2.5, we observe that under a *Lorentz transformation* from the laboratory frame F to a new coordinate frame F' moving with a velocity β in the z-direction, the rapidity y' of the particle in the new frame F' is related to the rapidity y in the old frame F by

$$y' = y - y_\beta, \tag{2.18}$$

where y_β is

$$y_\beta = \frac{1}{2}\ln\left(\frac{1+\beta}{1-\beta}\right). \tag{2.19}$$

According to Eq. (1) of Exercise 2.4, the quantity y_β is the rapidity a particle would have in the frame F, if it were traveling with the velocity β of the moving frame. The quantity y_β can conveniently be called the 'rapidity of the moving frame'. Thus, the rapidity of a particle in a moving frame is equal to the rapidity in the rest frame minus the rapidity of the moving frame, much like the subtraction of the velocity of the moving frame in the nonrelativistic case. This similarity is not surprising because nonrelativistically, y is equal to the longitudinal velocity β, as shown in Exercise 2.4. It is often useful to treat the rapidity variable as a relativistic measure of the 'velocity' of a particle. This simple property of the rapidity variable under a Lorentz transformation makes it a suitable choice to describe the dynamics of relativistic particles. To go from a frame of reference at rest to a moving frame of reference, it is only necessary to find the rapidity of the moving frame y_β and change the rapidity variables by subtracting this constant y_β.

•⟦ Exercise 2.6.
In the collision of a beam particle b with momentum b_z on a target particle a with momentum a_z, show that the initial rapidities of the particles are

$$y_a = \sinh^{-1}(a_z/m_a)$$

and

$$y_b = \sinh^{-1}(b_z/m_b)\,,$$

where m_a and m_b are the rest masses of particles a and b respectively. For the case when the rest mass of the projectile and the rest mass of the target particle are the same, show that the rapidity of the center-of-mass frame is given by

$$y_{cm} = (y_a + y_b)/2\,,$$

and, in the center-of-mass frame, the rapidities of a and b are

$$y_a^* = -(y_b - y_a)/2 \quad \text{and} \quad y_b^* = (y_b - y_a)/2\,.$$

◇ Solution:
The transverse momentum of the beam particle is zero. From Eq. (2.17), we have

$$b_z = m_b \sinh y_b,$$

or

$$y_b = \sinh^{-1}(b_z/m_b)\,.$$

From Eq. (2.16), we infer that the energy of the beam particle b in the laboratory frame is

$$b_0 = m_b \cosh y_b\,.$$

Similarly, because the target particle has no transverse momentum, the rapidity of the target particle a in the laboratory frame F is given by

$$y_a = \sinh^{-1}(a_z/m_a)\,,$$

where we have allowed the target particle to have a longitudinal momentum. The energy of the target particle a in the laboratory frame is

$$a_0 = m_a \cosh y_a\,.$$

The center-of-mass frame F_{CM} is obtained by boosting the laboratory frame F by a velocity of the center-of-mass frame β_{CM} such that the longitudinal momentum of the beam particle b_z^* and the longitudinal momentum of the target particle a_z^* are equal and opposite. Therefore, the velocity of the center-of-mass frame, β_{CM}, satisfies the following condition

$$a_z^* = \gamma_{CM}(a_z - \beta_{CM} a_0) = -b_z^* = -\gamma_{CM}(b_z - \beta_{CM} b_0)\,,$$

where

$$\gamma_{CM} = \frac{1}{\sqrt{1 - \beta_{CM}^2}} \, .$$

Hence, we have the velocity of the center-of-mass frame given by

$$\beta_{CM} = \frac{a_z + b_z}{a_0 + b_0} \, . \tag{1}$$

From Eq. (2.19), the rapidity of the center-of-mass frame is

$$y_{CM} = \frac{1}{2} \ln \left(\frac{1 + \beta_{CM}}{1 - \beta_{CM}} \right) . \tag{2}$$

Substituting the velocity of the center-of-mass frame from Eq. (1) into Eq. (2), we obtain

$$y_{CM} = \frac{1}{2} \ln \left[\frac{a_0 + a_z + b_0 + b_z}{a_0 - a_z + b_0 - b_z} \right] .$$

Writing the energies and the momenta in terms of the rapidity variables in the frame F, we can express the rapidity of the center-of-mass frame y_{CM} in terms of the rapidities of the colliding particles as

$$y_{CM} = \frac{1}{2} \ln \left[\frac{m_a e^{y_a} + m_b e^{y_b}}{m_a e^{-y_a} + m_b e^{-y_b}} \right] ,$$

$$= \frac{1}{2} (y_a + y_b) + \frac{1}{2} \ln \left[\frac{m_a e^{y_a} + m_b e^{y_b}}{m_a e^{y_b} + m_b e^{y_a}} \right] .$$

The above result is for the general case, when the rest masses of the beam particle b and the rest mass of the target particle a may be different. In the special case when the rest masses of a and b are equal, we have

$$y_{CM} = \frac{1}{2} (y_a + y_b) ,$$

and the rapidities of a and b in the center-of-mass frame are

$$y_a^* = y_a - y_{CM} = -\frac{1}{2} (y_b - y_a) ,$$

and

$$y_b^* = y_b - y_{CM} = \frac{1}{2} (y_b - y_a) . \qquad]\bullet$$

For a given incident energy, the rapidity of the projectile particles and the rapidity of the target particles can be determined easily (see Exercise 2.6). The greater the incident energy, the greater is the separation between the projectile rapidity and the target rapidity.

The region of rapidity about midway between the projectile rapidity and the target rapidity is called the *central rapidity* region. The rapidities of the produced particles lie mostly in this region. For example, in a *pp* collision at a laboratory momentum of 100 GeV/c, the beam rapidity y_b is 5.36 and the target rapidity y_a is 0. The central rapidity region is around $y \approx 2.7$.

While the light-cone variables x_+ or x_- are the ratios of two quantities which bear a (daughter)-(parent) relationship, the rapidity variable y of a particle is a kinematic variable of an individual particle in a given coordinate system. From the definitions of x_+ and y, we can relate x_+ to y. We consider a detected particle c which is found to have a rapidity y in a given frame of reference. Relative to the beam particle b, which has a beam rapidity y_b, the forward light-cone variable of c is x_+. From the definitions of x and y, we relate the forward light-cone variable x_+ to y by

$$x_+ = \frac{m_{c_T}}{m_b} e^{y - y_b}, \tag{2.20}$$

where m_{c_T} is the transverse mass of c. Conversely, we can express y in terms of the forward light-cone fraction x_+ as

$$y = y_b + \ln x_+ + \ln(m_b/m_{c_T}). \tag{2.21}$$

Similarly, relative to the target particle a with a target rapidity y_a, the backward light-cone variable of the detected particle c is x_-. The variable x_- is related to y by

$$x_- = \frac{m_{c_T}}{m_a} e^{y_a - y}, \tag{2.22}$$

and conversely,

$$y = y_a - \ln x_- - \ln(m_a/m_{c_T}). \tag{2.23}$$

We can illustrate the relationship between x_+ and y by using these quantities to describe a proton or a pion detected after a *pp* collision at an incident momentum of 100 GeV/c. These relations depend on the transverse momentum of the proton and the pion. For numerical purposes, we suppose that the detected proton acquires a transverse momentum with a magnitude 0.46 GeV/c, which is the average value for a proton detected after a collision at this incident momentum. Similarly, we suppose that the pion has a transverse momentum of magnitude 0.35 GeV/c, which is the average magnitude of p_T for produced pions. Figure 2.1 gives the light-cone variables x_+ for the

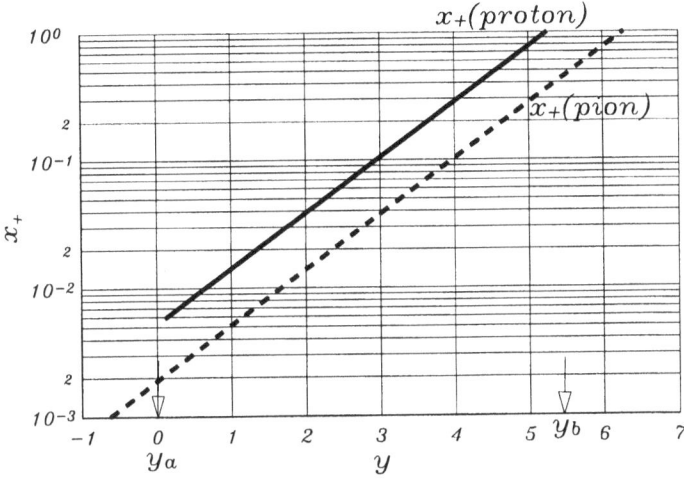

Fig. 2.1 The light-cone variables x_+ as a function of y, for a proton or a pion produced after a pp collision at an incident beam momentum of 100 GeV/c.

proton and the pion as a function of y. The arrows indicate the positions of the beam rapidity y_b and the target rapidity y_a. As one observes from Eq. (2.21) and Fig. 2.1, $\ln x_+$ is a linear function of y. A projectile fragmentation reaction is characterized by particles produced with light-cone variables close to unity. In this region of x_+ close to 1, a large fraction of the entire range of x_+ is covered by about one and a half units of rapidity, which comprise a small fraction of the whole range of the rapidity variable. Thus, for a reaction which leads to particles with a momentum close to the beam momentum, it is most appropriate to use the forward light-cone variable x_+ to describe those particles. Similarly, for reactions leading to particles with momentum close to the target momentum, the backward light-cone variable x_- is a better variable to describe those particles. On the other hand, for those particles detected in the central rapidity region, away from the beam rapidity and the target rapidity, a small region of the light-cone variable x_+ is mapped into a large region in the rapidity variable y, as is evident in Fig. 2.1. To examine particles in these regions, the rapidity variable y is a more appropriate kinematic variable. These relations indicate that a complete description of the full dynamical range of the produced particles requires $x_\pm$ and y variables.

§2.4 Pseudorapidity Variable

To characterize the rapidity of a particle, it is necessary to measure two quantities of the particle, such as its energy and its longitudinal momentum. In many experiments, it is only possible to measure the angle of the detected particle relative to the beam axis. In that case, it is convenient to utilize this information by using the *pseudorapidity variable* η to characterize the detected particle. The pseudorapidity variable of a particle c is defined as

$$\eta = -\ln[\tan(\theta/2)], \qquad (2.24)$$

where θ is the angle between the particle momentum $\boldsymbol{p}$ and the beam axis. In terms of the momentum, the pseudorapidity variable can be written as

$$\eta = \frac{1}{2}\ln\left(\frac{|\boldsymbol{p}| + p_z}{|\boldsymbol{p}| - p_z}\right). \qquad (2.25)$$

By comparing Eqs. (2.14) and (2.25), it is easy to see that the pseudorapidity variable coincides with the rapidity variable when the momentum is large, that is, when $|\boldsymbol{p}| \approx p_0$.

We consider the change of variables from $(y, \boldsymbol{p}_T)$ to $(\eta, \boldsymbol{p}_T)$. It is easy to express y as a function of η, and vice versa. From the definition of η, we have

$$e^\eta = \sqrt{\frac{|\boldsymbol{p}| + p_z}{|\boldsymbol{p}| - p_z}} \qquad (2.26)$$

and

$$e^{-\eta} = \sqrt{\frac{|\boldsymbol{p}| - p_z}{|\boldsymbol{p}| + p_z}}. \qquad (2.27)$$

Adding Eqs. (2.26) and (2.27), we obtain the relation

$$|\boldsymbol{p}| = p_T \cosh\eta,$$

where p_T is the magnitude of the transverse momentum:

$$p_T = \sqrt{\boldsymbol{p}^2 - p_z^2}.$$

Subtracting Eq. (2.27) from (2.26), we obtain

$$p_z = p_T \sinh\eta. \qquad (2.28)$$

Using these results, we can express the rapidity variable y in terms of the pseudorapidity variable η as

$$y = \frac{1}{2} \ln \left[\frac{\sqrt{p_T^2 \cosh^2 \eta + m^2} + p_T \sinh \eta}{\sqrt{p_T^2 \cosh^2 \eta + m^2} - p_T \sinh \eta} \right], \tag{2.29}$$

where m is the rest mass of the particle. Conversely, the pseudorapidity variable η can be expressed in terms of the rapidity variable y by

$$\eta = \frac{1}{2} \ln \left[\frac{\sqrt{m_T^2 \cosh^2 y - m^2} + m_T \sinh y}{\sqrt{m_T^2 \cosh^2 y - m^2} - m_T \sinh y} \right]. \tag{2.30}$$

If the particles have a distribution $dN/dy d\mathbf{p}_T$ in terms of the rapidity variable y, then the distribution in the pseudorapidity variable η is

$$\frac{dN}{d\eta d\mathbf{p}_T} = \sqrt{1 - \frac{m^2}{m_T^2 \cosh^2 y}} \, \frac{dN}{dy d\mathbf{p}_T}. \tag{2.31}$$

In many experiments, only the pseudorapidity variable of the detected particles is measured to give $dN/d\eta$, which is the integral of $dN/d\eta d\mathbf{p}_T$ with respect to the transverse momentum. One can compare this quantity with dN/dy, which is the integral of $dN/dy d\mathbf{p}_T$ with respect to the transverse momentum. From Eq. (2.31), we can infer that in the region of y much greater than zero, $dN/d\eta$ and dN/dy are approximately the same, but in the region of y close to zero, there is a small depression of the $dN/d\eta$ distribution relative to dN/dy due to the above transformation (2.31). In experiments at high energies where dN/dy has a plateau shape, this transformation gives a small dip in $dN/d\eta$ around $\eta \approx 0$.

The transformation (2.31) reveals the difference in the maximum magnitude of the pseudorapidity distribution for $dN/d\eta$, whether η is measured in the laboratory frame or in the center-of-mass frame. In the center-of-mass frame, the peak of the distribution is located around $y \approx \eta \approx 0$, and the peak value of $dN/d\eta$ is smaller than the peak value of dN/dy by approximately the factor $(1 - m^2/< m_T^2 >)^{1/2}$. In the laboratory frame, the peak of the distribution is located around half of the beam rapidity $\eta \approx y_b/2$ for which the factor $[1 - m^2/< m_T^2 > \cosh^2(y_b/2)]^{1/2}$ is about unity. The peak value of

$dN/d\eta$ is approximately equal to the peak value of dN/dy. Because the shape of the rapidity distribution dN/dy does not change when one goes from the center-of-mass to the laboratory frame, the peak value of the pseudorapidity distribution in the center-of-mass frame is lower than the peak value of the pseudorapidity distribution in the laboratory frame.

§Reference for Chapter 2

1. W. Pauli, *Theory of Relativity*, Pergamon Press, N.Y., 1958, p. 25.

3. Nucleon-Nucleon Collisions

A nucleus is a composite many-nucleon system. Nucleus-nucleus collisions involve the dynamics of colliding nucleons. Nucleon-nucleon reaction data provide valuable information for the discussion of nucleus-nucleus collisions. For example, an important question for nucleus-nucleus collisions at high energies is the extent to which the longitudinal kinetic energies carried initially by the colliding nuclei are dissipated by the collisions, with the release of energy into other degrees of freedom. One would like to know whether the longitudinal energy dissipated in a nucleus-nucleus collision leads to energy densities of matter high enough to allow the formation of a quark-gluon plasma. The search for any exotic behavior of quark-gluon plasma in nucleus-nucleus collisions also requires a comparison with what is expected from phenomenological nucleon-nucleon extrapolations. It is therefore necessary to determine what is expected from such extrapolations based on information from nucleon-nucleon data. We shall discuss these topics in later chapters. Here, we summarize the relevant information from nucleon-nucleon collisions.

§3.1 Particle Production in Nucleon-Nucleon Collisions

The nucleon-nucleon total cross section for a center-of-mass energy 3 GeV $< \sqrt{s} <$ 100 GeV is about 40 mb [1]. The reaction processes contributing to the total reaction cross section include the elastic scattering process, in which the colliding nucleons do not lose any energy, and the inelastic processes, in which the colliding nucleons lose varying amounts of their energies. For nucleon-nucleon collisions with a center-of-mass energy 3 GeV $< \sqrt{s} <$ 100 GeV, the inelastic cross section σ_{in} is about 30 mb [1]. For example, the measurement of the pp inelastic cross section at a laboratory momentum of 100 GeV/c gives $\sigma_{in} = 31.3 \pm 1.2$ mb [2]. Thus, the inelastic cross section is much greater than the elastic cross section.

An important property of a nucleon-nucleon collision at high energies is that the probability for the colliding nucleons to lose a large fraction of their energies is quite substantial. The nucleon energy loss is taken up to produce particles. An inelastic reaction process is therefore characterized by the production of particles.

The cross sections for many hadron reactions as a function of the laboratory momentum p of the incident hadron have been parametrized by the CERN-HERA group [3]. For pp reactions, the total cross section in units of mb is given by

$$\sigma_{\text{total}} = 48 + 0.522(\ln \ p)^2 + (-4.51) \ln \ p,\qquad (3.1)$$

and the elastic cross section by

$$\sigma_{\text{elastic}} = 11.9 + 26.9p^{-1.21} + 0.169(\ln \ p)^2 + (-1.85) \ln \ p,\qquad (3.2)$$

where p is in units of GeV/c. The difference of σ_{total} and σ_{elastic} gives the inelastic cross section.

Among the inelastic reaction processes, there is the *diffractive dissociation* process. In this process, one nucleon can be considered as a region of absorption, and the interference of the scattering amplitudes from different impact parameters gives rise to a diffraction pattern in the very forward or backward direction. After a nucleon suffers a diffractive scattering, it may become slightly excited and lose a relatively small amount of their energies. Only a very small number of particles are produced. On the other hand, in a nondiffractive inelastic event, the colliding nucleons lose a substantial fraction of their kinetic energies and a large number of particles are produced. The fraction of inelastic cross section which can be attributed to diffractive dissociation is not certain because the separation of diffractive and nondiffractive components is not without ambiguities. It is of the order of 10%. In our discussion of particle production and the 'stopping' of baryons, in which the nucleons sustain a substantial loss of their longitudinal kinetic energy, we focus our attention on nondiffractive inelastic collision processes, which constitute the dominant component of the inelastic events. For brevity, we shall often use simply the term 'inelastic nucleon-nucleon collision' to mean a nondiffractive inelastic nucleon-nucleon collision, except in cases where such an abbreviated term will lead to ambiguities.

Experimental data [4] of nucleon-nucleon collisions reveal that about 80-90% of the produced particles are pions; the rest consists of kaons, baryons, anti-baryons, and other particles. The total number of particles produced in a collision is called the *multiplicity* of the collision. Many measurements use detection methods which are only sensitive to ionizing particles so that only the charged particles are detected. It is therefore useful to speak of the total number of charged particles produced in the collision as the *charged multiplicity* of the collision. The charged multiplicity in a pp collision increases with the center-of-mass energy $\sqrt{s}$ approximately in a logarithmic way [5]. Figure 3.1 gives the average charged multiplicity in pp collisions as a function of $\sqrt{s}$ of the colliding nucleon-nucleon system [5]. On the

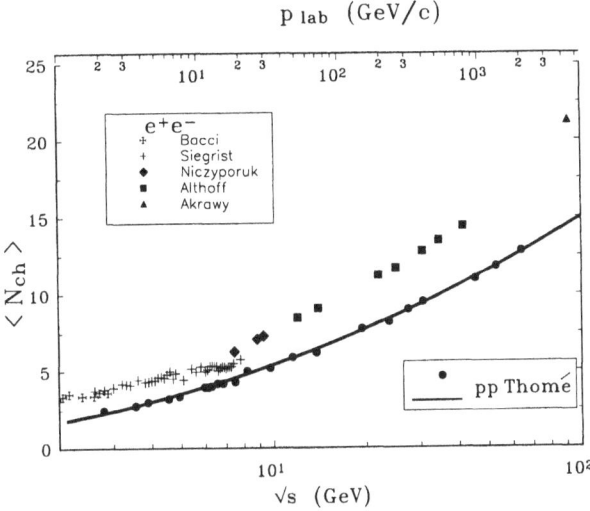

Fig. 3.1 Average charged multiplicity for pp reactions and e^+e^- annihilations. Data are from Refs. [4]-[6]. The solid curve is an extrapolation from Thomé [5].

upper horizontal axis in the same figure, we also give the corresponding laboratory momentum p_{lab} of the incident proton beam particles for collisions with a fixed proton target. The solid curve in Fig. 3.1 is the fit by Thomé *et al* to the pp data [5] parametrized in the form

$$< N_{ch} >= 0.88 + 0.44 \ln s + 0.118(\ln s)^2 , \qquad (3.3)$$

where s is in units of GeV^2. For orientation, we observe that at the incident beam momentum of 100 GeV/c, a nucleon-nucleon collision produces on the average about 6 charged particles. Neutral particles are also produced in the collision. Most of the neutral particles are π^0's. To estimate the number of neutral particles produced, we can assume that the detected charged particles are π^+ and π^-, and that equal numbers of π^+, π^-, and π^0 are produced. One can therefore estimate that the average total number of particles produced at an incident proton energy of 100 GeV is about $6 \times \frac{3}{2} = 9$.

In hadron-hadron collisions, there is a large probability of finding particles which closely resemble the incident particles in the respective fragmentation regions. They carry a substantial fraction of the incident center-of-mass energy and are among the fastest particles in the forward or the backward direction. Particles which resemble the incident particles in the fragmentation regions are called the leading particles. The phenomenon that these leading particles appear in the fragmentation regions is called the *leading particle effect*. A *pp* collision differs from an e^+e^- annihilation in that there is a *leading-particle effect* in a *pp* collision, but not in an e^+e^- annihilation. The colliding baryons in a *pp* collision retain about half of their momentum fractions after the collision (see Section 3.2 below). Only about half of the center-of-mass energy is used to produce particles. In consequence, the multiplicity of charged particles produced in a *pp* collision is lower than the corresponding multiplicity in an e^+e^- collision with the same center-of-mass energy. This is demonstrated in Fig. 3.1 where the average multiplicity of charged particles produced in e^+e^- annihilation is also plotted as a function of the center-of-mass energy of the colliding system [6].

The produced particles have a distribution of momenta. The longitudinal momentum distribution of the produced particles can be inferred from their distribution in rapidity y, or in pseudorapidity η. As discussed in Chap. 2, the pseudorapidity and the rapidity of a detected particle are approximately equal to each other when the energy of the particle is large. The pseudorapidity of a particle requires the measurement of only one kinematic quantity for the particle (such as the angle of emission of the particle relative to the beam axis), while the rapidity requires the measurement of two kinematic quantities. For this reason, in many experiments only the pseudorapidity variables of particles are measured, rather than the rapidity variables. As one can see from Fig. 3.2, in a *pp* collision for $p_{lab} = 100$ GeV/c [7], which corresponds to $\sqrt{s}$=13.8 GeV, the pseudorapidity distribution $dN_{ch}/d\eta_{lab}$ of the detected charged particles has the form of a bell-shaped curve. But at the higher energies of $\sqrt{s} =$ 23.6 and 62.8 GeV [5], the pseudorapidity distribution $dN_{ch}/d\eta_{CM}$ of charged particles assumes the shape of a plateau, reaching a value of about 2 in the central rapidity region for $\sqrt{s} = 62.4$ GeV. To exhibit the structure of the plateau in the $dN_{ch}/d\eta_{CM}$ case, we have plotted the distribution for negative η_{CM} by reflecting the data $dN_{ch}/d\eta_{CM}$ for positive η_{CM} with respect to $\eta_{CM} = 0$.

For many practical applications, it is desirable to parametrize the particle production data in terms of simple functions. In the $p + p \rightarrow \pi + X$ reaction, the distribution of the forward light-cone variable x_+ in the projectile fragmentation region follows a simple $(1 - x_+)^a$ relation, with $a \approx 3 - 4$ [2] which is related to the counting rules for constituents in a nucleon [8] (see Chap. 4). By symme-

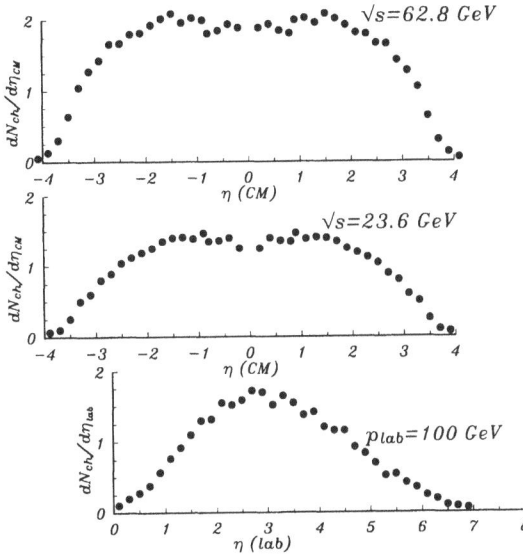

Fig. 3.2 The pseudorapidity distribution of charged particle $dN_{ch}/d\eta$ in pp reactions at various energies, from Refs. [5] and [7]. For the $p_{lab} = 100$ GeV/c case, the pseudorapidity variable η_{lab} is measured in the laboratory system. For the other two cases, the pseudorapidity variable η_{CM} is measured in the center-of-mass system.

try, there is an analogous distribution for the backward light-cone variable x_- for pions produced in the target fragmentation region. As the production of pions dominates the production process, it is convenient to parametrize the rapidity distribution of the produced charged particles in the form:

$$dN/dy = A(1 - x_+)^a(1 - x_-)^a \,, \qquad (3.4)$$

where x_+ and x_- are related to the rapidity y by Eqs. (2.20) and (2.22).

The average transverse momentum of the produced pions is about 350 MeV/c, which increases for other heavier produced particles and for increasing collision energies [4]. Because the invariant momentum element dp/E is equal to $dp_z \, dp_T/E$ we shall call the distribution $dN/dp_T (= dN/2\pi |p_T| d|p_T|)$ the *transverse momentum distribution* (not to be confused with the distribution $dN/d|p_T|$ sometimes used in the literature). The transverse momentum distribution dN/dp_T

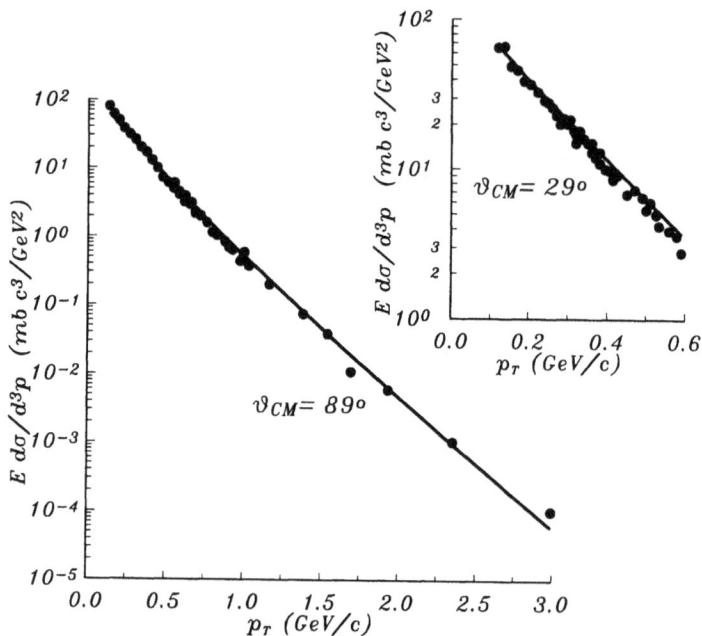

Fig. 3.3 The differential invariant cross section $E d\sigma/d\boldsymbol{p}$ for π^+ in pp reactions at $\sqrt{s}$=30.6 GeV. The solid points are the experimental data points from Ref. [4]. The solid curves are from the parametrization of Eq. (3.5).

of the produced pions has an approximately exponential shape. As an illustration, we show in Fig. 3.3 the differential invariant cross section $E d\sigma/dp_z d\boldsymbol{p}_T$ at $\theta_{\rm CM} = 89°$ for $p + p \rightarrow \pi^+ + X$ as a function of the transverse momentum $p_T (= |\boldsymbol{p}_T|)$ at a center-of-mass energy of 30.6 GeV [4]. The differential invariant cross section is the joint distribution with respect to the transverse momentum $\boldsymbol{p}_T$ and the longitudinal momentum p_z. The differential invariant cross section at $\theta_{\rm CM} \approx 90°$ gives the variation of the distribution with respect only to the transverse momentum, which is approximately an exponential function (Fig. 3.3). It can be represented by the sum of two exponential functions of p_T with different slope parameters. For example, the invariant cross section in Fig. 3.3 for $\sqrt{s} = 30.6$ GeV at center-of-mass

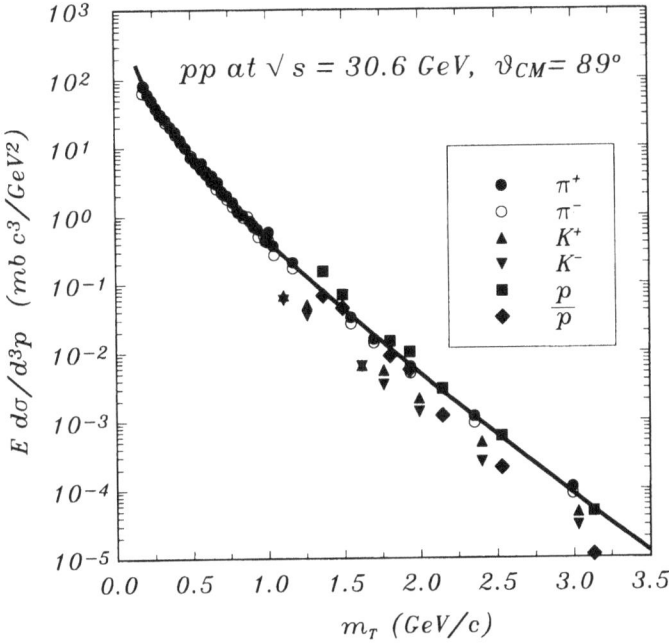

Fig. 3.4 Invariant cross section for the production of different particles in pp reactions at $\sqrt{s}$=30.6 GeV and $\theta_{CM} = 89°$, as a function of the transverse mass m_T [4]. The solid curve is from the parametrization of Eq. (3.7).

angles $\theta_{CM} = 29°$ and $89°$ can be represented by

$$E\frac{d\sigma}{d^3p} = [(1 - x_+)(1 - x_-)]^4[110e^{-6.6p_T} + 35e^{-4.2p_T}] \ \text{mb} \cdot \text{c}^3/\text{GeV}^2 ,$$

$$(3.5)$$

where p_T is given in units of GeV/c. The invariant cross section, as obtained by Eq. (3.5), is shown as the solid curve in Fig. 3.3. The general trend of the transverse momentum distribution is that the low transverse momentum part of the spectrum has a steep exponential slope which becomes less steep as the transverse momentum increases.

A useful way to exhibit the transverse momentum distribution of produced particle is to use the transverse mass variable

$$m_T^2 = m^2 + \boldsymbol{p}_T^2 , \qquad (3.6)$$

and to display the invariant differential cross section as a function of the transverse mass. We show in Fig. 3.4 the experimental data for the invariant cross section for the production of different kinds of particles, for pp collisions at $\sqrt{s} = 30.6$ GeV and $\theta_{CM} = 89°$, as a function of the transverse mass m_T [4]. The solid curve is the parametrization of the invariant cross section in the form [9]

$$E\frac{d\sigma}{d^3p} = A\frac{e^{-m_T/T}}{(m_T/\text{GeV})^\lambda}, \qquad (3.7)$$

with $T = 0.290$ GeV, $\lambda = 1.5$, and $A=13.9$ (mb·c³/GeV²). The data in Fig. 3.4 indicate that the invariant cross sections for the production of different types of particles as a function of m_T have similar shapes so that the parametrizations of these differential cross sections in terms of $e^{-m_T/T}/m_T^\lambda$ will have roughly the same T coefficients. This approximately universal behavior with respect to m_T is often called "m_T scaling".

Among the produced particles, one can refer to those produced particles with a transverse momentum p_T much below 1 GeV/c as 'soft' particles, and particles with transverse momentum far exceeding 1 GeV/c as 'hard' particles. The production mechanisms for these two types of particles are different. The production of soft particles belongs in the realm of nonperturbative QCD, while the production of hard particles can be described by perturbative QCD. As one can observe from Figs. 3.3 and 3.4, soft hadron production is an important process in hadron collisions. It occurs with a higher probability than the hard processes. However, because of the difficulty with nonperturbative studies of QCD dynamics, the production of soft hadrons can only be studied phenomenologically and qualitatively. Models based on the Schwinger mechanism [10], quantum electrodynamics in two dimensions [11,12], preconfinement [13], parton-hadron duality [14], cluster fragmentation [15], string-fragmentation [16], dual-partons [17], and topological domains [18] have been proposed and are successful in describing some aspects of the data. We shall describe some of these models later.

§3.2 Baryon Energy Loss in an Inelastic Collision

In any reaction, the net number of baryons (number of baryons minus the number of antibaryons) is a conserved number. Therefore, in a nucleon-nucleon (or a baryon-baryon) inelastic collision, there must be at least two baryons among the product hadrons because of this *law of baryon number conservation*. One baryon is likely to be found in the projectile fragmentation region and another in the target fragmentation region. Baryons found in these two regions are the *leading particles*. It should be noted that the term 'leading' only refers to the general fragmentation regions where the baryons are likely be found after a collision; it is not intended to imply that the leading particles must 'lead' the other produced particles as the most energetic, because there is a probability, no matter how small, that other produced particles may take up a large fraction of the light-cone momenta of the incident baryons. If one considers the leading baryons as related to the corresponding incident baryons, one can view an inelastic collision as a reaction in which the two energetic baryons suffer a degradation of their light-cone momenta after the collision and emerge as the leading baryons.

Because the characteristics of the target fragmentation process are identical to those of the projectile fragmentation, we shall focus our attention on the projectile fragmentation reactions and shall abbreviate x_+ by x for simplicity. In a baryon-baryon collision, an incident projectile baryon loses a substantial fraction of its light-cone momentum to emerge as a leading baryon. The degree of inelasticity can be characterized by the forward light-cone variable x, defined as the ratio of the (forward) light-cone momentum of the detected leading baryon to that of the incident parent baryon [see Eq. (2.10)].

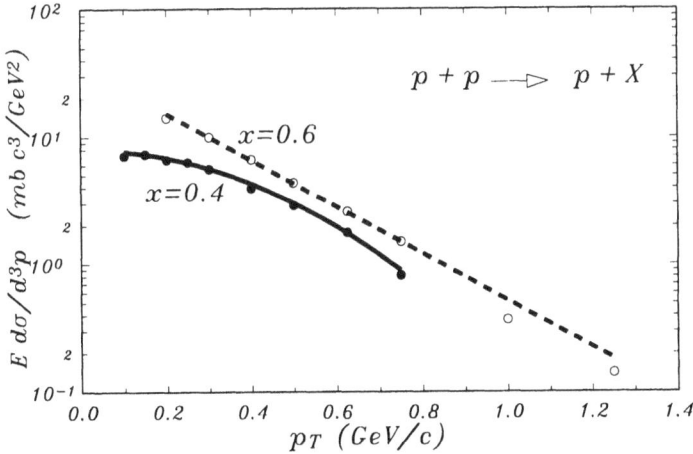

Fig. 3.5 The invariant cross section $E d\sigma/d^3 p$ for the reaction $p+p \to p+X$ at $p_{lab}=100$ GeV/c as a function of the transverse momentum p_T for $x = 0.6$ and $x = 0.4$. The data are from Ref. [2].

The transverse momentum distribution of the leading proton has a shape which depends on the degree of energy loss [2]. For a pp inelastic collision leading to a proton with x close to unity, the invariant cross section $E d\sigma/d^3p$ is in the form of an exponential $e^{-\alpha p_T}$, while for lower values of x it is in the form of a gaussian $e^{-\beta p_T^2}$. We show in Fig. 3.5 an example of data for $p + p \rightarrow p + X$ at an incident momentum of 100 GeV/c [2]. The invariant cross section at $x = 0.6$ can be approximated as the dashed curve in Fig. 3.5 given by

$$E\frac{d\sigma}{d^3p} = 35.5 e^{-4.22 p_T} \quad \text{mb} \cdot \text{c}^3/\text{GeV}^2 \,,$$

where p_T is in units of GeV/c. The average transverse momentum is

$$< p_T >= \frac{2}{4.22} \quad \text{GeV/c} = 0.47 \quad \text{GeV/c} \,. \tag{3.8}$$

At $x = 0.4$, the invariant cross section can be approximated as the solid curve which is parametrized as

$$E\frac{d\sigma}{d^3p} = 8 e^{-3.9 p_T^2} \quad \text{mb} \cdot \text{c}^3/\text{GeV}^2 \,.$$

The average transverse momentum for $x = 0.4$ is

$$< p_T >= \frac{\sqrt{\pi}}{2 \times 3.9} \quad \text{GeV/c} = 0.45 \quad \text{GeV/c} \,. \tag{3.9}$$

Even though the shapes of the distributions for different values of x are different, the average values of p_T are very close.

What is the distribution of the light-cone variable of the leading protons? We show in Fig. 3.6 the experimental inelastic cross section data $d\sigma/dx$ for the reaction $p + p \rightarrow p + X$ at an incident momentum of 100 GeV/c, obtained for each value of x by integrating over the transverse momentum. The cross section $d\sigma/dx$ is nearly independent of the incident energy. For x close to unity, the light-cone variable x and the Feynman scaling variable x_F nearly coincide (see Exercise 2.3). The phenomenon of *Feynman scaling* refers to the situation when the cross section plotted as a function of the Feynman scaling variable x_F or x of the detected particle, for x_F or $x \gtrsim 0.2\text{-}0.3$, is independent of the incident energy. Thus, the present case of $d\sigma/dx$ being nearly independent of the incident energy, for $x \gtrsim 0.2$, can be called the occurrence of Feynman scaling. Fig. 3.6 shows that there is Feynman scaling for the inelastic cross section when p_{lab} exceeds 100

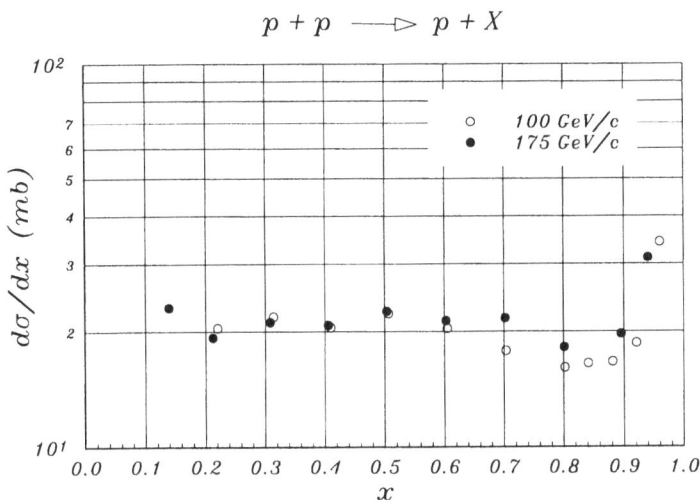

Fig. 3.6 The differential cross section $d\sigma/dx$ for the reaction $p+p \to p+X$ from Ref. [2].

GeV/c. A systematic examination of the nucleon-nucleon data shows that Feynman scaling commences at a lower energy of many tens of GeV in the laboratory system [19].

The independence of the measured cross section on incident energies in Feynman scaling implies that the cross section measures some intrinsic properties of the detected particle relative to the parent particle. The existence of such independence leads to the concept of partons as hadron constituents [20]. The detected particle arises from the hard scattering of the constituents of the colliding particles and their subsequent fragmentations, or from the direct fragmentation of one of the parent particles (see next Chapter). In the kinematic region with x close to unity, the cross section for these hard processes depends only on the structure function of the parton in the beam particle, which is the the momentum distribution of the parton in the parent beam particle, and on the fragmentation function of the parton, which describes the fragmentation distribution of the detected particle in the parton. If constituent partons are valid concepts, these two quantities should be the properties of the parent particle and of the fragmenting parton. They should not depend on the collision energy. Feynman scaling occurs as a consequence. The experimental observation of Feynman scaling provides support to the usefulness of the concept of constituent partons.

The shape of the distribution $d\sigma/dx$ as a function of the light-cone variable x reveals the degree of energy loss in an inelastic collision. Except for the diffractive dissociation region in the vicinity of x close to 1, the distribution is nearly flat. If one extrapolates to small regions of x but does not include diffractive dissociation in the description, one can approximate the differential cross section $d\sigma/dx$ as

$$d\sigma/dx \approx \sigma_{in}\theta(1-x)\theta(x-x_L)/(1-x_L), \qquad (3.10)$$

where σ_{in} is the total inelastic cross section, while x_L is the lower limit of x, as required by energy and momentum conservation. The value of x_L is approximately $e^{y_a-y_b}$, where y_a is the target rapidity and y_b is the projectile rapidity. The lower limit x_L is much smaller than unity. The distribution (3.10) shows that after an inelastic nucleon-nucleon collision, there is an equal probability of finding a product baryon in the whole range of the light-cone variable. The average value of x is

$$< x > \approx 1/2. \qquad (3.11)$$

Therefore, on the average, the product baryon carries about half of the initial light-cone momentum. This implies that about half of the initial light-cone momentum of the baryon is lost. Thus, if the incident baryon carries tens or hundreds of GeV, then the energy-momentum loss can be very substantial.

From the distribution (3.10), we can infer an approximate average rapidity loss for the incident baryon in a pp collision. We consider a leading proton with a transverse momentum of 0.46 GeV/c (Eqs. (3.8) and (3.9)), the average magnitude of the transverse momentum of the leading protons. The cross section $d\sigma/dy$ is

$$\frac{d\sigma}{dy} = \frac{d\sigma}{dx}\frac{dx}{dy}$$

$$= \frac{d\sigma}{dx}\frac{m_{cT}}{m_b}e^{y-y_b}.$$

With $d\sigma/dx$ approximately a constant, the rapidity distribution for the leading proton is approximately

$$\frac{d\sigma}{dy} \propto e^{y-y_b}. \qquad (3.12)$$

The average rapidity after a pp inelastic collision is then

$$< y > = \left\{\int y\,\frac{d\sigma}{dy}dy\right\}\Big/\left\{\int \frac{d\sigma}{dy}dy\right\} \qquad (3.13)$$

$$\approx y_b - 1.$$

Thus, the distribution of Eq. (3.10) implies that on the average, the incident proton loses about one unit of rapidity in a pp inelastic collision.

In nucleus-nucleus collisions, the nucleons of one nucleus suffer many collisions with nucleons of the other nucleus. In such a multiple-collision process, the loss of the incident energy and momentum can be quite large and may lead to the "stopping" of the baryons in the center-of-mass system. We shall return to this topic later.

The production of particles and the energy loss of the baryons are intimately related because the total energy of the system must be conserved. There is, in fact, an experimental correlation between the baryon energy loss and the multiplicity of particles produced [2,21]. The greater the energy loss of the baryons, the greater will be the number of particles produced.

§References for Chapter 3

1. For a general review of experimental data, see *Review of Particle Properties*, K. Hikasa *et al.*, Particle Data Group, Phys. Rev. D45, S1 (1992).
2. A. E. Brenner *et al.*, Phys. Rev. D26, 1497 (1982).
3. See Page III.83 of reference 1.
4. K. Alpgard *et al.*, Phys. Lett. 107B, 310 (1981); K. Alpgard *et al.*, Nucl. Phys. B87, 19 (1975).
5. W. Thomé *et al.*, Nucl. Phys. B129, 365 (1977).
6. ADONE: C. Bacci *et al.*, Phys. Lett. B76, 29 (1974); MARK II: J. L. Siegrist *et al.*, Phys. Rev. D26, 969 (1982); LENA: B. Niczyporuk *et al.*, Zeit. Phys. C9, 1 (1981); TASSO: M. Althoff *et al.*, Zeit. Phys. C22, 307 (1984); and OPAL: M. Akrawy *et al.*, Zeit. Phys. C47, 505 (1990).
7. Michigan-Rochester 100 GeV pp bubble-chamber exposure, as reported by J. E. Elias *et al.*, Phys. Rev. D22, 13 (1980).
8. S. J. Brodsky and G. R. Farrar, Phys. Rev. Lett. 31, 1153 (1973); Phys. Rev. D11, 1309 (1975); R. Blankenbecler and S. Brodsky, Phys. Rev. D10, 2973 (1974).
9. G. Gatoff and C. Y. Wong, Phys. Rev. D46, 997 (1992).
10. J. Schwinger, Phys. Rev. 82, 664 (1951).
11. J. Schwinger, Phys. Rev. 128, 2425 (1962); J. Schwinger, in *Theoretical Physics*, Trieste Lectures, 1962 (I.A.E.A., Vienna, 1963), p. 89.
12. A. Casher, J. Kogut, and L. Susskind, Phys. Rev. D10, 732 (1974).
13. S. Wolfram, Proc. 15th Recontre de Moriond (1980), ed. by Tran Thanh Van; G. C. Fox and S. Wolfram, Nucl. Phys. 238, 492 (1984).
14. L. Van Hove and A. Giovannini, Acta Phys. Polon. B19, 931 (1988); M. Garetto, A. Giovannini, T. Sjostrand, and L. van Hove, CERN Report CERN-TH-5252/88, Presented at Perugia Workshop on Multiparticle Dynamics, Perugia, Italy, June 21-28, 1988; Y. L. Dokshizer, V. A. Khoze and S. I. Troyan, in *Perturbative Quantum Chromodynamics*, ed. by A. H. Mueller, World Scientific Publishing, Singapore 1989, p. 241.
15. R. O'dorico, Nucl. Phys. B172, 157 (1980); R. O'dorico, Comp. Phys. Comm. 32, 139 (1984); G. Marchesini and B. R. Webber, Nucl. Phys. B238, 1 (1984);

B. R. Webber, Nucl. Phys. B238, 492 (1984); T. D. Gottschalk, Nucl. Phys. B239, 325 (1984).
16. B. Andersson, G. Gustafson, and T. Sjöstrand, Zeit. für Phys. C20, 317 (1983); B. Andersson, G. Gustafson, G. Ingelman, and T. Sjöstrand, Phys. Rep. 97, 31 (1983); T. Sjöstrand and M. Bengtsson, Comp. Phys. Comm. 43, 367 (1987); M. Gyulassy,"Attila", CERN Report CERN-TH.4794/87, 1987.
17. A. Capella and A. Krzywicki, Phys. Rev. D18, 3357 (1978); A. Capella and J. Tran Thanh Van, Zeit. Phys. C10, 249 (1981); A. Capella et al., Zeit. Phys. C33, 541 (1987).
18. J. Ellis, M. Karliner, and H. Kowalski, Phys. Lett. 235, 341 (1990); J. Ellis, and H. Kowalski, Phys. Lett. 214, 161 (1988).
19. F. E. Taylor et al., Phys. Rev. 14, 1217 (1976).
20. R. P. Feynman, Phys. Rev. Lett. 23, 1415 (1969).
21. M. Basile et al., Nuovo Cim. 65A, 400 (1981); Nuovo Cim. 67A, 244 (1981).

4. Hard Processes in Nucleon-Nucleon Collisions

We learned from the last chapter that the transverse momentum distribution dN/dp_T of particles produced in a nucleon-nucleon collision behaves approximately as $\exp\{-\alpha p_T\}$, and the average value of p_T is ~ 0.3 GeV/c. The region of transverse momentum with $p_T \sim 0.3$ GeV/c is called the soft p_T region. The rapidity distribution of the produced particles is the greatest in the mid-rapidity region. Processes which lead to the production of particles with a p_T in the soft p_T region and a rapidity in the mid-rapidity region are *soft processes*. They are associated with a length scale that is large in the context of quantum chromodynamics. They belong to the realm of nonperturbative QCD. There is a different kind of process, the *hard processes* [1-3], which are important in the production of particles with $p_T \gg 1$ GeV/c, or with a momentum fraction x close to unity. Since these processes involve large momentum transfers which are associated with a small coupling constant in QCD, the hard processes are examples of reactions in which the concept of partons is useful and the techniques of perturbative QCD may be applicable. We shall develop the tools for their study in this chapter. The concept of partons can also be applied to nucleus-nucleus collisions at high energies where the 'partons' of a nucleus are nucleons [2].

§4.1 The Infinite-Momentum Frame

We consider the collision of a beam particle B and a target particle A. For highly relativistic collisions, we represent the momentum of B and A by

$$B = (B_0, \boldsymbol{B}_T, B_z) = \left(P_1 + \frac{B^2 + B_T^2}{4P_1}, \boldsymbol{B}_T, P_1 - \frac{B^2 + B_T^2}{4P_1} \right) \quad (4.1)$$

and

$$A = (A_0, \boldsymbol{A}_T, A_z) = \left(P_2 + \frac{A^2 + A_T^2}{4P_2}, \boldsymbol{A}_T, -P_2 + \frac{A^2 + A_T^2}{4P_2} \right), \quad (4.2)$$

where we use the same symbol to represent a particle and its momenta. This is a rather general way to write down the momenta of particles, whether they are on the mass shell or not. One can easily check, by direct substitution, that

$$B_0^2 - B_z^2 - B_T^2 = B^2$$

and
$$A_0^2 - A_z^2 - A_T^2 = A^2 .$$

In our case of B colliding on A, the colliding particles are on the mass shell and we have $B^2 = m_B^2$ and $A^2 = m_A^2$.

There is much freedom to choose P_1 and P_2, representing different choices of the Lorentz frame. Without loss of generality, we can choose P_1 and P_2 to be positive and can arrange the coordinate system in such a way that the initial transverse momenta are zero:

$$B_T = A_T = \vec{0} .$$

Writing the momenta A and B in the form of Eqs. (4.1) and (4.2) is sometimes referred to as representing these momenta in the *infinite-momentum frame*. The term "infinite-momentum frame" refers to any frame in which the magnitudes of the longitudinal momenta of the particles are very large, corresponding to the situation where both P_1 and P_2 are much greater than the rest masses of A and B. This is possible when the beam momentum is much greater than the rest masses, as in high-energy collisions. The infinite-momentum frames are distinctly different from the target rest frame or the projectile rest frame, in which the longitudinal momentum of one of the two colliding particles is zero. An infinite-momentum frame can be obtained from the target frame or the projectile frame by a boost of the coordinate system in the longitudinal direction with a speed close to the speed of light.

Equations (4.1) and (4.2) represent a beam particle B moving in the positive z-direction, and a target particle A moving in the opposite direction. Among all the infinite-momentum frames, one can choose the *center-of-mass frame* for which A_z and B_z are equal and opposite. We can use this condition to determine the quantities P_1 and P_2 for the center-of-mass frame. From the relations (4.1) and (4.2) for the longitudinal momenta A_z and B_z, the center-of-mass frame is characterized by

$$P_1 - \frac{B^2}{4P_1} = P_2 - \frac{A^2}{4P_2} .$$

On the other hand, the center-of-mass energy of the colliding system $\sqrt{s}$ is given by

$$\sqrt{s} = \sqrt{(A+B)^2} = P_1 + \frac{B^2}{4P_1} + P_2 + \frac{A^2}{4P_2} .$$

Therefore, from these two equations, we obtain

$$P_1 + \frac{A^2}{4P_2} = \frac{\sqrt{s}}{2} \tag{4.3}$$

and

$$P_2 + \frac{B^2}{4P_1} = \frac{\sqrt{s}}{2}. \tag{4.4}$$

Upon substituting Eq. (4.4) into (4.3), we can solve for P_1. We obtain P_1 and P_2 for the center-of-mass system:

$$P_1 = [s + B^2 - A^2 + \lambda(s, A^2, B^2)]/4\sqrt{s} \tag{4.5a}$$

$$P_2 = [s + A^2 - B^2 + \lambda(s, A^2, B^2)]/4\sqrt{s}, \tag{4.5b}$$

where the function λ, introduced in Exercise 2.3, is defined by

$$\lambda(s, A^2, B^2) = [s^2 + A^4 + B^4 - 2(sA^2 + sB^2 + A^2B^2)]^{1/2}. \tag{4.5c}$$

In high-energy collisions with $s \gg A^2, B^2$, $\lambda \sim s$ and in the center-of-mass frame we have

$$P_1 \sim P_2 \sim \sqrt{s}/2. \tag{4.6}$$

§4.2 Relativistic Hard-Scattering Model

We consider the reaction

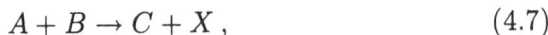

$$A + B \to C + X, \tag{4.7}$$

in which attention is focussed on the particle C while the occurrences of all other particles X are implicitly included but not sorted out. A reaction of the type is called an *inclusive reaction*. In many processes involving the collision of two composite particles B and A leading to the production of C, the reaction (4.7) can take place with the scattering of the constituent b of B and the constituent a of A as depicted in Fig. 4.1. The constituents b and a are called *partons*. The interaction of partons a and b through the basic process $b + a \to c + d$ leads to the parton c. The parton c subsequently fragments to give the detected particle C. The process depicted in 4.1 is a *relativistic hard-scattering* process [1,2]. (In the constituent interchange model [1], which is a special case of the hard-scattering model, the particles a, b, c, and d are generalized to include the possibility that they can be composite particles such as mesons. If the particle c is a hadron, it can be directly detected without the process of fragmentation.)

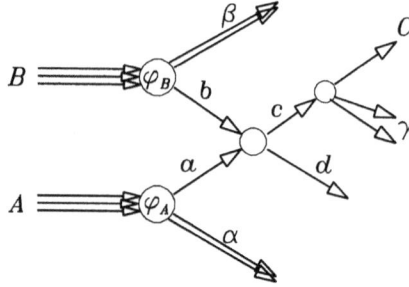

Fig. 4.1 Relativistic hard-scattering process.

We introduce the forward light-cone momentum fraction x_b of b relative to B:

$$x_b = \frac{b_0 + b_z}{B_0 + B_z}, \tag{4.8}$$

and the backward light-cone momentum fraction x_a of a relative to A:

$$x_a = \frac{a_0 - a_z}{A_0 - A_z}. \tag{4.9}$$

Using Eqs. (4.1) and (4.8), we can write the momentum b in terms of x_b and P_1,

$$b = \left(x_b P_1 + \frac{b^2 + b_T^2}{4x_b P_1}, b_T, x_b P_1 - \frac{b^2 + b_T^2}{4x_b P_1} \right). \tag{4.10}$$

Similarly, the momentum a can be written in the form

$$a = \left(x_a P_2 + \frac{a^2 + a_T^2}{4x_a P_2}, a_T, -x_a P_2 + \frac{a^2 + a_T^2}{4x_a P_2} \right). \tag{4.11}$$

As a consequence, the momenta $\beta = (B - b)$ and $\alpha = (A - a)$ can be written as

$$\beta = \left((1 - x_b)P_1 + \frac{\beta^2 + \beta_T^2}{4(1 - x_b)P_1}, -b_T, (1 - x_b)P_1 - \frac{\beta^2 + \beta_T^2}{4(1 - x_b)P_1} \right), \tag{4.12}$$

$$\alpha = \left((1 - x_a)P_2 + \frac{\alpha^2 + \alpha_T^2}{4(1 - x_a)P_2}, -a_T, (1 - x_a)P_2 + \frac{\alpha^2 + \alpha_T^2}{4(1 - x_a)P_2} \right). \tag{4.13}$$

Because $b = B - \beta$, we can form the scalar product of b with itself,

$$b^2 = B^2 - 2B \cdot \beta + \beta^2 .$$

Using Eq.(4.2) and (4.12) to get $B \cdot \beta$, we obtain

$$b^2 = [x_b(1 - x_b)B^2 - x_b\beta^2 - b_T^2]/(1 - x_b), \qquad (4.14)$$

and because $a = A - \alpha$, we obtain similarly

$$a^2 = [x_a(1 - x_a)A^2 - x_a\alpha^2 - a_T^2]/(1 - x_a) . \qquad (4.15)$$

We can treat β and α as having definite invariant on-shell masses complementary to b and a in particles B and A, respectively

$$\beta^2 = m_\beta^2$$

and

$$\alpha^2 = m_\alpha^2 .$$

The momenta a and b are off the mass-shell, because $a^2 \neq m_a^2$ and $b^2 \neq m_b^2$.

With simplifying assumptions, the relativistic hard-scattering process represented in Fig. 4.1 can be shown in Supplement 4.1 to lead to an inclusive invariant cross section given by

$$E_C \frac{d^3\sigma}{dC^3}\bigg|_{AB \to CX} = \sum_{ab} \int dx_b d\mathbf{b}_T dx_a d\mathbf{a}_T G_{b/B}(x_b, \mathbf{b}_T) G_{a/A}(x_a, \mathbf{a}_T)$$
$$\times r(s, s', x_b, x_a) E_C \frac{d^3\sigma}{dC^3}(ba \to CX') . \qquad (4.16)$$

Here, $r(s, s', x_b, x_a)$ is a kinematic factor defined by

$$r(s, s', x_b, x_a) = \frac{\lambda(s', a^2, b^2)}{x_b x_a \lambda(s, A^2, B^2)}, \qquad (4.17)$$

where $s' = (a + b)^2$. As $\lambda(s, A^2, B^2) \simeq s$, $\lambda(s', a^2, b^2) \simeq s'$, and $s' \simeq x_b x_a s$, the kinematic factor r is approximately 1. In Eq. (4.16), the symbol X represents the collection of all possible $\alpha + \beta + d + \gamma$ and the symbol X' represents the collection of all possible $d + \gamma$. The *structure function* $G_{b/B}(x_b, \mathbf{b}_T)$ is the probability of finding a constituent of type b in the particle B with a momentum fraction x_b and a transverse momentum $\mathbf{b}_T$. The other structure function $G_{a/A}(x_a, \mathbf{a}_T)$ is similarly defined as the probability of finding a constituent a in the particle A with a momentum fraction x_a and a transverse momentum $\mathbf{a}_T$. The quantity $E_C d^3\sigma/dC^3(ba \to CX')$ is the invariant cross section for the basic process $b + a \to C + X'$.

For the process depicted in Fig. 4.1, the detected particle C is the result of the fragmentation of the elementary particle c. In that case, the quantity $E_C d^3\sigma/dC^3(ba \to CX')$ can be further separated as the convolution of the constituent interaction cross section $E_c d^3\sigma/dc^3(ba \to cd)$ with the *fragmentation function* $G_{C/c}(x_C, C_T)$, which is the probability for the parton c to fragment into C. The inclusive invariant cross section for the process $B + A \to C + X$ is given by

$$E_C \frac{d^3\sigma}{dC^3}\bigg|_{AB \to CX} = \sum_{ab,cd} \int dx_b d\boldsymbol{b}_T dx_a d\boldsymbol{a}_T G_{b/B}(x_b, \boldsymbol{b}_T) G_{a/A}(x_a, \boldsymbol{a}_T)$$

$$\times r(s, s', x_b, x_a) \int dx_C d\boldsymbol{c}_T G_{C/c}(x_C, \boldsymbol{C}_T) E_c \frac{d^3\sigma}{dc^3}\bigg|_{ba \to cd} . \qquad (4.18)$$

$\bigoplus$〚 Supplement 4.1
We show in detail below how we can obtain the main results of the hard-scattering model. In this derivation, we follow the steps of Sievers, Brodsky, and Blankenbecler [1].

Using the rules for Feynman graphs [4], the differential cross section for $A + B \to C + X$ represented by Fig. 4.1 is

$$d\sigma = \frac{1}{2E_A E_B |V_A - V_B|} \left| M_{AB \to CX} \right|^2 dp, \qquad (1)$$

where E_i and V_i are respectively the energy and the velocity of particle i. To write down the matrix element $M_{AB \to CX}$, we associate different factors with different elements of graph 4.1. For simplicity, we shall assume that the appropriate spin summation and traces have been carried out so that the particles in diagram 4.1 can be treated effectively as spinless particles. We associate, for example, a vertex function $\phi_B(b)$ for the vertex $B \to b + \beta$, and a propagator $1/(b^2 - m_b^2)$ for the internal line b. The matrix element $M_{AB \to CX}$ is

$$M_{AB \to CX} = \sum_{ab,cd} \frac{\phi_A(a)}{(a^2 - m_a^2)} \frac{\phi_B(b)}{(b^2 - m_b^2)} \frac{\phi_c(C)}{(c^2 - m_c^2)} M_{ba \to cd} .$$

Assuming that the constituents $a, b, c,$ and d are distinct and 'localized' so that each distinct set of $a, b, c,$ and d leads to distinct final states, there is then no coherence in the decomposition and

$$\left| M_{AB \to CX} \right|^2 = \sum_{ab,cd} \frac{\phi_A^2(a)}{(a^2 - m_a^2)^2} \frac{\phi_B^2(b)}{(b^2 - m_b^2)^2} \frac{\phi_c^2(C)}{(c^2 - m_c^2)^2} \left| M_{ab \to cd} \right|^2 . \qquad (2)$$

The momentum volume element dp in Eq. (1) is

$$dp = \frac{d^4\beta}{(2\pi)^3} \delta(\beta^2 - m_\beta^2) \frac{d^4\alpha}{(2\pi)^3} \delta(\alpha^2 - m_\alpha^2) \frac{d^4\gamma}{(2\pi)^3} \delta(\gamma^2 - m_\gamma^2)$$

$$\times \frac{d^4C}{(2\pi)^3}\delta(C^2 - m_C^2)\frac{d^4d}{(2\pi)^3}\delta(d^2 - m_d^2)(2\pi)^4\delta(a + b - c - d).\qquad(3)$$

To obtain the inclusive cross section for $A + B \rightarrow C + X$, with X representing the collection β, α, γ and d, we need to sum over the final states. That is, we need to integrate over β, α, γ, and d (see Fig. 4.1). The integration over d can be easily carried out and gives just the constraint $a + b = c + d$. To integrate over β and α, we write them out explicitly. Following the procedures as in Exercise 2.2, we have

$$d^4\beta = \frac{d\boldsymbol{b}_T dx_b d\beta^2}{2(1 - x_b)}.$$

Therefore, we obtain from (2) and (3)

$$\int \frac{d^4\beta}{(2\pi)^3}\delta(\beta^2 - m_\beta^2)\frac{\phi_B^2(b)}{(b^2 - m_b^2)^2}\cdots = \int \frac{d\boldsymbol{b}_T}{(2\pi)^3}\frac{dx_b d\beta^2}{2(1 - x_b)}\delta(\beta^2 - m_\beta^2)\frac{\phi_B^2(b)}{(b^2 - m_b^2)^2}\cdots$$

$$= \int \frac{d\boldsymbol{b}_T}{(2\pi)^3}\frac{dx_b}{2(1 - x_b)}\frac{\phi_B^2(b)}{(b^2 - m_b^2)^2}\cdots .$$

$$(4)$$

We introduce the structure function $G_{b/B}(x_b, \boldsymbol{b}_T)$ defined as

$$G_{b/B}(x_b, \boldsymbol{b}_T) = \frac{1}{(2\pi)^3}\frac{x_b}{2(1 - x_b)}\frac{\phi_B^2(b)}{(b^2 - m_b^2)^2}.\qquad(5)$$

Equation (4) can be rewritten as

$$\int \frac{d^4\beta}{(2\pi)^3}\delta(\beta^2 - m_\beta^2)\frac{\phi_B^2(b)}{(b^2 - m_b^2)^2}\cdots = \int d\boldsymbol{b}_T\frac{dx_b}{x_b}G_{b/B}(x_b, \boldsymbol{b}_T)\cdots,$$

where the factor $1/x_b$ is introduced in anticipation that it will be needed later on. We shall see in Eqs. (8) and (9) below that this factor $1/x_b$ combines with $1/E_B$ to lead approximately to the factor $1/E_b$ for the basic cross section for $b + a \rightarrow c + d$. In a similar way, we introduce the structure function $G_{a/A}(x_a, \boldsymbol{a}_T)$ defined as

$$G_{a/A}(x_a, \boldsymbol{a}_T) = \frac{1}{(2\pi)^3}\frac{x_a}{2(1 - x_a)}\frac{\phi_A^2(a)}{(a^2 - m_a^2)^2}\qquad(6)$$

to give the integral over α as

$$\int \frac{d^4\alpha}{(2\pi)^3}\delta(\alpha^2 - m_\alpha^2)\frac{\phi_A^2(a)}{(a^2 - m_a^2)^2}\cdots = \int d\boldsymbol{a}_T\frac{dx_a}{x_a}G_{a/A}(x_a, \boldsymbol{a}_T)\cdots .$$

For the fragmentation of c into C and γ, the momentum c and γ are

$$c = \left(P_c + \frac{c^2 + c_T^2}{4P_c}, \boldsymbol{c}_T, P_c - \frac{c^2 + c_T^2}{4P_c}\right)$$

and

$$\gamma = \left((1 - x_C)P_c + \frac{\gamma^2 + \gamma_T^2}{4(1 - x_C)P_c}, \boldsymbol{c}_T - \boldsymbol{C}_T, (1 - x_C)P_c - \frac{\gamma^2 + \gamma_T^2}{4(1 - x_C)P_c}\right),$$

where $0 \leq x_C \leq 1$. The integral over γ becomes

$$\int \frac{d^4\gamma}{(2\pi)^3}\delta(\gamma^2 - m_\gamma^2)\frac{\phi_c^2(C)}{(c^2 - m_c^2)^2}\cdots = \int \frac{dc_T}{(2\pi)^3}\frac{dx_C}{2(1-x_C)}\frac{\phi_c^2(C)}{(c^2 - m_c^2)^2}\cdots$$

We introduce the fragmentation function $G_{C/c}(x_C, C_T)$ defined as

$$G_{C/c}(x_C, C_T) = \frac{1}{(2\pi)^3}\frac{1}{2(1-x_C)}\frac{\phi_c^2(C)}{(c^2 - m_c^2)^2}. \tag{7}$$

The integral over γ becomes

$$\int \frac{d^4\gamma}{(2\pi)^3}\delta(\gamma^2 - m_\gamma^2)\frac{\phi_c^2(C)}{(c^2 - m_c^2)^2}\cdots = \int dc_T dx_C G_{C/c}(x_C, C_T)\cdots$$

Comparing Eq. (7) and Eqs. (5) and (6), we note that the function $G_{C/c}$ is defined in a way slightly different from $G_{b/B}$ and $G_{a/A}$.

Collecting all the factors, the inclusive differential cross section $A + B \rightarrow C + X$ is

$$E_C\frac{d^3\sigma}{dC^3}\bigg|_{AB\rightarrow CX} = \sum_{ab,cd}\int dx_b d\mathbf{b}_T dx_a d\mathbf{a}_T G_{b/B}(x_b, \mathbf{b}_T)G_{a/A}(x_a, \mathbf{a}_T)$$

$$\times \int dx_C dc_T G_{C/c}(x_C, C_T)\frac{|M_{ab\rightarrow cd}|^2}{x_a x_b 2E_A 2E_B|V_A - V_B|}\frac{2\pi\delta(d^2 - m_d^2)}{2(2\pi)^3}. \tag{8}$$

On the other hand, if one considers only the basic process $b + a \rightarrow c + d$, the differential cross section is

$$E_c\frac{d^3\sigma}{dc^3}\bigg|_{ba\rightarrow cd} = \frac{|M_{ab\rightarrow cd}|^2}{2E_a 2E_b|V_a - V_b|}\frac{2\pi\delta(d^2 - m_d^2)}{2(2\pi)^3}. \tag{9}$$

Therefore, combining Eqs. (8) with (9), we have

$$E_C\frac{d^3\sigma}{dC^3}\bigg|_{AB\rightarrow CX} = \sum_{ab,cd}\int dx_b d\mathbf{b}_T dx_a d\mathbf{a}_T G_{a/A}(x_a, \mathbf{a}_T)G_{b/B}(x_b, \mathbf{b}_T)$$

$$\times r(s, s', x_b, x_a)\int dx_C dc_T G_{C/c}(x_C, C_T)E_c\frac{d^3\sigma}{dc^3}\bigg|_{ba\rightarrow cd}, \tag{10}$$

where

$$r(s, s', x_b, x_a) = \frac{2E_a 2E_b|V_a - V_b|}{x_a x_b 2E_A 2E_B|V_A - V_B|}. \tag{11}$$

As we discussed before, this kinematic factor is of the order of unity. We can write it in a simpler form. The quantity $E_A E_B|V_A - V_B|$ in the above equation is a Lorentz-invariant quantity. It is equal to

$$E_A E_B|V_A - V_B| = |E_A E_B V_A - E_A E_B V_B|$$
$$= |\ |\mathbf{p}_A|E_B - |\mathbf{p}_B|E_A\ |$$
$$= (\mathbf{p}_A^2 E_B^2 + \mathbf{p}_B^2 E_A^2 - 2|\mathbf{p}_A||\mathbf{p}_B|E_A E_B)^{1/2}.$$

We note that

$$-2|\mathbf{p}_A||\mathbf{p}_B|E_AE_B = (E_AE_B - |\mathbf{p}_A||\mathbf{p}_B|)^2 - E_A^2E_B^2 - \mathbf{p}_A^2\mathbf{p}_B^2$$
$$= (p_A \cdot p_B)^2 - E_A^2(\mathbf{p}_B^2 + m_B^2) - \mathbf{p}_A^2(\mathbf{p}_B^2 + m_B^2) + \mathbf{p}_A^2m_B^2$$
$$= (A \cdot B)^2 - E_A^2\mathbf{p}_B^2 - \mathbf{p}_A^2E_B^2 - (E_A^2 - \mathbf{p}_A^2)m_B^2$$
$$= (A \cdot B)^2 - \mathbf{p}_B^2E_A^2 - \mathbf{p}_A^2E_B^2 - m_A^2m_B^2 \,.$$

Therefore, we have

$$E_AE_B|V_A - V_B| = [(A \cdot B)^2 - A^2B^2]^{1/2}$$
$$= [(s - A^2 - B^2)^2/4 - A^2B^2]^{1/2}$$
$$= \lambda(s, A^2, B^2)/2 \,. \tag{12}$$

Similarly, we have

$$E_aE_b|V_a - V_b| = \lambda(s', a^2, b^2)/2 \,.$$

The kinematic factor r is therefore

$$r(s, s', x_b, x_a) = \frac{\lambda(s', a^2, b^2)}{x_a x_b \lambda(s, A^2, B^2)} \,. \tag{13}$$

Equations (10) and (13) are just Eqs. (4.18) and (4.17) respectively.

One can consider the inclusive basic process $ba \to CX'$ by integrating over all the momentum states of γ,

$$\left. E_C\frac{d^3\sigma}{dC^3}\right|_{ba \to CX'} = \int dx_C dc_T G_{C/c}(x_C, C_T) E_c \left.\frac{d^3\sigma}{dc^3}\right|_{ba \to cd} \,,$$

where X' represents d and γ. In terms of this inclusive basic cross section, we then have

$$\left. E_C\frac{d^3\sigma}{dC^3}\right|_{AB \to CX} = \sum_{ab} \int dx_b d\mathbf{b}_T dx_a d\mathbf{a}_T G_{b/B}(x_b, \mathbf{b}_T) G_{a/A}(x_a, \mathbf{a}_T)$$

$$\times \, r(s', s, x_b, x_a) E_C \left.\frac{d^3\sigma}{dC^3}\right|_{ba \to CX'} \,,$$

which is Eq. (4.16). $\mathbf{1}\oplus$

§4.3 Counting Rules

To make use of the hard-scattering model, we need to know the behavior of the structure function $G_{a/A}$ and the differential cross section $E_c d^3\sigma/dc^3(ba \to cd)$. We shall discuss here some simple estimates of these quantities for special cases.

From Eq. (6) of Supplement 4.1, the structure function $G_{a/A}(x, a_T)$ is given by

$$G_{a/A}(x, a_T) = \frac{1}{(2\pi)^3} \frac{x}{2(1-x)} \frac{\phi_A^2(a)}{(a^2 - m_a^2)^2}, \qquad (4.19)$$

where $\phi_A(a)$ is the vertex function for pulling a particle a out of A with the particle a off the mass shell. It is clear that the evaluation of the structure function for a parton in a hadron requires the solution of the hadron structure problem in QCD. This is not yet possible for a general case. However, the structure functions in the regions with x close to 1 and/or $a_T \gg 1$ GeV/c involve the interaction of the constituents at very short distances. As is well known, the QCD coupling constant depends on the distance scale of the interaction. For very short distances, the corresponding QCD coupling constant becomes small and the perturbation theory can be applicable. It is then possible to extract the leading behavior of the structure function from the lowest order Feynman diagram in the perturbation theory.

We consider a parent particle with N_A number of constituents. The fraction of the parent light-cone momentum each constituent carries is approximately $1/N_A$. Therefore, in a Feynman diagram which connects n number of constituent particles, the maximum fraction of forward light-cone momentum that a single constituent can carry is of the order of n/N_A. The occurrence of $x \sim 1$ means that a single constituent carries nearly all the forward light-cone momentum of the parent particle. The corresponding Feynman diagram requires that all or nearly all N_A number of constituent particle lines must be connected. The lowest order diagram which connects all the constituent lines is shown in Fig. 4.2. We label the constituents of A by i such that the momentum of the N_A-th constituent of A is a, and a is off the mass shell. In order for the constituent a in Fig. 4.2 to carry nearly all forward light-cone momentum of the parent particle, the other $N_A - 1$ constituents must carry almost no forward light-cone momentum. This case in which all the forward light-cone momentum is concentrated in a single constituent is illustrated in Fig. 4.2 where the intermediate light-cone momentum fraction of each constituent line is also indicated. Fig. 4.2 is a magnified microscopic picture of the vertex function part ϕ_A of diagram 4.1 for pulling a constituent a out of the composite particle A with x close to unity. Accordingly, we can take the Feynman diagram 4.2 as the leading diagram to extract the structure function $G_{a/A}(x)$, with x close to unity.

One associates a factor $1/(k_i^2 - m_i^2)$ for the constituent propagator with momentum k_i in Fig. 4.2. One also associates a propagator for the exchange of a particle (the gluon) between the constituents which

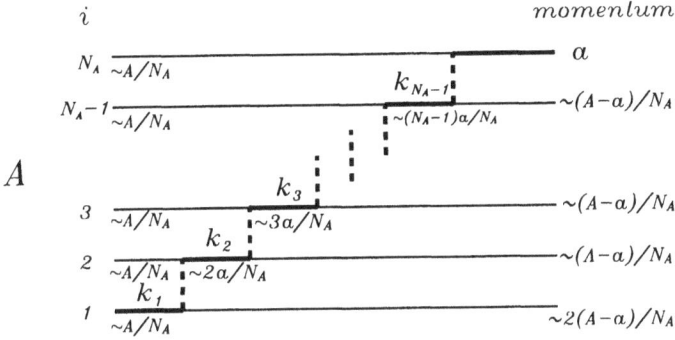

Fig. 4.2 The leading diagram to extract the structure function $G_{a/A}$ for x close to 1.

depends on the nature of the exchange interaction. For the present case involving a large momentum transfer, it is reasonable to assume a point interaction for the exchange of particles. Then the vertex function arises only from the constituent propagators, with momenta k_2, k_3, ..., k_{N_A-1}, as given by

$$\phi_A(a) \sim \frac{1}{(k_2^2 - m_2^2)} \frac{1}{(k_3^2 - m_3^2)} \cdots \frac{1}{(k_{N_A-1}^2 - m_{N_A-1}^2)}.$$

Because we are considering the case when a carries a large fraction of the momentum of A or a large transverse momentum, the dominant contribution to the diagram comes when the constituent momentum k_i in Fig. 4.2, which carries the flow of momenta of the constituents to a single constituent a, increases cumulatively as a function of i, leading eventually to a close to A. For example, by the conservation of momentum at every vertex of the diagram, k_i can increase with $i = 2, ..., N_A - 1$ as $k_i \sim (i/N_A)a$ so that as a result, a very substantial fraction of the momentum of A is concentrated in a. For this case, the quantities $|k_i^2|$ are much greater than the square of the constituent rest masses which can then be neglected. The vertex function becomes approximately

$$\phi_A(a) \sim \frac{1}{k_2^2} \frac{1}{k_3^2} \cdots \frac{1}{k_{N_A-1}^2},$$

$$\sim \frac{1}{[\frac{2}{N_A} \frac{3}{N_A} \cdots \frac{(N_A-1)}{N_A}]^2 (a^2)^{N_A-2}}.$$

On the other hand, in terms of the rest masses and the transverse momentum, the quantity a^2 is given by Eq. (4.15) as

$$a^2 = [x(1-x)A^2 - x\alpha^2 - a_T^2]/(1-x).$$

Hence, the vertex function is proportional to

$$\phi_A(a) \propto \frac{(1-x)^{N_A-2}}{[a_T^2 + x\alpha^2 - x(1-x)A^2]^{N_A-2}}.$$

With this vertex function, the structure function $G_{a/A}(x, a_T)$, as given from Eq. (4.19), is

$$G_{a/A}(x, a_T) \propto \frac{x(1-x)^{2(N_A-1)-1}}{[a_T^2 + x\alpha^2 - x(1-x)A^2]^{2(N_A-1)}}. \tag{4.20}$$

It is convenient to introduce a counting index g_a as

$$g_a = 2(N_A - 1) - 1 \tag{4.21}$$

and write Eq. (4.20) as

$$G_{a/A}(x, a_T) \propto \frac{x(1-x)^{g_a}}{[a_T^2 + x\alpha^2 - x(1-x)A^2]^{2(N_A-1)}}. \tag{4.20'}$$

When x is close to unity, the x-dependence of the structure function behaves as

$$G_{a/A}(x, a_T) \propto (1-x)^{g_a}. \tag{4.22}$$

When $|a_T|$ is much greater than the rest masses, the transverse momentum dependence of the structure function goes as

$$G_{a/A}(x, a_T) \propto \frac{1}{(a_T^2)^{g_a+1}}. \tag{4.23}$$

The above results were obtained when a is an elementary constituent. When a is a composite particle with n_a number of constituents, the same argument can be carried out. The structure function is given by Eqs. (4.22) and (4.23) where the index g_a is

$$\begin{aligned} g_a &= 2(N_A - n_a) - 1, \\ &= 2 \times \text{(number of spectators)} - 1. \end{aligned} \tag{4.24}$$

The index g_a involves the number of spectators. For this reason, this rule for g_a is called the *spectator counting rule* of the structure function, first derived by Blankenbecler and Brodsky [5].

There is another counting rule for the differential cross section of the basic collision process $b + a \rightarrow c + d$. This is the *dimensional counting rule* for large transverse momenta derived by Brodsky and Farrar, and by Matveev *et al.*, based on dimensional analysis [6]. One considers a basic process for which n active fields (particles) participate in the reaction $b + a \rightarrow c + d$, as shown in Fig. 4.3. The incident and the final particles can be composite. In that case, we need to integrate out the momenta of the constituents to give the momenta of the incident and the final particles.

From the rules for Feynman diagrams, the dimension of the cross section, (see Eq. (1) and Eq. (3) of Supplement 4.1), is equal to

$$[d\sigma] = \frac{1}{[p]^2}[|M|^2]\left[\frac{d^3p}{2\omega}\right]^{n-2}[\delta^4(p)] ,$$

where $[y]$ represents the dimension of the quantity y. From the above equation, the dimension of M is

$$[|M|^2] = [\text{momentum}]^{-2(n-4)} . \qquad (4.25)$$

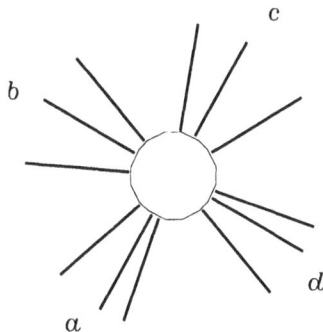

Fig. 4.3 The diagram for the basic process $b + a \rightarrow c + d$.

In the case when each of the elementary fields carries a fraction of the incident momentum, the relevant momentum scale is $\sqrt{s'}$ (or fractions thereof), where $s' = (a + b)^2$. Therefore, we have

$$|M|^2 \sim \frac{1}{(s')^{n-4}} . \qquad (4.26)$$

To get the cross section for $b + a \rightarrow c + d$, we need to integrate out the momenta of the constituents to obtain the momenta of a, b, c, and d. This integration can be carried out in the center-of-mass frame of each composite particle. It will not yield any dependence on the overall center-of-mass energy s'. The s'-dependence of the differential cross section $E_c d^3\sigma/dc^3$ comes from M, the flux factor $|E_a p_b - E_b p_a|$ which scales as s', and d^3c/E_c, which also scales as s'. Therefore, the basic invariant cross section varies with s' as

$$E_c \frac{d^3\sigma}{dc^3}\bigg|_{ba \rightarrow cd} \sim \frac{1}{(s')^2}|M|^2$$

$$\sim \frac{1}{(s')^{n-2}}\,.$$

This dimensional argument does not provide information on dimensionless quantities such as d^2/s' and θ_{CM} [or $t'/s' = (b-c)^2/(a+b)^2$]. There can be additional dependence on these variables. Thus, a more general expression for the invariant cross section is

$$E_c \frac{d^3\sigma}{dc^3}\bigg|_{ba \rightarrow cd} \sim \frac{1}{(s')^N} f(\theta_{CM}, \frac{d^2}{s'})\,, \qquad (4.27)$$

where the exponential index of s' is
$$N = n - 2$$
$$= (\text{number of active participants}) - 2\,. \qquad (4.28)$$

In the case when the transverse momentum of the scattered particle is large, and the relevant length scale is c_T^2 instead of s', the differential cross section is then

$$E_c \frac{d^3\sigma}{dc^3}\bigg|_{ba \rightarrow cd} \sim \frac{1}{(c_T^2)^N} f(\theta_{CM}, \frac{d^2}{s'})\,. \qquad (4.29)$$

Equation (4.28), which gives the exponential index of the transverse momentum for the basic cross section, is called the dimensional counting rule.

§4.4 Application of the Hard-Scattering Model

We shall use the counting rules for the structure function and the basic cross section to study the inclusive invariant cross section for $AB \rightarrow CX$ when C is observed with a large transverse momentum at $\theta_{CM} \sim 90°$. From Eq. (4.16), the inclusive invariant cross section for the process $AB \rightarrow CX$ involves the structure function $G_{a/A}(x, a_T)$,

$G_{b/B}(y, b_T)$, and the basic cross section $E_C d^3\sigma/dC^3(ba \to CX')$. A large transverse momentum is possible when the momentum fraction x (or y) is close to unity, for which the approximate expression Eq. (4.22) can be used for the structure function. This structure function $G_{a/A}(x, a_T)$ has a maximum at $a_T = 0$. The integration of $G_{a/A}$ over a_T can be carried out approximately by replacing the quantity a_T in Eq. (4.20) with an average quantity K_T. The average quantity K_T can be assumed to be much greater than the rest mass of α, so that the factor $K_T^2 + x\alpha^2 - x(1-x)A^2$ can be taken to be approximately a constant. We obtain then

$$\int da_T G_{a/A}(x, a_T) \propto x(1-x)^{g_a},$$

and similarly, we have

$$\int db_T G_{b/B}(y, b_T) \propto y(1-y)^{g_b}.$$

We shall parametrize the basic cross section for the process $ba \to CX'$, given by the dimensional counting rule (4.27) as

$$E_C \frac{d^3\sigma}{dC^3}(ba \to CX') = \frac{1}{(s')^{n-2}} \left[\frac{(X')^2}{s'} \right]^H,$$

where H is an index to be determined by comparison with data. Rewrite the ratio of the square of the 'missing mass' $(X')^2$ to s' as

$$(X')^2/s' = (a + b - C)^2/s'$$
$$\approx 1 - 2C_0'/\sqrt{s'}$$
$$= 1 - x_r',$$

where C_0' is the energy of the detected particle C, s' is $(a+b)^2$ the center-of-mass energy of a and b, and

$$x_r' = C_0'/(\sqrt{s'}/2). \tag{4.30}$$

The prime symbol is used to indicate that these quantities are taken in the center-of-mass frame of the system a and b of colliding constituents.

In the center-of-mass frame of the system A and B of parent particles, we can introduce the corresponding 'radial' variable x_r and transverse variable x_T as

$$x_r = \frac{C_0}{\sqrt{s}/2}, \tag{4.31}$$

and

$$x_T = \frac{C_T}{\sqrt{s}/2}, \tag{4.32}$$

where C_0 and C_T are the energy and the transverse momentum of the particle C in the the center-of-mass system of A and B. Because

$$s' \approx xys ,$$

we have

$$x_r \approx \frac{C_0'\sqrt{xy}}{\sqrt{s'}/2} = x_r'\sqrt{xy} .$$

The invariant cross section for large transverse momentum is therefore given approximately by

$$E_C \frac{d^3\sigma}{dC^3}\bigg|_{AB \to CX} \propto \int dx\, dy\, x(1-x)^{g_a} y(1-y)^{g_b} \frac{1}{(s')^{n-2}} \left(1 - \frac{x_r}{\sqrt{xy}}\right)^H .$$

When C has a large transverse momentum at $\theta_{CM} \sim 90^o$, we have $C_T \sim C_0 \sim C_0'$,

$$x_r \approx x_T \approx \frac{C_0'}{\sqrt{s}/2} .$$

and

$$s' \approx xys \approx \frac{4xyC_T^2}{x_r^2} .$$

Thus, we obtain

$$E_C \frac{d^3\sigma}{dC^3}\bigg|_{AB \to CX} \propto \frac{1}{(C_T^2)^{n-2}}$$

$$\times \int_{x_r}^1 dx \int_{x_r}^1 dy\, x(1-x)^{g_a} y(1-y)^{g_b} \left(\frac{x_r^2}{4xy}\right)^{n-2} \left(1 - \frac{x_r}{\sqrt{xy}}\right)^H .$$

By making use of the following changes of variables

$$x = x_r + \eta(1 - x_r) ,$$

and

$$y = x_r + \eta'(1 - x_r) ,$$

and further expanding the integrands in powers of $1 - x_r$,

$$1 - \frac{x_r}{\sqrt{xy}} = 0 + (1-x_r)\left\{\frac{d}{d(1-x_r)}\left(1 - \frac{x_r}{\sqrt{xy}}\right)\right\}_{1-x_r=0} + O[(1-x_r)^2]$$

$$= (1-x_r)[1 - \frac{1}{2}(1 - \eta + 1 - \eta')] + O[(1-x_r)^2] ,$$

we obtain

$$E_C \frac{d^3\sigma}{dC^3}\bigg|_{AB \to CX} \propto \frac{(1 - x_r)^{g_a+g_b+H+2}}{(C_T^2)^{n-2}}$$

$$\times \int_0^1 d\eta \int_0^1 d\eta' (1+\eta)^{g_a}(1+\eta')^{g_b} \left([1-\frac{1}{2}(2-\eta-\eta')]+O(1-x_r)\right)^H.$$

For $x \to 1$, the integral of η and η' gives a constant quantity, and thus

$$E_C \frac{d^3\sigma}{dC^3}\bigg|_{AB \to CX} \propto \frac{(1 - x_r)^{g_a+g_b+H+2}}{(C_T^2)^{n-2}}. \qquad (4.33)$$

A comparison with experimental pp data suggests that $H = -1$ gives the best fit to experimental data [3]. The inclusive cross section for large transverse momentum (around $\theta_{CM} \sim 90^\circ$) for the process $AB \to CX$ is

$$E_C \frac{d^3\sigma}{dC^3}\bigg|_{AB \to CX} \propto \frac{(1 - x_r)^{g_a+g_b+1}}{(C_T^2)^{n-2}}. \qquad (4.34)$$

The quantities g_a, g_b, and n in Eq. (4.34) are given by the counting rules which can be summarized as follows:

(i) Construct the diagram which allows a to be formed out of A. Count the total number of constituent particles N_A in A and the number of constituent particles N_a in a. Obtain the number of spectators $n_s = N_A - N_a$, by subtraction. The index g_a is then given by $2n_s - 1$. The quantity g_b can be obtained in a similar way.

(ii) Construct the diagram for the basic process $ba \to CX$. Count the total number of lines in this diagram. This gives the number of constituent particles n participating in the process.

(iii) The invariant cross section is proportional to $(1 - x_r)^F/(C_T^2)^N$, where $N = n - 2$ and $F = g_a + g_b + 1$.

We can use Eq. (4.34) and these counting rules to study the inclusive process $pp \to \pi^+ X$ for pions detected with a large transverse momentum at $\theta_{CM} = 90^\circ$ for which $x_r = x_T$. Experimentally, the data can be well represented by a formula of the form Eq. (4.34) with $2N = 2(n - 2) = 8.2$ and $F = g_a + g_b + 1 = 9$, as one can observe in Fig. 4.4 [3].

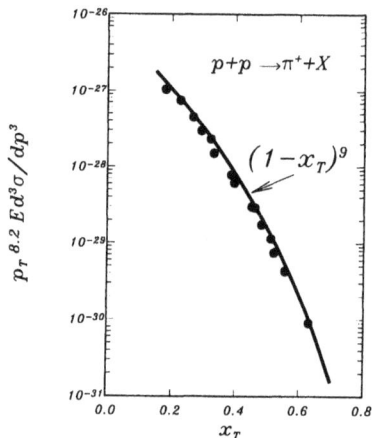

Fig. 4.4 The experimental invariant cross section $Ed^3\sigma/dp^3$ multiplied by $p_T^{8.2}$, where p_T is in GeV and σ in cm^2, for $pp \to \pi^+ X$, $\theta_{cm} = 90°$, $x_T > 0.18$, and beam energies from 200 to 400 GeV. Only data points for x_T near and above 0.2 are considered. The data from Ref. [7] are compared with the counting rule prediction $(1 - x_T)^9$ which is shown as a solid curve.

If one assumes that the dominant basic collision process in $pp \to \pi^+ X$ is $qq \to qq$, then the counting rule gives $g_b = g_a = 3$, and $n = 4$. Consequently, $2N = 2(n - 2) = 4$ and $F = g_a + g_b + 1 = 7$. These values differ from the observed power indices. On the other hand, if one assumes that the basic process (for $p_T \leq 10$ GeV/c), is $q + \text{meson} \to q + \text{meson}$, then g_b and g_a have the values of 5 and 3, and $n = 6$. This gives $2N = 8$ and $F = 9$, which agree well with the experimental indices. The basic process $q + \text{meson} \to q + \text{meson}$ involves the interchange of constituents. Such a model invoking the interchange of constituents is called the *constituent interchange* model [1-3]. It is further supported by other sets of experimental data related to the pion structure function, and by correlations of the momentum flow with the direction of the detected particle C.

The comparison indicates that for the region of $p_T < 10$ GeV/c, $q + \text{meson} \to q + \text{meson}$ is probably the dominant basic process, presumably because of the large hadron-quark coupling strength. The basic process $qq \to qq$ may still occur, but with a smaller probability in this transverse momentum region. It probably becomes dominant at still higher transverse momenta.

If the colliding particles involve a continuum of constituents instead of a finite number of particles, the invariant cross section would have to follow an exponential law instead of a power law. The power law behavior of the cross section behaving for high p_T as $(1 - x_r)^F/(p_T^2)^N$, with power indices N and F depending on the discrete number of constituent particles, indicates the constituent nature of hadrons.

§4.5 Direct Fragmentation Process

In many processes involving the collision of two composite particles B and A, the collision can take place with a constituent b of the particle B acting as a spectator, while the other constituents interact with the particle A. If the constituent b is a stable particle, then b can be detected directly as an outgoing particle, as depicted in Fig. 4.5. (If b is a quark or a gluon, then b will give rise to stable hadrons by fragmentation.) This process is different from the hard-scattering process, as the particle b has not undergone a collision. It is called a *direct fragmentation* of the particle b from B [8].

Following the same derivation of the hard-scattering cross section as in Supplement 4.1, we can show that the process represented by Fig. 4.5 leads to an inclusive invariant cross section given by

$$E_b \frac{d^3\sigma}{db^3}\bigg|_{BA \to bX} = x_b \tilde{G}_{b/B}(x_b, \boldsymbol{b}_T) \sum_i \int \frac{d^3 i}{E_i} E_i \frac{d^3\sigma}{di^3}(\beta + A \to i + X').$$

(4.35)

The structure function $\tilde{G}_{b/B}(x_b, \boldsymbol{b}_T)$ is the probability of finding a constituent of type b in the particle B with a fractional momentum x_b and transverse momentum $\boldsymbol{b}_T$. It is defined in terms of the vertex function with the particle b on the mass shell. The tilde symbol for $\tilde{G}_{b/B}$ indicates that it is, in principle, different from the structure function $G_{b/B}(x_b, \boldsymbol{b}_T)$ for which the vertex function is evaluated for an off-shell particle b. We shall derive the invariant cross section for the direct fragmentation process in Supplement 4.2.

To the extent that the structure function is a smooth function which may be continued smoothly from the off-shell region to the on-shell region, one can approximate $\tilde{G}$ by G. Furthermore, in Eq. (4.35), we expect that the sum over the total cross sections for all possible channels i is rather insensitive to the collision conditions when the collision energy is high enough. It is reasonable to approximate this by a constant. We obtain the approximate result for the direct

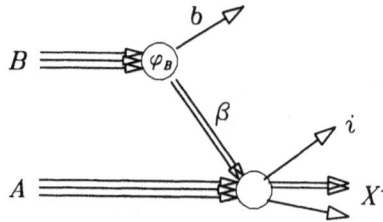

Fig. 4.5 Direct fragmentation process.

fragmentation of the projectile B directly into b, when x_b is close to unity, as given by

$$E_b \frac{d^3\sigma}{db^3}\bigg|_{BA \to bX} \propto x_b G_{b/B}(x_b, \boldsymbol{b}_T) \approx (1 - x_b)^{g_b}. \qquad (4.36)$$

There is a similar expression for the fragmentation of the target particle A.

We can compare the above results with the experimental data for the fragmentation of p into $\pi^{\pm}$ [9] at an incident momentum of 100 GeV/c and 175 GeV/c. The invariant cross section data for these fragmentation processes can be represented by $(1 - x)^n$, where n is 3.4-4.6. If we use the counting rule for the fragmentation of the proton into a pion, then the number of spectators is 3, and the counting index $g_b = 2 \times 3 - 1 = 5$. These counting numbers agree approximately with the experimental power indices.

$\oplus$ Supplement 4.2

We show in detail how we obtain the main results of the direct fragmentation process.

Using the rules for Feynman graphs, the differential cross section for $B + A \to b + X$, represented by Fig. 4.5, is

$$d\sigma = \frac{1}{2E_A E_B |V_A - V_B|} \sum_i \left| M_{BA \to biX'} \right|^2 dp. \qquad (1)$$

Again, we shall assume that the appropriate spin summation and traces have been carried out so that the particles in diagram 4.5 can be treated effectively as spinless particles. The matrix element $M_{BA \to biX}$ can be decomposed as

$$M_{BA \to biX'} = \frac{\tilde{\phi}_B^2(b)}{(\beta^2 - m_\beta^2)^2} M_{\beta A \to iX'} ,$$ (2)

where the tilde symbol in the vertex function $\tilde{\phi}$ indicates that the vertex function is evaluated when b is on the mass shell. In Eq. (1), dp is

$$dp = \frac{d^4b}{(2\pi)^3} \delta(b^2 - m_b^2) \frac{d^4i}{(2\pi)^3} \delta(i^2 - m_i^2) \frac{d^4X'}{(2\pi)^3} \delta(X' \cdot X' - m_{X'}^2)(2\pi)^4 \delta(\beta + A - i - X').$$ (3)

We have therefore

$$E_b \frac{d^3\sigma}{db^3} \bigg|_{BA \to biX'} \propto \frac{\tilde{\phi}_B^2(b)}{(\beta^2 - m_\beta^2)^2} \frac{1}{2E_A E_B |V_A - V_B|} \sum_i \left| M_{\beta A \to iX'} \right|^2 \frac{d^4i}{(2\pi)^3} \delta(i^2 - m_i^2).$$ (4)

We can identify

$$\frac{1}{(1 - x_b) 2 E_A E_B |V_A - V_B|} \left| M_{\beta A \to iX'} \right|^2 \frac{2\pi \delta(X'^2 - m_{X'}^2)}{2(2\pi)^3} = E_i \frac{d^3\sigma}{di^3} \bigg|_{\beta A \to iX'} ,$$

where x_b is

$$x_b = \frac{b_0 + b_z}{B_0 + B_z} .$$

The inclusive cross section for $BA \to bX$ is obtained from (4) by integrating over d^3i and summing over i. We obtain

$$E_b \frac{d^3\sigma}{db^3} \bigg|_{BA \to bX} = (1 - x_b) \frac{\tilde{\phi}_B^2(b)}{(\beta^2 - m_\beta^2)^2} \sum_i \int \frac{d^3i}{E_i} E_i \frac{d^3\sigma}{di^3} (\beta + A \to i + X') .$$

We note that the propagators of b and β are related. In fact, when b is not on the mass shell (but β is on shell), we have

$$m_b^2 - b^2 = \frac{1}{1 - x_b} \left[m_B^2 \left(x_b - \frac{m_B^2 + m_b^2 - m_\beta^2}{2m_B^2} \right) - \frac{1}{4m_B^2} \lambda^2(m_B^2, m_b^2, m_\beta^2) \right] .$$

And, when β is not on the mass shell (but b is on shell), we have

$$m_\beta^2 - \beta^2 = \frac{1}{x_b} \left[m_B^2 \left(x_b - \frac{m_B^2 + m_b^2 - m_\beta^2}{2m_B^2} \right) - \frac{1}{4m_B^2} \lambda^2(m_B^2, m_b^2, m_\beta^2) \right] .$$

Therefore the propagators of b and β are related by

$$\frac{1}{b^2 - m_b^2} : \frac{1}{\beta^2 - m_\beta^2} = (1 - x_b) : x_b .$$

We therefore have

$$(1 - x_b)\frac{\tilde{\phi}_B^2(b)}{(\beta^2 - m_\beta^2)^2} = x_b \frac{x_b}{(1 - x_b)} \frac{\tilde{\phi}_B^2(b)}{(b^2 - m_b^2)^2} .$$

The results suggest the usefulness of introducing the structure function

$$\tilde{G}_{b/B}(x_b, \boldsymbol{b}_T) = \frac{1}{(2\pi)^3} \frac{x_b}{2(1 - x_b)} \frac{\tilde{\phi}_B^2(b)}{(b^2 - m_b^2)^2} ,$$

which differs with $G_{b/B}$ of Eq. (5) of Supplement 4.1 only in the vertex function. The vertex function is evaluated when b is on the mass shell for $G_{b/B}(x_b, \boldsymbol{b}_T)$ and is evaluated when b is off shell for $\tilde{G}_{b/B}(x_b, \boldsymbol{b}_T)$. In terms of $\tilde{G}$, the invariant cross section is

$$E_b \frac{d^3\sigma}{db^3}\bigg|_{BA \to bX} = x_b \tilde{G}_{b/B}(x_b, \boldsymbol{b}_T) \sum_i \int \frac{d^3i}{E_i} E_i \frac{d^3\sigma}{di^3}(\beta + A \to i + X') ,$$

which is Eq. (4.35). $|\oplus$

§References for Chapter 4

1. An excellent review of the hard-scattering process can be found in D. Sivers, S. Brodsky, and R. Blankenbecler, Phys. Rep. 23C, 1 (1976).
2. E. A. Schmidt and R. Blankenbecler, Phys. Rev. D15, 332 (1977).
3. R. Blankenbecler, Lectures presented at Tübingen University, Germany, June 1977, SLAC-PUB-2077 (1977).
4. J.D. Bjorken and S. D. Drell, *Relativistic Quantum Mechanics*, McGraw-Hill Book Company, N.Y. 1964.
5. R. Blankenbecler and S. J. Brodsky, Phys. Rev. D16, 2973 (1974).
6. S. J. Brodsky and G. Farrar, Phys. Rev. Lett. 31, 1153 (1973); Phys. Rev. D11, 1309 (1975); V. Matveev, R. Muradyan, and A. Tavhelidze, Nuovo Cim. Lett. 7, 719 (1973).
7. D. Antreasyan *et al.*, Phys. Rev. Lett. 38, 112 (1977).
8. C. Y. Wong and R. Blankenbecler, Phys. Rev. D22, 2433 (1980).
9. A. E. Brenner *et al.*, Phys. Rev. D26, 1497 (1982).

5. Particle Production in a Strong Field

We saw in Chapter 3 that the production of soft particles is a very important process in high-energy nucleon-nucleon collisions. How are particles produced? What is the physical basis for their production? While the answers to these questions should be found in the theory of quantum chromodynamics (QCD), the quantitative analysis of the particle production process cannot yet be described from first principles of QCD. On the other hand, many qualitative aspects of the production process can be explained in terms of simple and analogous models. This chapter and several following chapters will discuss different models of particle production.

§5.1 Schwinger Particle Production Mechanism

The Schwinger particle production mechanism, first put forth to examine the production of electron-positron pairs in a strong and uniform electromagnetic field, has been applied to many problems in contemporary physics [1]. In particular, it has been invoked to study particle production in QCD [2]. As applied to nucleon-nucleon collisions or e^+e^- annihilations, the field between a quark and an antiquark is represented, as an approximation, by an Abelian gauge field in the same form as a constant electric field between two condenser plates in quantum electrodynamics. A particle-antiparticle pair is produced when a particle tunnels from the negative energy continuum to the positive energy continuum.

We can study how this approximate description arises. The interaction between a quark and an antiquark can be represented phenomenologically by a linear potential proportional to the separation between the two particles [3], which is an essential feature for quark confinement. The constant of proportionality is called the *string tension* κ, which has been estimated from the Regge slope parameter in hadron spectroscopy to be about 1 GeV/fm [3]. The linear potential suggests the usefulness of describing a $q\bar{q}$ pair in terms of the model of a confined color flux tube. In this model, one considers a quark q_0 with a color charge q at $z = 0$ and an antiquark $\bar{q}_0$ with a color charge $-q$ at $z = L$ (Fig. 5.1). The color electric field $\mathcal{E}$ inside the tube has a constant magnitude $\mathcal{E}$ and is directed in the positive z direction. The color flux is confined within a tube of length L and a cross sectional area A. The cross section of the tube is approximately a circle with

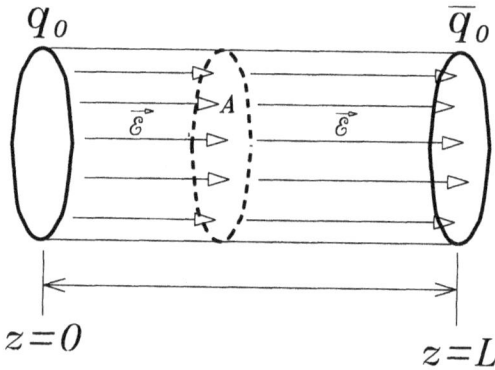

Fig. 5.1 The color electric field $\vec{E}$ between a quark q_0 at $z = 0$ and an antiquark $\bar{q}_0$ at $z = L$ in the color flux tube model.

a radius of about 0.5 fm [3]. Outside the flux tube, the color electric field is zero. The energy contained in the color flux tube is

$$\frac{1}{2}\mathcal{E}^2 \cdot A \cdot L,$$

which also gives the interaction energy between the quark q_0 and the antiquark $\bar{q}_0$ at a separation L. In this model, the interaction energy between a quark and an antiquark is linearly proportional to their separation L with a constant of proportionality $\mathcal{E}^2 A/2$. The model can be used to study interesting qualitative features involving the dynamics of quarks. One can identify the constant of proportionality of the linear interaction in the color flux tube model with the phenomenological string tension κ,

$$\frac{1}{2}\mathcal{E}^2 A = \kappa. \tag{5.1}$$

Because the color electric field is confined inside the tube and vanishes outside, Gauss's law can be used to relate $\mathcal{E} A$ to the color charge q of the quark q_0 at $z = 0$ by

$$\mathcal{E} A = q. \tag{5.2}$$

Using (5.1) and (5.2), one finds that the relation between the product $q\mathcal{E}$ and the string tension κ is

$$q\mathcal{E} = 2\kappa. \tag{5.3}$$

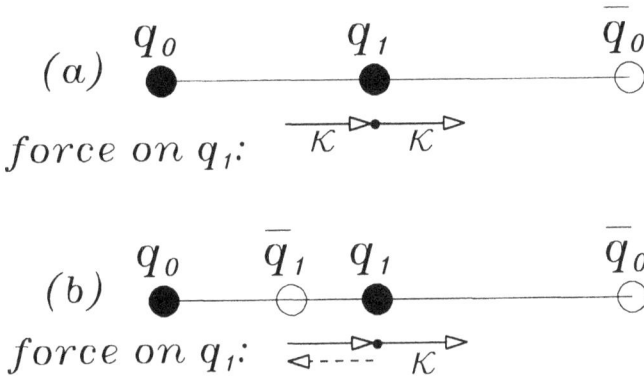

(a) q_0 q_1 $\bar{q}_0$

force on q_1: κ κ

(b) q_0 $\bar{q}_1$ q_1 $\bar{q}_0$

force on q_1: κ

Fig. 5.2 (a) The vector diagram shows the two forces acting on a single quark q_1 coming from q_0 and $\bar{q}_0$. (b) The production of the quark q_1 is accompanied by the production of $\bar{q}_1$. Here the vector diagram includes, in addition, the force acting on q_1 coming from the produced $\bar{q}_1$, shown as the dashed arrow.

When the separation L between the quark q_0 and the antiquark $\bar{q}_0$ exceeds a threshold value, a quark q_1 and an antiquark $\bar{q}_1$ can be produced in the flux tube. To study the production of q_1 and $\bar{q}_1$ in the tube, we treat the quark q_0 at $z = 0$ and the antiquark $\bar{q}_0$ at $z = L$ as static sources which generate an 'external' color electric field $\mathcal{E}$ for the production of q_1 and $\bar{q}_1$. For this picture, a $q_0\bar{q}_0$ pair acts as a pair of condenser plates placed at different potential levels with the potential difference being due to the color electric field from q_0 to $\bar{q}_0$. If a single quark q_1 with a color charge q were produced in the flux tube (Fig. 5.2(a)), it would experience a force $q\mathcal{E}$ and therefore a potential

$$V(z) = -q\mathcal{E}z$$
$$= -2\kappa z \quad (\text{ single } q_1 \text{ produced, unphysical}). \qquad (5.4)$$

The force $q\mathcal{E}$ acting on q_1 is equal to 2κ pointing in the positive z direction. This force can be divided into two parts, as illustrated in Fig. 5.2(a). One part κ is the repulsive 'push' force from the quark q_0 at $z = 0$; another part κ is from the attractive 'pull' force from the antiquark $\bar{q}_0$ at $z = L$.

However, a single quark is never produced all by itself, as this would violate charge conservation. The production of the quark q_1

is always accompanied by the production of an adjacent antiquark $\bar{q}_1$ [see Fig. 5.2(b)]. In consequence, the force acting on the quark q_1 due to the quark q_0 at $z = 0$ is cancelled by the force arising from the interaction due to the newly produced antiquark $\bar{q}_1$. The additional force coming from the additional interaction is shown by the dashed arrow in Fig. 5.2(b). It has a magnitude κ but points in the negative z direction. The cancellation of the two forces acting on q_1 due to q_0 and due to the the newly produced $\bar{q}_1$ is another manifestation of the phenomenon of *screening*. The quark q_0 is said to be screened by the newly produced $\bar{q}_1$ because q_0 now exerts no effect on the quark q_1. Therefore, after taking into account this additional interaction due to the produced antiquark $\bar{q}_1$, the magnitude of the force experienced by q_1 is κ, and not 2κ. Thus, when we take into account the fact that the production of q_1 is accompanied by the production of $\bar{q}_1$, the potential experienced by the produced quark q_1 is given not by Eq. (5.4), but by

$$V(z) = -\kappa z \quad (\text{ additional antiquark } \bar{q}_1 \text{ interaction included }). \quad (5.5)$$

This reduction of the potential due to the additional interaction was first pointed out by Glendenning and Matsui [4].

By treating the color charged pair q_0 and $\bar{q}_0$ as static external sources of the color electric field, we can identify the potential $V(z)$ as the time-like component $A_0(z)$ of an Abelian gauge field with vanishing space-like components $\mathbf{A}$. Equation (5.5) gives the potential that a quark will experience inside the flux tube. Outside the flux tube, the color electric field is zero, and so the potential is a constant. Let us set this constant at zero. Then the complete spatial dependence of the potential experienced by a quark is given by

$$A_0(z) = \begin{cases} 0 & \text{for } z \leq 0 \text{ (Region I)}; \\ -\kappa z & \text{for } 0 \leq z \leq L \text{ (Region II)}; \\ -\kappa L & \text{for } L \leq z \text{ (Region III)}. \end{cases} \quad (5.6)$$

The potential $A_0(z)$ is shown as the solid curve in Fig. 5.3(a).

Consider a quark with a rest mass m and a transverse mass $m_T = \sqrt{m^2 + p_T^2}$ in the above static vector potential $A = (A_0, \mathbf{A})$. In Fig. 5.3(a), the long-dashed curve shows $A_0(z) + m_T$, the potential A_0 plus the transverse mass, above which curve the positive energy continuum lies. The short-dashed curve shows $A_0(z) - m_T$, the potential A_0 minus the transverse mass, below which curve the negative energy continuum lies. According to Dirac's hole theory, the vacuum state is characterized by the configuration in which the single-particle states in the negative energy continuum are all occupied, and the states in the

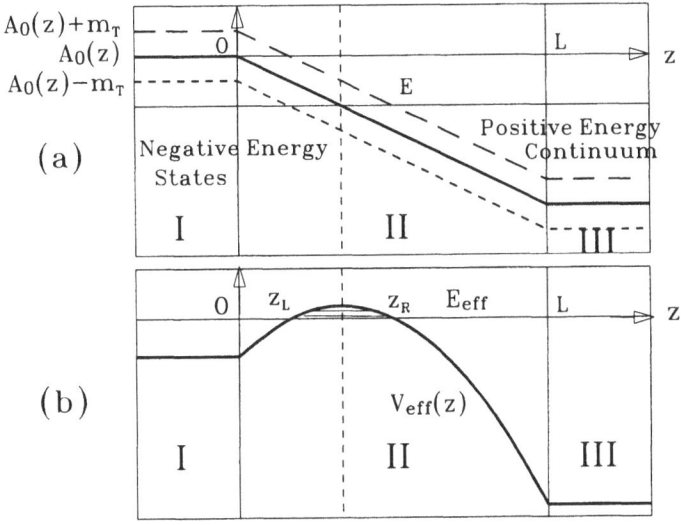

Fig. 5.3(a) The potential $A_0(z)$ for the Dirac or the Klein-Gordon equation. (b) The corresponding V_{eff} for the equivalent Schrödinger equation (5.13).

positive energy continuum are all empty. A quark from the negative energy sea in Region I can tunnel through a barrier and emerge as a quark in the positive energy continuum in Region III, leaving an empty hole in Region I and a quark in Region III. The hole has the characteristics of an antiquark. The tunneling process will lead to the production of a $q\bar{q}$ pair. How does one exhibit the nature of the tunneling mechanism explicitly?

To study the tunneling process, one should, in principle, use the Dirac equation to examine the behavior of a quark as a fermion in the external gauge potential A of Eq. (5.6). This was carried out in Refs. [2] and [5]. As is well known, the Dirac equation gives rise to a Klein-Gordon-like equation with an extra term involving $\alpha \cdot \nabla A_0$, where α is the Dirac α matrix [6]. This term comes from the spinor nature of the fermion wave function and gives only high-order effects in the tunneling probability; it does not affect the dominant term in the tunneling process. When this $\alpha \cdot \nabla A_0$ term is neglected, one has a Klein-Gordon equation for a quark of mass m in the vector potential of Eq. (5.6):

$$\left[(p - A)^2 - m^2\right]\psi = 0. \qquad (5.7)$$

The potential does not depend on the transverse coordinates x and y. (For a more realistic description of the transverse degrees of freedom, see Refs. [8]-[10]). The wave function in the transverse coordinates is just a plane wave with transverse momentum $\mathbf{p}_T = (p_x, p_y)$. The solution of the Klein-Gordon equation can be written in the form

$$\psi = \exp[i(p_x x + p_y y - Et)]f(z), \qquad (5.8)$$

where E is the energy of the quark under consideration. Written out explicitly, the Klein-Gordon equation becomes

$$\left\{[E - A_0(z)]^2 - p_z{}^2 - m_T^2\right\}f(z) = 0. \qquad (5.9)$$

For the tunneling process to occur, one observes from Fig. 5.3(a) that the quark state under consideration must lie in the negative energy continuum in Region I and must lie in the positive energy continuum in Region III. Therefore, only those quarks with an energy E lying within the range

$$-m_T \geq E \geq -\kappa L + m_T \qquad (5.10)$$

can tunnel from Region I to Region III, leaving a quark particle in Region III and an empty hole in Region I, which has the characteristics of an antiparticle $\bar{q}$. There is thus the production of a particle-antiparticle ($q\bar{q}$) pair in the strong field.

From the above energy requirement, we see that the production of a $q\bar{q}$ pair will not occur if the separation L between q_0 and $\bar{q}_0$ is not great enough to satisfy Eq. (5.10). The minimum separation $L_{\min}$ below which no pair will be produced can be inferred from Eq. (5.10) by equating its left-hand side with its right-hand side. We find

$$L_{\min} = 2m_T/\kappa. \qquad (5.11)$$

If one takes the quark mass to be about 0.325 MeV, and the string tension κ to be 1 GeV/fm, then the minimum length $L_{\min}$ required for the production of a $q\bar{q}$ pair is 0.7 fm for a particle with $p_T = 0$, but increases to 1 fm for a quark with a transverse momentum of 0.35 GeV.

To bring out the important concepts of tunneling through a barrier, it is simplest to write Eq. (5.9) in the form of the Schrödinger equation, for which techniques for its solution have been well developed.

Dividing Eq. (5.9) by $-2m_T$, we obtain the equivalent Schrödinger equation

$$\left\{ \frac{p_z^2}{2m_T} + \left[\frac{m_T}{2} - \frac{[E - A_0(z)]^2}{2m_T} \right] \right\} f(z) = 0 . \qquad (5.12)$$

This can be cast in the form

$$\left[\frac{p_z^2}{2m_T} + V_{\text{eff}}(z) - E_{\text{eff}} \right] f(z) = 0 , \qquad (5.13)$$

which is the Schrödinger equation for a particle with a mass m_T and an effective energy $E_{\text{eff}} = 0$, in an effective potential

$$V_{\text{eff}}(z) = \frac{m_T}{2} - \frac{[E - A_0(z)]^2}{2m_T} . \qquad (5.14)$$

A simple inspection shows that the effective potential $V_{\text{eff}}(z)$ experienced by the particle of mass m_T in the region between the plates (Region II) is quadratic in the spatial coordinate z. The quadratic dependence comes from the quadratic term $(p - A)^2$ in Eq. (5.7). In Region II, $V_{\text{eff}}(z)$ is in the form of an inverted parabola $m_T/2 - [E - A_0(z)]^2/2m_T$; this is joined onto a constant potential $m_T/2 - E^2/2m_T$ on the left of $z = 0$, and onto a constant potential $m_T/2 - (E + \kappa L)^2/2m_T$ on the right of $z = L$, as shown in Fig. 5.3(b). The system under consideration corresponds to a single particle with a mass m_T and an effective energy $E_{\text{eff}} = 0$. The particle must overcome a barrier to tunnel from Region I to Region III. The potential barrier is shown as the shaded region in Fig. 5.3(b). The top of the barrier is located at the point at which $E = A_0(z)$ in Fig. 5.3(a). It has a height of $m_T/2$. The left turning point is at

$$z_L = \frac{-E - m_T}{\kappa} ,$$

and the right turning point is at

$$z_R = \frac{-E + m_T}{\kappa} .$$

From the equivalent Schrödinger equation (5.13), and the semiclassical WKB method [6], the penetrability for the system to go from one side of the barrier to the other side is given by

$$P = e^{-I} , \qquad (5.15)$$

where

$$I = 2 \int_{z_L}^{z_R} \sqrt{2m_{\text{eff}}[V_{\text{eff}}(z) - E_{\text{eff}}]}dz \,,$$

and where m_{eff} is m_T. Then the integration of the above gives

$$I = 2 \int_{z_L}^{z_R} \sqrt{m_T^2 - [E - A_0(z)]^2}dz = \frac{\pi m_T^2}{\kappa} \,. \tag{5.16}$$

Therefore, the penetrability for the quark to tunnel from Region I to Region III is

$$P = \exp\left\{-\frac{\pi m_T^2}{\kappa}\right\} = \exp\left\{-\frac{\pi(m^2 + p_T^2)}{\kappa}\right\} \,. \tag{5.17}$$

This result indicates that the tunneling probability decreases with increasing particle mass. It increases with increasing string tension, which depends on the strength of the color electric field. The greater the strength of the color electric field, the greater will be the probability to produce a pair. The transverse momentum serves to provide an additional mass to the particle, and this effective mass must likewise tunnel through the barrier. It must work against the field with a strength κ; the result is a transverse momentum distribution of the form $\exp(-\pi p_T^2/\kappa)$.

§5.2 Particle Production Rate

We are now in a position to estimate the rate of particle production per unit time per unit volume. We consider an energy E in the interval given by Eq. (5.10) for which pair production can occur. The number of pairs produced in an elementary phase space volume, leading to the occurrence of positive energy particles at an energy E within a time interval Δt, is equal to the number of quantum states in that phase space volume element multiplied by the penetrability P given by Eq. (5.17).

The longitudinal length traveled by a particle with longitudinal momentum p_z and energy $E = (m_T^2 + p_z^2)^{1/2}$, in time Δt, is Δt multiplied by the longitudinal velocity p_z/E. The elementary phase space volume swept in time Δt is therefore $\Delta x \Delta y[\Delta t \times (p_z/E)]dp_z d\mathbf{p}_T$. The number of quantum states in the phase space volume is

$$\Delta x \Delta y[\Delta t \times (p_z/E)]dp_z d\mathbf{p}_T/(2\pi)^3 \,.$$

The number of positive energy outgoing particles which arise in this elementary phase space volume is

$$\Delta N = \Delta x \Delta y \Delta t \frac{p_z}{|E|} \frac{dp_z d\mathbf{p}_T}{(2\pi)^3} \exp\left\{-\frac{\pi(m^2 + \mathbf{p}_T^2)}{\kappa}\right\}.$$

Therefore, we have

$$\frac{\Delta N}{\Delta x \Delta y \Delta t \Delta E d\mathbf{p}_T} = \frac{1}{(2\pi)^3} \exp\left\{-\frac{\pi(m^2 + \mathbf{p}_T^2)}{\kappa}\right\}. \qquad (5.18)$$

In the process of particle production due to the tunneling of a particle from the negative energy state, a positive energy particle is created, leaving a hole in the negative energy state which can be interpreted as an antiparticle moving in the direction opposite to the direction of motion of the created particle. The created particle is characterized by the energy E. For such a pair, one cannot, strictly speaking, identify any particular point as the location where the pair is produced, although this pair of particles begins to emerge into the classically allowed region at the points $z_L = (-E - m_T)/\kappa$ and $z_R = (-E + m_T)/\kappa$. Nevertheless, one can approximately associate the midpoint between z_L and z_R,

$$z = -E/\kappa,$$

as the location in the vicinity of which the pair is created. With this association, the energy of the produced pair can be identified by the approximate location of production z, and the energy interval ΔE in which the particle states lie can be associated with a spatial interval Δz at which the pair of particles are produced:

$$\Delta E = \kappa \Delta z. \qquad (5.19)$$

From Eqs. (5.18-19), we obtain

$$\frac{\Delta N}{\Delta t \Delta x \Delta y \Delta z d\mathbf{p}_T} = \frac{\kappa}{(2\pi)^3} \exp\left\{-\frac{\pi(m^2 + \mathbf{p}_T^2)}{\kappa}\right\}. \qquad (5.20)$$

An integration over the transverse momentum leads to

$$\frac{\Delta N}{\Delta t \Delta x \Delta y \Delta z} = \frac{\kappa^2}{8\pi^3} \exp\left\{-\frac{\pi m^2}{\kappa}\right\}. \qquad (5.21)$$

When we include the additional spin degree of freedom, we need to multiply the above result by the spin degeneracy factor g_s which is

2 for fermions. Thus we obtain the rate of quark production in the strong color electric field to be

$$\frac{\Delta N}{\Delta t \Delta x \Delta y \Delta z} = g_s \frac{\kappa^2}{8\pi^3} \exp\left\{-\frac{\pi m^2}{\kappa}\right\}. \tag{5.22}$$

We can compare our result (5.22) with the following result given by Schwinger for fermions [1] and by Brezin and Itzykson for bosons [11]:

$$\frac{\Delta N}{\Delta t \Delta x \Delta y \Delta z} = g_s \frac{\kappa^2}{8\pi^3} \sum_{n=1}^{\infty} \frac{(\theta)^{n-1}}{n^2} \exp\left\{-\frac{n\pi m^2}{\kappa}\right\}, \tag{5.23}$$

where $\theta = 1$ for fermions and $\theta = -1$ for bosons. This result (5.23) is a series in powers of $e^{-\pi m^2/\kappa}$. For the case of quarks with m_u about 0.325 GeV and a string tension of 1 GeV/fm, the second term (i.e. the $n = 2$ term) is 0.0465 times the first term (i.e. the $n = 1$ term). We see that the production rate (5.22), obtained from the present simple analysis, gives the first term of the Schwinger series, which is the dominant term. We also note that the results for fermions and for bosons in Eq. (5.23) are the same for the dominant term with $n = 1$; they differ only in the higher order terms with $n \geq 2$. The use of the Klein-Gordon equation (5.7) to study the tunneling probability may be justified.

It should be noted that the Schwinger result (5.23) is for two condenser plates separated at an infinite distance. The derivation of the production probability for a finite separation requires the solution of the Dirac equation, or the Klein-Gordon equation, in the external potential of Eq. (5.6), which involves parabolic cylinder functions [2]. The solutions show that the penetrability has spatial oscillations for finite systems.

The model we have been discussing is one with the geometry of parallel plates. One may well ask on what basis can it be applied to nucleon-nucleon collisions which are better described by a flux tube of radius R. The generalization of the Schwinger result to include a limitation in the transverse degrees of freedom was carried out in Refs. [8], [9], and [10].

We can discuss the application of the Schwinger mechanism for particle production in QCD. As one can judge from Eq. (5.22), the rate of particle production depends on the mass of the produced particles. The masses of the quarks in the flux tube model are phenomenological constants. Using a set of *constituent masses* with $m_u = m_d = 0.325$ GeV for the up and down quarks, and $m_s = 0.45$ GeV for the strange quark, we find that Eq. (5.22) gives the ratio

$$\frac{\text{(rate of production of an } s\bar{s} \text{ pair)}}{\text{(rate of production of a } u\bar{u} \text{ pair)}} = 0.214 , \qquad (5.24a)$$

and

$$\frac{\text{(rate of production of an } s\bar{s} \text{ pair)}}{\text{(rate of production of a nonstrange } q\bar{q} \text{ pair)}} = 0.107 . \qquad (5.24b)$$

⊕[Supplement 5.1
We would like to relate the ratio $(s+\bar{s})/(u+\bar{u}+d+\bar{d})$ to the ratio K^+/π^+, in a simple qualitative model. We assume that the produced hadrons under consideration consists of pions and kaons and the symmetry of the production process leads to $N_{\pi^+} = N_{\pi^0} = N_{\pi^-}$ and $N_{K^+} = N_{K^0} = N_{K^-} = N_{\overline{K^0}}$, where N_i represent the number of the produced particle i.

From pages 15 and 16 of Ref. [12], the dominant valence quark components of the mesons are:

$$\pi^+ = u\bar{d}, \quad \pi^0 = (u\bar{u} - d\bar{d})/\sqrt{2}, \quad \pi^- = \bar{u}d ,$$

and

$$K^+ = u\bar{s}, \quad K^0 = d\bar{s}, \quad \overline{K^0} = \bar{d}s, \text{ and } K^- = \bar{u}s .$$

Therefore, upon counting the valence quarks in this hadron system, the number of quarks and the number of mesons are related by

$$N_u = N_{\pi^+} + N_{k^+} + \frac{1}{2}N_{\pi^0} ,$$

$$N_{\bar{u}} = N_{\pi^-} + N_{k^-} + \frac{1}{2}N_{\pi^0} ,$$

$$N_d = N_{\pi^-} + N_{k^0} + \frac{1}{2}N_{\pi^0} ,$$

$$N_{\bar{d}} = N_{\pi^+} + N_{\overline{K^0}} + \frac{1}{2}N_{\pi^0} ,$$

$$N_s = N_{\overline{K^0}} + N_{K^-} ,$$

and

$$N_{\bar{s}} = N_{K^+} + N_{K^0} .$$

From this set of equations, we have

$$N_s + N_{\bar{s}} = N_{\overline{K^0}} + N_{K^-} + N_{K^+} + N_{K^0}$$
$$= 4N_{K^+} , \qquad (1)$$

and

$$N_u + N_{\bar{u}} + N_d + N_{\bar{d}} = 4N_{K^+} + 2N_{\pi^+} + 2N_{\pi^-} + 2N_{\pi^0}$$
$$= 4N_{K^+} + 6N_{\pi^+} . \tag{2}$$

Therefore, upon taking the ratio of (1) and (2), we have

$$\frac{N_s + N_{\bar{s}}}{N_u + N_{\bar{u}} + N_d + N_{\bar{d}}} = \frac{N_{K^+}/N_{\pi^+}}{1.5 + N_{K^+}/N_{\pi^+}} . \tag{3}$$

The inverse relation is

$$\frac{N_{K^+}}{N_{\pi^+}} = \frac{1.5\,(N_s + N_{\bar{s}})/(N_u + N_{\bar{u}} + N_d + N_{\bar{d}})}{1 - (N_s + N_{\bar{s}})/(N_u + N_{\bar{u}} + N_d + N_{\bar{d}})} \tag{4}$$

Eq. (5.24b) and Eq. (4) of Supplement 5.1 lead to the estimate of the K^+/π^+ ratio equal to about 0.18. Experimentally, the observed K^+/π^+ ratio in e^+-e^- annihilation is $(1.3 \pm 0.2)/8.3 = 0.157 \pm 0.024$ for $\sqrt{s} = 10$ GeV and is $(1.48 \pm 0.09)/10.3 = 0.144 \pm 0.009$ for $\sqrt{s} = 29$ GeV [13].

In nucleon-Be collisions at 14.5 GeV from the E802 Collaboration, K^+/π^+ ratio is about 0.08 [14], which corresponds to $(s\bar{s})/(u\bar{u}\,d\bar{d}) \sim 0.05$, as calculated by Eq. (3) of Supplement 5.1. However, in nucleon-nucleon collisions, a substantial fraction of the energy is carried by the leading particle and the energy available for particle production is about half of the incident center-of-mass energy, which may be the reason for the differences in the K^+/π^+ ratio in e^+e^- annihilations [13] and in p-Be collisions [14].

The suppression of the production of the K mesons can be qualitatively attributed to the difference in the masses of the up or down quarks and the mass of the strange quark. In a similar way, the suppression of the production of baryon-antibaryon pairs can also be attributed to the large value of mass involved. In this baryon-antibaryon production, a diquark-antidiquark pair is produced in the strong field. Subsequently, the produced diquark combines with an adjacent quark to form a baryon, while the produced antidiquark combines with an adjacent antiquark to form an antibaryon. The suppression of the production of baryon-antibaryon pairs may be attributed to the large value of the mass of the diquark, as compared to the mass of the up or down quarks.

Equation (5.17) shows that the transverse momentum contributes an additional mass which must work against the barrier for the tunneling process to occur. This leads to a Gaussian transverse momentum distribution for the quarks of the form

$$\frac{dN}{d\mathbf{p}_T} = (\text{const}) \; \exp\{-\pi p_T^2/\kappa\}\,, \qquad (5.25)$$

with a root-mean-square transverse momentum

$$\sqrt{<p_T^2>} = \sqrt{\frac{\kappa}{\pi}}\,. \qquad (5.26)$$

The root-mean-square transverse momentum of a produced pion, which consists of two quarks, is $\sqrt{2\kappa/\pi}$. Assuming a string tension of 1 GeV/fm, the root-mean-square transverse momentum is 0.25 GeV/c for a produced quark, and is 0.35 GeV/c for a produced pion. For the observed pion spectra in Fig. 3.3 with the distribution of Eq. (3.5), the observed root-mean-square transverse momentum of produced pions is about 0.37 GeV/c. Thus, the width of the pion transverse momentum distribution can be described qualitatively by the Schwinger production mechanism.

§References for Chapter 5

1. J. Schwinger, Phys. Rev. 82, 664 (1951).
2. R. C. Wang and C. Y. Wong, Phys. Rev. D38, 348 (1988) and references cited therein.
3. A. Casher, H. Neuberger, and S. Nussinov, Phys. Rev. D20, 179 (1979); H. Neuberger, Phys. Rev. D20, 2936 (1979); C. B. Chiu and S. Nussinov, *ibid.* 20, 945 (1979).
4. N. K. Glendenning and T. Matsui, Phys. Rev. D 28, 2890 (1983).
5. C. Martin and D. Vautherin, Phys. Rev. D 38, 3593 (1988); *ibid.* D40, 1667 (1989).
6. V. B. Berestetskii, L. D. Lifshitz, and L. P. Pitaevskii, *Quantum Electrodynamics*, Pergamon Press, Oxford, 1971.
7. L. D. Landau and L. D. Lifshitz, *Quantum Mechanics*, Pergamon Press, London, 1958.
8. K. Sailor *et al.*, Phys. Lett. 247, 5 (1990).
9. H.-P. Pavel and D. M. Brink, Zeit. Phys. C51, 119 (1991).
10. G. Gatoff and C. Y. Wong, Phys. Rev. D46, 997 (1992).
11. E. Brezin and C. Itzykson, Phys. Rev. D2, 1191 (1970).
12. G. P. Yost *et al.*, Particle Data Group, *Review of Particle Properties*, Phys. Lett. 204, 1 (1988).
13. Table 2 of W. Hoffmann, Ann. Rev. Nucl. Part. Sci. 38, 279 (1988).
14. S. Nagamiya, Nucl. Phys. A544, 5c (1992).

6. Particle Production in Two-Dimensional
Quantum Electrodynamics

The model of particle production we discussed in the last section comes from static considerations. The quark and the antiquark at the two ends of a flux tube are held fixed. On the other hand, in particle production processes in high-energy hadron collisions and e^+e^- annihilations, the quark and the antiquark move apart with a speed close to the speed of light. The concept of a static potential is therefore a crude approximation. Furthermore, the fermions and the antifermions produced in the Schwinger mechanism are not confined particles. They can travel individually to distant locations. The Schwinger mechanism itself gives no quantitative description of how the produced quarks and antiquarks are confined and combined together to form the observed composite objects (mesons and baryons).

It is of interest to study the physics of particle production with QCD-inspired nonperturbative models in which the property of quark confinement is a natural consequence of the dynamics of the model. The Schwinger model [1] of two-dimensional quantum electrodynamics (QED_2) furnishes an interesting arena for such a study.

We shall consider only QED_2 with massless fermions, which can be solved exactly [1-4]. This is a model in which the relevant fermion and antifermion particles are electrons and positrons interacting via the electromagnetic interaction. They reside in a two-dimensional space-time with one spatial coordinate and one temporal coordinate. The fermions are assumed to be massless, which is the proper limit when the electromagnetic coupling is large compared to the rest mass of the fermion. In this two-dimensional space-time and in the Coulomb gauge, the electrostatic potential between an electron and a positron separated at a distance r does not fall off as $1/r$, as in four-dimensional space-time. It is actually proportional to r, in a form much like the interaction between two particles joined by a string. Because of this property, massless QED_2 has many features similar to those of QCD in four-dimensional space-time [2]. In particular, in QED_2, the electrons and positrons are confined, and the interaction possesses the property of asymptotic freedom. It is a quantum mechanical system in which a neutral boson exists as a nonperturbative bound state, much as mesons are bound states in QCD. When a positively and a negatively charged pair are separated in such a system, the vacuum is so polarized that the positive and the negative charges are completely screened in a manner similar to the screening of color charges of quarks. It was demonstrated by Casher, Kogut, and Susskind [2] that the rapidity distribution of the produced particles in a system of two opposite charges separating at high relative momenta will exhibit a plateau structure in rapidity, which resembles the shape of the rapidity distribution in high-energy hadron collisions and e^+e^- annihilations (see Fig. 3.2). These desirable properties of confinement, charge screening, the existence of neutral bound states, and the proper high-energy behavior make it useful to study particle production using QED$_2$. We shall discuss some of these properties in this chapter.

§6.1 Qualitative Description of QED$_2$

In quantum electrodynamics, there is the *fermion field* ψ and the *electromagnetic field* A^μ. The fermion field and the electromagnetic field are subject to gauge transformations, and the dynamics is governed by the principle of local gauge invariance (see below). For these reasons, quantum electrodynamics is known as a gauge field theory and the electromagnetic field A^μ is also called the *gauge field*.

The gauge field A^μ depends on the fermion field ψ. The fermion field ψ, in turn, depends on the gauge field A^μ. The coupling is quite complicated and leads to a non-linear problem of great complexity. There are no known analytical solutions for the general QED problem in four-dimensional space-time. Remarkably, for the simple system with massless fermions in one space and one time dimension (QED$_2$), the fermion field can be represented in terms of a boson field, and QED$_2$ can be solved exactly [1,3,4]. Schwinger found that QED$_2$ involving massless fermions with the electromagnetic interaction is equivalent to a free boson field ϕ with a mass $m = e/\sqrt{\pi}$, where e is the coupling constant.

We can understand this remarkable property of massless QED$_2$ from the following line of argument. One starts with electrons occupying the negative-energy Dirac sea. If there is a density and/or current disturbance in some region of space, it will lead to an electromagnetic gauge field A^μ. This electromagnetic field A^μ will affect the fermion field operator. The resultant changes in the fermion field set all the electrons in all the states in motion and in excitation. This motion and excitation generate a physical charge density j^0 and charge current j^1. But how does the generated current j^μ depend on its source of disturbance, the gauge field A^μ ?

Before we write down such a dependence explicitly, it is worth noting the difference of the behavior of these two quantities under a *gauge transformation*. In quantum electrodynamics, physical quantities such as the current j^μ and the field strength tensor $F_{\mu\nu}$ are *gauge-invariant quantities*. That is, they do not change when one changes the gauge field from A^μ to $\bar{A}^\mu$, according to the gauge transformation

$$\bar{A}^\mu(x) = A^\mu(x) - \partial^\mu \lambda(x), \qquad (6.1a)$$

while the fermion field operator $\psi(x, A)$ transforms as

$$\psi(x, \bar{A}(x)) = e^{ie\lambda(x)}\psi(x, A(x)), \qquad (6.1b)$$

where $\lambda(x)$ is an arbitrary function of x. A different choice of the arbitrary function $\lambda(x)$ represents a different gauge, and the physical quantities must be independent of the gauge choice.

The generated current $j^\mu(x)$ depends on its source gauge field $A^\mu(x)$ but the properties of $j^\mu(x)$ and $A^\mu(x)$ differ with regard to a gauge transformation. While the current j^μ is gauge independent, the gauge field A^μ depends on the gauge choice. To be consistent, the generated current j^μ must depend on a particular combination of its source A^μ such that the combination does not depend on the gauge choice. The simplest such combination is $A^\mu - \partial^\mu (\partial^\lambda \partial_\lambda)^{-1} \partial_\nu A^\nu$. In fact, as shown in Section 6.2 below, when the gauge invariance property of the current is properly taken into account, the current $j^\mu(x)$ generated is related to the local electromagnetic disturbance $A^\mu(x)$ by

$$j^\mu(x) = -\frac{e^2}{\pi}[A^\mu(x) - \partial^\mu \frac{1}{\partial^\lambda \partial_\lambda} \partial_\nu A^\nu(x)]. \qquad (6.2)$$

It is easy to verify that $j^\mu(x)$ in Eq. (6.2) does not depend on the choice of a gauge for $A^\mu(x)$.

⊕[Supplement 6.1

We show in detail here that the current $j^\mu(x)$, as given by Eq. (6.2), is gauge invariant. According to Eq. (6.2), the current $j^\mu(x, A)$ for the gauge field $A^\mu(x)$ is

$$j^\mu(x, A) = -\frac{e^2}{\pi}[A^\mu(x) - \partial^\mu \frac{1}{\partial^\lambda \partial_\lambda} \partial_\nu A^\nu(x)]. \qquad (1)$$

We have kept the label A in the argument of $j^\mu(x, A)$ to indicate that this is the current for the gauge field A^μ.

Under a gauge transformation, the gauge field is changed from $A^\mu(x)$ to $\bar{A}^\mu(x)$ according to Eq. (6.1). From Eq. (6.2), the current $j^\mu(x, \bar{A})$ for the transformed gauge field $\bar{A}^\mu(x)$ is

$$j^\mu(x, \bar{A}) = -\frac{e^2}{\pi}[\bar{A}^\mu(x) - \partial^\mu \frac{1}{\partial^\lambda \partial_\lambda} \partial_\nu \bar{A}^\nu(x)], \qquad (2)$$

where the argument $\bar{A}$ in $j^\mu(x, \bar{A})$ indicates that the current is evaluated for the gauge field $\bar{A}^\mu$. To see how $j^\mu(x, A)$ and $j^\mu(x, \bar{A})$ are related, we can use the gauge transformation (6.1) to substitute $\bar{A}^\mu$ in terms of A^μ in Eq. (2), and we find

$$j^\mu(x, \bar{A}) = -\frac{e^2}{\pi}[\{A^\mu(x) - \partial^\mu\lambda(x)\} - \partial^\mu\frac{1}{\partial^\lambda\partial_\lambda}\partial_\nu\{A^\nu(x) - \partial^\nu\lambda(x)\}]$$
$$= j^\mu(x, A),\qquad\qquad (3)$$

where the terms with the arbitrary function $\lambda(x)$ cancel because

$$\frac{1}{\partial^\lambda\partial_\lambda}\partial_\nu\partial^\nu = 1.$$

The result of Eq. (3) indicates that the current, as given by Eq. (6.2), is the same for the gauge fields A^μ or $\bar{A}^\mu$, if A^μ and $\bar{A}^\mu$ are related by a gauge transformation. That is, $j^\mu(x)$ as given by Eq. (6.2) is gauge invariant. I⊕

The current j^μ generated by the gauge field A^μ is in turn the source of a gauge field $\mathcal{A}^\mu$. The gauge field $\mathcal{A}^\mu$ is determined from j^μ by the Maxwell equation:

$$\partial_\nu F^{\mu\nu} = \partial_\nu(\partial^\mu\mathcal{A}^\nu - \partial^\nu\mathcal{A}^\mu) = -j^\mu.\qquad\qquad (6.3)$$

The dynamics of the system can be found when this gauge field $\mathcal{A}^\mu$, generated by j^μ, is self-consistently the same as the electromagnetic field A^μ, which was first introduced in (6.2). Using this condition, we find an equation of motion to describe the dynamics of the gauge field A^μ:

$$\partial_\nu\partial^\mu A^\nu - \partial_\nu\partial^\nu A^\mu = \frac{e^2}{\pi}[A^\mu - \partial^\mu\frac{1}{\partial^\lambda\partial_\lambda}\partial_\nu A^\nu],\qquad\qquad (6.4)$$

which is satisfied if

$$-\Box A^\mu - \frac{e^2}{\pi}A^\mu = 0.\qquad\qquad (6.5)$$

Here, the operator $\Box$ stands for $\partial_\nu\partial^\nu$, the d'Alembertian operator. It is equal to the operator $-p^2$ in the coordinate representation.

⊕[Supplement 6.2
We would like to show that Eq. (6.4) is satisfied if Eq. (6.5) is true. We take the difference of the right-hand-side and the left hand side of Eq. (6.4) and use Eq.

(6.5). We get

$$\partial_\nu \partial^\mu A^\nu - \partial_\nu \partial^\nu A^\mu - \frac{e^2}{\pi}[A^\mu - \partial^\mu \frac{1}{\partial^\lambda \partial_\lambda} \partial_\nu A^\nu]$$

$$= \partial^\mu[\partial_\nu A^\nu + \frac{e^2}{\pi}\frac{1}{\partial^\lambda \partial_\lambda}\partial_\nu A^\nu]$$

$$= \partial^\mu[\partial_\nu A^\nu - \frac{1}{\partial^\lambda \partial_\lambda}\partial_\nu \partial^\kappa \partial_\kappa A^\nu]$$

$$= \partial^\mu[\partial_\nu A^\nu - \frac{1}{\partial^\lambda \partial_\lambda}\partial^\kappa \partial_\kappa \partial_\nu A^\nu]$$

$$= 0.$$

Therefore, Eq. (6.4) is satisfied if Eq. (6.5) is true.

In terms of p^2, Eq. (6.5) can be written as

$$p^2 A^\mu - \frac{e^2}{\pi}A^\mu = 0. \qquad (6.6)$$

We can compare this equation with the Klein-Gordon equation if A^μ is a free boson field with a mass m:

$$p^2 A^\mu - m^2 A^\mu = 0.$$

We see that the gauge field A^μ obeys the Klein-Gordon equation appropriate for a free boson with a mass m given by

$$m = \frac{e}{\sqrt{\pi}}. \qquad (6.7)$$

Therefore, QED$_2$ involving massless fermions is equivalent to a field of free bosons with a mass $e/\sqrt{\pi}$. In this way, a rather complicated non-linear problem for the field of massless interacting fermions can be written in terms of a simple, non-interacting field of bosons with a mass.

The above general results have been obtained for any choice of a gauge. The relation between the current and the gauge field can be seen more easily and the physical picture becomes even more transparent if one chooses the Lorentz gauge. In the Lorentz gauge with $\partial_\mu A^\mu = 0$, the current generated by the gauge field $A^\mu(x)$ is, according to Eq. (6.2), directly proportional to the gauge field itself:

$$j^\mu(x) = -\frac{e^2}{\pi}A^\mu(x). \qquad (6.8)$$

But in the Lorentz gauge, the Maxwell equation (6.3) gives $\Box A^\mu$ proportional to the current j^μ:

$$\Box A^\mu = j^\mu$$
$$= -\frac{e^2}{\pi} A^\mu \,.$$

Hence, A^μ is proportional to $\Box A^\mu$ and the gauge field A^μ propagates as if it were a boson field with a mass.

A field of bosons with a rest mass is called a *massive boson field*. As the dynamics of a free massive boson field is simple, it is convenient to transcribe many bilinear combinations of the massless fermion field in terms of functions of a massive boson field. Such a transcription is call a *bosonization*.

From the Maxwell equation (6.3), the fermion current j^μ satisfies the current conservation condition

$$\partial_\mu j^\mu = 0 \,. \tag{6.9}$$

The two components of current j^0 and j^1 are constrained by this equation. There is only one degree of freedom. It is useful to introduce a single boson field ϕ to represent the two components of the current. Equation (6.9) is easily solved by

$$j^\mu(x) = -m\epsilon^{\mu\nu}\partial_\nu\phi(x), \tag{6.10}$$

where $\epsilon^{\mu\nu}$ is the antisymmetric tensor with elements $\epsilon^{01} = -\epsilon^{10} = -1$. As the physical current is a gauge-invariant quantity, ϕ is also a gauge-invariant quantity.

Using this bosonization formula for j^μ, the gauge field A^μ can be written in terms of the boson field ϕ. This can be easily carried out first for the Lorentz gauge, and then generalized to any other gauge. In the Lorentz gauge with $\partial_\mu A^\mu = 0$, Eq. (6.8) gives A^μ in terms of ϕ:

$$A^\mu = \frac{1}{m}\epsilon^{\mu\nu}\partial_\nu\phi \,.$$

To obtain the relation between A^μ and ϕ for a general choice of a gauge, one performs a gauge transformation and obtains

$$A^\mu = \frac{1}{m}\epsilon^{\mu\nu}\partial_\nu\phi - \partial^\mu\lambda \,, \tag{6.11}$$

where λ is an arbitrary function of x.

Using the equation of motion for A^μ, we can get the equation of motion for ϕ. From the arbitrariness of the λ function, and the boundary condition that there are no current distributions at infinity, Eq. (6.5) leads to the equation of motion for ϕ:

$$-\Box\phi - \frac{e^2}{\pi}\phi = 0 \,. \tag{6.12}$$

Hence, the boson field ϕ, introduced to bosonize the fermion current j^μ and the gauge field A^μ, is a free massive boson field with a mass $m = e/\sqrt{\pi}$. It satisfies the Klein-Gordon equation.

We note here that because j^μ is a vector field and $\epsilon^{\mu\nu}$ is a pseudotensor, the field ϕ is a pseudoscalar field. The field ϕ is also a dipole field because its spatial derivative gives the charge density j^0.

We shall make use of the results obtained here to discuss particle production in Section 6.3. The next section deals with a derivation of the important relation of Eq. (6.2) between j^μ and A^μ. It may be by-passed by those who are not interested in the technical details of QED$_2$.

§6.2 Relation between j^μ and A^μ in QED$_2$

We would like to show how we can obtain Eq. (6.2) relating the current $j^\mu(x)$ and the gauge field $A^\mu(x)$. We consider a vacuum state in which all the negative energy states in the Dirac sea are occupied. A disturbance in density and/or current will generate a gauge field A^μ. We first study the charge density and the charge current induced by this gauge field A^μ. For this purpose, we study how the gauge field A_μ affects the fermion field operator ψ. The fermion field operator satisfies the Dirac equation

$$\gamma^\mu(i\partial_\mu - eA_\mu)\psi(x) = 0 \,. \tag{6.13}$$

The gamma matrices γ^μ can be chosen at will, provided they satisfy the anticommutation relation

$$\gamma^\mu\gamma^\nu + \gamma^\nu\gamma^\mu = 2g^{\mu\nu} \,.$$

It is convenient to choose

$$\gamma^0 = \begin{pmatrix} 0 & 1 \\ 1 & 0 \end{pmatrix} \tag{6.14}$$

and

$$\gamma^1 = i\sigma_2 = \begin{pmatrix} 0 & 1 \\ -1 & 0 \end{pmatrix} \,. \tag{6.15}$$

Thus, the matrix γ^5, the product of γ^0 and γ^1, is diagonal:

$$\gamma^5 = \gamma^0 \gamma^1 = \begin{pmatrix} -1 & 0 \\ 0 & 1 \end{pmatrix}. \tag{6.16}$$

We write the field operator $\psi(x)$ in terms of $\psi_F(x)$, the free fermion field operator for the case free of an electromagnetic field, and an unknown phase factor $\Phi(x)$:

$$\psi(x) = e^{ie\Phi(x)} \psi_F(x). \tag{6.17}$$

The phase factor $\Phi(x)$ is taken to contain diagonal matrices in the spinor space so that $\partial_\mu \Phi$ and Φ commute. The two independent diagonal matrices can be chosen to be the unit matrix and the γ^5 matrix. The phase factor $\Phi(x)$ is a linear combination of the unit matrix and γ^5:

$$\Phi(x) = \Phi_0(x) + \Phi_5(x)\gamma^5,$$

where $\Phi_0(x)$ and $\Phi_5(x)$ are functions of x. After substituting Eq. (6.17) into the Dirac equation (6.13), we obtain

$$[\gamma^\mu(-e\partial_\mu\Phi - eA_\mu)e^{ie(\Phi_0+\Phi_5\gamma^5)} + e^{ie(\Phi_0-\Phi_5\gamma^5)}\gamma^\mu i\partial_\mu]\psi_F(x) = 0. \tag{6.18}$$

The free fermion operator $\psi_F(x)$ for the case when there is no gauge field present satisfies the Dirac equation for a free particle:

$$\gamma^\mu i\partial_\mu \psi_F(x) = 0. \tag{6.19}$$

In the presence of the gauge field $A_\mu(x)$, the Dirac equation (6.18) will be satisfied if we choose $\Phi(x)$ such that

$$\gamma^\mu[\partial_\mu\Phi(x) + A_\mu(x)] = 0. \tag{6.20}$$

This equation gives a relation between $\Phi(x)$ and the gauge field $A_\mu(x)$, which we shall need later on.

We note here that the results of Eqs. (6.17) and (6.20) allow a 'geometrical' interpretation of the gauge field $A_\mu(x)$. According to these results, we can compare the fermion field operator $\psi(x)$ in the presence of a gauge field $A_\mu(x)$ with the free fermion field operator $\psi_F(x)$ when there is no gauge field. As indicated by Eq. (6.17), the effect of the gauge field A_μ is to modify the free fermion field operator $\psi_F(x)$ by the phase factor $e^{ie\Phi(x)}$ with the 'phase angle' $e\Phi(x)$. Therefore, the gauge field $A_\mu(x)$ can be interpreted as specifying, according to Eq. (6.20), the degree of local 'phase angle' $e\Phi(x)$ which modifies the free fermion field operator $\psi_F(x)$ at the space-time point x. In the presence of a gauge field, there will be a phase angle $e\Phi(x)$ associated with the gauge field $A_\mu(x)$ at each space-time point x.

The geometrical interpretation of the gauge field also provides a geometrical description of a gauge transformation and gauge invariance. The phase angle $e\Phi(x)$, due to the presence of the gauge field $A_\mu(x)$, will be changed if one rotates the orientation of the polar axis of the 'phase angle polar coordinate system', relative to which the phase angle $e\Phi(x)$ is measured. A rotation of the polar axis corresponds to a gauge transformation. However, the dynamics of the physical system should be independent of the choice of the polar axis used to carry out the measurement of the phase angle $e\Phi(x)$. Local gauge invariance can be interpreted as the independence of the Lagrangian when the polar axes used in the measurement of the phase angles $e\Phi(x)$ are rotated by different amounts at different space-time points x. Such an local gauge invariance is possible when the terms in the Langrangian are judiciously chosen to contain a product of fermion operators, the gauge field, and other operators in such a combination that at any space-time point the combined changes of the phase angles of various quantities due to any arbitrary rotation of the polar coordinate axis cancel out, and the cancellation occurs at all space-time points. The independence of the Lagrangian with respect to any choice of the orientation of the polar axis for the measurement of the phase angle $e\Phi(x)$ is the geometrical interpretation of *local gauge invariance*.

With the vacuum consisting of a sea of electrons occupying the negative energy states, the presence of the gauge field A_μ modifies the electron field operator, as given by Eq. (6.17). This change in the electron field operator leads to an electron current, which will depend on the gauge field A_μ. We can evaluate the electron current $j^\mu(x)$ by

$$j^\mu(x) = e < \bar{\psi}(x)\gamma^\mu\psi(x) >= e < \psi^\dagger(x)\alpha^\mu\psi(x) >,$$

where the expectation value is taken for the vacuum state in question. The α matrices are defined as $\alpha^\mu = \gamma^0\gamma^\mu$. For our choice of γ matrices in Eq. (6.14-16), we have

$$\alpha^0 = \gamma^0\gamma^0 = \begin{pmatrix} 1 & 0 \\ 0 & 1 \end{pmatrix}$$

and

$$\alpha^1 = \gamma^0\gamma^1 = \gamma^5 = \begin{pmatrix} -1 & 0 \\ 0 & 1 \end{pmatrix}.$$

The current $j^\mu(x)$ involves a bilinear combination of the field operators at the same point x. An evaluation of the expectation value for this bilinear combination will lead to a singular quantity. [See Eq. (6.29) and Supplement 6.4 below.] It is necessary to evaluate the expectation value by taking a suitable limiting procedure as x approaches x':

$$
\begin{aligned}
j^\mu(x) &= e \lim_{x \to x'} < T\big(\psi^\dagger(x')\alpha^\mu\psi(x)\big) > \\
&= e \lim_{x \to x'} < T\big(\psi^\dagger_\kappa(x')\alpha^\mu_{\kappa\lambda}\psi_\lambda(x)\big) >,
\end{aligned}
\tag{6.21}
$$

where the indices κ and λ label the spinor components of the fermion field operator. The symbol T in Eq. (6.21) denotes the time-ordered product. The time-ordered product of a product of fermion field operators is defined as the product of these fermion field operators arranged according to their time coordinates, a fermion field operator with an earlier time coordinate appearing on the right of a fermion field operator with a later time coordinate. There is an additional negative sign when the original product of operators reaches this ordering of the product of operators by an odd number of permutations. For example, the time-ordered product of $\psi^\dagger_\kappa(x')\psi_\lambda(x)$ is

$$
T\big(\psi^\dagger_\kappa(x')\psi_\lambda(x)\big) = \begin{cases} \psi^\dagger_\kappa(x')\psi_\lambda(x) & \text{if } x'^0 \geq x^0 \\ -\psi_\lambda(x)\psi^\dagger_\kappa(x') & \text{if } x^0 \geq x'^0. \end{cases}
$$

From the definition of the time-ordered product, it is easy to prove that

$$
T\big(\psi^\dagger_\kappa(x')\alpha^\mu_{\kappa\lambda}\psi_\lambda(x)\big) = -T\big(\alpha^\mu_{\kappa\lambda}\psi_\lambda(x)\psi^\dagger_\kappa(x')\big).
$$

Therefore, we have

$$
j^\mu(x) = -e \lim_{x \to x'} tr[\alpha^\mu < T\big(\psi(x)\psi^\dagger(x')\big) >].
\tag{6.22}
$$

We introduce the Green's function $G(x, x')$ as

$$
G(x, x') = < T\big(\psi(x)\psi^\dagger(x')\big) > .
\tag{6.23}
$$

Eq. (6.22) becomes

$$
j^\mu(x) = -e \lim_{x \to x'} tr[\alpha^\mu G(x, x')].
$$

In order to exhibit explicitly the properties of the Green's functions and the field operators under a gauge transformation, we include the gauge field dependence as part of the label of these quantities and write the above Green's function (6.23) as

$$G(x, x'; A) = < T(\psi(x, A)\psi^\dagger(x', A)) > ,\qquad (6.23')$$

where the label A indicates that they are evaluated for the gauge field A_μ. From Eq. (6.22), the current is related to the Green's function $G(x, x')$ as

$$j^\mu(x) = -e \lim_{x \to x'} tr[\alpha^\mu G(x, x'; A)] .\qquad (6.24)$$

However, the evaluation of the current must go through the procedure of taking the limit as $x \to x'$. In taking this limit while $x \neq x'$, the Green's function $G(x, x'; A)$ (defined by Eq. (6.23)) depends on the choice of a gauge and is not gauge invariant. The quantity $j^\mu(x)$ given by Eq. (6.24) using the Green's function $G(x, x'; A)$, will not be gauge invariant (see Supplement 6.3 below). On the other hand, the current $j^\mu(x)$ is a physical quantity and should be a gauge-invariant quantity. We must modify the definition of the current to insure that it is always a gauge-invariant quantity.

⊕[Supplement 6.3

We would like to show here that the current $j^\mu(x)$, as obtained by using Eqs. (6.22)-(6.24), is not gauge invariant.

We consider a gauge transformation from A to $\bar{A}$ of the following form:

$$\bar{A}_\mu(x) = A_\mu(x) - \partial_\mu \lambda(x) .\qquad (1)$$

The field operator $\psi(x, A)$ transforms as

$$\psi(x, \bar{A}(x)) = e^{ie\lambda(x)}\psi(x, A(x)) ,\qquad (2)$$

so that the Dirac equation is invariant under a gauge transformation

$$\gamma^\mu(i\partial_\mu - e\bar{A}_\mu)\psi(x, \bar{A}) = \gamma^\mu(i\partial_\mu - eA_\mu)\psi(x, A) = 0 .$$

Similarly, we have

$$\psi^\dagger(x', \bar{A}(x')) = e^{-ie\lambda(x')}\psi^\dagger(x', A(x')) .$$

Therefore, under the gauge transformation (1-2), $G(x, x'; A)$ transforms as

$$\begin{aligned} G(x, x'; \bar{A}) &= < T(\psi(x, \bar{A})\psi^\dagger(x', \bar{A})) > \\ &= e^{ie\{\lambda(x) - \lambda(x')\}} < T(\psi(x, A)\psi^\dagger(x', A)) > \\ &= e^{ie\{\lambda(x) - \lambda(x')\}} G(x, x'; A) .\end{aligned}\qquad (3)$$

If x differs from x', we observe from the above equation that $\lambda(x)$ need not equal $\lambda(x')$. Consequently, the Green's function $G(x, x'; \bar{A})$ need not equal $G(x, x'; A)$. The Green's function obtained with Eq. (6.23) depends on the gauge and is therefore not gauge invariant. Using Eq. (6.24) by taking the limit of this Green's function as x approaching but differing from x', the current $j^\mu(x)$ is also not gauge invariant.

$\boxed{\oplus}$

By multiplying $G(x, x'; A)$ by the factor $e^{ie \int_{x'}^x A_\mu(\xi) d\xi^\mu}$, Schwinger introduced the gauge-invariant Green's function $\widetilde{G}(x, x'; A)$ as

$$\widetilde{G}(x, x'; A) = e^{ie \int_{x'}^x A_\mu(\xi) d\xi^\mu} G(x, x'; A). \qquad (6.25)$$

It is easy to show that the Green's function $\widetilde{G}(x, x'; A)$ is gauge invariant. Suppose one changes the gauge field from A_μ to $\bar{A}_\mu$ by making the gauge transformation (6.1), the Green's function $\widetilde{G}(x, x'; \bar{A})$ for the new gauge field $\bar{A}_\mu$ is

$$\widetilde{G}(x, x'; \bar{A}) = e^{ie \int_{x'}^x \bar{A}_\mu(\xi) d\xi^\mu} G(x, x'; \bar{A})$$

$$= e^{ie \int_{x'}^x [A_\mu(\xi) - \partial_\mu \lambda(\xi)] d\xi^\mu} G(x, x'; \bar{A})$$

$$= e^{ie \int_{x'}^x A_\mu(\xi) d\xi^\mu} e^{-ie\{\lambda(x) - \lambda(x')\}} G(x, x'; \bar{A}).$$

When $G(x, x'; \bar{A})$ in the above equation is written out explicitly in terms of $G(x, x'; A)$ by using Eq. (3) of Supplement 6.3, the extra factors in λ cancel, and we obtain

$$\widetilde{G}(x, x'; \bar{A}) = e^{ie \int_{x'}^x A_\mu(\xi) d\xi^\mu} G(x, x'; A) = \widetilde{G}(x, x'; A).$$

Therefore, the Green's function $\widetilde{G}(x, x'; A)$ is a *gauge-invariant Green's function*. It is not necessary to label this gauge-invariant Green's function $\widetilde{G}(x, x')$ by the gauge field because it is the same for different gauge fields when they are related by a gauge transformation.

As we discussed before, the current j^μ, as defined by Eq. (6.24), is not gauge invariant because it is defined in terms of the Green's function $G(x, x'; A)$ of Eq. (6.23) which is not gauge invariant. On the other hand, quantities such as currents are physical quantities which must not depend on the choice of a gauge.

To define a gauge-invariant current, it is necessary to modify Eqs. (6.23) and (6.24) by replacing the gauge-noninvariant Green's function $G(x, x'; A)$ of Eq. (6.23) with the gauge-invariant Green's function

$\widetilde{G}(x, x')$ of Eq. (6.25). Furthermore, the Green's function is singular at $x = x'$, and the limit as x approaches x' depends on the direction of approach. We certainly would like to choose a space-like separation between x and x' because we want to evaluate the current without bringing in the dynamics of the operators. (The separation between the point (x^0, x^1) and the point $(x^{0'}, x^{1'})$ is space-like if $(x^1 - x^{1'})^2 - (x^0 - x^{0'})^2 \geq 0$.) We can take $x^0 = x^{0'}$ to make the separation between x and x' space-like. Furthermore, because of the singularity of the Green's function, it is necessary to average the limit $x^1 \to x^{1'}$ by approaching from both the left and the right. Accordingly, we modify the definition (6.24) and define a *gauge-invariant fermion current* j^μ as

$$ j^\mu(x) = \frac{-e}{2} \Big\{ \lim_{\substack{x^0=x^{0'} \\ x^1=x^{1'}-\epsilon}} tr[\alpha^\mu \widetilde{G}(x, x')] + \lim_{\substack{x^0=x^{0'} \\ x^1=x^{1'}+\epsilon}} tr[\alpha^\mu \widetilde{G}(x, x')] \Big\}. \quad (6.26) $$

Having defined a gauge-invariant current, we can put all other factors together to evaluate the current in the presence of the gauge field A_μ. The gauge field gives rise to the phase factor $e^{-ie\Phi(x)}$. From Eqs. (6.17) and (6.23), the Green's function $G(x, x')$ in the gauge field A_μ is

$$ G(x, x'; A) = e^{ie\Phi(x)} G_F(x, x') e^{-ie\Phi(x')}, \quad (6.27) $$

where $G_F(x, x')$ is the Green's function for free fermions:

$$ G_F(x, x') = <T(\psi_F(x)\psi_F^\dagger(x'))> . \quad (6.28) $$

It is easy to show in Supplement 6.4 below that the Green's function $G_F(x, x')$ at $x^0 = x^{0'}$ for free and massless fermions is given by

$$ G_F(x, x') \big|_{x^0=x^{0'}} = \frac{i\alpha^1}{2\pi(x^1 - x^{1'})} . \quad (6.29) $$

In terms of the free Green's function G_F, the gauge invariant Green's function (6.25) is

$$ \widetilde{G}(x, x') = e^{ie \int_{x'}^{x} d\xi^\mu A_\mu(\xi)} e^{ie\Phi(x)} G_F(x, x') e^{-ie\Phi(x')} . \quad (6.30) $$

As $\Phi(x')$ commutes with α^1, we can rewrite Eq. (6.30) as

$$ \widetilde{G}(x, x') = e^{ie \int_{x'}^{x} d\xi^\mu A_\mu(\xi)} e^{ie\{\Phi(x) - \Phi(x')\}} G_F(x, x') . $$

Expanding to first order in $x^1 - x^{1'}$, we have then

$$\tilde{G}(x, x')\,|_{x^0 = x^{0'}} = \left\{1 + ieA_1(x)(x^1 - x^{1'}) + ie[\Phi(x) - \Phi(x')]\right\}$$

$$\times \frac{i\alpha^1}{2\pi(x^1 - x^{1'})}. \qquad (6.31)$$

The first term inside the curly bracket of the above equal-time Green's function gives a singular quantity when $x^1 \to x^{1'}$. This singular term goes as $1/(x^1 - x^{1'})$ and depends on the direction from which x^1 approaches $x^{1'}$. The two singular terms cancel each other in the sum in Eq. (6.26). With this result, we obtain

$$j^\mu(x) = \frac{+e^2}{2\pi}tr\{\alpha^\mu\alpha^1[A_1(x) + \partial_1\Phi(x)]\}. \qquad (6.32)$$

We note here that one might naively think that the term $\int_{x'}^{x} d\xi^\mu A_\mu(\xi)$ would give a zero contribution when one takes the limit $x \to x'$. However, because of the singularity of the Green's function which goes as $1/(x - x')$ and the cancellation of the singularity by taking the limit from both directions, there is a finite contribution to the gauge-invariant Greens function from this term.

We shall first obtain j^1 and j^0 explicitly as a function of A_μ and Φ, and then eliminate Φ in terms of A_μ. Setting $\mu = 1$ in Eq. (6.32) and noting that $\alpha^1\alpha^1$ is equal to the unit matrix, we find

$$j^1(x) = \frac{e^2}{\pi}\{A_1(x) + \frac{1}{2}\partial_1[tr\Phi(x)]\}$$

$$= \frac{-e^2}{\pi}\{A^1(x) + \frac{1}{2}\partial^1[tr\Phi(x)]\}, \qquad (6.33)$$

where we have used $g^{11} = -1$. To obtain j^0, we use Eq. (6.20) to write down the following relation between Φ and A_μ:

$$\alpha^1(A_1 + \partial_1\Phi) = -\alpha^0(A_0 + \partial_0\Phi). \qquad (6.34)$$

We can substitute Eq. (6.34) into (6.32) to obtain

$$j^\mu(x) = \frac{-e^2}{2\pi}tr\{\alpha^\mu\alpha^0[A_0(x) + \partial_0\Phi(x)]\}.$$

Setting $\mu = 0$ in the above equation, we find

$$j^0(x) = \frac{-e^2}{\pi}\{A^0(x) + \frac{1}{2}\partial^0[tr\Phi(x)]\}. \tag{6.35}$$

We can eliminate Φ in Eqs. (6.33) and (6.35) in terms of A^μ by applying $\alpha^0\partial_0 - \alpha^1\partial_1$ to Eq. (6.34) and taking the trace. We obtain

$$\frac{1}{2}\partial^\lambda\partial_\lambda[tr\Phi(x)] = -\partial_\nu A^\nu. \tag{6.36}$$

Finally, using this result with Eqs. (6.33) and (6.35), the expression relating the gauge-invariant current j^μ with the gauge potential A_μ is found to be

$$j^\mu(x) = -\frac{e^2}{\pi}[A^\mu - \partial^\mu\frac{1}{\partial^\lambda\partial_\lambda}\partial_\nu A^\nu]. \tag{6.37}$$

The current, due to the presence of a gauge field, is related locally to a simple function of the gauge field. It leads to the remarkable property of QED$_2$ being equivalent to a free boson field with a mass, as discussed in the previous section.

$\oplus$[Supplement 6.4

We evaluate below the Green's function $G_F(x, x')$ at $x^0 = x^{0'}$ for free and massless fermions. The field operator $\psi_F(x)$ of free and massless fermions satisfies the following Dirac equation:

$$i\gamma^\mu\partial_\mu\psi_F(x) = 0. \tag{1}$$

We can introduce two independent spinors (u_L and u_R) defined by

$$u_L = \begin{pmatrix} 1 \\ 0 \end{pmatrix},$$

and

$$u_R = \begin{pmatrix} 0 \\ 1 \end{pmatrix}.$$

They have the property that they are the eigenstates of the chirality operator γ^5 of Eq. (6.16):

$$\gamma^5 u_L = -u_L$$

and

$$\gamma^5 u_R = u_R.$$

Thus, the spinor u_L describes states with a negative chirality, while the spinor u_R describes states with a positive chirality. We can expand the field operator $\psi_F(x)$ as the sum

$$\psi_F(x) = \psi_L(x)u_L + \psi_R(x)u_R. \tag{2}$$

The Dirac equation (1) becomes

$$i\gamma^\mu \partial_\mu \psi_F(x) = i\gamma^0(\partial_0 + \gamma^5 \partial_1)\psi_F(x)$$
$$= i\gamma^0[(\partial_0 - \partial_1)\psi_L(x)u_L] + (\partial_0 + \partial_1)\psi_R(x)u_R$$
$$= 0.$$

The solutions of the above equation can be explicitly written out. They are

$$\psi_L(x) = \frac{1}{\sqrt{2\pi}} \int_0^\infty dp \left[a_L(p)e^{-ip(x^0+x^1)} + b_L^\dagger(p)e^{+ip(x^0+x^1)}\right], \tag{3}$$

and

$$\psi_R(x) = \frac{1}{\sqrt{2\pi}} \int_0^\infty dp \left[a_R(p)e^{-ip(x^0-x^1)} + b_R^\dagger(p)e^{+ip(x^0-x^1)}\right]. \tag{4}$$

Thus, the left-moving fermions and antifermions, with the phases $\mp ip(x^0 + x^1)$, are associated with the negative chirality state u_L, and the right-moving fermions and antifermions are associated with the positive chirality state u_R. The canonical quantization rules for the fermion creation and annihilation operators a_λ, and $a_\lambda^\dagger$ are

$$\{a_\lambda(p), a_{\lambda'}^\dagger(p')\} \equiv a_\lambda(p)a_{\lambda'}^\dagger(p') + a_{\lambda'}^\dagger(p')a_\lambda(p) = \delta(p - p')\delta_{\lambda,\lambda'},$$

and similarly the quantization rules for the antifermion creation and annihilation operators b_λ, and $b_\lambda^\dagger$ are

$$\{b_\lambda(p), b_{\lambda'}^\dagger(p')\} = \delta(p - p')\delta_{\lambda,\lambda'},$$

where $\lambda, \lambda' = L, R$. All other anticommutators vanish. These quantization rules will lead to the equal-time anticommutation relations for the fermion field operators:

$$\left\{\psi_\lambda(x), \psi_{\lambda'}^\dagger(x')\right\}_{x^0=x^{0'}} = \delta_{\lambda\lambda'}\delta(x^1 - x^{1'}),$$

where $\lambda, \lambda' = L, R$.

We consider a vacuum state in which all the negative energy states in the Dirac sea are occupied. In terms of the fermion-antifermion picture, the absence of a fermion in the negative energy sea corresponds to the presence of an antifermion. In this equivalent fermion-antifermion description, the vacuum state $|0>$ is characterized by the absence of fermions or antifermions in the system. It is defined by

$$a_\lambda, b_\lambda|0> = 0.$$

Using this property of the vacuum state, substitution of Eqs. (2)-(4) into (6.28) leads to the Green's function

$$G_F(x, x') = < T(\psi_L(x)\psi_L^\dagger(x')) > u_L u_L^\dagger + < T(\psi_R(x)\psi_R^\dagger(x')) > u_R u_R^\dagger. \tag{5}$$

Those terms involving $u_L u_R^\dagger$ and $u_R u_L^\dagger$ vanish.

We can evaluate the first term of the above equation $< T(\psi_L(x)\psi_L^\dagger(x')) >$ by writing out the time-ordered product explicitly. This gives

$$< T(\psi_L(x)\psi_L^\dagger(x')) > = < \psi_L(x)\psi_L^\dagger(x') > \theta(x^0 - x^{0'}) - < \psi_L^\dagger(x')\psi_L(x) > \theta(x^{0'} - x^0).$$

Substituting Eq. (3) for $\psi_L(x)$ and making use of the anticommutation relations for the fermion creation and annihilation operators, we obtain

$$< T(\psi_L(x)\psi_L^\dagger(x')) > = \frac{1}{2\pi}\left\{ \int_0^\infty dp\, e^{-ip\{(x^0 + x^1) - (x^{0'} + x^{1'})\}}\theta(x^0 - x^{0'}) \right.$$

$$\left. - \int_0^\infty dp\, e^{-ip\{(x^{0'} + x^{1'}) - (x^0 + x^1)\}}\theta(x^{0'} - x^0) \right\}.$$

The integral over p can be easily carried out by introducing $\epsilon > 0$ and taking the limit $\epsilon \to 0$:

$$< T(\psi_L(x)\psi_L^\dagger(x')) > = \frac{1}{2\pi}\lim_{\epsilon \to +0}\left\{ \int_0^\infty dp\, e^{-ip\{(x^0 + x^1) - (x^{0'} + x^{1'}) - i\epsilon\}}\theta(x^0 - x^{0'}) \right.$$

$$\left. - \int_0^\infty dp\, e^{-ip\{(x^{0'} + x^{1'}) - (x^0 + x^1) - i\epsilon\}}\theta(x^{0'} - x^0) \right\}.$$

We get

$$< T(\psi_L(x)\psi_L^\dagger(x')) > = \frac{-i}{2\pi}\lim_{\epsilon \to +0}\left\{ \frac{\theta(x^0 - x^{0'})}{(x^0 + x^1) - (x^{0'} + x^{1'}) - i\epsilon} \right.$$

$$\left. + \frac{\theta(x^{0'} - x^0)}{(x^0 + x^1) - (x^{0'} + x^{1'}) + i\epsilon} \right\}. \qquad (6)$$

We can follow a similar procedure to obtain for the second term of Eq. (5):

$$< T(\psi_R(x)\psi_R^\dagger(x')) > = \frac{-i}{2\pi}\lim_{\epsilon \to +0}\left\{ \frac{\theta(x^0 - x^{0'})}{(x^0 - x^1) - (x^{0'} - x^{1'}) - i\epsilon} \right.$$

$$\left. + \frac{\theta(x^{0'} - x^0)}{(x^0 - x^1) - (x^{0'} - x^{1'}) + i\epsilon} \right\}. \qquad (7)$$

Substituting Eqs. (6) and (7) into Eq. (5) and noting that $\gamma^5 u_{R,L} = \pm u_{R,L}$, we have

$$G_F(x, x') = \frac{-i}{2\pi}\lim_{\epsilon \to +0}\left\{ \frac{\theta(x^0 - x^{0'})}{(x^0 - x^{0'}) - \gamma^5(x^1 - x^{1'}) - i\epsilon}(u_L u_L^\dagger + u_R u_R^\dagger) \right.$$

$$\left. + \frac{\theta(x^{0'} - x^0)}{(x^0 - x^{0'}) - \gamma^5(x^1 - x^{1'}) + i\epsilon}(u_L u_L^\dagger + u_R u_R^\dagger) \right\}.$$

$$= \frac{-i}{2\pi}\lim_{\epsilon \to +0}\left\{ \frac{\theta(x^0 - x^{0'})}{(x^0 - x^{0'}) - \gamma^5(x^1 - x^{1'}) - i\epsilon} + \frac{\theta(x^{0'} - x^0)}{(x^0 - x^{0'}) - \gamma^5(x^1 - x^{1'}) + i\epsilon} \right\}. \qquad (8)$$

We are interested in the equal-time Green's function at $x^0 = x^{0'}$. By setting $x^0 = x^{0'}$ in Eq. (8), we have the equal-time free-fermion Green's function given by

$$G_F(x, x')_{x^0 = x^{0'}} = \frac{-i}{2\pi} \lim_{\epsilon \to 0} \frac{1}{2} \left\{ \frac{1}{-\gamma^5(x^1 - x^{1'}) - i\epsilon} + \frac{1}{-\gamma^5(x^1 - x^{1'}) + i\epsilon} \right\}$$

$$= \frac{-i}{2\pi} \lim_{\epsilon \to 0} \frac{1}{2} \frac{-2\gamma^5(x^1 - x^{1'})}{(x^1 - x^{1'})^2 + \epsilon^2}$$

$$= \frac{i}{2\pi} \frac{\alpha^1}{x^1 - x^{1'}}.$$

which is the result of Eq. (6.29). The Green's function is singular as x approaches x'. We can also rewrite the Green's function in a more compact form. Eq. (8) can be rewritten as

$$G_F(x, x') = \frac{-i}{2\pi\gamma^0} \lim_{\epsilon \to +0} \left\{ \frac{\theta(x^0 - x^{0'})}{\not{x} - \not{x}' - i\epsilon\gamma^0} + \frac{\theta(x^{0'} - x^0)}{\not{x} - \not{x}' + i\epsilon\gamma^0} \right\},$$

where the slash symbol is defined by $\not{x} = \gamma^\mu x_\mu$.

$]\oplus$

§6.3 Inside-Outside Cascade

From the last two sections, we find that in two-dimensional quantum electrodynamics, a system of massless fermions, antifermions, and the electromagnetic gauge field is equivalent to a system of free bosons with a rest mass. The boson can be considered to be a fermion-antifermion bound state. This is analogous to the system of quarks and antiquarks in QCD in which the physical quanta are mesons, which are bound states each of which consists of a quark and an antiquark. Because massless QED$_2$ has many properties similar to those of QCD, we shall use QED$_2$ to simulate approximately the behavior of the dynamics when a quark and an antiquark separate from each other at high energies. In this qualitative simulation, the positively charged fermion and the negatively charged antifermion in QED$_2$ respectively play the roles of the quark and the antiquark in QCD. The physical boson in QED$_2$ can be considered the analogue of the pion in QCD.

The equation of motion for the boson field in QED$_2$ is the Klein-Gordon equation (6.12), and the relation between the fermion current and the boson field is given by Eq. (6.10). As the dynamics of free bosons is simple, it is convenient to describe the dynamics of a fermion system in terms of bosons. When the fermion current j^μ is initially known, then the boson field ϕ is also initially known, and the dynamics of the boson field can be inferred at later times. The knowledge of the boson field, in turn, allows one to obtain the fermion currents at later times.

Alternatively, the fermion currents are given as external sources with a prescribed space-time behavior. The relation between the fermion and the boson field quantities allows one to study the dynamics of the bosons produced by the external fermion currents.

We examine the case where the energy of the system is so large that the trajectories of the separating fermion and the antifermion are assumed known. We consider a positive charge e moving in the $+x^1$ direction with the speed of light, and a negative charge $-e$ moving in the $-x^1$ direction also with the speed of light. Both start at the space-time origin ($x^1 = 0, t = x^0 = 0$). In this case, as the trajectories of the fermion-antifermion pair are given, we can treat these two particles as providing an 'external' current j^μ_{ext} which will produce a boson field in a space-time domain. In the center-of-mass system, the space-time trajectories of the fermion-antifermion pair for $t \geq 0$ can be described by the 'external' charge density

$$j^0_{\text{ext}}(x^1, t) = -e\delta(x^1 + t) + e\delta(x^1 - t), \qquad (6.38a)$$

and the 'external' charge current

$$j^1_{\text{ext}}(x^1, t) = e\delta(x^1 + t) + e\delta(x^1 - t). \qquad (6.38b)$$

The fermion current j^μ_{ext} can be represented in terms of an equivalent time-dependent, 'external' boson field ϕ_{ext} which is related to j^μ_{ext} by Eq. (6.10),

$$j^\mu_{\text{ext}}(x^1, t) = -m\epsilon^{\mu\nu}\partial_\nu \phi_{\text{ext}}(x^1, t).$$

Thus, for the case of the external fermion currents (6.38), the corresponding external boson field ϕ_{ext} is

$$\phi_{\text{ext}}(x^1, t) = -\sqrt{\pi}[\theta(x^1 + t) - \theta(x^1 - t)]. \qquad (6.39)$$

As the external boson field changes in time, a boson field $\phi_{\text{pro}}(x^1, t)$ is produced. The spatial distribution of the produced boson field can be considered as that of the produced particles. The quanta of the produced boson field are bosons with a rest mass $m = e/\sqrt{\pi}$.

The total boson field $\phi(x^1, t)$ is the sum of the external boson field ϕ_{ext} and the produced boson field $\phi_{\text{pro}}(x^1, t)$:

$$\phi(x^1, t) = \phi_{\text{pro}}(x^1, t) + \phi_{\text{ext}}(x^1, t). \qquad (6.40)$$

It obeys the Klein-Gordon equation (6.12):

$$\Box\phi + m^2\phi = 0 \,.$$

From Eqs. (6.12) and (6.40), the equation for the produced boson field is

$$\Box\phi_{\text{pro}}(x^1, t) + m^2\phi_{\text{pro}}(x^1, t) = -m^2\phi_{\text{ext}}(x^1, t) \,.$$

The general solution to this equation, which can be verified by direct substitution, is

$$\phi_{\text{pro}}(x^1, t) = \frac{1}{\sqrt{2\pi}} \int d^2p \, \theta(p^0)\delta(p^2 - m^2)\left[c(p^1)e^{-ip\cdot x} + c^*(p^1)e^{+ip\cdot x}\right]$$

$$-\phi_{\text{ext}}(x^1, t) \,, \tag{6.41}$$

where $p^0 = \sqrt{(p^1)^2 + m^2}$ and $p \cdot x = p_\mu x^\mu = p^0 t - p^1 x^1$. The coefficients c and c^* can be fixed from the initial conditions.

We shall proceed to find the coefficients $c(p^1)$ and $c^*(p^1)$ for our problem of a fermion-antifermion pair separating at high energies. The initial conditions are that there are no produced charge densities and no produced charge currents at $t = 0$ when the 'quark' (the positive fermion charge) and the 'antiquark' (the negative fermion charge) begin to separate from each other :

$$j^0_{\text{pro}}(x^1, 0) = j^1_{\text{pro}}(x^1, 0) = 0 \,.$$

The initial condition on the charge density gives

$$j^0_{\text{pro}}(x^1, 0) = m \, \partial_1\phi_{\text{pro}}(x^1, t)\Big|_{t=0}$$

$$= \frac{m}{\sqrt{2\pi}} \int \frac{dp^1}{2p^0} ip^1 \left[c(p^1)e^{ip^1 x^1} - c^*(p^1)e^{-ip^1 x^1}\right] - j^0_{\text{ext}}(x^1, 0)$$

$$= 0 \,.$$

Because $j^0_{\text{ext}}(x^1, 0) = 0$, we have

$$\frac{m}{\sqrt{2\pi}} \int \frac{dp^1}{2p^0} ip^1 \left[c(p^1) + c^*(-p^1)\right]e^{+ip^1 x^1} = 0 \,,$$

which implies

$$c^*(-p^1) = -c(p^1) \,. \tag{6.42}$$

From the initial condition on the charge current, we find

$$
\begin{aligned}
j^1_{\text{pro}}(x^1, 0) &= -m \; \partial_0 \phi_{\text{pro}}(x^1, t)\Big|_{t=0} \\
&= \frac{m}{\sqrt{2\pi}} \int \frac{dp^1}{2} i \left[c(p^1) e^{ip^1 x^1} - c^*(p^1) e^{-ip^1 x^1} \right] - j^1_{\text{ext}}(x^1, 0) \\
&= \frac{e}{2\pi} \int dp^1 \left[\frac{ic(p^1)}{\sqrt{2}} e^{ip^1 x^1} - \frac{ic^*(p^1)}{\sqrt{2}} e^{-ip^1 x^1} \right] - e\delta(x^1) - e\delta(x^1) \\
&= 0 .
\end{aligned}
$$

Therefore, the condition of no initial produced charge current gives

$$
ic(p^1) = -ic^*(-p^1) = \sqrt{2} . \tag{6.43}
$$

With the coefficients c and c^* explicitly obtained from the initial conditions in Eq. (6.43), the dynamics of the produced boson field $\phi_{\text{ext}}(x^1, t)$ is completely known. Substituting c and c^* back into ϕ_{pro}, we get the produced boson field

$$
\begin{aligned}
\phi_{\text{pro}}(x^1, t) &= \frac{1}{\sqrt{2\pi}} \int \frac{dp^1}{2p^0} \frac{\sqrt{2}}{i} \left[e^{-ip \cdot x} - e^{+ip \cdot x} \right] - \phi_{\text{ext}}(x^1, t), \\
&= \frac{-1}{\sqrt{\pi}} \int \frac{dp^1}{p^0} \sin(p^0 t - p^1 x^1) + \sqrt{\pi}[\theta(x^1 + t) - \theta(x^1 - t)] .
\end{aligned}
$$
$$
\tag{6.44}
$$

We show in Fig. 6.1 the spatial distribution of the produced boson field $\phi_{\text{pro}}(x^1, t)/\sqrt{\pi}$ as a function of time for the early evolution of the system. In the numerical calculations, we fix the magnitude of e by choosing the mass of the boson m to be 0.40 GeV, corresponding to the average transverse mass of pions in high-energy processes. As one observes, the produced boson field ϕ_{pro} starts to emerge at the point of the separation of the fermion-antifermion pair and spreads from the 'inside' (small $|x|$) region to the 'outside' (large $|x|$) region, as time proceeds. The dynamics of the produced bosons, which can be considered to be that of the 'produced' particles, exhibits the *inside-outside cascade* picture of particle production, first suggested by Bjorken [5].

To see how the charge spreads out when a fermion-antifermion pair of 'external' charges separate from each other at high energies, we show in Fig. 6.2 the produced charge density $j^0_{\text{pro}}(x^1, t)$ as a function of the spatial coordinates for different times. The external positive

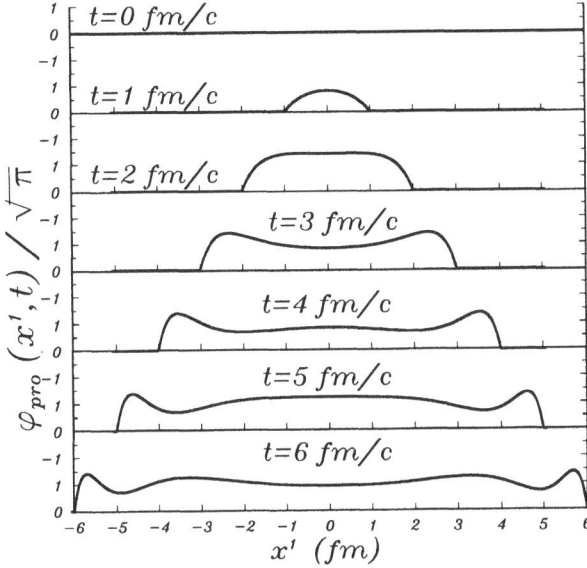

Fig. 6.1 The spatial distribution of the produced boson field $\phi_{\text{pro}}(x^1, t)/\sqrt{\pi}$ as a function of time t.

charge (the 'quark') is represented in Fig. 6.2 by a solid circle moving to the right, and the negative charge (the 'antiquark') is represented by an open circle moving to the left. The external charges are the source of the produced boson field. The spatial derivative of the produced boson field ϕ_{pro} gives the produced charge density $j^0_{\text{pro}}(x^1, t)$. Initially, the produced charge density $j^0_{\text{pro}}(x^1, 0)$ is zero. At time $t = 1$ fm/c, there are produced charges of opposite sign trailing behind each of the external charge sources. When these produced charges move outward, they in turn produce charges of opposite sign trailing behind them. The magnitudes of the secondary produced charges are smaller than the primary produced charges, but they spread out into a larger region. These secondary charges, in turn, produce later generations of charges of opposite sign as they move outward, but with diminished peak magnitudes. The space-time picture is that produced charges screen the charge sources which produce them, so that the later generation of particles have the signs of their charges governed by the signs of the most recently produced charged pair, and not by the signs of the earlier external sources.

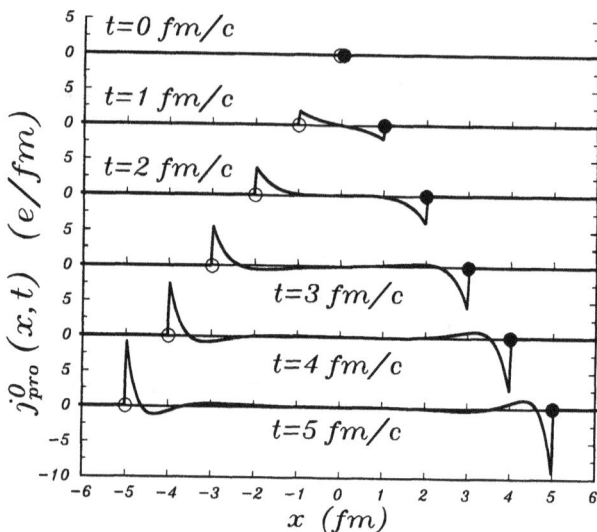

Fig. 6.2 The spatial distribution of the produced charged density $j^0_{\mathrm{pro}}(x^1, t)$ as a function of time t.

⊕[Supplement 6.5

We show below that the the produced boson field ϕ_{pro} is a function of the proper time.

We consider a fermion-antifermion pair of charges which begin to separate at the space-time origin ($x^1 = 0, t = 0$). From causality, we can limit our attention to the domain of time-like points (x^1, t) which have the property that $t^2 > (x^1)^2$. We would like to show that the produced boson field ϕ_{pro} at (x^1, t) is a function of the relativistic invariant quantity $\tau = \sqrt{t^2 - (x^1)^2}$, the proper time. To demonstrate this, we first show that the produced boson field $\phi_{\mathrm{pro}}(x^1, t)$ at (x^1, t) is the same as the produced boson field $\phi_{\mathrm{pro}}(x'^1, t')$ at (x'^1, t'), if (x^1, t) and (x'^1, t') are related by a Lorentz transformation. Because the proper time τ is an invariant quantity under a Lorentz transformation, the produced boson field is only a function of the proper time.

According to Eqs. (6.43) and (6.41), $\phi_{\mathrm{pro}}(x^1, t)$ at the point (x^1, t) is

$$\phi_{\mathrm{pro}}(x^1, t) = \frac{1}{i\sqrt{\pi}} \int d^2p\,\theta(p^0)\delta(p^2 - m^2)\left[e^{-ip\cdot x} - e^{+ip\cdot x}\right] + \sqrt{\pi}[\theta(x^1 + t) - \theta(x^1 - t)].$$

(1)

We consider another time-like space-time point $x'^\mu = (x'^1, t')$ which is related to $x^\mu = (x^1, t)$ by a Lorentz transformation $x'^\mu = \Lambda^\mu_\nu x^\nu$. Explicitly, the transformation is

$$t' = \gamma(t - \beta x^1), \tag{2}$$

and

$$x'^1 = \gamma(x^1 - \beta t), \tag{3}$$

where β is an arbitrary velocity parameter and $|\beta| < 1$. For this point (x'^1, t'), the boson field $\phi_{\text{pro}}(x'^1, t')$, according to Eqs. (6.43) and (6.41), is

$$\phi_{\text{pro}}(x'^1, t') = \frac{1}{i\sqrt{\pi}} \int d^2p\,\theta(p^0)\delta(p^2 - m^2) \left[e^{-ip\cdot x'} - e^{+ip\cdot x'} \right]$$

$$+ \sqrt{\pi}[\theta(x'^1 + t') - \theta(x'^1 - t')]. \tag{4}$$

We note that

$$p \cdot x' = p_\mu x'^\mu$$
$$= p_\mu \Lambda^\mu_\nu x^\nu.$$

This result suggests that to relate Eq. (4) to Eq. (1), it is useful to make a Lorentz transformation from p to p'

$$p_\nu' = \Lambda^\mu_\nu p_\mu. \tag{5}$$

Furthermore, because p^2 and d^2p are Lorentz-invariant quantities, Eq. (4) becomes

$$\phi_{\text{pro}}(x'^1, t') = \frac{1}{i\sqrt{\pi}} \int d^2p'\,\theta(p^0)\delta(p'^2 - m^2) \left[e^{-ip'\cdot x} - e^{+ip'\cdot x} \right]$$

$$+ \sqrt{\pi}[\theta(x'^1 + t') - \theta(x'^1 - t')]. \tag{6}$$

We make use of Eq. (5) to obtain

$$\theta(p'^0) = \theta(p'_0)$$
$$= \theta[\gamma(p_0 - \beta p_1)]$$
$$= \theta[\gamma(p^0 + \beta p^1)].$$

Because γ is always positive, $|\beta| < 1$ and $p^0 \geq |p^1|$, the above result leads to

$$\theta(p'^0) = \theta(p^0). \tag{7}$$

In addition, from Eqs. (2) and (3), we have

$$\theta(x'^1 + t') = \theta[\gamma(1 - \beta)(x^1 + t)]$$
$$= \theta(x^1 + t), \tag{8}$$

and

$$\theta(x'^1 - t') = \theta[\gamma(1 + \beta)(x^1 - t)]$$
$$= \theta(x^1 - t). \tag{9}$$

After substituting Eqs. (7) - (9) into (6) and relabeling p' by p, we have

$$\phi_{\text{pro}}(x'^1, t') = \frac{1}{2\sqrt{\pi}} \int d^2p\,\theta(p^0)\delta(p^2 - m^2)\left[e^{-ip\cdot x} - e^{ip\cdot x}\right]$$
$$+ \sqrt{\pi}[\theta(x^1 + t) - \theta(x^1 - t)]$$
$$= \phi_{\text{pro}}(x^1, t).$$

We have obtained the result that the produced boson field is the same for two time-like space-time points related by a Lorentz transformation. Under such a transformation, the only combination of coordinates which remain invariant is the proper time τ:

$$\tau = \sqrt{t^2 - (x^1)^2} = \sqrt{t'^2 - (x'^1)^2}.$$

Thus, the produced boson field can depend only on this combination of the space-time coordinates τ. That is,

$$\phi_{\text{pro}}(x^1, t) = \phi_{\text{pro}}(\tau). \qquad\qquad] \oplus$$

When a fermion is separated from an antifermion at high energies, the results from QED$_2$ show that the produced boson field ϕ_{pro} depends only on the proper time (see Supplement 6.5). One can obtain the explicit proper-time dependence by considering ϕ_{pro} at the space-time point $(x^1 = 0, t)$, for which the proper time is the same as the time coordinate t. We thus have from Eq. (6.44)

$$\phi_{\text{pro}}(\tau) = \phi_{\text{pro}}(x^1 = 0, t = \tau)$$
$$= \frac{-1}{\sqrt{\pi}} \int \frac{dp^1}{p^0} \sin p^0\tau + \sqrt{\pi}[\theta(\tau) - \theta(-\tau)], \qquad (6.45)$$
$$= -\sqrt{\pi} J_0(m\tau) + \sqrt{\pi},$$

where J_0 is the Bessel function of order 0. The produced boson field is an oscillating function of the proper time τ, as shown in Fig. 6.3. The boson field starts at zero for $\tau = 0$ and reaches a value of $\sqrt{\pi}$ at $2.405\hbar/mc^2 = 1.2$ fm/c. It oscillates about $\sqrt{\pi}$ for large values of the proper time. The space-time points, at which the produced boson field divided by $\sqrt{\pi}$ reaches the value of unity, give the locations at which the total charge produced is $+e$ on one side of the point and $-e$ on the other side (see Supplement 6.6). Because the production of a quark is accompanied by the production of an antiquark, the space-time locations, at which the total produced charge is $+e$ on one side and $-e$ on the other side, are the locations at which a quark and an antiquark are produced. Only after $q\bar{q}$ pairs are produced can a

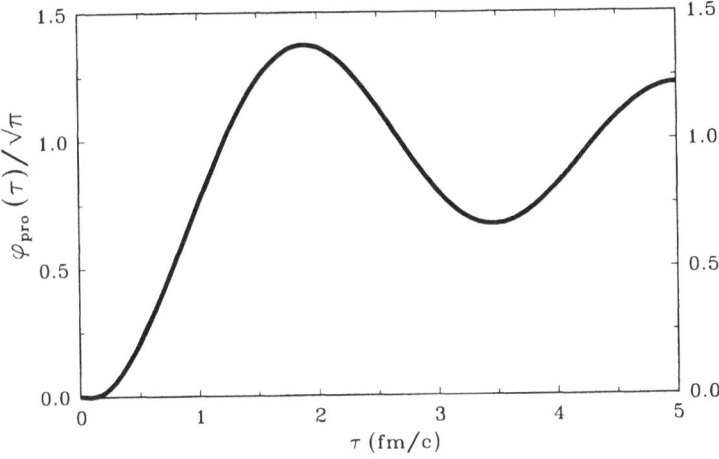

Fig. 6.3 The produced boson field $\phi_{pro}(\tau)$ as a function of the proper time τ.

quark from one vertex join with an antiquark in an adjacent vertex to form a boson. Thus, the proper time τ_{pro} at which the produced boson field ϕ_{pro} reaches the value of $\sqrt{\pi}$ can be called the 'particle production time'. For a proper time much less than τ_{pro}, the produced boson field is much less than $\sqrt{\pi}$ and the probability of producing a $q\bar{q}$ pair is small, and consequently the probability of producing a boson is also small. Numerically, as can be inferred from Eq. (6.45), τ_{pro} is determined by the location of the first zero of the Bessel function of order zero. The latter quantity is located at $m\tau_{pro} = 2.405$. Thus, *the particle production time τ_{pro} in QED$_2$ is given by*

$$\tau_{pro} = 2.405\hbar/mc^2. \tag{6.46}$$

We show in Fig. 6.4 a schematic description of the inside-outside cascade picture of particle production, as one can infer from what we learn in QED$_2$. We consider a positive charge q (the 'quark') which moves with the speed of light in the $+x^1$ direction, and a negative charge (the 'antiquark') which moves in the $-x^1$ direction, starting from the space-time origin. As the q and the $\bar{q}$ separate, many $q\bar{q}$ pairs are produced at vertices along the space-time points defined by $\tau = \tau_{pro}$. The locus of points with $\tau = \tau_{pro}$ are indicated in Fig. 6.4 by the dashed curve. The quark from one vertex combines with the antiquark from an adjacent vertex to form a physical boson with

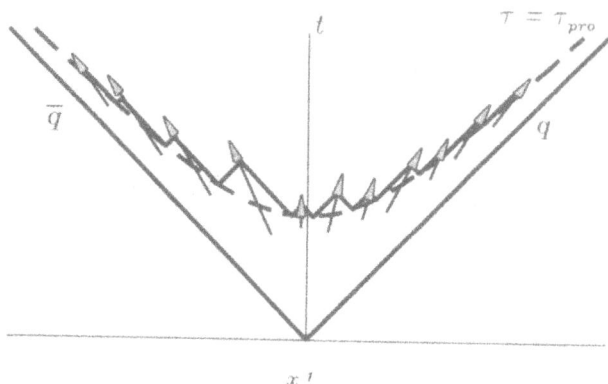

Fig. 6.4 A schematic picture of particle production in the *inside − outside cascade* picture.

a definite momentum, which is indicated by an arrow. Thus, the produced bosons begin to emerge at the proper time $\tau = \tau_{\text{pro}}$. Slow particles are the first to emerge, and the fastest particles are the last to emerge.

⊕Ⅱ Supplement 6.6

We would like to show that the total produced charge in the region $-\infty < x^1 \leq X^1$ is equal to $e\phi_{\text{pro}}(X^1, t)/\sqrt{\pi}$ and the total produced charge in the region $X^1 \leq x^1 < \infty$ is equal to $-e\phi_{\text{pro}}(X^1, t)/\sqrt{\pi}$.

The relation between the produced boson field ϕ_{pro} and the produced charge density j^0_{pro} is given by

$$j^0_{\text{pro}}(x^1, t) = m\, \partial_1 \phi_{\text{pro}}(x^1, t).$$

Therefore, the total produced charge in the region $-\infty < x^1 \leq X^1$ is

$$Q_< = \int_{-\infty}^{X^1} j^0_{\text{pro}}(x^1, t)dx^1$$

$$= \int_{-\infty}^{X^1} \frac{e}{\sqrt{\pi}}\, \partial_1 \phi_{\text{pro}}(x^1, t)dx^1$$

$$= e\phi_{\text{pro}}(X^1, t)/\sqrt{\pi}.$$

Similarly, the total produced charge in the region $X^1 \leq x^1 < \infty$ is equal to

$$Q_> = \int_{X^1}^{+\infty} j^0_{\text{pro}}(x^1, t)dx^1$$

$$= \int_{X^1}^{+\infty} \frac{e}{\sqrt{\pi}} \partial_1 \phi_{\text{pro}}(x^1, t) dx^1$$

$$= -e\phi_{\text{pro}}(X^1, t)/\sqrt{\pi} \,.$$

1 ⊕

§6.4 Rapidity Distribution from QED$_2$

The separation of the boson field into an 'external' part and a 'produced' part is useful to illustrate the space-time distribution of the produced particles in the presence of a polarizing 'external' field. It is a reasonable description in the limit in which the energy of the system is so high that the motion of the leading charges can be taken to follow simple, prescribed trajectories as given by Eq. (6.38). In an isolated QED$_2$ system with a finite energy, such a separation is not possible as the produced charges have a tendency to neutralize the leading charges. It is more useful to study the dynamics of the total boson field [6]. The dynamics of the system can be represented by the evolution of a real massive pseudoscalar boson field ϕ. The most general solution for ϕ is

$$\phi(x^1, t) = \frac{1}{\sqrt{2\pi}} \int d^2 p \, \theta(p^0) \delta(p^2 - m^2) \left[c(p^1) e^{-ip \cdot x} + c^*(p^1) e^{+ip \cdot x} \right],$$
(6.47)

which satisfies the Klein-Gordon equation (6.12). The coefficient $c(p^1)$ and its complex conjugation $c^*(p^1)$ can be obtained from the initial data of $j^0(x^1, t)$, and $j^1(x^1, t)$ at $t = 0$, by using Eq. (6.10).

As a simple example, we shall show how this initial-value problem can be formulated to yield the dynamics of the system and the rapidity distribution of the bosons. We start at $t = 0$ with the quark-antiquark pair superimposed so that the total charge density of the system at $t = 0$ is zero:

$$j^0(x^1, 0) = 0 \,.$$
(6.48)

As in Eq. (6.42), this initial condition leads to

$$c(p^1) = -c^*(-p^1) \,.$$
(6.49)

The initial charge current $j^1(x^1, 0)$, is nonzero, however. In the following discussion, we shall work in the center-of-mass system. In this frame, the initial current $j^1(x^1, 0)$ is a symmetric function of x^1 as the separating pair of opposite charges are moving in opposite directions.

We introduce the function $\tilde{j}^\mu(p^1)$, the Fourier transform of $j^\mu(x^1, 0)$ as

$$\tilde{j}^\mu(p^1) = \frac{1}{\sqrt{2\pi}} \int dx^1 e^{-ip^1 x^1} j^\mu(x^1, 0) \,. \tag{6.50}$$

From the initial condition on the charge current, we find

$$j^1(x^1, 0) = -m\, \partial_0 \phi(x^1, t)\Big|_{t=0}$$

$$= \frac{m}{\sqrt{2\pi}} \int \frac{dp^1}{2} i \left[c(p^1) e^{ip^1 x^1} - c^*(p^1) e^{-ip^1 x^1} \right]$$

$$= \frac{e}{2\pi} \int dp^1 \left[\frac{ic(p^1)}{\sqrt{2}} e^{ip^1 x^1} - \frac{ic^*(p^1)}{\sqrt{2}} e^{-ip^1 x^1} \right]$$

$$= \frac{1}{\sqrt{2\pi}} \int dp^1 \tilde{j}^1(p^1) e^{ip^1 x^1} \,.$$

Therefore, using Eq. (6.49) and the above equation, the initial charge current gives

$$c(p^1) = -\frac{i\sqrt{\pi}}{e} \tilde{j}^1(p^1) \,. \tag{6.51}$$

With the coefficients c and c^* thus determined, the energy P^0 of the system can be determined from

$$P^0 = \int \left[\frac{1}{2} \left(\frac{\partial \phi}{\partial t} \right)^2 + \frac{1}{2} \left(\frac{\partial \phi}{\partial x^1} \right)^2 + \frac{1}{2} m^2 \phi^2 \right] dx^1 \,. \tag{6.52}$$

With $\phi(x^1, t)$ given by Eq. (6.47), this leads to the energy

$$P^0 = \int dp^1 c^*(p^1) c(p^1)/2 \,,$$

which is clearly a time-independent quantity. From the above equation, we have

$$P^0 = \int dp^1 p^0 \frac{c^*(p^1) c(p^1)}{2p^0} \,,$$

and we obtain the momentum distribution of the bosons given by

$$\frac{dN}{dp^1} = \frac{c^*(p^1) c(p^1)}{2p^0}$$

$$= \frac{\pi}{2p^0 e^2} |\tilde{j}^1(p^1)|^2 \,. \tag{6.53}$$

By relating p^1 with the rapidity as $p^1 = m \sinh y$, the rapidity distribution of the particles is

$$\frac{dN}{dy} = \frac{\pi}{2e^2}|\tilde{j}^1(p^1)|^2 = \frac{\pi}{2e^2}|\tilde{j}^1(m \sinh y)|^2 . \qquad (6.54)$$

We obtain the result that the rapidity distribution of the boson particles in the fragmentation of a fermion-antifermion pair is given by the square of the Fourier transform of the initial fermion current. When the energy of the system is so large that the initial current can be represented by a delta function in the coordinate x^1, the Fourier transform $\tilde{j}^1(p^1)$ is a constant and the rapidity distribution is a constant, independent of rapidity.

⊕[Supplement 6.7
    As an example of the relation between the rapidity distribution and the initial fermion current, we can consider the initial density distribution

$$j^0(x^1, 0) = 0$$

and an initial current which arises from a charge νe moving in the positive x direction and another charge $-\nu e$ moving in the negative x direction given by [6]

$$j^1(x^1, 0) = \frac{\nu e}{\sigma \cosh^2(x^1/\sigma)} .$$

In the limit as σ approaches zero, the right side of the above equation is proportional to a delta function. The diffusivity σ is related to the total invariant mass $\sqrt{s}$ of the system; using Eq. (6.52) to calculate the energy $P^0 = \sqrt{s}$ at the initial time $t = 0$, we obtain a relation between σ and $\sqrt{s}$:

$$\sigma = \frac{2\pi\nu^2}{3\sqrt{s}} .$$

For this current distribution $j^\mu(x, 0)$ the Fourier transform of $j^1(x, 0)$ is [6]

$$\tilde{j}^1(p^1) = \frac{iec(p^1)}{\sqrt{\pi}} = -\frac{iec^*(-p^1)}{\sqrt{\pi}} = \frac{\nu e \pi p^1 \sigma}{\sqrt{2\pi} \sinh(\pi p^1 \sigma/2)} ,$$

and the rapidity distribution is

$$\frac{dN}{dy} = \frac{\nu^2 \xi^2}{\sinh^2 \xi} ,$$

where

$$\xi = \frac{\nu^2 \pi^2 m \sinh y}{3\sqrt{s}} .$$

The rapidity distribution therefore shows a plateau structure around $y \sim 0$. In the limit of very high energy, the rapidity distribution is $dN/dy = \nu^2$.   ]⊕

§**References for Chapter 6**

1. J. Schwinger, Phys. Rev. 128, 2425 (1962); J. Schwinger, in *Theoretical Physics,* Lecture Presented at Seminar on Theoretical Physics, Trieste, 1962, published by IAEA, Vienna, 1963, p. 89.
2. A. Casher, J. Kogut, and L. Susskind, Phys. Rev. D10, 732 (1974).
3. J. H. Lowenstein and J. A. Swieca, Ann. Phys. (N.Y.) 68, 172 (1971).
4. T. Eller, H-C. Pauli, and S. Brodsky, Phys. Rev. D35, 1493 (1987); T. Eller and H-C. Pauli, Zeit. für Phys. 42, 59 (1989).
5. J. D. Bjorken, Lectures presented in the 1973 Proceedings of the Summer Institute on Particle Physics, edited by Zipt, SLAC-167 (1973).
6. C. Y. Wong, R. C. Wang, and C. C. Shih, Phys. Rev. D44, 257 (1991).

7. Classical String Model

In the last two chapters, particle production was considered as a tunneling process in a strong field, or as a dynamical excitation of an external polarizing field in two-dimensional QED. Although these basic considerations provide much insight into the dynamics of the production process, their applications to practical problems have been limited by the inability to describe many different produced mesons as stable states and to incorporate the transverse degrees of freedom. For phenomenological applications, there are some merits in the classical string model which describes mesons as string segments executing longitudinal expansion and contraction, and the process of particle production is represented by the fragmentation of a stretching string formed by a separating quark and antiquark pair. This type of model [1-3] has enjoyed a great deal of success and is useful for many applications. We shall describe the classical string model for mesons and for particle production.

§7.1 The String Model of Hadrons

The notion that hadrons can be described as strings originated in the late 1960's from the study of the dual resonance model [4] which was later interpreted in terms of the scattering of strings [5-7]. In the dual resonance model, one calculates the amplitude for the elastic scattering of mesons with initial momenta p_1 and p_2 to end up as mesons with final momenta p_3 and p_4: $p_1 + p_2 \rightarrow p_3 + p_4$. In the lowest-order Feynman diagram, this can be done in two different ways, as shown in Fig. (7.1a) and (7.1b). In one way, as indicated in Fig. (7.1a), one allows p_1 and p_2 to interact to lead to a particle (or a resonance) p_J which subsequently breaks up into p_3 and p_4. As the momentum of p_J in Fig. (7.1a) is equal to the sum $p_1 + p_2$ and $(p_1 + p_2)^2 = s$, a diagram of the type (7.1a) is called an s-channel diagram. There is a diagram (7.1a) for each value of the spin J of the intermediate particle p_J. The s-channel scattering amplitude is the sum of all the amplitudes coming from all the diagrams of the type in Fig. (7.1a) with different values of the spin J. Another way of calculating the amplitude describes the process as the exchange of a particle p_J from the particle p_1 to p_2 as they propagate forward in time, leading to the final particles p_3 and p_4, as indicated by Fig. (7.1b). Because the momentum p_J is equal to $p_1 - p_3$ and $(p_1 - p_3)^2 = t$,

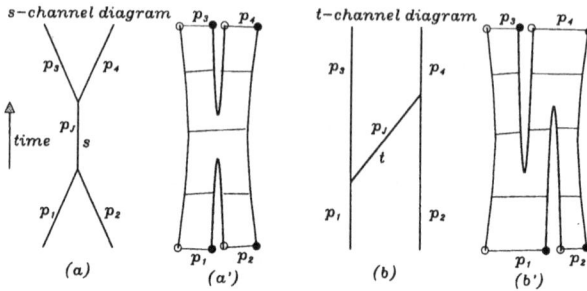

Fig. 7.1 (a) The s-channel diagrams for elastic meson-meson scattering. (a') The s-channel diagram for elastic meson-meson scattering when mesons are assumed to be strings. The configuration of the strings (mesons) at different proper times is represented by thin lines. The solid and the open circles represent a quark and an antiquark respectively. (b) The t-channel diagram for elastic meson-meson scattering. (b') The t-channel diagram for elastic meson-meson scattering when mesons are assumed to be strings.

a diagram of the type in $(7.1b)$ is called a t-channel diagram. There is a t-channel diagram $(7.1b)$ for each value of the spin J of the exchanged particle p_J. The t-channel scattering amplitude is the sum of all the amplitudes coming from all the t-channel diagrams with different values of J.

In the dual resonance model, it was found that the s-channel amplitude and the t-channel amplitude are the same. The s- and the t-channel diagrams give two alternative, or 'dual', descriptions of the same physics. This 'duality' property is the origin of the name 'dual resonance model'. A quantitative introduction to the dual resonance model will be presented in Chapter 8.

A very simple way the duality property can be realized is to assume that a meson is not a point object but an extended object, with the structure of a string connecting a quark and an antiquark as shown in Fig. $(7.1a')$ and Fig. $(7.1b')$. In terms of strings, the s-channel diagram Fig. $(7.1a)$ for mesons as point particles is actually a configuration of the form in Fig. $(7.1a')$ if the mesons are strings. Similarly, the t-channel diagram in Fig. $(7.1b)$ for point-like mesons becomes Fig. $(7.1b')$ if the mesons are described as strings. Because Figs. $(7.1a')$ and $(7.1b')$ have the same topology, they are just the same diagram in terms of quarks and antiquarks. The differences of the s-channel and t-channel diagrams in terms of point-like mesons are merely two manifestations of the same quark-antiquark diagram (Fig. $(7.1a')$ or

Fig. $(7.1b')$) in two different Lorentz frames. The s-channel amplitude is therefore the same as the t-channel amplitude. It is in this way that the dual resonance model led to the suggestion of a string description of mesons.

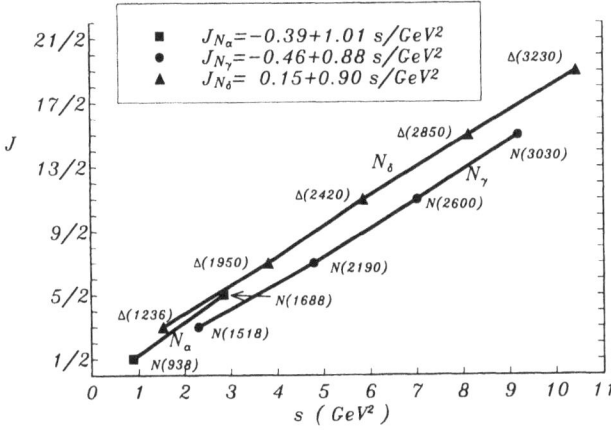

Fig. 7.2 The relation between J and $s(= M^2)$ for baryons (from Ref. 8).

A string picture of hadrons also gives the appropriate relationship between the mass M of the hadron and its spin J. It was found empirically that hadron resonances with the same internal symmetry quantum numbers but different spins fall into straight lines when one plots J as a function of $s(= M^2)$. These straight lines are called *Regge trajectories* (Fig. 7.2). That is, J is related to M by

$$J(M) = \alpha(0) + \alpha' M^2 , \qquad (7.1a)$$

where $\alpha(0)$ is called the *Regge intercept* and α' the *Regge slope*. The Regge slopes for three different sets of baryons can be read off from the relation between J and s in Fig. 7.2. It appears that the Regge slope is approximately a universal constant with a value

$$\alpha' \approx 1 \;\; \text{GeV}^{-2} . \qquad (7.1b)$$

Such a relation between the spin and the mass can be obtained in the model of a spinning string. We can consider a string of length $2L$ rotating about the axis passing through its center O (see Fig. 7.3) and perpendicular to the rotating string. The end points of the string are attached to a massless quark (solid circle) and a massless antiquark (open circle) traveling with the speed of light in the direction perpendicular to the string. We assume a rigid-body rotation. The string segment dx at the point x has a speed $v = x/L$. If the string were at rest, the energy stored in the segment dx would be $\kappa\,dx$, where κ is the string tension coefficient. For the string in rotation, the energy content in the segment dx is $\gamma\kappa dx$, where $\gamma = 1/\sqrt{1-v^2}$ is the gamma factor for this string segment in motion. The total energy content of the string is therefore

$$
\begin{aligned}
M &= 2\int_0^L \gamma\kappa dx \\
  &= 2\int_0^L \frac{\kappa dx}{\sqrt{1-(x/L)^2}} \\
  &= \pi\kappa L \, .
\end{aligned}
$$

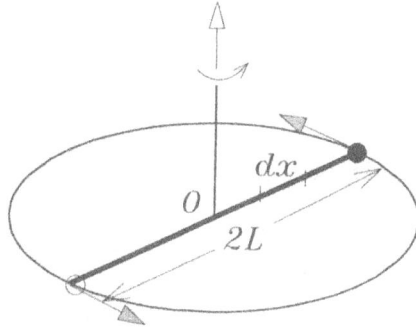

Fig. 7.3 A rotating relativistic string.

On the other hand, the angular momentum content of the segment dx is its energy content multiplied by vx. Hence, the total angular momentum of the string is

$$
\begin{aligned}
J &= 2\int_0^L \gamma\kappa vx dx \\
  &= 2\int_0^L \frac{\kappa vx dx}{\sqrt{1-(x/L)^2}} \, . \\
  &= \kappa L^2 \pi/2
\end{aligned}
$$

From these relationships, we have

$$J = \frac{1}{2\pi\kappa} M^2 . \tag{7.2}$$

Thus, a string description of the structure of hadrons gives the proper relationship between the angular momentum and the mass as in Eq. (7.1a). In this picture, a comparison of Eqs. (7.1a) and (7.2) shows that the string tension κ is related to the Regge slope α' by

$$\kappa = \frac{1}{2\pi\alpha'} \approx 1 \text{ GeV/fm} . \tag{7.3}$$

It should be pointed out that, although strings may be used as an *effective* description of strong interaction phenomena in connection with the large-distance structure of hadrons, we now know that the string theory cannot be used as a *fundamental* model of the strong interaction. There is a general consensus that the fundamental theory for the strong interaction is the theory of quantum chromodynamics. The field theory of strings for strong interactions has the problem that it predicts a massless vector particle ρ and a massless tensor particle. Any attempt to move these states to positive masses leads to other types of difficulties such as negative norms. The string theory of strong interaction also has difficulties in explaining many hard processes with large transverse momentum transfers observed in deep inelastic electron scattering experiments which led to the concept of partons in strong interactions. It has been suggested in the superstring theory that the field theory of strings may instead describe the structure of fundamental particles such as quarks, gluons, leptons, W$^{\pm}$, photons, and gravitons [9].

When we deal with the production of hadrons with a transverse momentum of a few hundred MeV/c, one is studying the strong interaction at a large distance scale. A phenomenological description in terms of strings provides much insight into the dynamics of the system. With the advent of the theory of QCD for strong interactions, and the realization that the color electric field between a quark and an antiquark, separated by a large distance, is constrained to reside mainly between them in the form of a flux tube [10,11], the string model of a meson is intuitively a reasonable concept. Much insight into the mechanism of particle production is provided by examining classical strings in one space and one time dimensions. They are useful as a simple representation of mesons, as well as highly excited $q\bar{q}$ systems. In this description, a meson at rest in its center-of-mass system is depicted as a string which has a quark and an antiquark at its two ends, the quark and the antiquark expanding and contracting in a yo-yo motion. A moving meson is described as a string with a translational motion superimposed on the intrinsic yo-yo motion. An excited $q\bar{q}$ system is described as a string whose quark and antiquark at its two ends move apart with a large momentum. The excited string fragments into hadrons with new $q\bar{q}$ pairs created between the separating quark and antiquark. The classical string model is capable of describing many aspects of the high-energy particle production process.

§7.2 The Yo-Yo Model of Stable Particles

We would like to begin our discussion of a simple string system which can be considered as a classical representation of a stable meson [1-3]. The description is simplest in the center-of-mass system in which the meson is at rest. The string is characterized by a quark and an antiquark located at the two ends of the string and interacting with a linear potential. It is convenient to take the quark and the antiquark to be massless so that their trajectories can be easily followed.

The Hamiltonian for this system of massless $q\bar{q}$ pair is

$$H(x_i, p_i) = |p_q| + |p_{\bar{q}}| + \kappa |x_q - x_{\bar{q}}| , \tag{7.4}$$

where κ is the string tension, and p_i and x_i denote, respectively, the (longitudinal) momentum and the (longitudinal) spatial coordinate of the particle i, with $i = q$ and $\bar{q}$. The motion of the quark q and the antiquark $\bar{q}$ can be determined from the Hamilton equations:

$$\dot{x}_i = \frac{\partial H}{\partial p_i} , \tag{7.5}$$

and

$$\dot{p}_i = -\frac{\partial H}{\partial x_i} . \tag{7.6}$$

From Eq. (7.5), we find that the velocity of the particle i (in units of the speed of light) is governed by

$$\dot{x}_i = \text{sign}(p_i) . \tag{7.7}$$

Thus, the direction of motion of the particle i is governed by the sign of its momentum. The spatial coordinate of the particle i is

$$x_i(t) = x_i(t_0) + \text{sign}(p_i)(t - t_0) , \tag{7.8}$$

where t_0 is the most recent time when the quark and the antiquark meet together at $x_i = 0$. From the Hamilton equation (7.6), the rate of change of the momentum of the particle i is determined by the sign of $x_i - x_{i'}$:

$$\dot{p}_i = -\text{sign}(x_i - x_{i'})\kappa , \tag{7.9}$$

where i' is different from i. The rate of change of the magnitude of the momentum of the quark or the antiquark is governed by the string tension κ. The solution for the momentum of the ith particle is

$$p_i(t) = p_i(t_0) - \text{sign}(x_i - x_{i'})\kappa(t - t_0). \qquad (7.10)$$

We consider the initial situation where both the quark and the antiquark are located at the spatial origin at $t = 0$. Based on the above results, we have the following description for the motion of the quark and the antiquark which characterize the two ends of the string, as shown in Fig. (7.4).

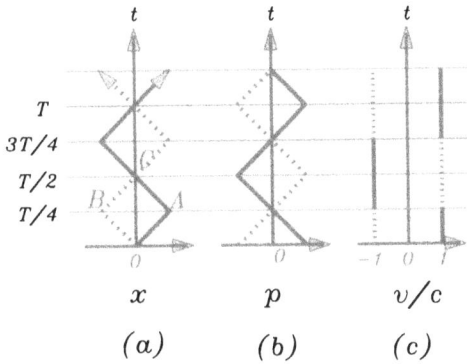

Fig. 7.4 (a). The space-time trajectories of the quark (the solid line) and the antiquark (the dashed line) for a stable meson (the yo-yo state) at rest. The time T, given by Eq. (7.13), is the period of the yo-yo state in its rest frame. (b) The time variation of the momentum of the quark (the solid line) and the antiquark (the dashed line) for the yo-yo state. (c) The time variation of the velocity of the quark (the solid line) and the antiquark (the dashed line) for the yo-yo state.

Without loss of generality, we consider initially that the quark is moving to the right with an initial momentum $p_q(0) > 0$, and the antiquark is moving to the left with an initial momentum $p_{\bar{q}}(0) < 0$ and $p_q(0) = -p_{\bar{q}}(0)$. During the initial stage, $p_q(t) > 0$ and $p_{\bar{q}}(t) < 0$, the Hamilton equation (7.5) gives

$$\dot{x}_q = \text{sign}(p_q) = +1 \qquad (7.11a)$$

and

$$\dot{x}_{\bar{q}} = \text{sign}(p_{\bar{q}}) = -1. \qquad (7.11b)$$

Thus, initially, the quark has a positive momentum and travels in the $+x$ direction, while the antiquark has a negative momentum and travels in the $-x$ direction, as shown in Figs. (7.4a) and (7.4c). Because these particles are massless, they travel with the speed of light.

In the initial stage, the motion of the quark and the antiquark pair leads to $x_q > x_{\bar{q}}$ (see Fig. (7.4a)). The Hamilton equation (7.6) for $\dot{p}_i$ then gives

$$\dot{p}_q = -\mathrm{sign}(x_q - x_{\bar{q}})\kappa = -\kappa \qquad (7.12a)$$

and

$$\dot{p}_{\bar{q}} = -\mathrm{sign}(x_{\bar{q}} - x_q)\kappa = +\kappa. \qquad (7.12b)$$

Therefore, as the quark and the antiquark separate from each other, the quark which has a positive momentum moves in the $+x$ direction, with its momentum diminishing linearly with time (see Fig. (7.4b)). It is pulled by the antiquark, by a force pointing in the $-x$ direction. As determined by Eq. (7.12a), the momentum of the quark is zero at the time $p_q(0)/\kappa$ which is one quarter of the period T for the motion of the quark or the antiquark in this center-of-mass frame. Thus, the period T is given by

$$T = 4p_q(0)/\kappa. \qquad (7.13)$$

After $t > T/4$, the momentum of the quark will be negative and its direction of motion will be reversed. It will turn around and travel in the $-x$ direction toward the antiquark. It experiences an attractive pull from the antiquark. The magnitude of its momentum increases linearly with time, from a value of zero at $t = T/4$ to reach its maximum value at $x = 0$ and $t = T/2$. After $t > T/2$, the quark moves away from the antiquark and experiences a pull from the antiquark to slow down. The magnitude of its momentum decreases linearly with time, until the momentum becomes zero at $t = 3T/4$. The behavior of the antiquark is just a mirror image of that of the quark. The time variations of the momentum of the quark and the antiquark are shown in Fig. (7.4b).

Combining the motion of the quark and the antiquark, we can describe the dynamics of the string as follows. Initially at $t = 0$, the quark and the antiquark are located at point O at $x = 0$ (Fig. (7.4a)). As time increases, the quark moves in the $+x$ direction, while the antiquark moves in the $-x$ direction. The string expands. The string reaches its maximum extension when the quark and the antiquark reach their respective turning points A and B at time $t = T/4$. Thereafter, the direction of motion of two particles reverses and the string contracts until the quark and the antiquark meet each other at point C, at $x = 0$ and $t = T/2$. This completes a half cycle of the expansion and contraction motion. The expansion and contraction motion will continue and will repeat themselves. This type of motion is called a yo-yo motion. A string which undergoes a yo-yo motion is said to be in a $yo - yo$ $state$.

In the center-of-mass system, the (t, x) coordinates for the points O, A, B, and C in Fig. 7.4 are given by

$$
\begin{aligned}
O &= (0,0) \\
A &= (T/4, T/4) \\
B &= (T/4, -T/4)
\end{aligned}
\tag{7.14}
$$

and

$$
C = (T/2, 0) \, .
$$

With the results of Eqs. (7.8) and (7.10) given above, it is easy to show that the total energy of the system H is a constant of motion, independent of time. We can evaluate the total energy H at time $t = 0$ at which the separation between the quark and the antiquark is zero. The total energy is

$$
H = 2|p_q(0)| \, .
\tag{7.15}
$$

We identify this energy H as the rest mass m of the stable meson. Since it is equal to the invariant mass $\sqrt{s}$ of the $q\bar{q}$ system at the moment when the quark q and the antiquark $\bar{q}$ meet each other, we shall also call this quantity the invariant mass of the yo-yo system.

It is convenient to introduce the forward light-cone space-time coordinate u, and the backward light-cone space-time coordinate v. They are related to coordinates t and x by

$$
(u, v) = (t + x, t - x) \, .
\tag{7.16}
$$

The (u, v) coordinates of the points O, A, B, and C are

$$
\begin{aligned}
O &= (\quad 0, \quad 0) \\
A &= (\ T/2, \quad 0) \\
B &= (\quad 0, \ T/2)
\end{aligned}
\tag{7.17}
$$

and

$$
C = (T/2, T/2) \, .
$$

That is, A lies on the axis $v = 0$ and B lies on the axis of $u = 0$. In the (u, v) coordinate system, the area $OACB$ enclosed by the quark and antiquark trajectories in a single half-cycle is

$$
\begin{aligned}
(\text{area } OACB) = OA \times OB &= T^2/4 \\
&= 4|p_q(0)|^2/\kappa^2 \\
&= s/\kappa^2 \, .
\end{aligned}
\tag{7.18}
$$

In subsequent expansions and contractions, the area enclosed by the quark and the antiquark trajectories in each half cycle is the same as that of $OACB$. This area in the (u,v) coordinate system can be called an elementary area for the yo-yo state in question. From Eq. (7.18), we see that the elementary area enclosed by the trajectory of the quark and the antiquark lines is equal to the square of the invariant mass of the yo-yo state divided by κ^2. Therefore, in the classical string model, a meson is represented by a yo-yo state whose elementary area is equal to the square of the meson mass divided by κ^2.

§7.3 A Yo-Yo State in Motion

How do we describe a yo-yo state in translational motion? This is achieved by first following the yo-yo state in its center-of-mass system where the trajectories of the quark and the antiquark can be easily determined. This frame of reference is then boosted with a velocity β relative to the laboratory frame. The points O, A, B, and C for a yo-yo state at rest become O, A', B', and C' in the laboratory frame (see Fig. 7.5). The latter laboratory coordinates (t,x) can be obtained from the coordinates of Eq. (7.14) by a Lorentz transformation. They are

$$
\begin{aligned}
O &= (\quad\quad 0, \quad\quad\quad 0) \\
A' &= (\ \gamma(1+\beta)T/4, \quad \gamma(1+\beta)T/4) \\
B' &= (\ \gamma(1-\beta)T/4, \quad -\gamma(1-\beta)T/4)
\end{aligned}
\tag{7.19}
$$

and

$$
C' = (\gamma T/2,\ \gamma\beta T/2),
$$

where $\gamma = 1/\sqrt{1-\beta^2}$. The velocity of the yo-yo in the laboratory system is determined by the ratio of the $x-$coordinate of C' to the $t-$coordinate of C', which is given, as expected, by

$$
(\text{velocity of the yo}-\text{yo state}) = \frac{\gamma\beta T/2}{\gamma T/2} = \beta .
\tag{7.20}
$$

The trajectories for the boosted quark and the antiquark take on the form as shown in Fig. (7.5b).

In terms of the (u,v) coordinates, the points A' and B' for the yo-yo in motion are

$$
\begin{aligned}
O &= (\quad\quad 0, \quad\quad 0) \\
A' &= (\ \ \gamma(1+\beta)T/2, \quad\quad 0) \\
B' &= (\quad\quad 0, \ \ \gamma(1-\beta)T/2)
\end{aligned}
\tag{7.21}
$$

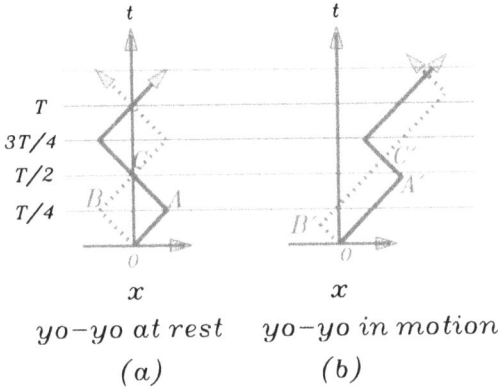

yo−yo at rest   yo−yo in motion

(a)   (b)

Fig. 7.5 (a). The space-time trajectories of the quark (the solid line) and the antiquark (the dashed line) for a yo-yo state at rest. (b) The space-time trajectories of the quark (the solid line) and the antiquark (the dashed line) for a yo-yo state in motion.

and
$$C' = (\gamma(1 + \beta)T/2, \gamma(1 - \beta)T/2) .$$

For the yo-yo state in motion, the elementary area $OA'C'B'$ enclosed by the quark and the antiquark trajectories in the (u, v) coordinate system is
$$\begin{aligned}(\text{area } OA'C'B') = OA' \times OB' &= T^2/4 \\ &= s/\kappa^2 ,\end{aligned} \qquad (7.22)$$

which is equal to the area $OACB$ for the yo-yo state at rest. Thus, the elementary area is an invariant quantity which does not depend on the motion of the yo-yo state. It is equal to the square of the invariant mass of the system divided by κ^2.

Strong interaction phenomena are characterized by a large number of stable particles and resonances. The ability to describe these stable particles as expanding and contracting strings is one of the advantages of the classical string model. The production of particles is represented in terms of the break-up of a string with a large invariant mass into smaller fragments with the masses of the stable mesons.

Knowing the velocity of the yo-yo state in the laboratory system, one can calculate the energy and the momentum of the yo-yo state as a single particle. These quantities give the forward and backward light-cone momenta p_+ and p_- of the yo-yo state. We find
$$p_+ = p_0 + p_x = \gamma(1 + \beta)\kappa T/2 \qquad (7.23a)$$

and

$$p_- = p_0 - p_x = \gamma(1 - \beta)\kappa T/2 \,. \tag{7.23b}$$

Comparing Eqs. (7.23) with (7.21), we note that (p_+, p_-) of the yo-yo state is given by the segment lengths of the trajectories of the quark and the antiquark $[OA'$ and OB' in (u, v) coordinates] multiplied by the string constant κ. The rapidity of the yo-yo state is

$$\begin{aligned} y &= \frac{1}{2}\ln\left(\frac{p_z + p_0}{p_z - p_0}\right) \\ &= \frac{1}{2}\ln\left(\frac{p_+}{p_-}\right) \\ &= \frac{1}{2}\ln\left(\frac{1 + \beta}{1 - \beta}\right), \end{aligned}$$

which is the same as the rapidity given by Eq. (1) of Exercise 2.4, for a single particle travelling with longitudinal velocity β.

§7.4 The Symmetric Lund Model

We shall examine here the symmetric Lund model [3] which gives the probability distribution function for the locations of the vertices at which the string is broken into separate pieces. The basic assumption of the symmetric Lund model is that the vertices at which the quark and antiquark pairs are produced lie approximately on a curve of constant proper time. This characteristics will lead to a rapidity distribution of the produced particles to be independent of the rapidity for an infinite-energy system (see Exercise 7.1) and is consistent with the experimental feature of a plateau shape for the rapidity distribution in finite-energy systems (Fig. 3.2). With additional simplifying assumptions on the functional dependence of the splitting function and the independence on the order of splitting, an explicit form of the distribution function for the locations of the vertices is then obtained.

To set up our framework, we consider a string system consisting of a pair of massless quark q and antiquark $\bar{q}$ with large forward and backward light-cone momenta respectively (Fig. 7.6). At the space-time origin $(t, x) = (0, 0)$, the quark q of the string meets with, and begins to separate from, the antiquark $\bar{q}$ of the string. We shall work in the center-of-mass frame in which the initial momentum of the quark $p_q(0)$ is equal and opposite to the initial momentum of the antiquark $p_{\bar{q}}(0)$. The invariant mass of the system is $\sqrt{s} =$

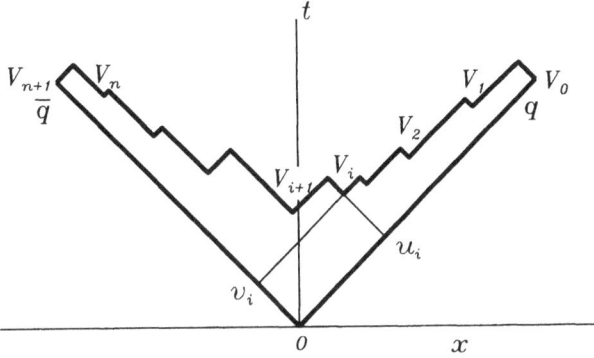

Fig. 7.6 The space-time trajectories of quarks and antiquarks from a string fragmentation.

$2|p_q(0)|$. Without loss of generality, we can take $p_q(0) > 0$. As time proceeds, the massless quark q traverses to the positive-x direction, while the massless antiquark $\bar{q}$ traverses to the negative-x direction. A string is formed joining the quark and the antiquark. When the $q\bar{q}$ pair separates, the string is stretched and new $q_i\bar{q}_i$ pairs will be spontaneously produced at various vertices V_i (see Fig. 7.6). A quark q_i produced at vertex V_i will subsequently combine with an antiquark $\bar{q}_{i+1}$ from a neighboring vertex V_{i+1} to form a yo-yo state, which can be considered to represent a hadron in the inside-outside cascade description of particle production, as shown previously in Fig. (6.4).

A vertex V is specified by the light-cone space-time coordinates $(u, v) = (t + x, t - x)$. From the classical motion of a string fragment discussed in the last section, we note from Eqs. (7.23) and (7.21) that the forward and the backward light-cone momenta of a yo-yo state are directly proportional to the segment lengths of the trajectories of the quark and the antiquark in the (u, v) coordinates. It is therefore convenient to label the location of the vertex V alternatively by the forward and the backward light-cone momenta (p_+, p_-) defined by

$$(p_+, p_-) \equiv (\kappa u, \kappa v) , \tag{7.24}$$

which differ from (u, v) only by the multiplicative scale factor κ. These quantities can be written in terms of the square of the proper momentum Γ, and the rapidity y as

$$\Gamma = p_+ p_- = \kappa^2 uv = \kappa^2(t^2 - x^2) , \tag{7.25a}$$

and

$$y = \frac{1}{2} \ln \frac{p_+}{p_-} = \frac{1}{2} \ln \frac{u}{v}. \tag{7.25b}$$

There is thus one more way to label the location of the vertex V by the proper momentum and the rapidity, (Γ, y). From Eq. (7.25a), the proper momentum variable Γ is related to the proper time coordinate, $\tau = \sqrt{t^2 - x^2}$, by

$$\sqrt{\Gamma} = \kappa\tau. \tag{7.26}$$

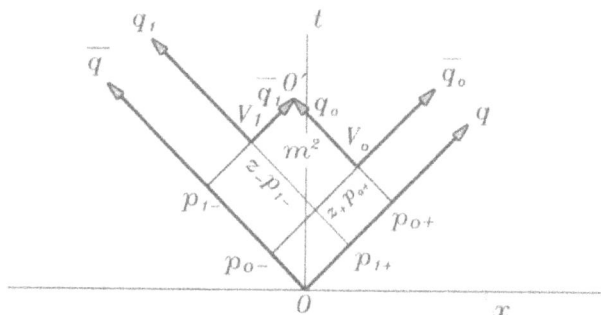

Fig. 7.7 The space-time trajectories of quarks and antiquarks from a string fragmentation at vertices V_0 and V_1.

We consider the occurrence of two adjacent vertices V_0 and V_1 as shown in Fig. (7.7). The order of occurrence of the vertices depends on the frame of reference. One can boost the system to a frame of reference such that the vertex V_0 precedes V_1, or to another frame of reference in which the vertex V_1 precedes V_0.

We shall label the location of the vertex V_0 by (Γ_0, y_0). The probability element for the occurrence of a vertex producing the pair q_0 and $\bar{q}_0$ at the point V_0, specified by the proper momentum Γ_0 and the rapidity y_0, is given by

$$dP_0 = \rho(\Gamma_0, y_0) d\Gamma_0 dy_0. \tag{7.27}$$

In particle production processes at very high energies, it is observed experimentally that the rapidity distribution of the produced particles dN/dy has a plateau shape (Fig. 3.2). On the other hand, it is expected theoretically that if all the production vertices fall on a curve of a constant proper time τ_0 (or proper momentum $\Gamma_0 = \kappa\tau_0$), the rapidity distribution of the produced particles will be independent of the rapidity, and dN/dy will be a constant (see Exercise 7.1).

Therefore, in high-energy processes, it is reasonable to assume that the probability function $\rho(\Gamma_0, y_0)$ is a function of Γ_0 only.

We can show in passing that if all the production vertices fall on a curve of a constant proper time τ_0, the produced particles are ordered in space-time (see Exercise 7.2). The ordering of produced particles means that in the center-of-mass system, particles with a greater magnitude of rapidity $|y|$ are produced at a distance farther away from the point of collision and at a time later than and more separated from the time of collision. We shall come back to discuss these properties of particle production in connection with pion interferometry in e^+-e^- annihilation in Section 17.4 of Chapter 17.

After the appearance of the vertex V_0, the string is further fragmented into two smaller strings with the production of the quark-antiquark pair q_1 and $\bar{q}_1$ at the vertex $V_1 = (p_{1+}, p_{1-})$. In the configuration of Fig. (7.7), the set of vertices V_0, and V_1 occur in the sequence from right to left. Before the quark-antiquark pair is produced at the vertex V_1, the coordinates (Γ_0, y_0) or the light-cone momentum corrdinates (p_{0+}, p_{0-}) of vertex V_0 are presumed known. Thus, the coordinates of V_1 can be specified by the forward light-cone momentum fraction z_+

$$z_+ = (p_{0+} - p_{1+})/p_{0+}, \tag{7.28}$$

and the square of the invariant mass of the produced hadron:

$$M_{01}^2 = (p_{0+} - p_{1+})(p_{1-} - p_{0-}).$$

We can use the quantities z_+ and M_{01}^2 to specify the location (p_{0+}, p_{0-}) of vertex V_1.

We introduce the probability distribution f_{01} so that the probability element for the occurrence of this vertex at V_1 is

$$dP_1 = f_{01}(z_+, M_{01}^2)g(M_{01}^2)dz_+dM_{01}^2, \tag{7.29}$$

where the ordering of the subscript 01 in f_{01} and M_{01} indicates the first occurrence of vertex V_0, to be followed by the occurrence of the vertex V_1. In general, f_{01} depends on z_+, M_{01}^2, and Γ_0 [12]. The assumption that f depends only on z_+ and M_{01}^2 allows one to obtain an explicit form of the distribution by solving well-defined equations [3].

The invariant mass M_{01} formed between the vertices V_0 and V_1 is given by a distribution $g(M_{01}^2)$ which depends on the dynamics of the problem. We can consider the case where M_{01}^2 is determined from some other considerations to have a certain set of values m_λ^2, with probabilities c_λ for the occurrence of the mass m_λ:

$$g(M_{01}^2) = \sum_\lambda c_\lambda \delta(M_{01}^2 - m_\lambda^2). \tag{7.30}$$

For example, in the Lund model, as implemented in the computer program JETSET [13], one finds from a given probability distribution what particle will be produced next. From a transverse momentum distribution, the transverse momentum and the transverse mass of the produced particle is determined. The transverse mass is then identified as the mass m of the particle produced by the occurrence of that vertex.

Combining Eqs. (7.27) and (7.29), the joint probability element for the occurrence of both vertices V_0 and V_1 is

$$dP = dP_0 \cdot dP_1 = \rho(\Gamma_0) f_{01}(z_+, M_{01}^2) g(M_{01}^2) d\Gamma_0 dy_0 dz_+ dM_{01}^2 . \quad (7.31)$$

On the other hand, the occurrence of these two vertices in Fig. 7.7 can proceed by the occurrence of the vertex V_1 first, and the vertex V_0 subsequently. Following the preceding arguments as before and interchanging the labels 0 and 1, the combined probability element for this sequence of events is

$$dP' = dP_1' \cdot dP_0' = \rho(\Gamma_1) f_{10}(z_-, M_{10}^2) g(M_{10}^2) d\Gamma_1 dy_1 dz_- dM_{10}^2 , \quad (7.32)$$

where (Γ_1, y_1) are respectively the square of the proper momentum and the rapidity of V_1, and

$$z_- = (p_{1-} - p_{0-})/p_{1-} . \quad (7.33)$$

The quantity M_{01} is equal to M_{10}, as the invariant mass is independent of the order of the occurrence of the two vertices. The symmetric Lund model makes the observation that if the vertices are indistinguishable, the different ordering of the occurrences of the two vertices represent the same physical process in different Lorentz frames, and the joint probability should be symmetrical with respect to the ordering of the vertices V_0 and V_1. The joint probability should fulfill the following "left-right symmetry" condition:

$$dP = dP' . \quad (7.34)$$

From this condition, Andersson et al. obtained the splitting function [3]

$$f(z_+, m^2) = N z_+^{-1} (1 - z_+)^a e^{-bm^2/z_+} \quad (7.35)$$

and

$$\rho(\Gamma) = C \Gamma^a e^{-b\Gamma} , \quad (7.36)$$

where the constant N and C are normalization constants, and we have used the notation m for M_{10}, (see Exercise 7.3).

If the vertices are distributed according to (7.36), the average proper time at which the vertices occur is then

$$< \tau > = \frac{\Gamma(a + 1 + \frac{1}{2})}{\kappa \sqrt{b} \Gamma(a + 1)},$$ (7.37)

where Γ in the above equation is the gamma function (not to be confused with the Γ's in Eq. (7.36)). The standard deviation of the proper formation time for the distribution (7.36) is

$$\sigma_\tau = \sqrt{\frac{(a + 1)}{\kappa^2 b} - < \tau >^2}.$$ (7.38)

Thus, the values of the parameters a and b give information about the vertex occurrence time and the particle formation time.

•⟦ Exercise 7.1
Show that if all the pair-production vertices of a fragmenting string fall on the curve of the proper time $\tau = \tau_0$, the rapidity distribution of the produced particles is a constant given by [12]

$$\frac{dN}{dy} = \frac{\kappa \tau_0}{m_T}.$$

◇Solution:
We consider vertices V_{i-1} and V_i. The light-cone space-time coordinates of the vertices are $V_{i-1} = (u_{i-1}, v_{i-1})$ and $V_i = (u_i, v_i)$ (see Fig. 7.6). A particle i is formed by combining the antiquark from vertex V_{i-1} with the quark from vertex V_i. This particle is a yo-yo state which has a forward light-cone momentum $p_{i+} = \kappa(u_{i-1} - u_i)$ and a backward light-cone momentum $p_{i-} = \kappa(v_i - v_{i-1})$. The rapidity of this yo-yo particle i is given according to Eq. (7.25b) by

$$y_i = \frac{1}{2} \ln\left(\frac{p_{i+}}{p_{i-}}\right)$$

$$= \frac{1}{2} \ln\left(\frac{u_{i-1} - u_i}{v_i - v_{i-1}}\right).$$ (1)

Similarly, the rapidity of the adjacent yo-yo particle $i + 1$ formed by combining the antiquark from vertex V_i with the quark from vertex V_{i+1} is

$$y_{i+1} = \frac{1}{2} \ln\left(\frac{u_i - u_{i+1}}{v_{i+1} - v_i}\right).$$

There is the addition of one produced particle in the interval between y_i and y_{i+1}. Therefore, in the neighborhood of the vertex V_i, the rapidity density $\Delta N / \Delta y$ is

$$\frac{\Delta N}{\Delta y} = \frac{1}{y_{i+1} - y_i}$$

$$= 2 \left/ \left[\ln\left(\frac{u_i - u_{i+1}}{v_{i+1} - v_i}\right) - \ln\left(\frac{u_{i-1} - u_i}{v_i - v_{i-1}}\right)\right].\right.$$

Taking the continuum limit, we have

$$
\frac{\Delta N}{\Delta y} = 2 \bigg/ \left[\frac{\Delta \ln(-\frac{du}{dv})}{\Delta v} \Delta v \right]
$$
$$
= -2 \frac{du}{dv} \bigg/ \left(\frac{d^2 u}{dv^2} \Delta v \right) ,
$$

(2)

where $\Delta v = v_{i+1} - v_i$. On the other hand, the mass of the produced particle is given by

$$
(u_i - u_{i+1})(v_{i+1} - v_i) = m_T^2/\kappa^2 .
$$

Hence, we have

$$
-\frac{du}{dv}(\Delta v)^2 = m_T^2/\kappa^2 ,
$$

and

$$
\Delta v = \frac{m_T}{\kappa} \bigg/ \sqrt{-\frac{du}{dv}} .
$$

(3)

Therefore, substituting (3) into (2), we obtain the rapidity distribution:

$$
\frac{dN}{dy} = 2 \frac{\kappa}{m_T} \left(-\frac{du}{dv}\right)^{3/2} \left(\frac{d^2 u}{dv^2}\right)^{-1} .
$$

(4)

For the simple case when all the vertices fall on the curve of a proper time τ_0,

$$
\tau_0^2 = uv ,
$$

(5)

then we obtain from Eq. (4)

$$
\frac{dN}{dy} = \frac{\kappa \tau_0}{m_T} .
$$

Thus, in this simplifying example of vertices occurring at a constant proper time τ_0, the rapidity distribution is flat and the magnitude of the rapidity plateau provides an estimate of the particle formation time τ_0.

We can use the above formula to estimate the particle production time from dN/dy. From Fig. 3.2, we observed that the rapidity plateau for charged particles $dN_{ch}/d\eta$ at $\sqrt{s}=62.8$ GeV has the value ≈ 2. If we include the neutral particles, the plateau value is

$$
dN/dy \approx dN/d\eta \approx 3 .
$$

Assuming that after a nucleon-nucleon collision two quark-diquark strings are formed and taking $m_T \approx 0.4$ GeV and $\kappa = 1$ GeV/fm, we find

$$
\tau_0 \approx 0.6 \ \text{fm}/c .
$$

]•

•⟦ Exercise 7.2
Show that if all the pair-production vertices of a fragmenting string fall on the curve of the proper time $\tau = \tau_0$, then in the center-of-mass system particles with a greater

magnitude of rapidity $|y|$ are produced at a distance farther away from the point of collision and at a time later than and more separated from the time of collision.

◇Solution:
From Eq. (1) of Exercise 7.1, the rapidity of the particle produced between the vertex at (u_{i-1}, v_{i-1}) and the vertex at (u_i, v_i) is related to coordinates of the vertices by

$$y_i = \frac{1}{2} \ln\left(\frac{u_{i-1} - u_i}{v_i - v_{i-1}}\right), \tag{1}$$

which can be written as

$$y_i = \frac{1}{2} \ln\left(-\frac{\Delta u}{\Delta v}\right). \tag{2}$$

Taking the continuum limit of the above equation as $(u_{i-1}, v_{i-1}) \to (u_i, v_i) \equiv (u, v)$, we can write the particle rapidity y_i in terms of the derivative of u with respect to v,

$$y_i = \frac{1}{2} \ln\left(-\frac{du}{dv}\right). \tag{3}$$

For the case when all the vertices fall on the curve of a proper time τ_0,

$$\tau_0^2 = uv, \tag{4}$$

we obtain from Eq. (3)

$$\begin{aligned} y_i &= \frac{1}{2} \ln\left(\frac{\tau_0^2}{v^2}\right) \\ &= \frac{1}{2} \ln\left(\frac{\tau_0^2}{(t-x)^2}\right). \end{aligned} \tag{5}$$

The coordinates t and x are related to the coordinate rapidity variable y of Eq. (7.25b) by

$$t = \tau_0 \cosh y,$$

and

$$x = \tau_0 \sinh y.$$

Therefore, from Eq. (5), we have

$$\begin{aligned} y_i &= \frac{1}{2} \ln\left(\frac{\tau_0^2}{\tau_0^2 (e^{-y})^2}\right) \\ &= y. \end{aligned} \tag{6}$$

This means that the rapidity of the produced particle coincide with the coordinate rapidity at that point, as defined by $\ln\{(t+x)/(t-x)\}/2$ in Eq. (7.25b).
    We reach the conclusion that the particle with the rapidity y is produced at the coordinates

$$(t = \tau_0 \cosh y, \quad x = \tau_0 \sinh y).$$

Because $|\sinh y|$ is a monotonically increasing function of $|y|$, we deduce that particles with a greater magnitude of rapidity $|y|$ are produced at a greater longitudinal distance $|x|$ from the point of collision. Furthermore, because $\cosh y$ is a monotonically increasing function of $|y|$, particles with a greater magnitude of rapidity $|y|$ are produced at a time t later and more separated from the time of collision.. ▮•

•⟦ Exercise 7.3
Prove that Eqs. (7.35) and (7.36) satisfy the symmetry condition (7.34).

◇Solution: .
    The validity of Eqs. (7.35) and (7.36) can be readily verified by substituting them into Eqs. (7.31) and (7.32). A change of variables and some simple geometrical considerations from Fig. 7.7 give the following relations for the independent variables:

$$\Gamma_0 = M_{01}^2 (1 - z_-)/(z_+ z_-), \tag{1}$$

$$\Gamma_1 = M_{01}^2 (1 - z_+)/(z_+ z_-), \tag{2}$$

and

$$\frac{dz_+ dM_{01}^2}{z_+} = dp_{1+} dp_{1-} = d\Gamma_1 dy_1, \tag{3}$$

$$\frac{dz_- dM_{10}^2}{z_-} = dp_{0+} dp_{0-} = d\Gamma_0 dy_0. \tag{4}$$

After substituting Eqs. (1) and (3) into (7.31), we get

$$dP = NCM_{01}^{2a} \exp\left\{-b\frac{M_{01}^2}{z_+ z_-}\right\} \left[\frac{(1 - z_-)}{z_+}\right]^a \left[\frac{(1 - z_+)}{z_-}\right]^a d\Gamma_0 dy_0 d\Gamma_1 dy_1 g(M_{10}^2). \tag{5}$$

After substituting Eqs. (2) and (4) into (7.32), we obtain

$$dP' = NCM_{01}^{2a} \exp\left\{-b\frac{M_{01}^2}{z_+ z_-}\right\} \left[\frac{(1 - z_-)}{z_+}\right]^a \left[\frac{(1 - z_+)}{z_-}\right]^a d\Gamma_0 dy_0 d\Gamma_1 dy_1 g(M_{10}^2)$$
$$= dP.$$

Therefore, Eqs. (7.35) and (7.36) satisfy the symmetry condition (7.34).
    We note that $M_{01}^2/z_+ z_-$ in Eq. (5) is κ^2 times the area formed by the rectangle with the diagonal OO' (Fig. 7.7). The probability for the occurrence of the vertices is proportional to the exponential of this area.   ▮•

    In practical Monte Carlo calculations with the Lund model, an iterative scheme is introduced [13] with an additional assumption. It is stipulated that a subsystem fragments independently, and the same splitting function $f(z, m^2)$ is used iteratively to generate all the vertices, starting from the known right-most vertex V_0 and the left-most vertex V_{n+1} (see Fig. 7.6). The assumption of the same splitting function for all fragmentations is the same as that in the outside-inside cascade picture of multiperipheral parton fragmentation, as was recognized earlier by Artru [14].

It should be pointed out that when one uses the splitting function $f(z_+)$ of Eq. (7.35) iteratively, starting with the left-most (or the right-most) vertex for which $\tau = 0$, the proper time for the first few vertices will not distribute themselves according to $\rho(\Gamma)$ of Eq. (7.36). Only by using $f(z_+)$ of Eq. (7.35) iteratively many times will one obtain vertices whose proper times coordinates will be distributed according to $\rho(\Gamma)$ of Eq. (7.36) and whose average vertices occurrence proper times are given by Eq. (7.37). Other types of fragmentation functions similar to Eqs. (7.35) and (7.36) have been proposed [2,12] in the literature.

§7.5 Lund Model for Nuclear Collisions

In the Lund model for hadron-hadron collisions [15,16], it is assumed that the hadrons are not transversely excited but are only longitudinally stretched. The collision of a beam hadron b with a target hadron a will result in two excited hadrons b' and a':

$$b + a \rightarrow b' + a'. \tag{7.39}$$

The longitudinally excited hadrons b' and a' subsequently decay. They are the sources of particle production. The decay of these stretched hadrons is analyzed in the same way as in the fragmentation of a string stretched between a q and a $\bar{q}$ which has been discussed in the last section.

It is convenient to represent the energy-momentum of a hadron by the light-cone momenta. We introduce

$$(b_+, b_-) = (b_0 + b_z, b_0 - b_z), \tag{7.40}$$

where the forward light-cone momentum b_+ is the sum of its energy and its longitudinal momentum, while the backward light-cone momentum b_- is the difference (Eq. (2.9)). Without loss of generality, we can take the beam particle b to be moving in the positive z direction. The initial light-cone momenta of the beam particle b are

$$b = (b_+, b_-) = (b_+, \frac{m_b^2}{b_+}), \tag{7.41}$$

where $b_+ \gg b_-$. The initial light-cone momenta of the target particle a are

$$a = (a_+, a_-) = (\frac{m_a^2}{a_-}, a_-), \tag{7.42}$$

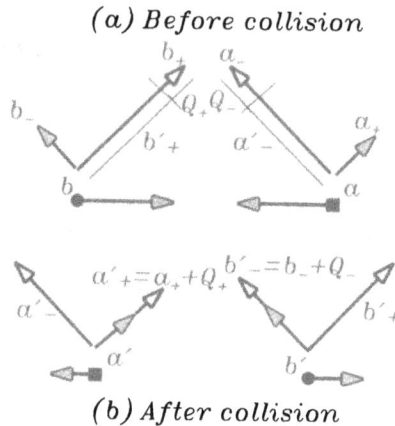

Fig. 7.8 (a) The forward and backward light-cone momenta of particles b and a before the collision. (b) The forward and backward light-cone momenta of particles b' and a' after the collision.

where $a_+ \ll a_-$ (see Fig. 7.8).

In the Lund model, it is assumed that collisions between the hadrons will lead to a flow of their light-cone momenta from one hadron to another. The flow of light-cone momenta is from the hadron with the larger amount to the other hadron with the lesser amount. After the collision, the beam hadron with the forward light-cone momentum b_+ will transfer an amount Q_+ of its forward light-cone momentum to the target hadron. In return, the target hadron with the backward light-cone momentum a_- will transfer an amount Q_- of its backward light-cone momentum to the beam hadron (see Fig. 7.8). The resultant momentum of the beam hadron after the collision is

$$b' = (b_+ - Q_+, \; \frac{m_b^2}{b_+} + Q_-),\qquad(7.43)$$

and the momentum of the target hadron is

$$a' = (\frac{m_a^2}{a_-} + Q_+, \; a_- - Q_-).\qquad(7.44)$$

It is assumed that the momentum transfers Q_+ and Q_- obey a probability distribution given by

$$d^2 P \propto \frac{dQ_+}{(Q_+ + m_b^2/b_+)} \frac{dQ_-}{(Q_- + m_a^2/a_-)},$$
(7.45)

where the mass term comes from the lower limits of momentum transfer, while the dQ/Q type of probability distribution comes from the Field-Feynman parameterization of the momentum distribution of the wee-partons [17].

After the hadrons collide and become excited, each excited hadron decays to yield the produced particles. It is assumed that the decay of a final hadron with a momentum (p_+, p_-) is the same as the fragmentation of a string formed in q-$\bar{q}$ system in which a massless quark travels in the forward light-cone direction with a forward light-cone momentum p_+, and an antiquark travels in the backward light-cone direction with a backward light-cone momentum p_-. The Lund model for the fragmentation of a stretched string developed in the last section can be used for that purpose. With a set of well-adjusted parameters and other additional assumptions on the nature of fragmentation, the Lund model is found to give a good description of many observable quantities in nucleon-nucleon collisions such as the multiplicity distributions, the rapidity distributions, and the transverse momentum distributions of the produced particles [15,16]. The parameters a and b are found to be $a = 0.5$ and $b = 0.9$ GeV^{-2}.

The Lund model for hadron-hadron collisions has been generalized for hadron-nucleus and nucleus-nucleus collisions [16]. In a hadron-nucleus or nucleus-nucleus collision, it is assumed that a Glauber-type multiple collision model is valid, so that one can either follow the trajectory of the colliding baryons by the Monte-Carlo method or decompose a nucleus-nucleus collision as a combination of a set of collisions of nucleons of one nucleus with many nucleons of the other nucleus (see Chapter 12 below).

When one considers the collision of the projectile hadron with a sequence of n target hadrons in a hadron-nucleus collision, the initial light-cone momenta of the projectile hadron before the collision is

$$(P_{1+}, \frac{m_1^2}{P_{1+}}),$$
(7.46a)

and the initial light-cone momenta of the n target nucleons are

$$(\frac{m_i^2}{P_{i-}}, P_{i-}) \quad \text{for} \quad i = 2, ..., n+1.$$
(7.46b)

In the Lund model for hadron-nucleus collision, each binary collision between the hadrons will lead to the transfer of their light-cone momenta between them. The transfer is assumed to go from the hadron with the greater light-cone momentum to the hadron with the lesser light-cone momentum. There is otherwise no quantum number flow from one baryon to another. After the sequence of collisions, the ith target baryons in the set $2, ..., n+1$ gain a momentum Q_{i+} in the forward light-cone direction, but lose a momentum Q_{i-} in the backward light-cone direction. The resultant momenta are

$$\left(\frac{m_i^2}{P_{i-}} + Q_{i+}, P_{i-} - Q_{i-}\right) \quad \text{for} \quad i = 2, ..., n+1, \qquad (7.47a)$$

and the momentum of the incident hadron is

$$\left(P_{1+} - \sum_{i=2}^{n+1} Q_{i+}, \frac{m_1^2}{P_{1+}} + \sum_{i=2}^{n+1} Q_{i-}\right). \qquad (7.47b)$$

It is assumed that the momentum transfer obeys a probability distribution similar to Eq. (7.45) given by

$$d^2P \propto \frac{dQ_{i+}}{(Q_{i+} + m_1^2/P_{1+})} \frac{dQ_{i-}}{(Q_{i-} + m_i^2/P_{i-})}. \qquad (7.48)$$

In the Lund model for nucleus-nucleus collisions (implemented with the computer code FRITIOF [16]), the above procedure is similarly generalized. That is, in each binary collision of the projectile baryon with momentum (P_{b+}, P_{b-}) and a target baryon with momentum (P_{a+}, P_{a-}), there is an exchange of momentum Q_+ and Q_- to result in the momentum $(P_{b+} - Q_+, P_{b-} + Q_-)$ for the projectile baryon and the momentum $(P_{a+} + Q_+, P_{a-} - Q_-)$ for the target nucleon. The only restriction is that Q_+ and Q_- must be greater than zero so that the dominant components of the momentum, $P_{b+} - Q_+$ for the projectile baryon and $P_{a-} - Q_-$ for the target baryon, will always decrease in their magnitudes after collision. The probability distribution of the exchange momentum is taken to be Eq. (7.45). After the baryons complete their collisions, the decay of each excited baryon is then studied to obtain the spectrum of the produced particles. It is again assumed that the decay of the baryon with a momentum (P_{i+}, P_{i-}) is equivalent to the decay of a string formed in the q-$\bar{q}$ system with a massless quark traveling in the forward light-cone direction with a light-cone momentum P_{i+}, and an antiquark traveling in the backward light-cone direction with momentum P_{i-}. The Lund model for q-$\bar{q}$

fragmentation can be used for that purpose. The Lund model for nucleus-nucleus collisions [16] has been successfully applied to the study of relativistic nucleus-nucleus reactions.

§References for Chapter 7

1. A comprehensive review of the application of the string fragmentation model for nucleon-nucleon collisions and for $e^+ - e^-$ annihilation can be found in B. Andersson, G. Gustafson, G. Ingelman, and T. Sjöstrand, Phys. Rep. 97, 31 (1983) and X. Artru, Phys. Rep. 97, 147 (1983).
2. X. Artru and G. Mennessier, Nucl. Phys. B70, 93 (1974).
3. B. Andersson, G. Gustafson and B. Söderberg, Z. Phys. C20, 317 (1983).
4. G. Veneziano, Nuovo Cim. 57A, 190 (1968).
5. Y. Nambu, in *Symmetry and Quark Model*, ed. R. Chand (Gordon and Breach), 1970; Lectures at the Copenhagen Symposium, 1970.
6. L. Susskind, Nuovo Cim. 69A, 457 (1970).
7. T. Goto, Prog. Theo. Phys. 46, 1560 (1971).
8. M. B. Green, Physica Scripta T15, 7 (1987).
9. M. B. Green, J. H. Schwarz and E. Witten, *Superstring Theory*, Cambridge University Press, Cambridge, U.K., Volumes 1 and 2, 1987.
10. S. L. Adler and T. Piran, Rev. Mod. Phys. 56, 1 (1984).
11. R. Sommer, Nucl. Phys. B291, 673 (1987).
12. C. Y. Wong and R. C. Wang, Phys. Rev. D44, 679 (1991).
13. T. Sjöstrand, Comp. Phys. Comm. 39, 347 (1986); T. Sjöstrand, and M. Bengtsson, Comp. Phys. Comm. 43, 367 (1987).
14. X. Artru, Nuovo Cim. 93A, 69 (1986).
15. B. Andersson, G. Gustavson, and B. Nilsson-Almqvist, Nucl. Phys. B281, 289 (1987).
16. B. Nilsson-Almqvist and E. Stenlund, Comp. Phys. Comm. 43, 387 (1987).
17. R. D. Field and R. P. Feynman, Phys. Rev. B136, 1 (1978).

8. Dual Parton Model

In the last few chapters, we discussed different descriptions of the particle production process. We shall give here an alternative description of particle production based on the dual resonance model, which describes many characteristics of soft hadron physics. We shall briefly review the dual resonance model and the concepts of a Reggeon and a Pomeron, at an elementary level.

§8.1 Dual Resonance Model

In Section 7.1, we gave a simple qualitative discussion of the dual resonance model in order to introduce the concept of the string model of hadrons. We shall present here a more quantitative discussion of the dual resonance model [1,2]. We recall that the basic assumption of the dual resonance model is that hadrons interact through the formation of intermediate states (resonances) as described by Fig. 8.1a, which is called an *s-channel scattering process*. They can also interact through the exchange of particles (or resonances) as depicted in Fig. 8.1b, which is called a *t-channel scattering process*. The variables $s = (p_1 + p_2)^2$ and $t = (p_1 - p_3)^2$ are the usual Mandelstam variables.

The intermediate particle p_J in Fig. 8.1a or the exchanged particle p_J in Fig. 8.1b can possess a spin. There are problems when this intermediate particle or exchange particle possesses a large spin. In a t-channel process (Fig. 8.1b), if the exchange particle has a spin J, the scattering amplitude A at high energies grows as s^J (see Supplement 8.1). The scattering amplitude and the total cross section grow without bounds if there is only the exchange of a single particle with a high spin J at very high energies.

Similarly, in the s-channel process of Fig. 8.1a, if the intermediate state p_J in a hadron-hadron scattering is a single particle (or a resonance) with a spin J, the scattering amplitude will likewise grow as t^J, which diverges for large values of J and t. These types of asymptotic behavior for large s and t are not desirable.

Remarkably, there is a way to write down a scattering amplitude $A(s, t)$ which is asymptotically convergent for large values of s (and fixed t) and represents the sum of amplitudes for the exchange of a whole set of particles with all possible spins. This is the Veneziano

s-channel   t-channel

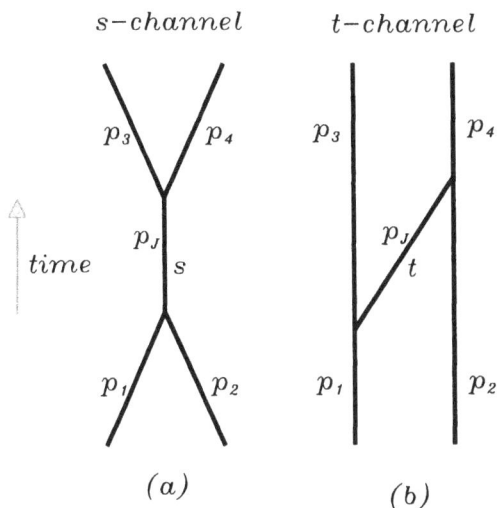

(a)   (b)

Fig. 8.1 (a) An s-channel diagram for elastic meson-meson scattering. (b) A t-channel diagram for elastic meson-meson scattering.

scattering amplitude [1] given by

$$A(s,t) = g^2 \frac{\Gamma(-\alpha(s))\Gamma(-\alpha(t))}{\Gamma(-\alpha(s) - \alpha(t))} , \qquad (8.1)$$

where $\alpha(x)$ is taken to depend linearly on x as

$$\alpha(x) = \alpha_0 + \alpha'x \qquad (8.2)$$

and Γ is the gamma function.

This amplitude has many remarkable properties (see Supplement 8.2). If one keeps t to be fixed and expands the amplitude (8.1) as a power series of s, then each term of the expansion corresponds to a t-channel Feynman diagram (8.1b) for the exchange of a particle with a spin J and a mass M_J, which are related to each other by

$$J = \alpha_0 + \alpha'M_J^2 . \qquad (8.3)$$

The set of particles whose spin J and mass M_J obey the relation (8.3) is said to lie on a *Regge trajectory* with a *Regge intercept* α_0 and a *Regge slope* α'. The amplitude (8.1) then represents the sum of amplitudes for the exchange of the set of particles lying on the Regge trajectory with intercept α_0 and slope α'. For brevity of notation, we often simply use the term 'the Regge trajectory' to represent the whole set of particles lying on the trajectory, characterized by a Regge intercept and a Regge slope. Accordingly, we speak of the exchange of the whole set of particles lying on a *Regge trajectory* as the '*exchange of a Regge trajectory*' or simply as '*the exchange of a Reggeon*.'

The Veneziano amplitude (8.1) is invariant with respect to the interchange of s with t. There is thus a symmetry between the s- and the t-channel processes, the duality property of the Veneziano scattering amplitude. To be specific, if one fixes s and expands the amplitude (8.1) as a power series of t, then each term of the expansion corresponds to an s-channel process for having an intermediate resonance of spin J and mass M_J which obey the linear relation (8.3) between J and M_J^2. The Veneziano amplitude is the sum of all the amplitudes for which the intermediate resonances consist of the set of particles lying on the Regge trajectory with an intercept α_0 and a slope α'. Therefore, the s-channel description and the t-channel description give two alternative or 'dual' descriptions of the same physics. This 'duality' property is the origin of the name 'dual resonance model.'

⊕【 Supplement 8.1
We would like to study here the Veneziano scattering amplitude for the t-channel diagram (8.1b) in perturbation theory, when the exchange particle possesses a spin J. The spin J field is specified by $\sigma_{\mu_1\mu_2...\mu_J}$, and the simplest coupling of the spin J field to the particle field ϕ is

$$-g_J' \phi \partial^{\mu_1} \partial^{\mu_2}...\partial^{\mu_J} \phi \sigma_{\mu_1\mu_2...\mu_J} , \tag{1}$$

where g_J' is a coupling constant which may depend on J. The scattering amplitude for the process $p_1 + p_2 \to p_3 + p_4$ is obtained by taking the matrix element of the product of an operator of the form (1) associated with the vertex joining p_1 and p_3 to p_J and another operator of the same form associated with the vertex joining p_2 and p_4 to p_J (Fig. 8.1b). If there were no derivative operators in Eq. (1), the scattering amplitude would be simply $g_J^2/(t - M_J^2)$ [3], where M_J is the rest mass of the exchanged particle with spin J. For each of the two vertices, each operator ∂_μ brings down a factor of the momentum $-ip_{1\mu}$ or $-ip_{2\mu}$. Hence for high energies, the scattering amplitude is proportional to s^J,

$$A_J(s,t) \propto -\frac{(g_J')^2(-s)^J}{t - M_J^2} . \tag{2}$$

When one includes the exchanges of particles with all possible spins, we have

$$A(s,t) = -\sum_J \frac{g_J^2(-s)^J}{t - M_J^2} , \tag{3}$$

where g_J is related to g'_J by a constant. In a similar way, one can consider the s-channel scattering amplitude. If one considers an intermediate state with spin J, the s-channel amplitude is

$$A_J(s,t) = -\frac{(g'_J)^2(-t)^J}{s - M_J^2}.$$

When intermediate states of all J values are included, the total amplitude is

$$A(s,t) = -\sum_J \frac{g_J^2(-t)^J}{s - M_J^2}. \qquad (4)$$

⊕[Supplement 8.2

We would like to expand the Veneziano amplitude (8.1) as a sum of t-channel amplitudes involving the poles of the exchange particles. We first note that the gamma function $\Gamma(-\alpha)$ obeys the relation

$$\Gamma(-\alpha) = \frac{\Gamma(-\alpha + 1)}{-\alpha} = \frac{\Gamma(-\alpha + 2)}{-\alpha(-\alpha + 1)} = \frac{\Gamma(-\alpha + J + 1)}{-\alpha(-\alpha + 1)...(-\alpha + J)}.$$

In the neighborhood of $-\alpha \simeq -J$ where J is an integer, the gamma function can be represented by

$$\Gamma(-\alpha) = \frac{\Gamma(1)}{(-J)(-J + 1)...(-\alpha + J)}$$

$$= \frac{(-1)^J}{J!(-\alpha + J)}. \qquad (1)$$

Thus, $\Gamma(-\alpha)$ has poles at $-\alpha = 0, -1, -2, ..., -J, ..., $.
We consider now the Veneziano amplitude (8.1)

$$A(s,t) = g^2 \frac{\Gamma(-\alpha(s))\Gamma(-\alpha(t))}{\Gamma(-\alpha(s) - \alpha(t))}.$$

For fixed s and when $-\alpha(t)$ approaches $-J$, then according to Eq. (1)

$$A(s,t) = g^2 \frac{\Gamma(-\alpha(s))}{\Gamma(-\alpha(s) - J)} \frac{(-1)^J}{J![-\alpha(t) + J]}$$

$$= g^2[-\alpha(s) - 1][-\alpha(s) - 2]...[-\alpha(s) - J]\frac{(-1)^J}{J![-\alpha(t) + J]}$$

$$= g^2 \frac{[\alpha(s) + 1][\alpha(s) + 2]...[\alpha(s) + J]}{J![-\alpha(t) + J]}.$$

If we include all the singularities of $-\alpha(t) = 0, -1, -2, ... - J, ...$, we have

$$A(s,t) = \sum_{J=0}^{\infty} g^2 \frac{[\alpha(s) + 1][\alpha(s) + 2]...[\alpha(s) + J]}{J![-\alpha(t) + J]}.$$

When the function α is taken to be a linear form,

$$\alpha(x) = \alpha_0 + \alpha' x,$$

then the Veneziano amplitude for large s is

$$A(s,t) = \sum_{J=0}^{\infty} g^2 \frac{(\alpha' s)^J}{J! \, [-\alpha_0 - \alpha' t + J]},$$

which can be written as

$$A(s,t) = -\sum_{J=0}^{\infty} \frac{g^2 (\alpha')^{J-1}}{J!} \frac{s^J}{t - \frac{J-\alpha_0}{\alpha'}}. \tag{2}$$

We can now compare this amplitude with the amplitude (3) in Supplement 8.1 for the exchange of a set of particles with different spins J. We see that the amplitude $A(s,t)$ in Eq. (2) is a sum of amplitudes for the exchange of particles with spin $J = 0, 1, 2, ..., \infty$ and mass M_J given by

$$M_J^2 = \frac{J - \alpha_0}{\alpha'}. \tag{3}$$

The amplitude (8.1) then represents the sum of amplitudes for the exchange of the set of particles whose spin J and mass M_J are related to each other by

$$J = \alpha_0 + \alpha' M_J^2. \qquad\qquad \text{⏛⊕}$$

The Veneziano amplitude $A(s,t)$ is symmetric with respect to the interchange of s and t. A complete crossing symmetry requires the amplitude to be invariant under the interchange of s, t, and u, where $u = (p_1 - p_4)^2$. We would like to use a scattering amplitude that has such a complete *crossing symmetry* which is given by

$$A_4 = A(s,t) + A(t,u) + A(u,s). \tag{8.4}$$

It is easy to show that for large values of s and a fixed value of t, the crossing-symmetric amplitude A_4 behaves as

$$A_4 = -g^2 \pi e^{-\alpha(t)} \frac{(1 + e^{-i\pi\alpha(t)})}{\Gamma(1 + \alpha(t))} \frac{(\alpha' s)^{\alpha(t)}}{\sin \pi\alpha(t)} \tag{8.5}$$

(see Supplement 8.3).

It is instructive to compare the behavior of the above scattering amplitude with the behavior of the scattering amplitude for the exchange of a particle with spin J as given by Eq. (2) of Supplement

8.1. One notes from the power of s that the scattering amplitude A_4 behaves as if the hadrons had exchanged a *single* particle with a spin J, which needs not be an integer, and the spin J is related to t by

$$J = \alpha(t) = \alpha_0 + \alpha't. \tag{8.6}$$

It is thus interesting that an *infinite* sum of the amplitudes for the exchange of a set of particles lying on a Regge trajectory can be effectively described as the exchange of a *single* fictitious particle with an effective spin $J = \alpha(t)$.

The optical theorem states that the imaginary part of the scattering amplitude at zero degree is related to the total cross section by [4]

$$\mathcal{I}m\, A_4\big|_{t=0} = \lambda(s, m^2, m^2)\sigma_{\text{tot}}, \tag{8.7}$$

where $\lambda(s, a^2, b^2)$ is given by Eq. (4.5c). From Eq. (8.5), the imaginary part of A_4 is

$$\mathcal{I}m\, A_4 = g^2\pi e^{-\alpha(t)}\frac{(\alpha's)^{\alpha(t)}}{\Gamma(1 + \alpha(t))}. \tag{8.8}$$

From Eqs. (8.7) and (8.8), the total cross section for a hadron-hadron collision in the dual resonance model then depends on s according to

$$\sigma_{\text{tot}} \propto s^{\alpha_0 - 1}. \tag{8.9}$$

$\oplus\!\mathbb{[}$ Supplement 8.3

We would like to study the behavior of the crossing-symmetric amplitude A_4 as defined by Eq. (8.4), for large values of s and a fixed value of t. We first examine the amplitude $A(s, t)$, which is the first term in Eq. (8.4). Using the formula [5]

$$\Gamma(-\alpha) = \frac{-\pi}{\Gamma(1 + \alpha)\sin\pi\alpha}, \tag{1}$$

we can write

$$\frac{\Gamma(-\alpha(s))}{\Gamma(-\alpha(s) - \alpha(t))} = \left[\frac{\Gamma(1 + \alpha(s) + \alpha(t))}{\Gamma(1 + \alpha(s))}\right]\left[\frac{\sin\pi(\alpha(s) + \alpha(t))}{\sin\pi\alpha(s)}\right]. \tag{2}$$

Upon using Eq. (1) and Eq. (2), the scattering amplitude (8.1) becomes

$$A(s, t) = g^2\frac{-\pi}{\Gamma(1 + \alpha(t))\sin\pi\alpha(t)}\left[\frac{\Gamma(1 + \alpha(s) + \alpha(t))}{\Gamma(1 + \alpha(s))}\right]\left[\frac{\sin\pi(\alpha(s) + \alpha(t))}{\sin\pi\alpha(s)}\right]. \tag{3}$$

We shall use Stirling's formula [5]

$$\Gamma(x) \simeq \sqrt{2\pi}x^{x+\frac{1}{2}}e^{-x},$$

which gives an approximate representation of $\Gamma(x)$ for large values of $|x|$ away from the negative x-axis. Applying Stirling's formula to the first factor in (2), we obtain

$$\frac{\Gamma(1 + \alpha(s) + \alpha(t))}{\Gamma(1 + \alpha(s))} \simeq \frac{[1 + \alpha(s) + \alpha(t)]^{1+\alpha(s)+\alpha(t)+\frac{1}{2}} e^{-1-\alpha(s)-\alpha(t)}}{[1 + \alpha(s)]^{1+\alpha(s)+\frac{1}{2}} e^{-1-\alpha(s)}}$$

$$\simeq [\alpha(s)]^{\alpha(t)} e^{-\alpha(t)} . \tag{4}$$

The second factor in Eq. (2) is

$$\frac{\sin \pi(\alpha(s) + \alpha(t))}{\sin \pi \alpha(s)} = \cos \pi \alpha(t) + \cot \pi \alpha(s) \sin \pi \alpha(t) . \tag{5}$$

To simplify this factor, we note that when s is large, this factor is a rapidly varying function of s. From Eq. (3), we see that $A(s,t)$ passes through many poles whenever $\alpha(s)$ is an integer and it passes through zeros whenever $\alpha(s) + \alpha(t)$ is an integer. One can average over these poles and zeros by giving a positive imaginary part to s, corresponding to providing a width to the resonances. One expects that as s increases, the width of the resonances also increases. So, this imaginary part of s is large compared to the separation between the poles (although it is still small compared to the real part of s). By giving an imaginary part to s, the function $\alpha(s)$, which is linear in s with real coefficients, will have a real part $\alpha_R(s)$ and an imaginary part $\alpha_I(s)$. The large widths of the resonances correspond to the situation $\pi \alpha_I(s) \gg 1$. We can evaluate the limit $\pi \alpha_I(s) \gg 1$,

$$\lim_{\pi \alpha_I(s) \gg 1} \cot \pi \alpha(s) = i \lim_{\pi \alpha_I(s) \gg 1} \frac{e^{i\pi(\alpha_R(s)+i\alpha_I(s))} + e^{-i\pi(\alpha_R(s)+i\alpha_I(s))}}{e^{i\pi(\alpha_R(s)+i\alpha_I(s))} - e^{-i\pi(\alpha_R(s)+i\alpha_I(s))}}$$

$$= -i .$$

We then have

$$\lim_{\pi \alpha_I(s) \gg 1} \frac{\sin \pi(\alpha(s) + \alpha(t))}{\sin \pi \alpha(s)} = e^{-i\pi \alpha(t)} . \tag{6}$$

Eqs. (2), (4), (5) and (6) lead to

$$\frac{\Gamma(-\alpha(s))}{\Gamma(-\alpha(s) - \alpha(t))} = (\alpha_0 + \alpha' s)^{\alpha(t)} e^{-\alpha(t)} e^{-i\pi \alpha(t)} .$$

Using the above equation and Eq. (1), we find from Eq. (8.1) the amplitude $A(s,t)$ for large s and fixed t as given by

$$A(s,t) \simeq -g^2 \frac{\pi (\alpha' s)^{\alpha(t)} e^{-\alpha(t)} e^{-i\pi \alpha(t)}}{\Gamma(1 + \alpha(t)) \sin \pi \alpha(t)} . \tag{7}$$

We consider next the amplitude $A(t,u)$, which is the second term in Eq. (8.4),

$$A(t,u) = g^2 \frac{\Gamma(-\alpha(t))\Gamma(-\alpha(u))}{\Gamma(-\alpha(t) - \alpha(u))} .$$

We note that $\alpha(s), \alpha(t)$, and $\alpha(u)$ are related by

$$\begin{aligned} \alpha(s) + \alpha(t) + \alpha(u) &= 3\alpha_0 + \alpha'(s + t + u) \\ &= 3\alpha_0 + \alpha'(4m^2), \end{aligned}$$

where m is the rest mass of the colliding (equal) hadrons. Using this relation, we can express $\alpha(u)$ in terms of $\alpha(s)$ and $\alpha(t)$:

$$\alpha(u) = -\alpha(s) - \alpha(t) + 3\alpha_0 + 4\alpha'm^2,$$

and

$$\alpha(t) + \alpha(u) = -\alpha(s) + 3\alpha_0 + 4\alpha'm^2.$$

The amplitude $A(t, u)$ then becomes

$$A(t, u) = g^2 \frac{\Gamma(\alpha(s) + \alpha(t) - 3\alpha_0 - 4\alpha'm^2)\Gamma(-\alpha(t))}{\Gamma(\alpha(s) - 3\alpha_0 - 4\alpha'm^2)}.$$

Applying Stirling's formula and using Eq. (1), we find

$$A(t, u) = -g^2\pi e^{-\alpha(t)} \frac{1}{\Gamma(1 + \alpha(t))} \frac{(\alpha's)^{\alpha(t)}}{\sin \pi\alpha(t)}. \tag{8}$$

We consider finally the amplitude $A(u, s)$, which is the last term in Eq. (8.4). Writing $\alpha(u)$ again in terms of $\alpha(s)$ and $\alpha(t)$, we have

$$\begin{aligned} A(u, s) &= g^2 \frac{\Gamma(-\alpha(u))\Gamma(-\alpha(s))}{\Gamma(-\alpha(u) - \alpha(s))} \\ &= -g^2 \frac{\Gamma(\alpha(s) + \alpha(t) - 3\alpha_0 - 4\alpha'm^2)\pi}{\Gamma(\alpha(t) - 3\alpha_0 - 4\alpha'm^2)\Gamma(1 + \alpha(s)) \sin \pi\alpha(s)} \\ &\simeq -g^2 \frac{(\alpha(s))^{\alpha(t)-1-3\alpha_0-4\alpha'm^2}}{\Gamma(\alpha(t) - 3\alpha_0 - 4\alpha'm^2)} \frac{2\pi i}{e^{i\pi(\alpha_R(s)+i\alpha_I(s))} - e^{-i\pi(\alpha_R(s)+i\alpha_I(s))}}. \end{aligned}$$

This amplitude $A(u, s)$ vanishes when $\pi\alpha_I(s) \gg 1$.

Summing up all the terms, the crossing-symmetric amplitude A_4 for large s is given by

$$A_4 = -g^2\pi e^{-\alpha(t)} \frac{(1 + e^{-i\pi\alpha(t)})}{\Gamma(1 + \alpha(t))} \frac{(\alpha's)^{\alpha(t)}}{\sin \pi\alpha(t)},$$

which is Eq. (8.5). $\mathbb{I}\oplus$

One can use the scattering amplitude (8.5) to analyze experimental scattering data for elastic πN reactions for large s and fixed t. In principle, there can be many contributions coming from the exchange of many different Regge trajectories. Among the trajectories, the *leading trajectory* is the one which gives the dominant contribution

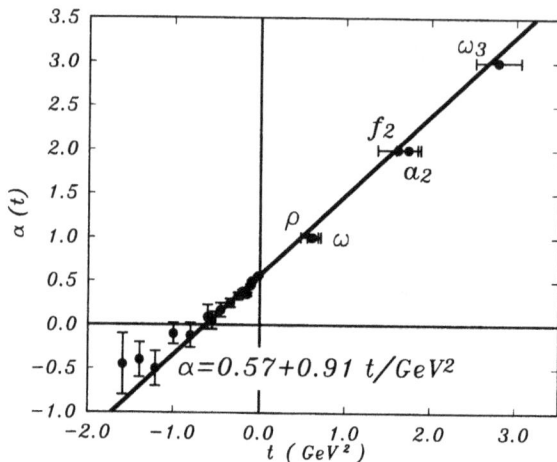

Fig. 8.2 The function $\alpha(t)$ from elastic πN data (for negative t) and from resonance masses for positive t, from Ref. [6].

to the scattering amplitude. From Eq. (8.5), one observes that the trajectory with the largest value of $\mathcal{R}e\ \alpha(t)$ will be dominant asymptotically; the leading trajectory is the one with the largest value of $\mathcal{R}e\ \alpha(t)$.

From the experimental data, one can extract the quantity $\alpha(t)$ for the leading trajectory for a fixed value of t, which is negative for elastic πN reactions. We show in Fig. 8.2 the data points from such an analysis. We also show $\alpha(t)$ as a function of t for positive values of t, which gives the relation between the mass of a resonance and its spin $J = \alpha$. It is remarkable that $\alpha(t)$ extracted from elastic πN data joins on smoothly to $\alpha(t)$ as determined from the masses of resonances for positive values of t. Thus, the leading Regge trajectory from the elastic data coincides with the $\rho - \omega$ meson trajectory from meson spectra. The intercept of the leading Regge trajectory is $\alpha_R(0) \sim 0.5$ and the slope parameter $\alpha'_R \sim 1$ GeV^{-2}, where the subscript R indicates that these quantities are for the leading Regge trajectory. The Regge slope for mesons is, within the experimental uncertainty, the same as the Regge slope for the baryons, as seen by comparison of Figs. 8.2 and 7.2. These comparisons indicate the approximate validity of the dual resonance model as a description of some aspects of hadron physics.

§8.2 Pomeron Exchange

The concept of a *Pomeron* arises from the consideration of many special features of hadronic reactions at very high energies [7]. It was found experimentally that for hadron-hadron reactions at high energies

(i) inelastic processes with no quantum number flow appear to dominate the reactions in the fragmentation regions,

(ii) the forward scattering amplitudes are approximately purely imaginary,

and (iii) the total cross section approaches nearly a constant value at very high energies.

Property (i) suggests that the exchange of an object with the quantum numbers of the vacuum can be a useful concept for forward reactions. Features (ii) and (iii) can be understood if the exchanged object is a Regge trajectory having a Regge intercept $\alpha_0 = 1$ [7].
It is easy to show that the exchange of such a Regge trajectory will lead to the characteristics (ii) and (iii). From the scattering amplitude A_4 in (8.5), the real part of A_4 is given by

$$\mathcal{R}e \, A_4 = -g^2 \pi e^{-\alpha(t)} \frac{(1 + \cos \pi\alpha(t)) \, (\alpha' s)^{\alpha(t)}}{\Gamma(1 + \alpha(t))} \frac{}{\sin \pi\alpha(t)}. \tag{8.10}$$

Forward scattering is characterized by $t \to 0$. Therefore, if the Regge intercept α_0 is equal to unity, then

$$\mathcal{R}e \, A_4 \big|_{t \to 0} = - \lim_{\alpha(t) \to 1} g^2 \pi e^{-\alpha(t)} \frac{(1 + \cos \pi\alpha(t)) \, (\alpha' s)^{\alpha(t)}}{\Gamma(1 + \alpha(t))} \frac{}{\sin \pi\alpha(t)} \to 0.$$

On the other hand, the imaginary part of A_4, as given by Eq. (8.8) is

$$\mathcal{I}m \, A_4 \big|_{t \to 0} = g^2 \pi e^{-1} \alpha' s,$$

which is a nonzero quantity. The scattering amplitude for the exchange of a Regge trajectory with an intercept $\alpha_0 = 1$ is approximately a purely imaginary quantity, in agreement with property (ii).
From Eq. (8.9), the total cross section is proportional to $s^{\alpha_0 - 1}$. The exchange of a Regge trajectory with $\alpha_0 = 1$ gives a total cross section that is independent of energy, in agreement with property (iii).

Properties (i), (ii), and (iii) have been used by Pomeranchuk to discuss many features of hadron reactions at very high energies [8] and are known as the Pomeranchuk properties. The Regge trajectory with the intercept $\alpha_0 = 1$, which is used to explain these properties, is therefore named the *Pomeranchuk trajectory* or simply the *Pomeron*. Comparison with the experimental elastic and quasi-elastic data with the assumption of the exchange of a Pomeron gives $\alpha_P(0) \sim 1$ and $\alpha'_P \sim 0.4$ GeV^{-2} [2], where the subscript P is to denote that these are the parameters for the Pomeron.

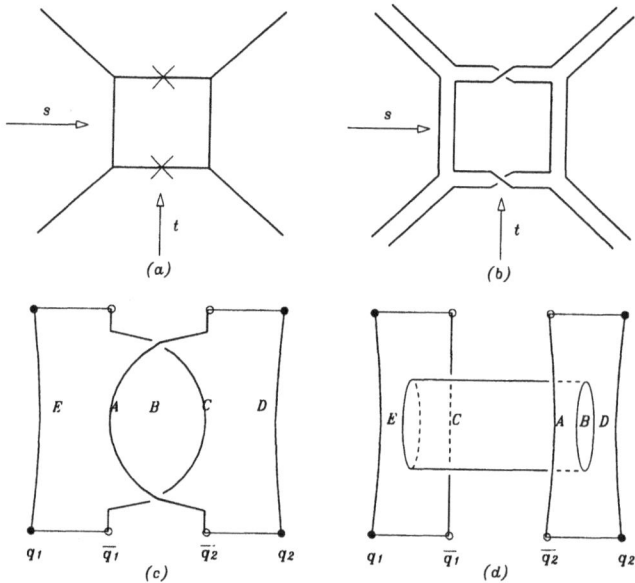

Fig. 8.3 (a) The hadron Feynman diagram which gives rise to Pomeranchuk properties. The crosses represent twists. (b) The Feynman diagram of Figure (a) expressed in terms of quark lines and antiquark lines. (c) The dual diagram of Figure (b) is redrawn to label the prominent parts and features. The straight lines joining a quark (solid circle) and an antiquark (open circle) represent strings, or stable hadrons. (d) The same dual diagram as Figure (c), redrawn in such a way to bring out the features of two world sheets joined by a cylinder.

We have discussed the Pomeron exchange as the exchange of a set of particles lying on the Pomeranchuk trajectory in a t-channel process. According to duality, the exchange of this trajectory in the t-channel implies that if one looks at the s-channel, one would expect the presence of intermediate resonances whose masses and spins place them on the Pomeranchuk trajectory. These particles are however not observed experimentally. Their absence leads to the suggestion that Pomeron exchange represents a different type of process. It turns out that the higher-order hadron Feynman diagram with one loop and two twists, Fig. 8.3a, gives rise to new singularities in channels that are naturally identified as having vacuum quantum numbers [9]. It is natural to relate these singularities to the 'particles' on the Pomeranchuk trajectory. Accordingly, one identifies this diagram as the one responsible for the Pomeron exchange and treats the dual of the Pomeron exchange in the t-channel as the nonresonant background scattering amplitudes in the s-channel [10]. This hadron Feynman diagram 8.3a can be represented in a different way in terms of a dual diagram of quarks and antiquarks. A cross in the Feynman diagram 8.3a represents a twist, of a quark line and an antiquark line, as shown in Fig. 8.3b. We redraw Fig. 8.3b to change it into Fig. 8.3c and label different prominent parts of the diagram with letters $A, B, ..., E$. We further redraw Fig. 8.3c to change it into Fig. 8.3d. By comparing the labelled parts in Fig. 8.3c and in 8.3d, one can visualize, with a little imagination, that Fig. 8.3d is the same figure as Fig. 8.3c. Thus, the process of a Pomeron exchange can be represented by a nonplanar diagram of two world sheets joined together by a cylinder. Since the cross section of a cylinder is a closed circle, Pomeron exchange is then the exchange of a *closed string*, which is distinctly different from the *open string* discussed in Chapter 7. The Pomeron exchange process has a nonplanar topology, in contrast to the planar topology for the exchange of a Reggeon (Fig. 7.1a' and 7.1b').

§8.3 Particle Production Processes

In this Chapter, we have so far focussed on the amplitude for elastic hadron-hadron scattering. One of our main objectives is to understand particle production, which is an inelastic process. We shall utilize the property of the unitarity of the S matrix (or equivalently the optical theorem) to provide a link between the forward scattering amplitude and the probability for inelastic processes. The unitarity condition gives [4]

$$2 \, \mathcal{I}m \, T_{fi} = \sum_n (2\pi)^4 \delta^4(P_n - P_i) T^*_{nf} T_{ni}, \qquad (8.11)$$

where the matrix T is related to the S matrix by $iT = S - 1$, $|i>$ and $|f>$ are the initial and the final two-body states and $|n>$ is a general intermediate state. Thus, a transition matrix element T_{ni} multiplied by its Hermitian conjugation T_{ni}^* gives one of the contributions to the imaginary part of the elastic scattering amplitude. Conversely, one can start by writing down the set of diagrams which contribute to the elastic scattering amplitude. In each of the diagrams of this set, one can cut the diagram into two parts: one part corresponding to the transition of the initial channel i to a channel n, T_{ni}, and another piece representing its complex conjugate T_{ni}^*. The diagram after the splitting then gives the amplitude for the transition from the initial state $|i>$ to a multiparticle state $|n>$. This is the principle of the cutting procedure [11] for cutting an elastic scattering diagram to obtain an inelastic scattering diagram. The dual resonance model provides information as to how one writes down the important diagrams to represent the elastic process, and the cutting procedure then gives the dominant diagrams which represent the inelastic processes. In particular, in the *dual topological unitarization* scheme [12,13], the importance of various diagrams is classified according to topology: planar diagrams are the most important, cylindrical non-planar diagrams next, and so on. For example, in meson-meson collisions, the dominant elastic amplitude can be represented as a sum of many planar diagrams. A typical diagram involves the exchange of many Reggeons, as shown for example in Fig. 8.4a. From a multi-Reggeon exchange planar diagram, the amplitude for the process of multiparticle production can be obtained by a single cut to give a diagram for an inelastic process. In this case, there is only a single 'chain' of produced particles as shown in Fig. 8.4b.

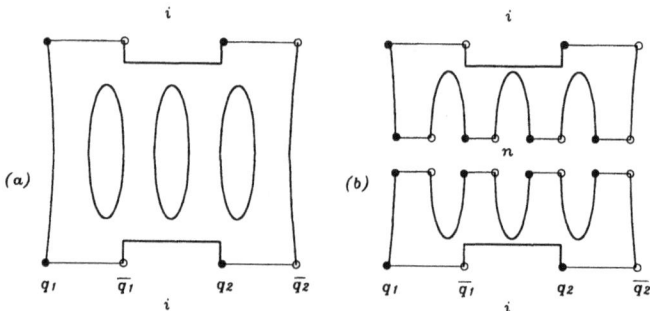

Fig. 8.4 (a) A planar diagram for elastic meson-meson scattering. (b) A cut of the elastic meson-meson scattering diagram (a) to give a diagram for an inelastic process.

In a nucleon-nucleon collision, we can treat a nucleon as composed of a quark q and a diquark qq system. A diquark behaves like an antiquark, except that unlike an antiquark which can be annihilated by a quark, a diquark cannot be annihilated by a quark. Thus, there is no planar diagram to represent a nucleon-nucleon collision. The simplest diagram is that of a nonplanar cylinder diagram, as in Fig. 8.3d. The elastic scattering diagram for nucleon-nucleon collisions can be represented by many diagrams of the type 8.3d and 8.5a. A cut of the cylinder diagram 8.5a for elastic scattering gives the diagram for multiparticle production 8.5b. There are two 'chains' of produced particles, as can be observed in Fig. 5b. If we label the quark and the diquark in one nucleon by i and jk, and the quark and diquark of the other nucleon by l and mn, then the quark i and the diquark mn form a chain while the diquark jk and the quark l form another chain (see Fig. 8.5). Using a nonplanar diagram for a nucleon-nucleon collision and the properties of the Reggeon and Pomeron trajectories, the dual parton model of Capella and Tran Thanh Van [14-16] gives the distribution of the momentum fraction x_q of the valence quarks in a baryon as

$$f_{q/B}(x_q) \propto x_q^{-1/2}(1 - x_q)^{1.5} \qquad (8.12)$$

(see Supplement 8.4). The momentum fraction of the diquark x_{qq} is related to the momentum fraction of the quark q by

$$x_{qq} = 1 - x_q . \qquad (8.13)$$

From distribution (8.12), the momentum distribution of the diquark in a nucleon is given as

$$f_{qq/B}(x_{qq}) \propto (1 - x_{qq})^{-1/2} x_{qq}^{1.5} . \qquad (8.14)$$

We note that the momentum distribution of the valence quark is large at small values of x_q, where it behaves as $x_q^{-1/2}$. The momentum distribution of the diquark is large near $x_{qq} \sim 1$, where it varies as $(1 - x_{qq})^{-\frac{1}{2}}$.

⊕[Supplement 8.4
    We show how the distribution of the valance quark momentum fraction (8.12) is obtained from the dual resonance model, following the arguments of Ref. [15]. Without loss of generality, we can take the nucleon containing the quark i and the diquark jk to be the target nucleon and the nucleon containing the quark l and the diquark mn to be the projectile nucleon. Because of the leading particle effect (Chapter 2), cross sections for the leading particle appear to dominate in the

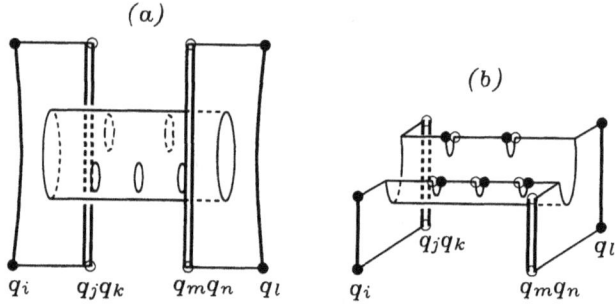

Fig. 8.5 (a) A nonplanar diagram for elastic nucleon-nucleon scattering. (b) A cut of the elastic nucleon-nucleon scattering diagram to give the diagram for particle production.

forward directions; it is reasonable to assume that a diquark carries a large fraction of the momentum of the parent nucleon (on the average). The rapidities of the diquarks are then approximately $y_{jk} \sim y_{\text{target}}$ and $y_{mn} \sim y_{\text{projectile}}$.

We consider the diagram 8.5b and study first the probability distribution for the valence quarks i and l when they carry small momentum fractions x_i and x_l. The rapidity of the valence quark i is y_i and the rapidity of the valence quark l is y_l.

The probability for the occurrence of the quark rapidity gap $y_i - y_{jk}$ is proportional to the probability for the occurrence of any reaction when a hadron, characterized by the rapidity y_i and having the quark i as one of its constituents, interacts with another hadron characterized by the rapidity y_{jk} and having the diquark jk as one of its constituents. To describe the interaction between these two hadrons at rapidities y_i and y_{jk}, we assume that the dual resonance model holds locally so that the interaction between these two hadrons can be represented by the exchange of a trajectory. We further assume that the rapidity gap between y_i and y_{jk} is large so that asymptotic results for high energies hold. Then, the probability for the hadron at y_i to interact with the hadron at y_{jk}, as inferred from Eqs. (8.7) and (8.8), is proportional to $(s_{jk,i})^{\alpha_{jk,i}(0)}$, where $s_{jk,i}$ is the square of the center-of-mass energy of relative motion between the hadron with the rapidity y_i and the hadron with the rapidity y_{jk}, and $\alpha_{i,i}(0)$ is the intercept of the exchanged trajectory. Similarly, for the hadron with the quark i and another hadron with the quark l, there is an exchange of a trajectory with an intercept $\alpha_{i,l}(0)$ and the probability for the occurrence of the quark rapidity gap between y_i and y_l is proportional to $(s_{i,l})^{\alpha_{i,l}(0)}$. Finally, for the hadron with the quark l and another hadron with the quark mn, there is an exchange of a trajectory with an intercept $\alpha_{l,mn}(0)$, and the probability for the occurrence of the rapidity gap between y_l and y_{mn} is proportional to $(s_{l,mn})^{\alpha_{l,mn}(0)}$.

Putting all these factors together, we have the distribution for quark rapidities y_i and y_l given by

$$d^2n/dy_i dy_l \propto (s_{jk,i})^{\alpha_{jk,i}(0)} (s_{i,l})^{\alpha_{i,l}(0)} (s_{l,mn})^{\alpha_{l,mn}(0)} .$$

The square of the center-of-mass energy $s_{a,b}$ can be expressed in terms of the rapidities y_a and y_b as

$$s_{a,b} \propto \exp\{y_b - y_a\}.$$

Therefore, we have

$$d^2n/dy_i dy_l \propto \exp\{\alpha_{jk,i}(0)(y_i - y_{jk}) + \alpha_{i,l}(0)(y_l - y_i) + \alpha_{l,mn}(0)(y_{mn} - y_l)\}.$$

Upon making the transformation

$$x_i \propto \exp\{-(y_i - y_{\text{target}})\}$$

and

$$x_l \propto \exp\{y_l - y_{\text{projectile}}\},$$

we have

$$d^2n/dx_i dx_l \propto x_i^{(-\alpha_{jk,i}(0)+\alpha_{i,l}(0)-1)} x_l^{(+\alpha_{i,l}(0)-\alpha_{l,mn}(0)-1)}. \tag{1}$$

The probability for the momentum fractions depends on the intercepts of the trajectories.

Because the quark i and the diquark jk lie on the same plane (Fig. 8.5), the trajectory which is exchanged between i and jk is a Regge trajectory. Similarly, the quark l and the diquark mn lie on the same plane, and the trajectory which is exchanged between l and mn is a Regge trajectory. The intercepts for these exchanged trajectories are

$$\alpha_{jk,i}(0) = \alpha_{l,mn}(0) = \alpha_R(0) = 1/2.$$

On the other hand, the trajectory between i and l is a Pomeron (Fig. 8.5), for which the intercept is

$$\alpha_{i,l}(0) = \alpha_P(0) = 1.$$

With these intercepts which determine the probability distribution (1), we get for small values of x_l of the valence quark l in a baryon

$$dn/dx_l = f_{q/B}(x_l) \propto x_l^{-\frac{1}{2}} \tag{2a}$$

and similarly

$$dn/dx_i = f_{q/B}(x_i) \propto x_i^{-\frac{1}{2}}. \tag{2b}$$

We consider next the case in which the valence quark l has a large momentum fraction. That is, x_l is close to unity. Because of the relation between the momentum fraction of the diquark and the quark, this corresponds to the occurrence of the diquark mn at a small momentum fraction. In contrast, another diquark $m'n'$ (the leading diquark) is expected to occur and to have a momentum fraction $x_{m'n'}$ close to 1. The occurrence of the diquark mn at small momentum fraction x_{mn} implies that the diquark mn is created out of the vacuum in the form of $mn\bar{m}\bar{n}$ joining on with $\bar{q}q...\bar{q}q$ to the leading diquark $m'n'$ (which has a rapidity close to the rapidity of the projectile or the rapidity of the fast valence quark l). The trajectory which is exchanged between the diquark mn with l is a $qq\bar{q}\bar{q}$ exotic trajectory. Repeating the same argument as before, we have then the distribution for the momentum fractions x_i and x_{mn}

$$d^2n/dx_i dx_{mn} \propto x_i^{(-\alpha_{jk,i}(0)+\alpha_{i,mn}(0)-1)} x_{mn}^{(+\alpha_{i,mn}(0)-\alpha_{mn,l}(0)-1)}. \tag{3}$$

Thus, for small values of x_{mn}, we have

$$dn/dx_{mn} = f_{qq/B}(x_{mn}) \propto x_{mn}^{(+\alpha_{i,mn}(0)-\alpha_{mn,l}(0)-1)}.$$

Because $x_{mn} = 1 - x_l$, therefore, for x_l close to 1, we have

$$dn/dx_l = f_{q/B} \propto (1 - x_l)^{(+\alpha_{i,mn}(0)-\alpha_{mn,l}(0)-1)}.$$

Taking the trajectory exchanged between mn and l to be an exotic $qq\bar{q}\bar{q}$ trajectory with an intercept -1.5 [15], and identifying the trajectory for the exchange between i and mn to be the Pomeron, we obtain for $x_{mn} \sim 0$

$$f_{qq/B}(x_{mn}) \propto x_{mn}^{1.5} \tag{4a}$$

and for $x_l \sim 1$

$$f_{q/B}(x_l) \propto (1 - x_l)^{1.5}. \tag{4b}$$

Combining Eqs. (2) and (4) for the region $x_l \sim 0$ and the region $x_l \sim 1$, we have

$$f_{q/B}(x_q) = x_q^{-1/2}(1 - x_q)^{1.5},$$

which is Eq. (8.12).

⊕

Equations (8.12) and (8.14) determine the momentum distribution of the quarks and the diquarks which participate in a nucleon-nucleon collision. These participants form two chains of produced particles. In the dual-parton model, the produced particles in each chain are assumed to be determined by phenomenological fragmentation functions, which are parametrized in Ref. [15] and [16]. The resultant momentum distribution of the produced particles is just a folding of the momentum distribution of the constituents with the momentum distribution of the fragments produced from the constituents. Upon choosing the structure functions and the fragmentation functions to fit experimental hadron-hadron data, the model is then used to study hadron-nucleus and nucleus-nucleus collisions. In a hadron-nucleus or nucleus-nucleus collision, it is assumed that a Glauber-type multiple collision model is valid, so that a nucleus-nucleus collision can be decomposed as a combination of a set of collisions of nucleons of one nucleus with many nucleons of the other nucleus (see Chapter 12 below). It is necessary to provide a description for the dynamics of the quark and diquarks when a nucleon collides with a sequence of other nucleons. In the dual-parton model, we again invoke the elastic process to write down the dominant diagram. Because diquarks cannot be annihilated by a quark, there is no planar diagram for the collision of, for example, one projectile nucleon with n other target nucleons. From the Dual Topological Unitarization scheme [12,13], the diagrams are classified according to topology. The simplest and the most important diagram for elastic scattering is the diagram with the projectile nucleon exchanging a Pomeron with each of the target nucleons. The diagram for elastic scattering consists of the world sheet of the projectile nucleon joined by a cylinder to each of the world sheets of the n participant target nucleons. There are then n cylinders. When we cut the elastic diagram to obtain the diagram for particle production, we cut across the cylinders to obtain $2n$ chains of produced particles. For example, in the collision of one projectile nucleon with two target nucleons, the diagram of quark and antiquark lines for the process of particle production can be represented by Fig. 8.6. One observes that if there were only one collision, as the quark-diquark system of the incident projectile baryon proceeds forward, a chain is formed between the diquark $q_p q_p$ of the projectile nucleon with the valence quark q_1 of the target nucleon. A similar chain is formed between the valence quark of the projectile nucleon q_p and the diquark $q_1 q_1$ of the target nucleon. (The chains can be identified as the strings in the Lund model for nucleon-nucleon collisions.) The diagram is just a simpler representation of Fig. 8.5b. When there is a second collision, the diquark $q_p q_p$ of the projectile nucleon proceed forward to form a chain with the valence quark q_2 of the second target nucleon, and there is a chain between the quark q_{sea} and the diquark $q_2 q_2$ of the second nucleon (Fig. 8.6). The quark q_{sea} is called a *sea quark* because it arises from the vacuum (sea) as one of the two members of the $q_{\text{sea}} \bar{q}_{\text{sea}}$ pair. There are therefore two types of chains: those not involving sea quarks and those involving sea-quarks. In a collision of a baryon with n target nucleons, there are two chains not involving sea quarks and $2n - 2$ chains involving sea quarks. It is assumed that each chain leads to an independent production of particles.

The momentum distribution of the sea-quarks is assumed to be given by a distribution of the form which goes as $\sim 1/x$ as $x \to 0$,

$$f_s(x) = (x^2 + \mu^2/p^2)^{-1/2},$$

where the parameter μ is introduced to cut off the singular point at $x = 0$. The sea quarks fragment into hadrons according to a prescribed fragmentation function. For example, the fragmentation function for a sea quark to go into a pion is assumed to be of the form

$$x D(s \to \pi) = F(x).$$

With the introduction of the various momentum distributions and fragmentation functions, the momentum distribution of the produced

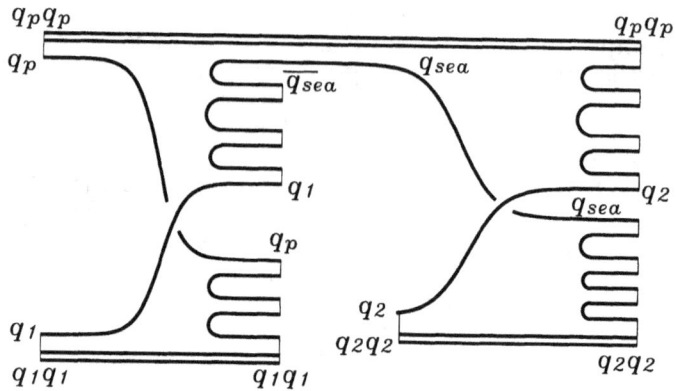

Fig. 8.6 The diagram for the collision of one projectile nucleon with two target nucleons in the dual-parton model.

particles can be obtained by convolution. The dual-parton model has been applied successfully to analyze a large body of experimental data in hadron-nucleus and nucleus-nucleus collisions [13-16].

It is worth mentioning some other variations and modifications of the dual-parton model and concepts. In the multistring model of Werner (with the computer code VENUS)[17], the dominant diagram for a nucleon-nucleon collision includes the same diagram as in Fig. 8.5b and 8.6. In addition, it also includes the probability for finding an antiquark in a nucleon and the collision of this antiquark with the other nucleon. For nucleus-nucleus collisions, there are thus additional contributions from the collision of the antiquark with nucleons of the other nucleus. In contrast to the dual-parton model, for which the momentum distribution of the partons are inferred from the dual resonance model, the momentum distribution function of quarks and antiquarks in the multistring model of Werner are taken to be the same as those determined by deep-inelastic collision experiments, represented in terms of the functions and parameters as given by Duke and Owens [18]. The momentum distribution of the diquark is then determined by momentum conservation. The partons form strings by interacting with the partons of other nucleons. The fragmentation of these strings is described, as in the dual-parton model, by using fragmentation functions which are a generalization of the Field-Feynman distribution functions [19]. For example, in reference [17], for the fragmentation of a pion and a baryon from a quark, the fragmentation functions are

$$xD_{\pi/q}(x) = (1 - x),$$

$$xD_{b/q}(x) = x^{\alpha}(1 - x)^3.$$

For the fragmentation of a pion and a baryon from a diquark, the fragmentation functions are

$$xD_{\pi/q}(x) = (1 - x)^3,$$

$$xD_{b/q}(x) = x^{\alpha}(1 - x),$$

where α is a parameter. A later version of the model (VENUS 3) [20] uses a different fragmentation scheme, which is similar to the string fragmentation scheme of the Lund model (Chapter 7), treating the quark and the diquark pair in a string as the two ends of a relativistic string. The probability for the breaking of the string at the space-time point x_+, x_- is given by the probability function

$$dP \propto e^{-\text{const} \times x_+ x_-},$$

which was first suggested by Artru and Mennessier [21]. The model has been applied successfully to analyze a large set of experimental data in nucleus-nucleus collisions [17,20].

A different multiple-collision model, the multi-chain model, was proposed by Kinoshita, Minaka, and Sumiyoshi [22]. In this model for a baryon-nucleus collision, incident baryons make collisions with many nucleons in the target nucleus. Each collision leads to the formation of a chain which later evolves into produced particles. The leading cluster degrades its momentum as the collision process proceeds. How the leading cluster loses energy is taken to be an unknown parametrized function $P(x)$ to be determined by experiment. A convenient parametrization of the probability function $P(x)$ is taken to be

$$P(x) = \alpha x^{\alpha - 1},$$

where x is the light-cone variable of the leading cluster. Other methods of partitioning the light-cone momentum have also been attempted [22]. The momentum degradation of the leading cluster is assumed to be sequential so that the momentum distribution after n collisions is the folding of the momentum distribution after $n - 1$ collisions and the probability function $P(x)$. After each collision, each chain also acquires a momentum fraction, and it decays into hadrons, depending on a fragmentation function G which is a function of the light-cone momentum of the chain. For example, for charged secondary particles, the fragmentation function was assumed to have the form

$$G(z, x) = (1 - z)^\beta (1 - x)^\alpha,$$

where z is the light-cone variable of the detected hadron relative to the chain, x is the light-cone variable of the detected hadron relative to the target nucleon, and α and β are parameters.

Another multiple collision model [23] combines the effects of the momentum distribution of the quark and diquarks with the fragmentation function to come up with an assumed law of energy loss for the leading baryon in the form

$$P(x) = \alpha \left(\frac{x - x_L}{1 - x_L} \right)^{\alpha - 1},$$

where

$$x_L = m_N / (b_0 + b_z),$$

and m_N is the rest mass of the nucleon, b_0 and b_z are the laboratory energy and momentum of the incident nucleon. The quantity x_L is introduced to satisfy the kinematic lower bound of the light-cone variable x. In a nucleus-nucleus collision, the leading baryon loses energy, but continues to make further collisions with the nucleons of the other nucleus. The energy loss is used to produce particles which are taken to be given by the particle production law

$$dn/dy \propto [(1 - x_+)(1 - x_-)]^a,$$

where the index a is parametrized by fitting nucleon-nucleon collision data, is taken to be

$$a = 3.5 + 0.7 \ln \sqrt{s/\text{GeV}^2},$$

where s is the squared center-of-mass energy available for particle production. The model has been applied and found to reproduce the general features of nucleus-nucleus collisions.

A different approach has been taken by Ludlam and his collaborators, in the HIJET computer program [24]. It is assumed in a similar way, that the leading clusters in nucleus-nucleus collisions make successive collisions with nucleons of the other nucleus. In each collision, the energy loss and the spectrum of produced particles are calculated with the ISAJET program, which has been written to reproduce the nucleon-nucleon collision data. While the leading clusters can continue to make more collisions as they pass through the other nucleus, the produced particles emerge only after the nuclei are far apart and do not participate in secondary collisions in the first approximation. A later version of the program includes the effects of secondary collisions. The HIJET program has been used as an event generator to aid the design of experimental detectors and has also been used to compare with preliminary experimental data.

There is another model of nucleus-nucleus collision, the Relativistic Molecular Dynamics (RQMD) model [25], in which the trajectories of all hadrons, including the produced particles, are followed as a function of space and time. The interaction of the hadrons at high energies produces strings as in the Lund model of hadron collisions, and the Lund model program JETSET [26] for string fragmentation is used to produce hadrons. The model has been applied successfully to analyze many pieces of experimental data.

Another model, the parton cascade model [27], assumes that the nucleons in a nucleus consist completely of partons, with parton distribution given by the structure functions from deep-inelastic scattering, the Drell-Yan process and direct photon production. The dynamics of the partons are followed to provide a description of the status of the system as a function of space and time. Various interesting physical results are obtained by following the dynamics.

Combining the parton model and the string model is the string-parton model [28] which was put forth to simulate nuclear collisions. In this model, dynamical quark and antiquark are attached to the end points and are allowed to execute dynamical motion leading to various parton trajectories. Parton intrinsic momentum distribution is specified in terms of a linear combination of a set of parton trajectories. The interaction between a parton-string and another parton-string is described by parton-parton scattering and parton exchange. The model has been applied successfully to analyze many pieces of experimental data.

§References for Chapter 8

1. G. Veneziano, Phys. Lett. 30B, 351 (1969).
2. Excellent reviews of the dual resonance model can be found in P. Frampton, *Dual Resonance Model*, W. A. Benjamin, Inc., Reading, Mass., 1974; P. D. B. Collins, Phys. Rep. 1, 103 (1971), P. D. B. Collins, *Regge Theory and High Energy Physics* (Cambridge University Press, Cambridge, 1977); and G. Veneziano, Phys. Rep. 9, 199 (1974).
3. J. D. Bjorken and S. Drell, *Relativistic Quantum Mechanics*, (McGraw-Hill Book Company, N.Y. 1964).
4. C. Itzykson and J.-B. Zuber, *Quantum Field Theory*, McGraw-Hill Book Com-

pany, N.Y., 1980, p. 242. Note that the function λ defined by these authors is the square of the function λ defined by Eq. (4.5c) in Chapter 2.

5. M. Abramowitz and I. Stegun, *Handbook of Mathematical Functions*, Dover Publications, N.Y., 1965.
6. M. B. Green, Phys. Scr. T15, 7 (1987).
7. V. N. Gribov, Soviet Phys.-JETP 14, 478 and 1395 (1961); and G. F. Chew, and S. Frautschi, Phys. Rev. Lett. 7, 394 (1961); an early review of the Pomeron can be found in G. F. Chew, Comments Nucl. Part. Phys. 1, 121 (1967).
8. I. Ya. Pomeranchuk, Soviet Phys. JETP 3, 306 (1956); *ibid* 7, 499 (1958); I. Ya. Pomeranchuk and L. B. Okun, Soviet Phys.-JETP 3, 307 (1956).
9. D. J. Gross, A. Neveu, J. Scherk, and J. H. Schwarz, Phys. Rev. D2, 697 (1970).
10. H. Lee, Phys. Rev. Lett. 30, 719 (1973).
11. V. N. Gribov, Soviet Phys.- JETP 30, 709 (1970); A. Capella and A. Krzywicki, Phys. Rev. D 18, 3357 (1978).
12. G. Veneziano, Nucl. Phys. B74, 365 (1974).
13. Chan Hong-Mo, J. E. Paton, and Tsou Sheung Tsun, Nucl. Phys. B86, 479 (1975).
14. A. Capella, U. Sukhatame, C-I. Tan, and J. Tran Thanh Van, Phys. Lett. 81B, 68 (1979); A. Capelle, U. Sukhatame, and J. Tran Thanh Van, Zeit. Phys. C3, 329 (1980).
15. A. Capella and J. Tran Thanh Van, Zeit. Phys. C10, 249 (1981).
16. A. Capella, C. Pajares, and A. V. Ramallo, Nucl. Phys. B241, 75 (1984); A. Capella, A. Staar, and J. Tran Thanh Van, Phys. Rev. D32, 2933 (1985); A. Capella, A. V. Ramallo and J. Tran Thanh Van, Z. Phys. C33, 541 (1987); A. Capella and J. Tran Thanh Van, Zeit. Phys. C38, (1988) 177; P. Aurenche, F. W. Bopp, A. Capella, J. Kwiecinski, M. Maire, J. Ranft, and J. Tran Thanh Van, Phys. Rev. D45, 92 (1992).
17. K. Werner, Phys. Rev. D39, 780 (1989)
18. D. W. Duke and J. F. Owens, Phys. Rev. D30, 49 (1984).
19. R. D. Field and R. P. Feynman, Phys. Rev. D15, 2590 (1977); R. D. Field and R. P. Feynman, Nucl. Phys. B136, 1 (1978); R. P. Feynman, R. D. Field and G. C. Cox, Phys. Rev. D18, 3320 (1978).
20. K. Werner, Lectures presented at Cape Town Summer School, 1990, p. 133.
21. X. Artru and G. Mennessier, Nucl. Phys. B70, 93 (1974).
22. K. Kinoshita, A. Minaka, and H. Sumiyoshi, Prog. Theo. Phys. 63, 1268 (1980).
23. C. Y. Wong, Phys. Rev. Lett. 52, 1393 (1984); C. Y. Wong, Phys. Rev. D30, 972 (1984); C. Y. Wong Phys. Rev. D32, 94 (1985); C. Y. Wong and Z. D. Lu, Phys. Rev. D39, 2606 (1989).
24. T. W. Ludlam (unpublished).
25. H. Sorge, H.Stöcker, and W. Greiner, Nucl. Phys. A498, 567c (1989); H. Sorge, A. von Keitz, R. Mattiello, H. Stöcker, and W. Greiner, Phys. Lett. B243, 7 (1990); R. Matiello *et al*, Nucl. Phys. B24, 221 (1991).
26. T. Sjöstrand, Comp. Phys. Comm. 39, 347 (1986); T. Sjöstrand, and M. Bengtsson, Comp. Phys. Comm. 43, 367 (1987).
27. K. Geiger and B. Müller, Nucl. Phys. A544, 467c (1992); K. Geiger, Phys. Rev. D46, 4965 (1992); K. Geiger, Phys. Rev. D47, 133 (1993).
28. D. J. Dean, A. S. Umar, J.-S. Wu, and M. Strayer, Phys. Rev. C45, 400 (1992).

9. Quarks, Gluons, and Quark-Gluon Plasma

It is generally held that the physics of the strong interaction is described by the theory of *quantum chromodynamics* (QCD) [1]. In this theory, the relevant fields are quark fields and gluon fields with the associated particles of quarks and gluons. There is an internal degree of freedom, the color degree of freedom, which provides the arena for the interaction of these particles. The dynamics in this *color* degree of freedom gives rise to the term '*chromo*dynamics' for the strong interaction. The equations governing quantum chromodynamics can be written out explicitly. Since the coupling constant for the interaction depends on the circumstances of the interaction, the study of systems of quarks and gluons falls into the categories of perturbative QCD and nonperturbative QCD. The study of new phases of quark matter is a problem in nonperturbative QCD, as it involves describing the system on a large spatial scale. While the techniques for studying perturbative QCD have been well developed, analytical and numerical solutions of problems in nonperturbative QCD are rather difficult to obtain. Phenomenological models, such as the bag model, can be useful to provide a qualitative guide to understand some aspects of the strong interaction phenomena. We shall use the bag model to motivate our discussions of a system of quarks and gluons under unusual circumstances of high temperature and density. We shall discuss the more quantitative descriptions of the system in the lattice gauge theory in the next chapter.

§9.1 Quarks and Gluons

Quarks have *spin* 1/2. They are fermions. Other properties of the quarks are listed in Table 9.1. Quarks are characterized by their *flavors*. The discovery of the flavor degree of freedom of quarks was one of the great triumphs of modern physics [2]. It led to the concept that quarks are basic fundamental particles out of which many other particles can be built up. Up to the present time, quarks with five different flavors are known to exist. They are denoted $u, d, c, s,$ and b which are respectively, the *up, down, charm, strange* and *bottom* quarks. A sixth quark with the *top* flavor is expected to exist by consideration of the unification of quarks and leptons. In this unification scheme, the (u, d) quark doublet and the (ν_e, e) lepton doublet form the *first generation* of elementary particles, and the (c, s)

quark doublet and the (ν_μ, μ) doublet form the *second generation* of elementary particles [3]. If the pattern repeats itself for another generation, then one would expect the existence of a quark doublet (t, b) to combine with the lepton doublet (ν_τ, e_τ) to form the *third generation* of particles. Of the twelve particles in the three generations, only the top quark t is yet to be discovered. Furthermore, the possibility of a third generation of particles has the attractive feature that it would provide a natural framework to describe CP violation [4]. Therefore, the top quark has been the subject of current research. The mass of the top quark was recently predicted to be $m_t \simeq 132^{+31}_{-37}$ GeV [5]. The discovery of such a particle may require the use of particle accelerators yet to be constructed.

In Table 9.1, Q is the electric charge, I_z is the z-component of the isospin, C is the charm quantum number, S is the strangeness, T is the topness, and B is the bottomness. We adopt the convention that the sign of the quantum numbers C, S, T, and B follow the same sign as the electric charge of the quark [6].

Table 9.1

Quarks	Q	I_z	C	S	T	B	mass [7] (MeV)
u	$\frac{2}{3}$	$\frac{1}{2}$	0	0	0	0	5.6 ± 1.1
d	$-\frac{1}{3}$	$-\frac{1}{2}$	0	0	0	0	9.9 ± 1.1
c	$\frac{2}{3}$	0	1	0	0	0	1350 ± 50
s	$-\frac{1}{3}$	0	0	-1	0	0	199 ± 33
t	$\frac{2}{3}$	0	0	0	1	0	> 90000
b	$-\frac{1}{3}$	0	0	0	0	-1	$\simeq 5000$

The mass of a quark listed in Table 9.1 is the *current mass* of the quark, which is the mass of the quark in the absence of confinement. When the quark is confined in a hadron, the quark may acquire an effective mass which includes the effect of the zero-point energy of the quark in the confining potential. The effective mass of a confined quark in a hadron is known as the *constituent mass* of the quark and is typically a few hundred MeV in magnitude.

Each quark carries a *baryon number* $1/3$ and a *color*. There are three different colors a quark can carry. The interaction between the quarks depends on the colors of the interacting quarks, similar to the interaction between electric charges. For this reason, the color of a quark is sometimes called its *color charge*. By the exchange of a *gluon*, a quark with one color can interact with another quark of any other color.

It is generally held that the field theory for gluons and quarks belongs to a special class of field theories known as *gauge field theories*. That is, the interaction in the field theory can be represented as arising from the requirement that the Lagrangian is invariant under a local gauge transformation. This invariance is called a local gauge invariance, or simply '*gauge invariance*.' If the quanta of the gauge field have a rest mass, then the Lagrangian will not be invariant under a local gauge transformation. To maintain gauge invariance, the quanta of the gauge field, the gluons, must be massless. In this respect, the gluons resemble photons, which are the quanta for the gauge field of electromagnetic interactions. They are spin-1 particles and are therefore bosons.

In a quark-antiquark interaction, a particle with three types of color charges interacts with another particle with three types of color charges. There are in principle nine possible types of gluons belonging to a color singlet state in the $U(1)$ group and a color octet state in $SU(3)$ group. However, the gluon in the singlet state would not carry a color charge and therefore would be colorless. A colorless and massless gluon will lead to a long-range strong interaction between colorless hadron states. The existence of a color singlet gluon is however ruled out by the absence of this long-range color interaction between color singlet hadrons. There are thus only *eight gluons* as members of the color octet, all of which carry color charges. In the color space, the internal symmetry group which describes the gluons and quarks is that of the $SU(3)_c$ group, with the quarks residing in the fundamental representation of the group. The subscript c denotes the color degree of freedom. Because the gluons also carry color charge, they interact with quarks and gluons with the exchange of other gluons. The theory which describes the interaction of the color charges of quarks and gluons is called quantum chromodynamics (QCD). Since the gluons do not carry flavor and the interaction does not depend on the flavor degree of freedom, the flavor labels and the flavor group are often not explicitly written out.

In phenomenological quark models, mesons can be described as quark-antiquark bound states, and baryons can be considered as three-quark bound states [8]. Up to now, it is found that all the hadron states which can be observed in isolation are *color singlet* states which are completely antisymmetric with respect to the exchange of any two quarks of the hadron. Experimentally, no single quark, which is described by a color triplet state, has ever been isolated. Therefore, it is held that only hadrons in the color singlet state can be isolated and observed. The absence of the observation of a single quark in isolation also suggests that the interaction between quarks and gluons must be strong on large distance scales.

In the other extreme, much insight into the nature of the interaction of quarks and gluons on short distance scales was provided by deep-inelastic scattering experiments. In these experiments, an incident electron interacts with a quark within a hadron and is accompanied by a transfer of momentum from the electron to the quark. The measurement of the momentum of the electron before and after the collision allows a probe of the momentum distribution of the quarks (or partons) inside the nucleon. It was found that with very large momentum transfers, the quarks inside the hadron behave as if they were almost free, as demonstrated by the success of Bjorken scaling [9] and the parton model [10].

A non-Abelian gauge theory can describe a system which is weakly interacting on short distance scales but very strongly interacting on a large distance scale [11]. The non-Abelian gauge theory is essentially unique in this respect [12]. In a non-Abelian gauge field theory, the gauge field operators do not commute. This is in contrast to the Abelian gauge field theories for electromagnetic interactions, in which they do commute. It was found that non-Abelian gauge field theories possess the property of 'asymptotic freedom' [11]. The effective strength of the interaction between quarks and gluons depends on the conditions of their interaction. The QCD coupling constant α is related to the scale of the momentum transfer q by the relation [11]

$$\alpha(q^2) = \frac{\alpha_0}{1 + \alpha_0 \frac{(33-2n_f)}{12\pi} \ln(\frac{-q^2}{\mu^2})}, \qquad (9.1)$$

where α_0 is the coupling constant for the momentum transfer μ and n_f is the number of flavors. When the distance scale of the interaction is small, as for example when one probes the high momentum component of the distribution of the quarks, the coupling constant is small. This is the case of 'asymptotic freedom'. For that case, since the coupling constant is small, a perturbative treatment is a good description of the process and the parton model [8-10] is a useful concept. Much knowledge of the interaction of quarks and gluons has been derived from such studies. On the other hand, when the distance scale is large, as for example in the study of the structure of the ground state of a hadron, then the interaction strength is large. A perturbative treatment, based on an expansion in powers of the coupling constant, is no longer applicable. The constituent quarks are subject to confining forces and a nonperturbative treatment is needed.

§9.2 Bag Model of a Hadron

A very useful method to study nonperturbative quark-gluon systems is to examine the system on a discrete lattice of space and time. The formulation of QCD on a lattice is known as the QCD lattice gauge theory which we shall discuss in the next chapter. Theoretical results from lattice gauge theory indicate that when the distance scale is comparable to the size of a hadron, quarks interact with an effective interaction which goes approximately linearly with the spatial distance. Experimentally, no single quark has been isolated. This leads to the concept that the large scale behavior of quarks in hadrons is characterized by their confinement inside the hadrons. An intuitive way to visualize confinement is the picture of a linear tube between a quark and an antiquark (or a diquark) in which the color electric flux is restricted (see Chapters 5 and 7). As the separation between the quark and the antiquark becomes large, it becomes energetically more favorable to produce another quark-antiquark pair at a point along the tube such that the produced quark is connected to the antiquark while the produced antiquark is connected to the quark. To isolate a quark by separating it from its antiquark partner would then be an impossible task. The confinement of quarks when one considers the large distance behavior of QCD is sometimes referred to as the 'infrared slavery' of quarks. The term 'infrared' comes from analogy to optics where the wavelengths in the infrared region are greater than those in the optical spectrum. It is used here to denote the circumstances involving a large spatial distance, or small momentum transfer. Likewise, it is customary to use the term 'ultraviolet' to denote the circumstances involving a small distance scale, or a large momentum transfer.

With quarks being confined inside a hadron, a useful phenomenological description of quarks in hadrons is provided by the bag model [13,14]. While there are many different versions of the model, the MIT bag model [13] contains the essential characteristics of the phenomenology of quark confinement; we shall use it to understand the circumstances of how quarks can become deconfined in new phases of quark matter.

In the MIT bag model, quarks are treated as massless particles inside a bag of finite dimension, and are infinitely massive outside the bag. Confinement in the model is the result of the balance of the *bag pressure B*, which is directed inward, and the stress arising from the kinetic energy of the quarks. Here, the bag pressure B is a phenomenological quantity introduced to take into account the nonperturbative effects of QCD. If the quarks are confined in the bag, the gluons should also be confined in the bag. In this description, the total color charge of the matter inside the bag must be colorless, by virtue of the Gauss's Law. As there are three different types of color, the bag model would imply that the allowable hadronic bags should include colorless qqq and $q\bar{q}$ states.

⊕⟦ Supplement 9.1

We can get an estimate of the magnitude of the bag pressure by considering massless free fermions in a spherical cavity of radius R. The Dirac equation for a massless fermion in the cavity is

$$\gamma \cdot p\psi = 0. \tag{1}$$

We use the Dirac representation of the γ matrices,

$$\gamma^0 = \begin{pmatrix} I & 0 \\ 0 & -I \end{pmatrix},$$

and

$$\gamma = \begin{pmatrix} 0 & \sigma \\ -\sigma & 0 \end{pmatrix},$$

where I is a 2×2 unit matrix and σ are the Pauli matrices. We write the four-component wave function for the massless fermion ψ as

$$\psi = \begin{pmatrix} \psi_+ \\ \psi_- \end{pmatrix},$$

where ψ_+ and ψ_- are two dimensional spinors. Equation (1) becomes

$$\begin{pmatrix} p^0 & -\sigma \cdot \mathbf{p} \\ +\sigma \cdot \mathbf{p} & -p^0 \end{pmatrix} \begin{pmatrix} \psi_+ \\ \psi_- \end{pmatrix} = 0. \tag{2}$$

Writing the above matrix equation into two equations and eliminating the ψ_- component, we obtain

$$[\mathbf{p}^2 - (p^0)^2]\psi_+ = 0.$$

The lowest energy solution for the above equation is the $s_{1/2}$ state given by

$$\psi_+(\mathbf{r}, t) = \mathcal{N}e^{-ip^0t}j_0(p^0r)\chi_+,$$

where j_0 is the spherical Bessel function of order zero, χ_+ is a two dimensional spinor, and $\mathcal{N}$ is a normalization constant. From Eq. (2), the other component is

$$\psi_-(\mathbf{r}, t) = \mathcal{N} e^{-ip^0 t} \, \boldsymbol{\sigma} \cdot \hat{\mathbf{r}} \, j_1(p^0 r) \, \chi_-.$$

The confinement of the quarks is equivalent to the requirement that the normal component of the vector current $J_\mu = \bar{\psi} \gamma_\mu \psi$ vanishes at the surface. This condition is the same as the requirement that the scalar density $\bar{\psi}\psi$ of the quark vanishes at the bag surface $r = R$ (see p. 412 of Ref. [8]). This leads to

$$\bar{\psi}\psi \bigg|_{r=R} = [j_0(p^0 R)]^2 - \boldsymbol{\sigma} \cdot \hat{\mathbf{r}} \, \boldsymbol{\sigma} \cdot \hat{\mathbf{r}} [j_1(p^0 R)]^2 = 0$$

or

$$[j_0(p^0 R)]^2 - [j_1(p^0 R)]^2 = 0.$$

From the tabulated values of the spherical Bessel functions [15], this equation is satisfied for

$$p^0 R = 2.04, \quad \text{or} \quad p^0 = \frac{2.04}{R}.$$

For a system of N quarks in a bag, the total kinetic energy of the confined quarks is inversely proportional to R. It decreases with an increase in the bag radius. In the bag model, the nonperturbative effects of confinement are represented phenomenologically by the presence of a bag pressure directed from the region outside the bag toward the region inside the bag. The energy density of the 'vacuum' inside the bag is higher than that outside the bag, the difference being the bag pressure B. The energy of a system of N confined quarks in a bag of radius R is

$$E = \frac{2.04N}{R} + \frac{4\pi}{3} R^3 B.$$

We observe that the tendency to increase the radius due to the kinetic energy of the quarks is counterbalanced by this inward pressure B directed from the region outside the bag towards the region inside the bag. The equilibrium radius of the system is located at the radius R determined by $dE/dR = 0$, which leads to a bag pressure constant B related to the radius by

$$B^{1/4} = \left(\frac{2.04N}{4\pi}\right)^{1/4} \frac{1}{R}.$$

If we take the confinement radius to be 0.8 fm for a 3 quark system in a baryon, we obtain an estimate of the bag pressure constant

$$B^{1/4} = 206 \text{ MeV}.$$

(We have chosen to use the units $\hbar = c = 1$. To include $\hbar$ and c explicitly, the bag pressure constant should be $B^{1/4} = 206\text{MeV}/(\hbar c)^{3/4}$.) The value of the bag pressure $B^{1/4}$ is in the range between 145 MeV (Ref. 16) and 235 MeV (Ref. 17).

§9.3 Quark-Gluon Plasma

We have outlined the essence of the bag model in the last section. More refined descriptions taking into account chiral symmetry and perturbative gluon interactions have been quite successful in describing many experimental data, such as hadron masses and magnetic moments [18]. For our purposes, it is useful to use it as a heuristic model to discuss matter under extreme conditions.

We can interpret the bag model as indicating that the essential effect of nonperturbative QCD is to give rise to an inward bag pressure of magnitude B. For a hadron in which the quarks can be considered to be confined to be in the lowest $s_{1/2}$ state inside the bag, the inward bag pressure is balanced by the quantum stress [19] arising from the wave function of the quarks. This balance of opposing pressures allows a simple intuitive description of the competing forces leading to a stable system. It also provides a simple intuitive understanding of why new phases of quark matter are expected.

It is clear that if the pressure of the quark matter inside the bag is increased, there will be a point when the pressure directing outward is greater than the inward bag pressure. When that happens, the bag pressure cannot balance the outward quark matter pressure and the bag cannot confine the quark matter contained inside. A new phase of matter containing the quarks and gluons in an unconfined state is then possible. It is this situation which leads to the possible existence of different phases of the quark matter. The main condition for a new phase of quark matter is the occurrence of a large pressure exceeding the bag pressure B.

A large pressure of quark matter arises 1) when the temperature of the matter is high, and/or 2) when the baryon number density is large. New phases of quark matter are then expected [20]. We shall discuss these two extreme cases separately.

§9.3.1 Quark-Gluon Plasma at High Temperatures

We consider the case of a quark-gluon system in thermal equilibrium at a high temperature T, within a volume V. For simplicity, we examine the case where quarks and gluons are idealized to be noninteracting and massless and there is no net baryon number. The number of quarks and the number of antiquarks in the system are equal.

The partial pressures arising from the quarks, antiquarks, and the gluons can be obtained from standard textbooks [21]. For completeness, we shall describe the steps in Supplement 9.2 below.

Adding the contributions from quarks, antiquarks and gluons together (see Eqs. (3a), (3b), and (8) in Supplement 9.2), we have the total pressure of an ideal quark-gluon plasma given by

$$P = g_{total} \frac{\pi^2}{90} T^4, \tag{9.2}$$

where

$$g_{total} = [g_g + \frac{7}{8} \times (g_q + g_{\bar{q}})], \tag{9.3}$$

where g_g, g_q, and $q_{\bar{q}}$ are respectively the degeneracy number of the gluons, the quarks and the antiquarks. If the quarks and gluons are confined in a finite volume, the total pressure should include additional contributions from the quantum kinetic energies of the particles, which is inversely proportional to the radius of the confined volume. We are interested in the case when the quark matter may be deconfined in a large volume, and we can neglect this contribution.

To evaluate the degeneracy numbers, we note that there are 8 gluons, each having two possible polarizations. Therefore, we have

$$g_g = 8 \times 2. \tag{9.4}$$

The degeneracy number g_q of the quarks depends on the number of flavors under consideration. In order to find the critical temperature for a transition into a quark-gluon plasma with two flavors, we shall take the flavor degeneracy to be two. The degeneracy numbers g_q and $g_{\bar{q}}$ are

$$g_q = g_{\bar{q}} = N_c N_s N_f. \tag{9.5}$$

where $N_c(= 3)$ is the number of colors, $N_s(= 2)$ is the number of spins, and $N_f(= 2 \text{ or } 3)$ is the number of flavors. From Eq. (9.3), the total number of degrees of freedom g_{total} is 37. The pressure of a quark-gluon plasma at a temperature T is therefore

$$P = 37 \frac{\pi^2}{90} T^4, \tag{9.6}$$

and the energy density of the quark-gluon matter at a temperature T is

$$\epsilon = 37 \frac{\pi^2}{30} T^4, \tag{9.7}$$

which gives an energy density of 2.54 GeV/fm^3 at a temperature of 200 MeV.

From Eq. (9.6), the *critical temperature* at which the quark-gluon pressure is equal to the bag pressure B is given by

$$T_c = (\frac{90}{37\pi^2})^{1/4} B^{1/4}. \tag{9.8}$$

Thus, for $B^{1/4} = 206$ MeV as estimated in Section 9.2, we have $T_c \sim 144$ MeV. If the quark matter in a bag are heated up to a high temperature greater than the critical temperature, the quark matter inside the bag will have a pressure which is greater than that of the bag pressure. When this happens, the bag will not be able to hold the quark matter in the bag and the quark matter will be deconfined. The deconfined phase of the quark matter is given the general name 'quark-gluon plasma'. Thus, a quark-gluon plasma may arise when the temperature of the quark matter is very high.

⊕[Supplement 9.2

We would like to obtain the pressure arising from a relativistic massless quark gas at temperature T. We first determine the energy density of the quark gas. The phase space volume of quarks in a spatial volume V with momentum p in the momentum interval dp is $4\pi p^2 dp V$. As each state occupies a phase space volume of $(2\pi\hbar)^3$, the number of states characterized by a momentum p in the interval dp is $4\pi p^2 dp V/(2\pi)^3$. (In our notation, we use $\hbar = 1$.) At the specified temperature T, not all of the states are occupied. The occupation probability for the state with a momentum p is given by the Fermi-Dirac distribution appropriate for the temperature T. The number of quarks in a volume V with momentum p within the interval dp is

$$dN_q = \frac{g_q V 4\pi p^2 dp}{(2\pi)^3} \left\{ \frac{1}{1 + e^{(p-\mu_q)/T}} \right\}$$

where the factor in the curly brackets is the Fermi-Dirac distribution, μ_q is the quark Fermi energy (or chemical potential), and $g_q(= N_c N_s N_f)$ is the degeneracy of quarks.

Given the quark chemical potential μ_q, we can obtain the density of the antiquarks. The presence of antiquarks corresponds to the absence of quarks in the negative energy states. The number density of antiquarks is therefore

$$n_{\bar{q}}(\mu) = \frac{g_q}{(2\pi)^3} \int_{-\infty}^{0} 4\pi p_0^2 dp_0 \left[1 - \frac{1}{1 + e^{(p_0 - \mu_q)/T}} \right]$$

$$= \frac{g_q}{(2\pi)^3} \int_{0}^{\infty} 4\pi p_0^2 dp_0 \frac{1}{1 + e^{(p_0 + \mu_q)/T}} .$$

Thus, for our case when the number density of the quarks is the same as that of the antiquark, we have $\mu_q = 0$.

After setting $\mu_q = 0$, the energy of the massless quarks in the system of volume

V and temperature T is

$$
\begin{aligned}
E_q &= \frac{g_q V}{2\pi^2} \int_0^\infty \frac{p^3 dp}{1 + e^{p/T}} \\
&= \frac{g_q V}{2\pi^2} T^4 \int_0^\infty \frac{z^3 dz}{1 + e^z} \\
&= \frac{g_q V}{2\pi^2} T^4 \int_0^\infty z^3 dz e^{-z} \sum_{n=0}^\infty (-1)^n e^{-nz} \\
&= \frac{g_q V}{2\pi^2} T^4 \Gamma(4) \sum_{n=0}^\infty (-1)^n \frac{1}{(n+1)^4},
\end{aligned}
$$

where Γ is the gamma function. It is easy to show that

$$
\begin{aligned}
\sum_{n=0}^\infty (-1)^n \frac{1}{(n+1)^4} &= \sum_{m=1,3,5..} \frac{1}{m^4} - \sum_{m=2,4,6..} \frac{1}{m^4} \\
&= \sum_{m=1,2,3,..} \frac{1}{m^4} - 2 \sum_{m=2,4,6..} \frac{1}{m^4} \\
&= \sum_{m=1,2,3,..} \frac{1}{m^4} - 2 \sum_{m=1,2,3..} \frac{1}{(2m)^4} \\
&= (1 - 2^{-3})\zeta(4),
\end{aligned}
$$

where $\zeta(x)$ is the Riemann zeta function defined by [15]

$$
\zeta(x) = \sum_{m=1,2,3,..} \frac{1}{m^x}.
$$

The function $\zeta(4)$ has the value $\pi^4/90$ [15]. The energy of the system due to quarks is therefore

$$
E_q = \frac{7}{8} g_q V \frac{\pi^2}{30} T^4. \tag{1}
$$

For massless fermions and bosons, it can be shown [21] that the pressure P is related to the energy density E/V by

$$
P = \frac{1}{3} \frac{E}{V}. \tag{2}
$$

The pressure due to quarks is

$$
P_q = \frac{7}{8} g_q \frac{\pi^2}{90} T^4. \tag{3a}
$$

Similarly, the pressure due to antiquarks is

$$
P_{\bar{q}} = \frac{7}{8} g_{\bar{q}} \frac{\pi^2}{90} T^4. \tag{3b}
$$

Hence the pressure of the quark-antiquark gas at a temperature T is

$$P_q + P_{\bar{q}} = \frac{7}{8}(g_q + g_{\bar{q}})\frac{\pi^2}{90}T^4. \tag{4}$$

The number density of the quarks and antiquarks are

$$n_q = n_{\bar{q}} = \frac{g_q}{2\pi^2}\int_0^\infty \frac{p^2 dp}{1 + e^{p/T}},$$

$$= \frac{g_q}{2\pi^2}T^3 \frac{3}{2}\zeta(3). \tag{5}$$

The function $\zeta(3)$ has the value of 1.20205 [15].

It is instructive to have a general idea on the magnitude of the densities of quarks and antiquarks we deals with in a quark-gluon plasma. For a quark-gluon plasma with a temperature of 200 MeV, treated as a free gas of massless quarks with 3 flavors, Eq. (5) gives

$$n_q = n_{\bar{q}} = 1.71/\text{fm}^3. \tag{6}$$

Our next step is to evaluate the pressure due to the gluons at temperature T. The energy of the gluons in the system of volume V and temperature T is

$$E_g = \frac{g_g V}{2\pi^2}\int_0^\infty p^3 dp \left\{\frac{1}{e^{p/T} - 1}\right\},$$

where the factor in brackets is the Bose-Einstein distribution for bosons and g_g is the gluon degeneracy,

$$g_g = (\text{number of different gluons, 8}) \times (\text{number of polarizations, 2}).$$

The energy of the gluons is

$$E_g = \frac{g_g V}{2\pi^2}T^4 \int_0^\infty \frac{z^3 dz}{e^z - 1}$$

$$= \frac{g_g V}{2\pi^2}T^4 \int_0^\infty z^3 dz e^{-z} \sum_{n=0}^\infty e^{-nz}$$

$$= \frac{g_g V}{2\pi^2}T^4 \Gamma(4) \sum_{n=0}^\infty \frac{1}{(n+1)^4}$$

$$= \frac{g_g V}{2\pi^2}T^4 \Gamma(4)\zeta(4)$$

$$= g_g V \frac{\pi^2}{30}T^4. \tag{7}$$

Using the relation (2) between the energy density and the pressure, $P = \frac{1}{3}E/V$, we have

$$P_g = g_g \frac{\pi^2}{90}T^4. \tag{8}$$

The number density of the gluons are

$$n_g = \frac{g_g}{2\pi^2} \int_0^\infty p^2 dp \left\{ \frac{1}{e^{p/T} - 1} \right\}$$
$$= \frac{g_g}{2\pi^2} T^3 \Gamma(3) \zeta(3)$$
$$= \frac{g_g}{\pi^2} 1.202 \ T^3 \ . \tag{9}$$

For a quark-gluon plasma at a temperature of 200 MeV treated as a free gas of massless bosons, Eq. (9) gives

$$n_g = 2.03 \frac{\text{gluons}}{\text{fm}^3} \ . \tag{10}$$

From the results of Eqs. (6), (9), and Eq. (9.7), we concludes that as an order of magnitude estimate for a a quark-gluon plasma at a temperature of 200 MeV, (which is of the order of the quark-gluon-plasma transition temperature), there are approximately 1.7 quarks, 1.7 antiquarks and 2 gluons in a fm³, with an energy density of about 2.5 GeV/ fm³.]⊕

⊕[Supplement 9.3
     For a relativistic pion gas in thermal equilibrium, treated approximately as a massless boson gas, the number of the degrees of freedom g_h is 3, corresponding to the possibilities of π^+, π^-, and π^0. For this pion gas, the energy density ϵ_π is given according to Eq. (7) of Supplement 9.2 as

$$\epsilon_\pi = \frac{E_\pi}{V} = 3 \frac{\pi^2}{30} T^4 \ . \tag{1}$$

The pressure of the pion gas is given according to Eq. (8) of Supplement 9.2 as

$$P_\pi = \frac{\pi^2}{30} T^4 . \tag{2}$$

The number density of the pion gas is given by Eq. (9) of Supplement 9.2 as

$$n_\pi = 3 \frac{1.202}{\pi^2} T^3 \ . \tag{3}$$

Thus, at a temperature of $T = 200$ MeV, the pion energy density ϵ_π and the pion number density n_π of a pion gas consisting of π^+, π^- and π^0 are

$$\epsilon_\pi = 0.21 \ \text{GeV/fm}^3 \ ,$$

and

$$n_\pi = 0.38 \frac{\text{pions}}{\text{fm}^3} \ .$$

Under a constant temperature and pressure, the variation of the energy, the volume and the entropy of the hadron gas is given by Eq. (12.3) of Ref. 21,

$$dE_\pi = -P_\pi dV + TdS_\pi .$$

The above equation gives the relation for entropy density dS_π/dV as

$$\frac{dS_\pi}{dV} = \frac{\epsilon_\pi + P_\pi}{T} . \tag{4}$$

Therefore, from Eqs. (1-4), the entropy per pion in the pion gas, under a constant temperature and pressure, is

$$\begin{aligned}
\frac{dS_\pi}{dN_\pi} &= \frac{dS_\pi}{n_\pi dV} \\
&= \frac{\epsilon_\pi + P_\pi}{n_\pi T} \\
&= \frac{4\pi^2}{90} \times \frac{\pi^2}{1.202} \\
&= 3.6 .
\end{aligned} \tag{5}$$

This means that for a pion gas approximated to be a massless boson gas, it takes 3.6 unit of entropy to produce a pion. That is, the ratio of the entropy of the system to the number of pions in the system is a constant, independent of temperature and pressure. The property that the number of pions is proportional to the entropy is often used in statistical and hydrodynamical models [9,22], where one infers the number of produced pions from the entropy of reaction products.

It is instructive to find out the amount of entropy needed to produce a quark, an antiquark and a gluon in a quark-gluon plasma. Following the same arguments leading to Eqs. (4) and (5), the entropy per quark or antiquark in a quark-gluon plasma, under a constant temperature and pressure is

$$\begin{aligned}
\frac{dS_q}{dN_q} &= \frac{dS_{\bar{q}}}{dN_{\bar{q}}} \\
&= \frac{dS_q}{n_q dV} \\
&= \frac{\epsilon_q + P_q}{n_q T} .
\end{aligned} \tag{6}$$

For a massless quark-gluon plasma, this becomes

$$\frac{dS_q}{dN_q} = \frac{4P_q}{3n_q T} . \tag{7}$$

Using the results of Eqs. (3) and (5) in Supplement (9.2), we obtain

$$\begin{aligned}
\frac{dS_q}{dN_q} &= \frac{dS_q}{dN_q} \\
&= \frac{4}{3}\frac{7}{8}g_q\frac{\pi^2}{90}T^4 \frac{1}{g_q T\, T^3\, \frac{3}{2}\, 1.202/2\pi^2} \\
&= 1.40 .
\end{aligned} \tag{8}$$

Similarly, the entropy per gluon in the quark-gluon plasma is

$$\frac{dS_g}{dN_g} = \frac{4P_g}{3n_gT}.$$ (9)

Using the results of Eqs. (8) and (9) in Supplement (9.2), we obtain

$$\frac{dS_g}{dN_g} = \frac{4}{3}g_g\frac{\pi^2}{90}T^4\frac{1}{Tg_g 1.202\ T^3/\pi^2}$$
$$= 1.2.$$ (10)

Thus, in a quark-gluon plasma approximated as a massless gas of quarks, antiquarks and gluons, it takes 1.4 unit of entropy to produce a quark or an antiquark, and an entropy of 1.2 unit to produce a gluon. The amount of entropy needed to produce a particle in a quark gluon plasma is slightly less than the amount of entropy needed to produce a pion. However, there are about 37 degree of freedom (which can be roughly considered as different kinds of quark, antiquark and gluon particles) in the quark-gluon plasma but only three degrees of freedom in the pion gas. Consequently, the entropy needed to produce all types of particles in the quark-gluon plasma is much greater than the entropy needed to produce all types of particles in the pion gas. The quark-gluon plasma stores a greater amount of entropy per unit volume than does the pion gas.]⊕

It should be noted that the quark-gluon plasma is a continuum concept. The results discussed in this section is based on treating the quark-gluon plasma as a continuum. On the other hand, quark matter or quark-gluon plasma which may be produced in the laboratory must be a spatially finite system with a boundary. Even though the quark-gluon plasma is expected to be deconfined at high temperature, the deconfinement extends only over the region within the boundary of the hot quark matter. Because of this boundary, the pressure as given by Eq. (9.2) in the plasma will be subject to the bag pressure B arising from the presence of the boundary:

$$P_{\text{quark-gluon plasma with boundary}} = P_{\text{quark-gluon plasma continuum}} - B$$
$$= g\frac{\pi^2}{90}T^4 - B,$$ (9.9a)

and the energy density is correspondingly altered,

$$\epsilon_{\text{quark-gluon plasma with boundary}} = \epsilon_{\text{quark-gluon plasma continuum}} + B$$
$$= g\frac{\pi^2}{30}T^4 + B.$$ (9.9b)

Eqs. (9.9a) and (9.9b) are often used in the literature.

§9.3.2 Quark-Gluon Plasma with a High Baryon Density

We shall now consider another possibility in which the pressure inside a bag can be large enough to lead to the deconfinement of the quark matter, even at $T = 0$. We examine a situation in which the quark matter inside a bag consists of quarks of very high baryon density. Because of the Pauli exclusion principle, no more than one fermion can populate a state with a definite set of quantum numbers. The requirement that the quarks must populate states with different quantum numbers leads to the restriction that as the density of the quarks increases, the quarks must populate states of greater momentum. Thus, the quark gas acquires a pressure due to the degeneracy of the quark gas and this pressure increases with quark density. However, if the density of the quark matter inside the bag increases, there will be a point when the pressure from the degenerate quark gas exceeds that of the bag pressure. When that happens, the bag pressure will not be able to hold the bag together. A state of quark matter in which the quarks becomes unconfined is then possible. As each quark carries a baryon number 1/3, the high quark density corresponds to a high baryon density. Consequently, a new phase of quark matter, with a non-zero net baryon content will be possible for a large baryon density. What is the order of magnitude of the baryon density at which this state of quark matter is expected? We shall estimate this critical baryon density at $T = 0$.

We need to determine the pressure arising from a relativistic degenerate quark gas. For simplicity, we shall neglect the contributions from antiquarks and gluons. The number of state in a volume V with momentum p within the momentum interval dp is

$$\frac{g_q V}{(2\pi)^3} 4\pi p^2 dp.$$

As each state is occupied by one quark, the total number of quarks up to the quark Fermi momentum μ_q is

$$N_q = \frac{g_q V}{(2\pi)^3} \int_0^{\mu_q} 4\pi p^2 dp$$

$$= \frac{g_q V}{6\pi^2} \mu_q^3.$$

The number density of the quark gas is given by

$$n_q = \frac{N_q}{V} = \frac{g_q}{6\pi^2} \mu_q^3. \tag{9.10}$$

The energy of the quark gas in a volume V is

$$E_q = \frac{g_q V}{(2\pi)^3} \int_0^{\mu_q} 4\pi p^3 dp$$

$$= \frac{g_q V}{8\pi^2} \mu_q^4. \tag{9.11}$$

The energy density of the quark gas is therefore

$$\epsilon_q = \frac{E_q}{V} = \frac{g_q}{8\pi^2} \mu_q^4. \tag{9.12}$$

From the relation between the pressure and the energy density, we have

$$P_q = \frac{1}{3}\frac{E}{V} = \frac{g_q}{24\pi^2} \mu_q^4. \tag{9.13}$$

Critical change of the state of the matter will take place when the pressure from the degenerate quark matter becomes equal to the bag pressure, $P_q = B$. This leads to

$$\mu_q = \left(\frac{24\pi^2}{g_q} B\right)^{1/4}, \tag{9.14}$$

which corresponds to a critical quark number density n_{qc} given by

$$n_q(\text{quark} - \text{gluon plasma}) = 4\left(\frac{g_q}{24\pi^2}\right)^{1/4} B^{3/4}. \tag{9.15a}$$

The corresponding critical baryon number density is

$$n_B(\text{quark} - \text{gluon plasma}) = \frac{4}{3}\left(\frac{g_q}{24\pi^2}\right)^{1/4} B^{3/4}. \tag{9.15b}$$

To get some numerical values to aid our intuitive understanding of the nature of the plasma with a high baryon content, we consider the plasma to arise by compressing ordinary nuclear matter, which has only u and d valence quarks. The degeneracy number for such quark matter is

$$g_q = (3\text{ colors}) \times (2\text{ spins}) \times (2\text{ flavors}).$$

For a bag pressure of $B^{1/4} = 206$ MeV, the *critical baryon number density* at which the compressed hadron matter becomes a quark-gluon plasma with a high baryon content at $T = 0$ is

$$n_B(\text{quark} - \text{gluon plasma}) = 0.72/\text{fm}^3, \tag{9.16a}$$

corresponding to a quark Fermi momentum μ_q with the value

$$\mu_{u,d} = 434 \text{ MeV} . \qquad (9.16b)$$

These values for the quark-gluon plasma should be compared with the nucleon number density $n_B = 0.14/fm^3$ and a nucleon Fermi momentum of 251 MeV for the normal nuclear matter at equilibrium. Thus, the critical baryon density is about 5 times the normal nuclear matter density. When the density of baryons exceeds this density, the baryon bag pressure is not strong enough to withstand the pressure due to the degeneracy of quarks, and the confinement of quarks within individual baryon bags will not be possible, leading to formation of a state of deconfined quarks.

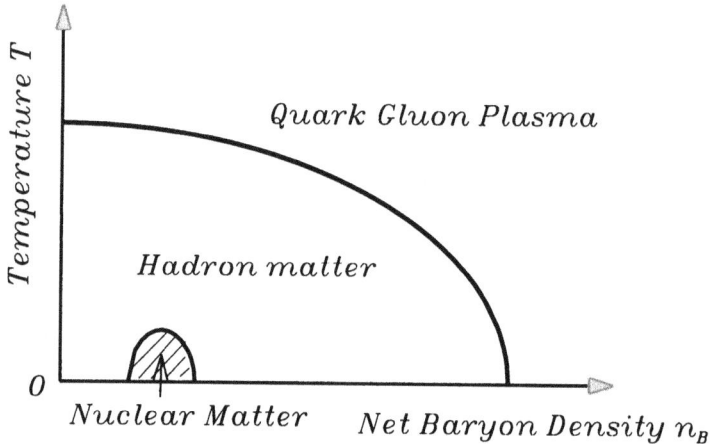

Fig. 9.1 Phase diagram in the plane of the temperature and the net baryon density. The region of nuclear matter is indicated by the shaded area.

We have discussed the possible occurrence of deconfined phases of quark matter in two extreme situations. The first situation occurs when the temperature is very high and the net baryon density (the density of baryons minus the density of the antibaryons) is zero. We estimate that the critical temperature is about 140 MeV. The second situation occurs when the temperature is zero and the baryon density is about 5 times the equilibrium nuclear matter density. For a system in between these two limits, there is a pressure arising from the thermal motion of the particles, there is also a pressure arising from the degeneracy of the fermion gas. The total pressure is the sum of the two contributions. Thus, for a system whose temperature and net baryon density is not zero, the critical temperature at which the quark matter becomes deconfined, will lie between the limits of $T = 0$ for a degenerate quark gas, and the other limit of T_c for a pure plasma with no net baryon density. The phase diagram one can construct, in the space of temperature and baryon density can be shown schematically as in Fig. 9.1. A important objective of relativistic heavy ion physics is to explore the phase diagram of quark matter in various regions of temperature and baryon density so as to confirm the existence of the new phase of quark matter.

It should be pointed out that we have used the bag model to provide a heuristic understanding of the possibility of the occurrence of a new phase of quark matter. The bag model has nevertheless limitations. There have been many predictions of the existence of exotic states [13] using the bag model. These exotic states are multiquark and multi-gluon color singlet bag states which do not occur in the simplest quark models, of $q\bar{q}$ and qqq. Many of these are quark-gluon systems with a large number of quarks (greater than 3). Experimental observation of such exotic multiquark systems is still rather elusive. To gain a better understanding of the possibility of a new phase of quark matter, it is necessary to examine the quark matter system within the framework of the lattice gauge theory. This will be followed in the next chapter.

§References for Chapter 9

1. For a general review of quantum chromodynamics, see F. Wilczek, Ann. Rev. Nucl. and Part. Sci. 32, 177 (1982).
2. M. Gell-Mann, Phys. Lett. 8, 214 (1964); G. Zweig, CERN Report, Th 401 and Th 412, 1964 (unpublished); Y. Ne'eman, Nucl. Phys. 26, 222 (1961).
3. S. L. Glashow, J. Iliopolous, and L. Maiani, Phys. Rev. D2, 1285 (1970).
4. M. Kobayashi and M. Maskawa, Prog. Theo. Phys. 49, 652 (1973).
5. J. Ellis and G. L. Fogli, Phys. Lett. 249B, 543 (1990).
6. K. Hisaka et al., Particle Data Group, *Review of Particle Properties*, Phys. Rev. D45, S1 (1992), p. III.68.
7. The masses of quarks in Table 9.1 were estimated for the standard model of electroweak interactions, as reported in Ref. 6, p. III.59. They were estimated by C. A. Dominguez and E. de Rafael, Ann. Phys. 174, 372 (1987), J. Gasser and H. Leutwyler, Phys. Rep. 87, 77 (1982), S. Narison, Phys. Lett. B216, 191 (1989), and J. F. Donoghne, Ann. Rev. Nucl. Part. Sci. 39, 1 (1989). See also Ref. 5.
8. See for example, F. E. Close, *Introduction to Quarks and Partons*, Academic Press, London, 1979.
9. J. D. Bjorken, Phys. Rev. 179, 1547 (1969).

10. R. P. Feynman, Phys. Rev. Lett. 23, 1415 (1969).
11. D. J. Gross and F. Wilczek, Phys. Rev. Lett. 30, 1343 (1973); H. D. Politzer, Phys. Rev. Lett. 30 1346 (1973).
12. S. Coleman and D. J. Gross, Phys. Rev. Lett. 31, 1343 (1973).
13. A. Chodos, R. L. Jaffe, K. Johnson, C. B. Thorn, and V. F. Weisskopf, Phys. Rev. D9, 3471 (1974);
14. For a review of the Bag Model, see C. D. DeTar and J. F. Donoghue, Ann. Rev. Nucl. Part. Sci. 33, 235 (1983); and L. Wilets, *Bag Model of Nucleus*, World Scientific, 1989.
15. M. Abramowitz and I. A. Stegun, *Handbook of Mathematical Tables*, Dover Publications, New York, 1965.
16. W. C. Haxton and L. Heller, Phys. Rev. D22, 1198 (1980).
17. P. Hasenfratz, R. R. Horgan, J. Kuti, and J. M. Richard, Phys. Lett. 95B, 199 (1981).
18. J. C. Collins and M. J. Perry, Phys. Rev. Lett. 34, 1353 (1975); L. D. McLerran and B. Svetitisky, Phys. Lett. 98B, 195 (1981); J. Kuti, J. Polonyi, K. Szlachanyi, Phys. Lett. 98B, 199 (1981); R. Hagedorn and J. Rafelski Phys. Lett. 97B, 180 (1980); K. Kajantie, C. Montonen, and C. Pietarinen, Zeit. Phys. C9, 253 (1981).
19. C. Y. Wong, J. Math. Phys. 17, 1008 (1976).
20. For a review of the physics of quark-gluon plasma, see L. McLerran, Rev. Mod. Phys. 58, 1021 (1986), and B. Müller, Lecture Notes in Physics, Vol. 225, Springer Verlag, 1985.
21. L. D. Landau and E. M. Lifshitz, *Statistical Physics*, Pergamon Press, Oxford, 1980, Third Edition.
22. L. D. Landau, Izv. Akad. Nauk. SSSR Ser. Fiz. 17, 51 (1953); S. Z. Belenkji and L. D. Landau, Suppl. Nuovo Cim. 3, 15 (1956).

10. Lattice Gauge Theory

The discussions in the last chapter were based on a simple model to provide qualitative arguments for the possible existence of a new phase of quark matter. Such an approach is useful as a guide. For a more quantitative understanding of the new phases of quark matter, we need to use a better theoretical model.

As the gauge theory of QCD is generally considered to be the correct theory for the interaction of quarks and gluons, it is necessary to study the phases of quark matter using the gauge theory of QCD. Furthermore, since the equilibrium phases and the phase transition involve quarks and gluons interacting over a large distance scale, they need to be studied within the framework of nonperturbative QCD.

Lattice gauge theory is a nonperturbative treatment of quantum chromodynamics formulated on a discrete lattice of space-time coordinates [1,2]. The discretization of the space-time continuum provides two main advantages. In nonperturbative QCD, while the main interest lies in the infrared (large-distance) behavior of the system, divergences associated with the ultraviolet (short-distance) behavior need to be avoided. In the evaluation of many Feynman diagrams in perturbation theory, one runs into momentum integrals which become divergent when the upper limit of the momentum variable is taken to go to infinity. Such a divergence, associated with a large magnitude of the momentum, is called an *ultraviolet divergence*. The term "ultraviolet" is used because a large momentum scale corresponds to a small wavelength scale, by the De Broglie momentum-wavelength relation, and the ultraviolet radiation has wavelengths shorter than those of the visible spectrum. In perturbation theory, the ultraviolet divergences may be handled by the procedure of *renormalization* [3]. This is done order by order by identifying the infinities into well defined forms and by using the redefinition of observable quantities, such as the mass and the coupling constant, in terms of unobservable bare quantities. This method of regulating the ultraviolet divergences is based on a perturbative expansion. In nonperturbative QCD, a perturbation expansion is no longer valid and a different procedure is needed.

The finite space-time intervals in the lattice gauge theory for nonperturbative QCD provide a natural way to regularize the ultraviolet divergences. There is a shortest distance scale, the spacing between the nearest lattice points. This shortest distance scale defines the maximum momentum scale for the problem in question and gives a momentum cut-off for the calculation. Terms which diverge for very large momenta will be rendered finite.

Another advantage of the lattice gauge theory is that by discretizing the space-time coordinates and going to the domain of imaginary time, the partition function in statistical mechanics can be written in the form of a path integral. Monte-Carlo methods of importance sampling can then be employed to find the equilibrium states of the system.

In practical calculations, because of the limitation in computer memory and computer speed, the number of lattice points is limited and the size of the lattice spacing cannot be too small. It is then necessary to perform calculations for different numbers of lattice points (different lattice spacings) until 'scaling' behavior occurs. That is, the relationship between the coupling constant and the scale of the spatial spacing is in accordance with what is expected from perturbative QCD. Only when scaling behavior is achieved can one relate the lattice spacing to physical scales, to allow the extraction of physical quantities from lattice QCD calculations.

We shall outline the basic ideas of the path integral method and the lattice gauge theory.

§10.1 Path Integral Method for a Single-Particle System

We first consider a single-particle system with the particle located initially at the starting point x_a at time t_a and at the end point x_b at time t_b. A 'path' of the system can be described by specifying the coordinate x as a function of t, for t between t_a and t_b. One can construct many paths from the starting point to the end point. Corresponding to each path, there is a value of the action S_M given by

$$S_M = \int_{t_a}^{t_b} dt \, L\left(x(t), \dot{x}(t) \right), \qquad (10.1)$$

where L is the Lagrangian. To distinguish the above action from the Euclidean action, which we shall introduce later in Eqs. (10.12) and (10.27), we shall call this the *Minkowski action* because it uses the Minkowski metric tensor $g_{\mu\nu}$ with diagonal elements $(g^{11}, g^{22}, g^{33}, g^{44})$ $= (1, -1, -1, -1)$.

As is well known, the *classical trajectory* is the path which leads to the least Minkowski action (see for example [4]). However, in quantum mechanics, all of the paths contribute to the probability amplitude and a quantum mechanical description can be obtained by the *path integral method* [4]. One would like to determine the quantum mechanical amplitude for the system to be initially located at (x_a, t_a) and to end up at (x_b, t_b). According to Feynman's path integral method, the amplitude is given by summing the phase factor e^{iS_M} over all paths,

$$\text{Amplitude } [(x_a, t_a) \to (x_b, t_b)] = \sum_{all\ paths} e^{iS_M}. \qquad (10.2)$$

We can write down an alternative expression for the probability amplitude in a quantum mechanical description. We consider the basis states in which the position operator is diagonal. The initial state is then

$$(\text{ initial state at time } t_a) = \mid x_a > .$$

The evolution of the state from time t_a to t_b is obtained by applying the operator $e^{-iH(t_b-t_a)}$ to the initial state,

$$(\text{ state at time } t_b) = e^{-iH(t_b-t_a)} \mid x_a >,$$

where H is the Hamiltonian operator of the system. The probability amplitude of finding the system to begin at x_a at time t_a and to end up at x_b at time t_b is the scalar product of the final state $\mid x_b >$ with the evolved state,

$$\text{Amplitude } [(x_a, t_a) \to (x_b, t_b)] = < x_b \mid e^{-iH(t_b-t_a)} \mid x_a > . \qquad (10.3)$$

Thus, Eq. (10.2) can be expressed as

$$< x_b \mid e^{-iH(t_b-t_a)} \mid x_a > = \sum_{all\ paths} e^{iS_M}. \qquad (10.4)$$

To make the concept of path integral operationally meaningful, we need a way to perform the operation 'summing over paths'. A 'path' consists of the locus of points (x, t) which start at the point (x_a, t_a) and ends up at (x_b, t_b). To sum over the paths, we treat the spatial coordinates of the intermediate points along the path as variables. By allowing these spatial coordinates to vary, we obtain the set of paths to be summed over. Accordingly, we divide the time separation between t_a and t_b into n_t evenly-spaced intervals, with t_a being the zeroth time coordinate and t_b being the n_t-th time coordinate, as in Fig. 10.1. The discrete time coordinates are $t_0, t_1, t_2, ..., t_{n_t-1}$, and t_{n_t}, with $t_0 = t_a$ and $t_{n_t} = t_b$. At the time coordinate t_1, the intermediate point (x_1, t_1) joins onto the starting point (x_a, t_a) to form a part of

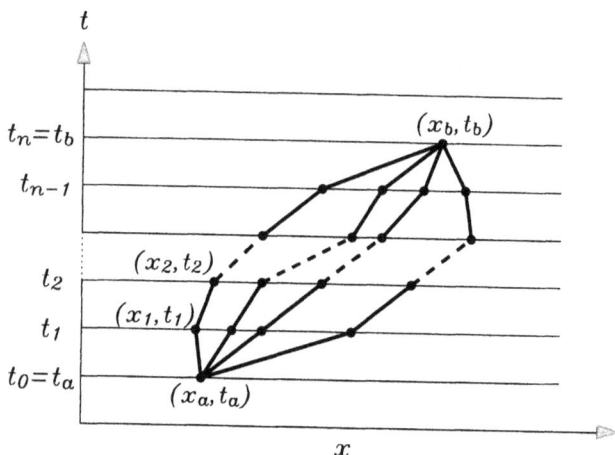

Fig. 10.1 Various paths from the space-time point (x_a, t_a) to the point (x_b, t_b). Different paths can be obtained by varying x_1 at t_1, x_2 at t_2, and so on.

the path from (x_a, t_a) to the end point (x_b, t_b). The variation of the paths which pass through the time t_1 can be specified as a variation of x_1. Therefore, to generate the set of paths passing through t_1, we treat the spatial coordinate x_1 as a variable, which has the range from $-\infty$ to $+\infty$. The sum (10.4) over all paths passing through t_1 can be transcribed into the integral over all possible values of x_1, with $-\infty < x_1 < +\infty$:

$$(\text{sum over paths which go through } t_1) \rightarrow \int dx_1 \, e^{iS_M}.$$

The same argument can be made for the other intermediate time coordinates $t_2, t_3, ...,$ and t_{n_t-1}. Associated with each time coordinate t_i, there is a corresponding integral over dx_i to specify the sum over the paths passing through the time t_i. We have therefore a direct transcription of the 'sum over paths' (10.4) in terms of an explicit $(n_t - 1)$-dimensional integral over the coordinates $x_1, x_2, ..., x_{n_t-1}$:

$$\begin{pmatrix} \text{sum over paths which go} \\ \text{from } (x_a, t_a) \text{ to} (x_b, t_b) \end{pmatrix} \rightarrow \int \prod_{i=1}^{n_t-1} dx_i \, e^{iS_M}.$$

From the above equation and Eq. (10.4), Feynman's path integral method gives the *probability amplitude* for the system to be located at x_a at time t_a and to end up at x_b at time t_b as given by

$$< x_b \mid e^{-iH(t_b-t_a)} \mid x_a > = \int \prod_{i=1}^{n_t-1} dx_i \; e^{iS_M(x_a,x_1,x_2,\ldots x_{n_t-1},x_b)}. \quad (10.5)$$

§10.2 Partition Function for a Single-Particle System

The partition function is an important quantity in statistical mechanics. From the partition function, many thermodynamic functions of the system can be obtained. The definition of the partition function and the path integral probability amplitude we just discussed are so similar that the partition function can be formulated in terms of a path integral. Such a representation of the partition function will facilitate the evaluation of the partition function by the Monte Carlo method.

We first study the partition function for a single-particle system. The *partition function* is the sum of the expectation value of the operator $e^{-\beta_B H}$, with the sum to be carried out over all states of the system [5]. Here, H is the Hamiltonian operator, $\beta_B = 1/kT$, k is the *Boltzmann* constant, and T is the temperature. In statistical mechanics involving strongly interacting particles, it is more convenient to absorb the Boltzmann constant k into the definition of the temperature T so that the temperature T is measured in energy units, (for example, in MeV or GeV), and β_B is just the inverse of the tempertaure,

$$\beta_B = 1/T. \quad (10.6)$$

When we choose the eigenstates to be diagonal in the coordinate x, the partition function is given by

$$Z = \sum_{x_a} < x_a | e^{-\beta_B H} | x_a > . \quad (10.7)$$

We can compare the above definition of the partition function (10.7) with the lefthand side of the path-integral probability amplitude Eq. (10.5). We see that they differ in three aspects. First, the operator in the partition function is of the form $e^{-\beta_B H}$ whereas the operator in the probability amplitude is of the form $e^{-iH(t_b-t_a)}$. To write the partition function in the form of a path integral, we introduce the imaginary time coordinate τ:

$$t = -i\tau, \quad (10.8a)$$

and allow τ to range from

$$\tau_a = 0 \quad \text{to go to} \quad \tau_b = \beta_B. \tag{10.8b}$$

By going to the domain of imaginary time, the operator in the partition function becomes identical to the operator in the path-integral method. The inverse of the temperature β_B then gives the interval of the 'time coordinate' τ for the problem in question.

Secondly, in the partition function, we evaluate the matrix element of the operator $e^{-\beta_B H}$. The state at the initial imaginary time τ_a is at x_a and the state at the final imaginary time τ_b is also at x_a. In contrast, in the path integral probability amplitude, the initial state at t_a is at x_a and the final state at t_b can be different. To write the partition function in the form of a path integral, we need to impose the *periodic boundary condition* that the configuration at the final imaginary time τ_b should be the same as the configuration at the initial imaginary time τ_a. As in Section 10.1, we likewise divide the imaginary time separation from τ_a to τ_b into n_t evenly-spaced intervals, with τ_a being the zeroth time coordinate and τ_b being the n_t-th time coordinate, then the discrete time coordinates are $\tau_0, \tau_1, \tau_2, ..., \tau_{n_t-1}$, and τ_{n_t}, with $\tau_0 = \tau_a$ and $\tau_{n_t} = \tau_b$. The periodic boundary condition becomes

$$x(\tau_{n_t}) = x(\tau_0). \tag{10.9}$$

Finally, in the partition function of Eq. (10.7) there is a summation over the states x_a, whereas the probability amplitude (10.5) involves only a single term. Accordingly, the partition function can be obtained from Eq. (10.5) by performing an additional summation (integration) over $x_a = x(\tau_0) = x(\tau_{n_t})$, which can be labeled x_{n_t}. From these considerations, the partition function for a single-particle system is given by

$$Z = \int dx_{n_t} \prod_{i=1}^{n_t-1} dx_i \; e^{iS_M|_{t=-i\tau}} \; = \; \int \prod_{i=1}^{n_t} dx_i \; e^{iS_M|_{t=-i\tau}}. \tag{10.10}$$

What does the Minkowski action S_M become under the transformation of t into $-i\tau$? We would like to consider the action

$$S_M = \int_{t_a}^{t_b} dt \; L(x, \dot{x}), \tag{10.11}$$

where

$$L(x, \dot{x}) = \left\{ \frac{m}{2} \left(\frac{dx}{dt} \right)^2 - V(x) \right\},$$

for a single-particle in a potential $V(x)$. When we make the transformation $t = -i\tau$, we obtain

$$iS_M \mid_{t=-i\tau} = \int_{\tau_a}^{\tau_b} d\tau \left\{ -\frac{m}{2} \left(\frac{dx}{d\tau} \right)^2 - V(x) \right\}.$$

In order to transform e^{iS_M} to e^{-S_E}, we introduce the *Euclidean action* S_E by the relation

$$iS_M \mid_{t=-i\tau} = -S_E. \tag{10.12}$$

The Euclidean action is then given explicitly by

$$S_E = \int_{\tau_a}^{\tau_b} d\tau \left\{ \frac{m}{2} \left(\frac{dx}{d\tau} \right)^2 + V(x) \right\}. \tag{10.13}$$

Note that the sign for the term $V(x)$ in the Euclidean action differs from that in the Minkowski action.

From Eqs. (10.10) and (10.12), the partition function is given in terms of the Euclidean action by

$$Z = \int \prod_{i=1}^{n_t} dx_i \, e^{-S_E(x_1, x_2, \ldots x_{n_t-1}, x_{n_t})}. \tag{10.14}$$

It is convenient to introduce the abbreviated symbol $\int [dx]$ to stand for the n_t-dimensional integration of the variables $x_1, x_2, \ldots, x_{n_t}$:

$$\int [dx] \equiv \int \prod_{i=1}^{n_t} dx_i.$$

The partition function is then given by

$$Z = \int [dx] \, e^{-S_E(x)}. \tag{10.15}$$

§10.3 Path Integral Method for a Scalar Field

We shall now generalize the above results for a single particle to the case for a scalar field ϕ. We can imagine that initially at the time t_a, the field configuration is given by $\phi(x, t_a)$. We would like to find out the probability amplitude that the field configuration is $\phi(x, t_b)$ at time t_b. Between t_a and t_b, there are many possible field configurations at the intermediate time steps. Each possible set of field configurations $\phi(x,t)$ for $t_a < t < t_b$ constitutes a 'path'. Corresponding to each path, there is a value of the *Minkowski action* S_M given by

$$S_M = \int_{t_a}^{t_b} dt \int dx \ \mathcal{L}[\phi(x,t), \partial_\mu \phi(x,t)], \tag{10.16}$$

where $\mathcal{L}$ is the Lagrangian density. The term 'the Lagrangian density' occurs so often in field theory that for brevity of notation, it is often called simply 'the Lagrangian' in the literature [6]. We shall likewise use the term 'the Lagrangian' to represent both 'the Lagrangian' $\int dx \mathcal{L}$ and 'the Lagrangian density' $\mathcal{L}$. Any ambiguity which may arise can easily be resolved from the context by dimensional analysis.

In order to distinguish the Lagrangian density defined in Eq. (10.16) from the Euclidean Lagrangian density which we shall later introduce in Eq. (10.30), we shall call the Lagrangian in the above equation the Minkowski Lagrangian $\mathcal{L}_M$ denoted with subscript M.

According to Feynman's path integral method, the probability amplitude for the field configuration to be $\phi(x,t_a)$ at t_a and to emerge as $\phi(x,t_b)$ at t_b is given by summing the phase factor e^{iS_M} over all paths:

$$\text{Amplitude } [\phi(x,t_a) \rightarrow \phi(x,t_b)] = \sum_{all \ paths} e^{iS_M}. \tag{10.17}$$

The probability amplitude can be expressed in a slightly different way in the quantum field description. We work with states for which the field operator $\widehat{\phi}(x,t_a)$ is diagonal,

$$\widehat{\phi}(x,t_a) \mid \phi(x,t_a) > = \phi(x,t_a) \mid \phi(x,t_a) > . \tag{10.18}$$

The initial state is then $\mid \phi(x,t_a) >$ and the evolution of the initial state from time t_a to t_b leads to the state vector

$$(\text{ state at time } t_b \ ) = e^{-iH(t_b-t_a)} \mid \phi(x,t_a) > .$$

The probability amplitude for the field configuration to be $\phi(x,t_a)$ at t_a and to emerge as $\phi(x,t_b)$ at t_b is given by

$$\text{Amplitude } [\phi(x,t_a) \rightarrow \phi(x,t_b)] = < \phi(x,t_b) \mid e^{-iH(t_b-t_a)} \mid \phi(x,t_a) > . \tag{10.19}$$

Thus, Eq. (10.17) can be expressed as

$$< \phi(x, t_b) \mid e^{-iH(t_b - t_a)} \mid \phi(x, t_a) >= \sum_{all \ paths} e^{iS_M}. \qquad (10.20)$$

How do we perform the 'sum over paths' for the right hand side of Eq. (10.20)? Again, we discretize the time coordinates between the initial time t_a and the end point time t_b into n_t equally-spaced time coordinates $t_0(= t_a), t_1, t_2, ..., t_{n_t-1}$, and $t_{n_t}(= t_b)$. The configuration of the field at the time t_j is specified by the field quantity $\phi(x, t_j)$ at all the spatial points x as in Fig. 10.2. It is convenient to discretize also the spatial dimension into n_x equally-spaced coordinates $x_1, x_2, x_3, ..., x_{n_x}$. Periodic boundary conditions are often imposed when we consider cases with an infinite spatial extension.

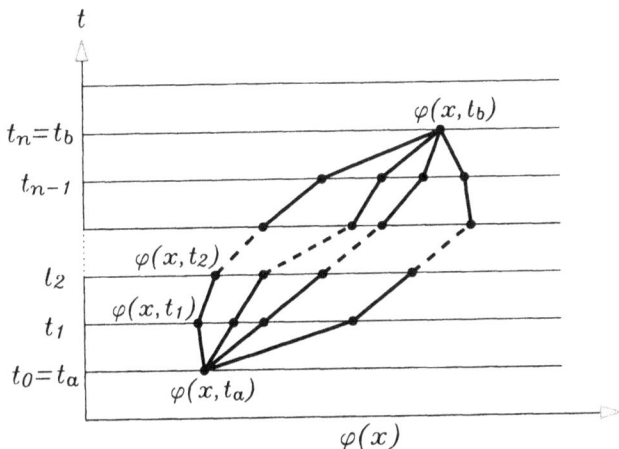

Fig. 10.2 Paths from the field configuration $\phi(x, t_a)$ at time t_a to the field configuration $\phi(x, t_b)$ at time t_b.

With the spatial points discretized, the field $\phi(x, t_j)$ at time t_j is specified by a set of n_x numbers: $\phi(x_1, t_j), \phi(x_2, t_j), ...\phi(x_{n_x}, t_j)$. Each of these field quantities is a degree of freedom. A 'path' from $\phi(x, t_a)$ at time t_a to $\phi(x, t_b)$ at time t_b consists of specifying all the field configurations $\phi(x, t)$ between the intial time t_a and the final time t_b. A variation of the path associated with the spatial coordinate x_i and the time coordinate t_j corresponds to a variation in the field quantity $\phi(x_i, t_j)$, which must be treated as a variable to generate the set of paths. The sum (10.20) over the paths associated with (x_i, t_j) becomes a single integral over all possible values of $\phi(x_i, t_j)$:

$$\left(\begin{array}{l} \text{sum over paths associated with } t_j \\ \text{and the spatial point } x_i \end{array} \right) \rightarrow \int d\phi(x_i, t_j) \, e^{iS_M}.$$

The sum (10.20) over all paths associated with t_j and all the spatial points $x_1, x_2, ..., x_{n_x}$ can be transcribed to be an n_x-dimensional integral over all possible values of the field configurations $\phi(x_i, t_j)$ at all the spatial points $x_1, x_2, x_3, ..., x_{n_x}$:

$$\left(\begin{array}{l} \text{sum over paths associated with } t_j \\ \text{and } all \text{ spatial points} \end{array} \right) \rightarrow \int \prod_{i=1}^{n_x} d\phi(x_i, t_j) \, e^{iS_M}.$$

The same argument can be made for the other time coordinates. When all the $n_t - 1$ time coordinates are included, there is a corresponding $(n_t - 1) \times n_x$-dimensional integral of the variables $\phi(x_1, t_1)$, $\phi(x_2, t_1), ..., \phi(x_{n_x}, t_{n_t-1})$ to specify the sum (10.20) over the paths passing through all the intermediate time coordinates $t_1, t_2, t_3, ..., t_{n_t-1}$:

$$\left(\begin{array}{l} \text{sum over paths which go} \\ \text{from } \phi(x, t_a) \text{ to } \phi(x, t_b) \end{array} \right) \rightarrow \int \prod_{j=1}^{n_t-1} \prod_{i=1}^{n_x} d\phi(x_i, t_j) \, e^{iS_M}.$$

Therefore, Feynman's path integral method gives the probability amplitude for the field configuration to be $\phi(x, t_a)$ at t_a and to emerge as $\phi(x, t_b)$ at t_b as

$$< \phi(x, t_b) \mid e^{-iH(t_b - t_a)} \mid \phi(x, t_a) > = \int \prod_{j=1}^{n_t-1} \prod_{i=1}^{n_x} d\phi(x_i, t_j) \, e^{iS_M}.$$

$$(10.21)$$

§10.4 Partition Function for a Scalar Field

We have shown how we can express the partition function for a single-particle system in terms of a path integral. In a similar way, we can express the partition function for the system of a scalar field in terms of a path integral. The basic idea which allows this representation is the similarity between the definition of the partition function and the path integral probability amplitude. The partition function is the sum of the expectation value of the operator $e^{-\beta_B H}$ over all states, where H is the Hamiltonian operator. Using states for which the field operator $\hat{\phi}(x)$ in the Heisenberg picture is diagonal, the partition function is given by

$$Z = \sum_{\phi(x)} < \phi(x)|e^{-\beta_B H}|\phi(x) > . \qquad (10.22)$$

To write the partition function in the form of a path integral, we need to introduce the imaginary time coordinate τ:

$$t = -i\tau. \qquad (10.23)$$

We allow τ to start at $\tau_a = 0$ to end at $\tau_b = \beta_B$. By going to the domain of imaginary time, the operator in the partition function in Eq. (10.22) becomes identical to the operator in the path-integral probability amplitude, Eq. (10.21).

In the definition of the partition function, we take the matrix element between two states which are characterised by the same field configuration, $\phi(x)$. In the language of the path integral method, the state at the initial imaginary time τ_a is $|\phi(x) >$ and the state at the final imaginary time τ_b is also $|\phi(x) >$. This requires that in order to represent the partition function as a path integral, we need to impose the periodic boundary condition

$$\phi(x, \tau_b) = \phi(x, \tau_a). \qquad (10.24)$$

As in Section 10.3, we discretize the imaginary time coordinate between the initial time τ_a and the final time τ_b into n_t equally-spaced intervals, with the discretized imaginary time coordinates given by $\tau_0(= \tau_a), \tau_1, \tau_2, ..., \tau_{n_t-1}, \tau_{n_t}(= \tau_b)$. A comparison of the definition of the partition function, Eq. (10.21), and the path integral probability amplitude, Eq. (10.22), shows that the partition function involves an additional summation over the field quantities at the time τ_0, $\phi(x, \tau_0)$, which is equal to $\phi(x, \tau_{n_t})$ due to the periodic boundary condition. Therefore, from Eqs. (10.21) and (10.22), we can write the partition function as

$$Z = \int d\phi(x_1, \tau_{n_t})...d\phi(x_{n_x}, \tau_{n_t}) \prod_{j=1}^{n_t-1} \prod_{i=1}^{n_x} d\phi(x_i, \tau_j) \, e^{iS_M|_{t=-i\tau}}$$

$$= \int \prod_{j=1}^{n_t} \prod_{i=1}^{n_x} d\phi(x_i, \tau_j) \, e^{iS_M|_{t=-i\tau}}. \qquad (10.25)$$

We need to write down the Minkowski action S_M under the transformation of t into $-i\tau$ for the above equation. We consider the following action for a scalar field ϕ:

$$S_M = \int dt dx \; \mathcal{L}_M(\phi(x,t), \partial_\mu \phi(x,t)), \qquad (10.26a)$$

where $\mathcal{L}_M$ is the Minkowski Lagrangian given by

$$\mathcal{L}_M[\phi(x,t), \partial_\mu \phi(x,t)] = \left\{ \frac{1}{2}\left(\frac{\partial \phi}{\partial t}\right)^2 - \frac{1}{2}\left(\frac{\partial \phi}{\partial x}\right)^2 - V(\phi)\right\}. \qquad (10.26b)$$

If we make the transformation $t = -i\tau$, then

$$iS_M \mid_{t=-i\tau} \; = \; \int_{\tau_a}^{\tau_b} d\tau dx \left\{ -\frac{1}{2}\left(\frac{\partial \phi}{\partial \tau}\right)^2 - \frac{1}{2}\left(\frac{\partial \phi}{\partial x}\right)^2 - V(\phi)\right\}.$$

Again, in order to transform e^{iS_M} to e^{-S_E}, we define the Euclidean action S_E by the relation

$$iS_M \mid_{t=-i\tau} \; = \; -S_E. \qquad (10.27)$$

Then the Euclidean action is given explicitly by

$$S_E = \int d\tau dx \left\{ \frac{1}{2}\left(\frac{\partial \phi}{\partial \tau}\right)^2 + \frac{1}{2}\left(\frac{\partial \phi}{\partial x}\right)^2 + V(\phi)\right\}.$$

This can be written as

$$S_E = \int d\tau dx \left\{ \frac{1}{2}\delta^{\mu\nu}\partial_\mu \phi \partial_\nu \phi + V(\phi)\right\}, \qquad (10.28)$$

which corresponds to using the Euclidean metric $\delta^{\mu\nu}$ for the Euclidean 'space-time' coordinates (τ, x, y, z). The quantity S_E is therefore called the *Euclidean action*. The sign for the term $V(x)$ in the Euclidean action (10.28) also differs from that in the Minkowski action.

Along with the Euclidean action, one can introduce the Euclidean Lagrangian (density) $\mathcal{L}_E$ by

$$S_E = \int d\tau dx \; \mathcal{L}_E(\phi(x,\tau), \partial_\mu \phi(x,\tau)). \qquad (10.29)$$

From this definition, the Euclidean Lagrangian $\mathcal{L}_E$ is related to the Minkowski Lagrangian $\mathcal{L}_M$ by

$$\mathcal{L}_E = -\mathcal{L}_M \mid_{t=-i\tau} . \qquad (10.30)$$

For the Minkowski Lagrangian $\mathcal{L}_M$ given by (10.26b), Eq. (10.30) gives the Euclidean Lagrangian $\mathcal{L}_E$ as

$$\mathcal{L}_E = \left\{ \frac{1}{2} \delta^{\mu\nu} \partial_\mu \phi \partial_\nu \phi + V(\phi) \right\},$$

which can also be obtained from (10.28). In the literature of lattice gauge calculations, the Lagrangian density employed is often implicitly taken to be the Euclidean Lagrangian density.

From Eqs. (10.25-27), the partition function is given in terms of the Euclidean action by

$$Z = \int \prod_{i=1}^{n_x} \prod_{j=1}^{n_t} d\phi(x_i, \tau_j) \, e^{-S_E(\phi)}. \qquad (10.31)$$

It is convenient to introduce the abbreviated symbol $\int [d\phi]$ to stand for the multiple dimensional integration of $d\phi(x_i, \tau_j)$:

$$\int [d\phi(x, \tau)] \equiv \int \prod_{i=1}^{n_x} \prod_{j=1}^{n_t} d\phi(x_i, \tau_j). \qquad (10.32)$$

Equation (10.31) can be abbreviated in the form:

$$Z = \int [d\phi(x, \tau)] e^{-S_E[\phi(x, \tau)]}. \qquad (10.31')$$

The partition function is the sum of the 'weighting factors' e^{-S_E}, evaluated for all field quantity $\phi(x, \tau)$, at all space-time points between $\tau = 0$ and $\tau = \beta_B$.

§10.5 Basic Quantities in Lattice Gauge Theory

The results obtained for the scalar field can be generalized to the gauge theory of quantum chromodynamics. One sets up a lattice of space-time points, with the time coordinate taken to be the imaginary time τ. For convenience of notation, we denote a lattice point by n, which is characterised by an ordered set of d numbers in a d-dimensional space-time coordinate system. The vectors between n and the adjacent lattice points in different space-time directions are labeled by μ, ν, κ, ... The lattice points adjacent to n are then $n + \mu$,

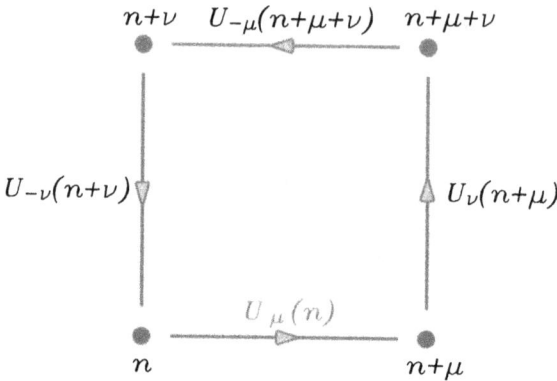

Fig. 10.3 Lattice points and the link variable U in a plaquette.

$n + \nu$, $n + \kappa$, ... For simplicity, the lattice spacings in all directions can be taken to have the same value, a. We show in Fig. 10.3 the points in the ν and μ directions.

A field configuration consists of specifying classically the independent fermion fields (quark fields), and the vector gauge field (gluon field) at the lattice points. The fermion field ψ and its conjugate momentum $\bar{\psi}(= \partial\mathcal{L}/\partial(\partial_0\psi))$ are independent. They are not ordinary numbers as they need to satisfy the fermion anticommutation relations. They can be represented by Grassmann variables which obey different rules for integration and differentiation (see Chapter 11, Supplement 11.2). Besides ψ and $\bar{\psi}$, it is necessary to include only the gauge field A_μ to specify the field configuration for lattice gauge theory as the degree of freedom involving the momentum conjugate of the gauge field can be effectively integrated out [7]. The gauge field $A_\mu(n)$ is a vector in configuration space with space-time components labeled by the subscript μ, with $\mu = 0, 1, 2$, and 3. The gauge field A_μ interacts with the fermion field through its coupling with the fermion current. The interaction Lagrangian is $\bar{\psi}\gamma^\mu A_\mu\psi$.

At each space-time point in the lattice, there are internal degrees of freedom, the color degrees of freedom. One associates a color coordinate frame of reference at each space-time point. The fermion field $\psi(n)$ at the lattice point n is represented by a 3-dimensional complex column vector in color space. One can also give a geometrical interpretation of the gauge field $A_\mu(n)$ as specifying, at the spatial location n, the degree of phase rotation for the fermion field in color space. [See the analogous geometrical interpretation of the gauge field $A_\mu(x)$ for two-dimensional quantum electrodynamics following Eqs. (6.17) and (6.20) in Chapter 6.] The interaction Lagrangian depends on this degree of rotation but should be independent of the orientation of the local frame of reference for the color coordinate. The property that the Lagrangian is independent of the orientation of the color frame of reference is a simple way of stating the *gauge invariance* of the Lagrangian. As one color frame of reference is related to another frame by a rotation operation, gauge invariance means that the Lagrangian is invariant under an arbitrary rotation in color space.

The QCD system we wish to study possesses $SU(3)_C$ symmetry. This means that the rotation operation in color space is described by an element of the $SU(3)$ group and that the gauge field $A_\mu(n)$ at the point n can in general be expanded in terms of the generators of the $SU(3)$ group.

We would like to review briefly the properties of the $SU(3)$ group, the *special unitary group in 3 dimensions*. An element U of the $SU(3)$ group is a 3×3 matrix which satisfies the unitary condition

$$U^\dagger U = U U^\dagger = 1, \tag{10.33a}$$

and the term 'special' signifies that the matrix U has a unit determinant,

$$\det U = 1. \tag{10.33b}$$

When we represent an element U of the $SU(3)$ group in the exponential form e^{iH}, it is easy to show below that the matrix H is a Hermitian matrix and the trace of H is zero.

•[Exercise 10.1
Show that an element U of the $SU(3)$ group can be represented by $U = e^{iH}$, where H is a 3×3 traceless Hermitian matrix.

◇Solution:
A group element of the $SU(3)$ is a 3×3 matrix. The matrix e^{iH}, which comes from the exponentiation of a 3×3 matrix H, is also a 3×3 matrix. One can introduce the matrix H such that a group element U of the $SU(3)$ group can be represented by e^{iH}. From the unitary property of U, one has

$$e^{iH} e^{-iH^\dagger} = 1.$$

Upon multiplication of the above equation by $e^{iH^\dagger}$ from the right, one obtains

$$e^{iH} = e^{iH^\dagger}$$

which implies that H satisfies

$$H = H^\dagger,$$

and H is therefore a Hermitian matrix.

We can show that the trace of the matrix H is zero by the requirement that the determinant of U is unity. To demonstrate this, we need to make use of the theorem that the product of the determinants of two square matrices is equal to the determinant of the product of the two matrices. This can be shown as follows. The product of the determinants of A and B is

$$
\begin{aligned}
(\det A)(\det B) &= \frac{1}{3!} \sum_{\alpha,\beta,\gamma\, a,b,c} \epsilon_{abc}^{\alpha\beta\gamma} A_{a\alpha} A_{b\beta} A_{c\gamma} \sum_{l,m,n} \epsilon_{123}^{lmn} B_{1l} B_{2m} B_{3n} \\
&= \frac{1}{3!} \sum_{\alpha,\beta,\gamma\, a,b,c\, l,m,n} \epsilon_{abc}^{lmn} A_{a\alpha} A_{b\beta} A_{c\gamma} B_{\alpha l} B_{\beta m} B_{\gamma n} \\
&= \frac{1}{3!} \sum_{a,b,c\, l,m,n} \epsilon_{abc}^{lmn} (AB)_{al} (AB)_{bm} (AB)_{cn} \\
&= \det (AB).
\end{aligned}
$$

Thus, the determinant of the product of matrices is equal to the product of the determinants of the matrices. Making use of this theorem, we have

$$
\det(e^{iH}) = \det(C e^{iH} C^{-1})
$$

The matrix on the righthand side of the above equation can be written in a different form. We have the equality

$$
\begin{aligned}
C e^{iH} C^{-1} &= C \sum_{n=0}^{\infty} \frac{(iH)^n}{n!} C^{-1} \\
&= \sum_{n=0}^{\infty} \frac{(iCHC^{-1})^n}{n!} \\
&= e^{iCHC^{-1}}.
\end{aligned}
$$

Therefore, we have

$$
\det(e^{iH}) = \det(e^{iCHC^{-1}}).
$$

The matrix C can be any matrix. In particular, we can always find a matrix C to diagonalize the matrix H. We then have

$$
\begin{aligned}
\det(e^{iH}) &= \det \begin{pmatrix} e^{i(CHC^{-1})_{11}} & 0 & 0 \\ 0 & e^{i(CHC^{-1})_{22}} & 0 \\ 0 & 0 & e^{i(CHC^{-1})_{33}} \end{pmatrix} \\
&= e^{i\, tr\{CHC^{-1}\}} \\
&= e^{i\, tr\{H\}}.
\end{aligned}
$$

Therefore, the requirement that the determinant of e^{iH} is unity implies

$$
tr\{H\} = 0.
$$

In general, one needs 18 real quantities to specify a 3×3 matrix containing complex number matrix elements. However, the Hermitian property of H reduces the number to 9 and the condition of a zero trace means that an ordered set of only eight real quantities is sufficient to specify the matrix H. One can introduce eight independent *generators*, each of which is a 3×3 traceless Hermitian matrix, and expands the traceless Hermitian matrix H in terms of these generators λ_i. One can choose these eight generators to be the 3×3 λ_i matrices of Gell-Mann [8], given explicitly by

$$\lambda_1 = \begin{pmatrix} 0 & 1 & 0 \\ 1 & 0 & 0 \\ 0 & 0 & 0 \end{pmatrix} \qquad \lambda_2 = \begin{pmatrix} 0 & -i & 0 \\ i & 0 & 0 \\ 0 & 0 & 0 \end{pmatrix}$$

$$\lambda_3 = \begin{pmatrix} 1 & 0 & 0 \\ 0 & -1 & 0 \\ 0 & 0 & 0 \end{pmatrix} \qquad \lambda_4 = \begin{pmatrix} 0 & 0 & 1 \\ 0 & 0 & 0 \\ 1 & 0 & 0 \end{pmatrix} \qquad (10.34)$$

$$\lambda_5 = \begin{pmatrix} 0 & 0 & -i \\ 0 & 0 & 0 \\ i & 0 & 0 \end{pmatrix} \qquad \lambda_6 = \begin{pmatrix} 0 & 0 & 0 \\ 0 & 0 & 1 \\ 0 & 1 & 0 \end{pmatrix}$$

$$\lambda_7 = \begin{pmatrix} 0 & 0 & 0 \\ 0 & 0 & -i \\ 0 & i & 0 \end{pmatrix} \quad \text{and} \quad \lambda_8 = \begin{pmatrix} \frac{1}{\sqrt{3}} & 0 & 0 \\ 0 & \frac{1}{\sqrt{3}} & 0 \\ 0 & 0 & -\frac{2}{\sqrt{3}} \end{pmatrix}.$$

Following conventional notation [9], when we expand a traceless Hermitian matrix H as a linear combination of the generators, we shall expand these matrices with respect to $\lambda_i/2$ (instead of λ_i) with coefficients H^i:

$$H = (\frac{\lambda_i}{2})H^i. \qquad (10.35a)$$

The $SU(3)$ element U associated with H can then be expressed as

$$U = \exp(i\frac{\lambda_i}{2}H^i). \qquad (10.35b)$$

⊕[Supplement 10.1
    One defines the scalar product of two vectors in the generators space as

$$(a,\ b) = tr\{ab\}.$$

One can form the scalar product among the generators in Eq. (10.34) and finds

$$\left(\lambda_i, \frac{\lambda_j}{2}\right) = \delta_{ij}.$$

Because of this relation, we have

$$H^i = (\lambda_i, H),$$

which is the reason to expand H as a linear combination of $\lambda_i/2$, instead of λ_i, in Eq. (10.35a).

$1 \oplus$

Equation (10.35) means that the set of eight λ_i matrices can be used to *generate* any element of the $SU(3)$ group. This set of eight λ_i matrices are therefore called the *generators of the $SU(3)$ group*. With these eight generators, an element U of the $SU(3)$ group can be specified by an ordered set of eight real numbers H^i with $i = 1, 2, ..8$, as given by Eq. (10.35).

As the QCD system we wish to study possesses $SU(3)_c$ symmetry, a rotation operation in color space is described by an element of the $SU(3)$ group. The gauge field $A_\mu(n)$ at the space-time point n specifies the degree of rotation of the fermion field operator in color space at that space-time point n. $A_\mu(n)$ is associated with an element of the $SU(3)$ group. Therefore, $A_\mu(n)$ can be expanded as a linear combination of the eight generators λ_i with real coefficients $A^i_\mu(n)$,

$$A_\mu(n) = \frac{\lambda_i}{2} A^i_\mu(n). \tag{10.36}$$

The gauge field $A_\mu(n)$ at the space-time point n can be considered as having eight degrees of freedom in color space with eight components $A^i_\mu(n)$, $i = 1, 2, ..., 8$.

One could use the gauge field components $A^i_\mu(n)$ as basic variables. However, in lattice gauge theory, it is more convenient to use the *link variable*, $U_\mu(n)$, instead of the gauge field variables $A^i_\mu(n)$. The advantage of the link variable is that it allows one to take into account explicitly the $SU(3)_c$ symmetry of the color degrees of freedom. Furthermore, the Euclidean action written in terms of the link variable is also quite simple. For this purpose, one chooses $U_\mu(n)$ to be an element of the $SU(3)_c$ group, which can be represented as a 3×3 $SU(3)$ matrix [1]. The link variable $U_\mu(n)$ is specified at the link *between* the lattice points n and $n + \mu$, and not at the lattice sites

n or $n + \mu$ (Fig. 10.3). Explicitly, it is defined in terms of the path integral of A_μ^i between n and $n + \mu$ as

$$U_\mu(n) = P \; \exp\left\{ ig \int_n^{n+\mu} A_\mu(x) dx^\mu \right\}, \qquad (10.37)$$

where P stands for *path ordering*, with operators occurring earlier along the path placed to the right of operators appearing later along the path, and g is the coupling constant. In the above equation, as an exception to the summation convention, no summation over μ is implied by the repeated indices of μ in $A_\mu(x) dx^\mu$.

It is instructive to clarify the differences between the gauge field variables A_μ in Minkowski space and in Euclidean space. The coordinates (x^0, x^1, x^2, x^3) in Minkowski space are (t, x, y, z), and the components (A_0, A_1, A_2, A_3) of the gauge field A_μ are (A_t, A_x, A_y, A_z). In Euclidean space, the coordinates (x^0, x^1, x^2, x^3) are (τ, x, y, z), and the gauge field components (A_0, A_1, A_2, A_3) are (A_τ, A_x, A_y, A_z). We choose the component A_t along the t direction in Minkowski space to be related to the component A_τ along the τ direction in Euclidean space by

$$A_t = i A_\tau. \qquad (10.38)$$

Then, because $t = -i\tau$, the line integral $\int A_0 dx^0$ of Eq. (10.37) is the same in Minkowski space and in Euclidean space:

$$\int A_0 dx^0 = \int A_t dt \quad (\text{ Minkowski space }),$$

$$= \int A_\tau d\tau \quad (\text{ Euclidean space }).$$

The spatial coordinates (x^1, x^2, x^3) and the gauge field components (A_1, A_2, A_3) are the same in Minkowski space and in Euclidean space. Therefore, the other line integrals $\int A_i dx^i$ with $i = 1, 2$, and 3 are also the same in Minkowski space and in Euclidean space.

The link variable $U_\mu(n)$ is associated with the point n. It points in the μ direction and links the point n to the adjacent point $n + \mu$. Linking the same two points but pointing in the opposite direction is the link variable $U_{-\mu}(n + \mu)$. It is easy to show (see Exercise 10.2 below) that the link variable $U_{-\mu}(n + \mu)$ is the inverse of $U_\mu(n)$:

$$U_{-\mu}(n + \mu) = U_\mu^{-1}(n). \qquad (10.39)$$

•[Exercise 10.2 :
Show Eq. (10.39) by using the definition of the link variable (10.37).

◇Solution:
We can write down explicitly the link variables $U_\mu(n)$ and $U_{-\mu}(n+\mu)$ in terms of
the gauge field variable A_μ along the path. We divide the path from n to $n+\mu$ into
N equally-spaced intervals with a small spacing $\epsilon = \mu/N$ such that $\epsilon << a$ where
a is the spacing between n and $n+\mu$. The link variable becomes

$$U_\mu(n) = P \ \exp\left\{ ig\epsilon \left[\frac{1}{2}A_\mu(n) + A_\mu(n+\epsilon) + ... + A_\mu(n+(N-1)\epsilon) + \frac{1}{2}A_\mu(n+N\epsilon) \right] \right\},$$
(1)

where we have used the Extended Trapezoidal Rule to write down the integral in
terms of a summation. Path ordering of the operators requires that those operators
appearing earlier along the path from n to $n+\mu$ are placed to the right of those
operators appearing later along the path. Therefore, by definition of path ordering,
Eq. (1) can be expressed as

$$U_\mu(n) = e^{ig\frac{\epsilon}{2}A_\mu(n+N\epsilon)} e^{ig\epsilon A_\mu(n+(N-1)\epsilon)} \cdots e^{ig\epsilon A_\mu(n+\epsilon)} e^{ig\frac{\epsilon}{2}A_\mu(n)}.$$
(2)

On the other hand, the link variable $U_{-\mu}(n+\mu)$ linking $n+\mu$ to n and pointing
from $n+\mu$ to n is defined, according to (10.37), by

$$U_{-\mu}(n+\mu) = P \ \exp\left\{ ig \int_{n+\mu}^{n} A_\mu(x)dx^\mu \right\}.$$

Writing out the path explicitly, we have

$$U_{-\mu}(n+\mu) = P \ \exp\{ -ig\epsilon[\frac{1}{2}A_\mu(n+N\epsilon) + ... + A_\mu(n+\epsilon) + \frac{1}{2}A_\mu(n)] \}$$
$$= e^{-ig\frac{\epsilon}{2}A_\mu(n)} e^{-ig\epsilon A_\mu(n+\epsilon)} ... e^{-ig\epsilon A_\mu(n+(N-1)\epsilon)} e^{-ig\frac{\epsilon}{2}A_\mu(n+N\epsilon)}.$$
(3)

Hence, by comparing the above equation with Eq. (2), we have

$$U_\mu(n)U_{-\mu}(n+\mu) = 1$$

and

$$U_{-\mu}(n+\mu) = U_\mu^{-1}(n). \qquad \qquad]•$$

The results of the last section for a scalar field indicate that the
partition function is obtained by summing the weighting factor e^{-S_E}
for all possible values of the field quantities at all space-time points
between $\tau = 0$ and $\tau = \beta$. The generalization of this result to the
gauge theory of QCD gives the partition function as the sum of the
weighting factor e^{-S_E}, for all values of the independent components

of the field quantities ψ, $\bar{\psi}$ and U, at all space-time points between $\tau = 0$ and $\tau = \beta$:

$$Z = \int [d\bar{\psi}] \, [d\psi] \, [dU] \, e^{-S_E}, \qquad (10.40a)$$

where for simplicity the abbreviation $[d\bar{\psi}] \, [d\psi] \, [dU]$ is used to represent the product of the differential elements at all space-time points,

$$[d\bar{\psi}] \, [d\psi] \, [dU] = \prod_n d\bar{\psi}(n) \, d\psi(n) \, dU_\mu(n), \qquad (10.40b)$$

and $d\psi(n)$, $d\bar{\psi}(n)$ and $dU_\mu(n)$ represent further the differential elements of the independent quantities characterizing $\psi(n)$, $\bar{\psi}(n)$ and $U_\mu(n)$ respectively. For example, $U_\mu(n)$ is a unitary 3×3 matrix with unit determinant. It is specified by nine independent quantities subject to the constraint that the determinant of the matrix is unity and $dU_\mu(n)$ represents a nine-dimensional differential element with a constraint.

The expectation value of an operator $\mathcal{O}$ is then given by

$$< \mathcal{O} >= \int [d\bar{\psi}] \, [d\psi] \, [dU] \, \mathcal{O} e^{-S_E}/Z, \qquad (10.41)$$

which allows the evaluation of many quantities of physical interest.

§10.6 The Lagrangian and the Euclidean Action for QCD

We would like to write down the Lagrangian and the Euclidean action appropriate for QCD. The type of theory we wish to construct for QCD is a non-Abelian gauge field theory. One can imagine that associated with every point in space-time, there is an additional, internal 3-dimensional coordinate frame, the color coordinate frame. The term "non-Abelian" refers to the property that the symmetry group which is associated with QCD in color space is a non-commutative group. The group elements do not commute and the generators of the group do not commute. The symmetry group for QCD in color space is generally taken to be the $SU(3)_C$ group so that a rotation in color space is carried out by using a unitary 3×3 $SU(3)$ matrix.

The field theory of QCD is a gauge field theory. This means that in the color space, QCD possesses the property that its Lagrangian is invariant under a rotation of the color coordinate frame of reference, with the amount of color frame rotation allowed to be arbitrary at different space-time points. The property that the Lagrangian is invariant under arbitrary color frame rotations at different space-time points is called *local gauge invariance*. The term 'local' here refers to different color frame rotations at different space-time points, in contrast to 'global' rotations in which the amount of rotation are the same at all space-time points.

We fix our attention at the space-time point n and consider the consequence of a gauge transformation, which is an arbitrary rotation of the color coordinate frame at n. As the gauge theory of QCD possesses $SU(3)_c$ symmetry, an arbitrary rotation in color space can be carried out by the operation of a unitary 3×3 $SU(3)$ matrix $G(n)$. In accordance with Eq. (10.35b), the unitary matrix $G(n)$ can be generated by using the generators λ_i and an ordered set of eight real numbers $\chi_i(n)$ with $i = 1, 2, ..., 8$,

$$G(n) = \exp[i\frac{\lambda_i}{2}\chi_i(n)]. \tag{10.42}$$

Treating the set of eight generators λ_i as components of a vector $\boldsymbol{\lambda}$ and the the set of eight real numbers $\chi_i(n)$ as components of a vector $\boldsymbol{\chi}(n)$, the above equation can be written in the form:

$$G(n) = \exp[i\frac{\boldsymbol{\lambda}}{2} \cdot \boldsymbol{\chi}(n)]. \tag{10.42'}$$

Under the rotation of the color coordinate frame by an arbitrary 'angle' $\boldsymbol{\chi}(n)$, a vector V in color space will be transformed into $G(n)V$.

The fermion field $\psi(n)$ at the space-time point n is a vector in color space which can be represented as a three component column matrix. Accordingly, under the rotation of the color coordinate frame by an arbitrary angle $\boldsymbol{\chi}(n)$, the fermion field $\psi(n)$ is transformed to $\psi'(n)$ according to

$$\psi(n) \rightarrow \psi'(n) = G(n)\psi(n). \tag{10.43a}$$

The fermion field $\bar{\psi}(n)$ is transformed to $\bar{\psi}'(n)$

$$\bar{\psi}(n) \rightarrow \bar{\psi}'(n) = \bar{\psi}(n)G^{-1}(n). \tag{10.43b}$$

Eqs. (10.42-43) give the rules of gauge transformation for $\psi(n)$ and $\bar{\psi}(n)$. To determine the rule of gauge transformation for the gauge field $A_\mu(n)$, we follow the procedure familiar in Abelian gauge field theory. We define a gauge covariant-derivative D_μ to be applied to the fermion field ψ:

$$D_\mu \psi = (\partial_\mu - ig A_\mu)\psi. \tag{10.44}$$

The 'minimal-coupling' term $i\bar{\psi}\gamma^\mu D_\mu \psi$ can be introduced in the Lagrangian to represent the kinetic energy part of the fermion field and the coupling of the fermion field ψ with the gauge field A_μ. We require that this term $i\bar{\psi}\gamma^\mu D_\mu \psi$ be invariant under local gauge transformations (10.42-43). This requirement leads to the rule of gauge transformation for the gauge field A_μ given by

$$A_\mu(n) \rightarrow A'_\mu(n) = G(n)A_\mu(n)G^{-1}(n) - \frac{i}{g}[\partial_\mu G(n)]G^{-1}(n). \tag{10.45}$$

• [Exercise 10.3 :
Show that the rule of gauge transformation for the gauge field A_μ follows from the invariance of the term $i\bar{\psi}\gamma^\mu D_\mu \psi$ under the local gauge transformation (10.42-43).

◇Solution:
Under the gauge transformation (10.42-43), if $D_\mu \psi$ transforms in the same way as ψ, that is , if

$$\{D_\mu \psi\} \rightarrow \{D_\mu \psi\}' = G\{D_\mu \psi\}, \tag{1}$$

then $i\bar{\psi}\gamma^\mu D_\mu \psi$ is invariant under the gauge transformation. This requirement can be explicitly written out. We focus our attention at the point n to write out the content of Eq. (1) explicitly:

$$\{\partial_\mu - ig A'_\mu(n)\}\psi'(n) = G(n)\{\partial_\mu - ig A_\mu(n)\}\psi(n). \tag{2}$$

Substituting Eq. (10.43a) into the above equation, we have

$$\{\partial_\mu - ig A'_\mu(n)\}G(n)\psi(n) = G(n)\{\partial_\mu - ig A_\mu(n)\}\psi(n).$$

We therefore have

$$\{G(n)\partial_\mu + [\partial_\mu G(n)] - ig A'_\mu(n)G(n)\}\psi(n) = G(n)\{\partial_\mu - ig A_\mu(n)\}\psi(n).$$

This relation holds true for all arbitrary ψ. Hence, we have

$$\{[\partial_\mu G(n)] - ig A'_\mu(n)G(n)\} = G(n)\{-ig A_\mu(n)\}.$$

Multiplication with $G^{-1}(n)$ from the right gives the rule of gauge transformation for the gauge field:

$$A_\mu(n) \rightarrow A'_\mu(n) = G(n)A_\mu(n)G^{-1}(n) - \frac{i}{g}[\partial_\mu G(n)]G^{-1}(n). \qquad]•$$

Besides the term $i\bar{\psi}\gamma^\mu D_\mu\psi$, the QCD Lagrangian should contain a term which depends on the mass m of the fermion. The fermion mass term is simply $-m\bar{\psi}\psi$, which is gauge invariant, as one can easily see from the definition of the gauge transformation (10.43). (When the flavor degrees of freedom are taken into account explicitly, this term should be generalized to include different masses m_f for fermions with different flavors.) The fermion part of the (Minkowski) Lagrangian $\mathcal{L}_{Mq}$ is therefore [6]

$$\mathcal{L}_{Mq} = i\bar{\psi}\gamma^\mu D_\mu\psi - m\bar{\psi}\psi. \qquad (10.46)$$

In addition to $\mathcal{L}_{Mq}$, the Lagrangian should contain a term which depends on the field strength tensor $F_{\mu\nu}$. To express $F_{\mu\nu}$ as a function of A_μ, we again follow a procedure similar to that in Abelian gauge field theory. We construct the operator $F_{\mu\nu}$ as the commutator of the gauge-covariant derivatives D_μ and D_ν :

$$-igF_{\mu\nu} = D_\mu D_\nu - D_\nu D_\mu. \qquad (10.47a)$$

The type of gauge field theory we wish to construct for QCD is a non-Abelian gauge field theory. The gauge field operators A_μ and A_ν do not commute when $\mu \neq \nu$. After substituting the gauge-covariant derivatives into the above equation, we obtain

$$F_{\mu\nu} = \partial_\mu A_\nu - \partial_\nu A_\mu - ig[A_\mu, A_\nu]. \qquad (10.47b)$$

To see how the field strength tensor $F_{\mu\nu}$ transforms under a gauge transformation, we first determine the gauge transformation of the covariant derivative D_μ. From Eq. (1) of Exercise (10.3), we have the following property for the gauge-covariant derivative D_μ under a gauge transformation:

$$\begin{aligned} D_\mu(n)\psi(n) \rightarrow\ & [D_\mu(n)\psi(n)]' \\ & = D'_\mu(n)\psi'(n) \\ & = D'_\mu(n)G(n)\psi(n). \end{aligned}$$

On the other hand, Eq. (2) of Exercise 10.3 gives

$$[D_\mu(n)\psi(n)]' = G(n)D_\mu(n)\psi(n).$$

Therefore, under the gauge transformation (10.42-43), D_μ transforms as

$$D'_\mu(n) = G(n)D_\mu(n)G^{-1}(n). \qquad (10.48)$$

With this transformation property for D_μ, the field strength tensor $F_{\mu\nu}$ at the point n transforms as

$$F_{\mu\nu}(n) \rightarrow F'_{\mu\nu}(n) = G(n)F_{\mu\nu}G^{-1}(n), \tag{10.49}$$

and the quantity $tr\{F_{\mu\nu}F^{\mu\nu}\}$ transforms as

$$
\begin{aligned}
tr\{F_{\mu\nu}F^{\mu\nu}\} \rightarrow \ & tr\{F'_{\mu\nu}F^{\mu\nu\prime}\} \\
= \ & tr\{G(n)F_{\mu\nu}G^{-1}(n)G(n)F^{\mu\nu}G^{-1}(n)\} \\
= \ & tr\{F_{\mu\nu}G^{-1}(n)G(n)F^{\mu\nu}G^{-1}(n)G(n)\} \\
= \ & tr\{F_{\mu\nu}F^{\mu\nu}\} \qquad \text{(no sum over } \mu \text{ or } \nu) ,
\end{aligned}
$$

where we have used the cyclic property of the trace of matrices. Thus, the quantity $tr\{F_{\mu\nu}F^{\mu\nu}\}$ is invariant under a local gauge transformation. It can be used for the gauge field strength term in the QCD Lagrangian.

We have defined $F_{\mu\nu}$ in terms of A_μ, which can be expanded as a linear combination of the generators λ^i of the $SU(3)_c$ group with coefficients A^i_μ as in Eq. (10.36). We can likewise expand $F_{\mu\nu}$ as a linear combination of the generators of the $SU(3)_c$ group with coefficients $F^i_{\mu\nu}$ by the definition:

$$F_{\mu\nu} = \frac{\lambda_i}{2}F^i_{\mu\nu}. \tag{10.50}$$

Then, in terms of A^i_μ, the field strength tensor $F^i_{\mu\nu}$ is

$$F^i_{\mu\nu} = \partial_\mu A^i_\nu - \partial_\nu A^i_\mu + g f_{ijk}A^j_\mu A^k_\nu, \tag{10.51}$$

where we have used the commutator for the Gell-Mann λ matrices:

$$[\frac{\lambda_j}{2}, \frac{\lambda_k}{2}] = i f_{ijk}\frac{\lambda_i}{2}, \tag{10.52}$$

and f_{ijk}'s are the structure constants of $SU(3)$ which are tabulated in Ref. [6]. By using the relation for the trace of the λ matrices :

$$tr\{\frac{\lambda_i}{2}\frac{\lambda_j}{2}\} = \frac{1}{2}\delta_{ij}, \tag{10.53}$$

the quantity $tr\{F_{\mu\nu}F^{\mu\nu}\}$ can be rewritten as

$$tr\{F_{\mu\nu}F^{\mu\nu}\} = \frac{1}{2}F^i_{\mu\nu}F^{\mu\nu,i} \quad \text{(no sum over } \mu \text{ or } \nu). \tag{10.54}$$

As the left-hand side quantity is a gauge invariant quantity, the right-hand side quantity is also gauge-invariant.

With the fermion part of the (Minkowski) Lagrangian as given by Eq. (10.46), the gauge-invariant QCD (Minkowski) Lagrangian is therefore

$$\mathcal{L}_M = i\bar\psi\gamma^\mu D_\mu\psi - m\bar\psi\psi - \frac{1}{4}F^i_{\mu\nu}F^{\mu\nu,i} \quad \text{(sum over repeated indices)}$$

(10.55)

where the coefficient of the $F^i_{\mu\nu}F^{\mu\nu,i}$ term is chosen by analogy with the Abelian gauge theory of QED.

We shall now study how we can construct a gauge-invariant Euclidean action for QCD. As in previous considerations, we can associate a color coordinate frame with each space-time point n. Gauge invariance here means that the action is independent of the angle of phase rotation of the color coordinate frame, which is allowed to be different at different space-time points. For this purpose, we need to know the law of gauge transformation for the link variable $U_\mu(n)$. The link variable is transformed as

$$U_\mu(n) \rightarrow G(n+\mu)U_\mu(n)G^{-1}(n).$$

(10.56)

• [Exercise 10.4
Show Eq. (10.56) by using the definition of the link variable (10.37).

◇Solution:
From the definition of the link variable (10.37), we can write down explicitly the link variable $U_\mu(n)$ in terms of the gauge field variable $A_\mu(x)$ along the path from n to $n+\mu$. We divide the path from n to $n+\mu$ into N equally-spaced intervals with a small spacing $\epsilon = \mu/N$ such that $\epsilon \ll a$. Following the arguments presented in Exercise 10.2, the path-ordered product for $U_\mu(n)$ is given by Eq. (2) of Exercise 10.2:

$$U_\mu(n) = e^{ig\frac{\epsilon}{2}A_\mu(N)}e^{ig\epsilon A_\mu(N-1)}\dots e^{ig\epsilon A_\mu(1)}e^{ig\frac{\epsilon}{2}A_\mu(0)}.$$

(1)

where for simplicity of notation, we have abbreviated the location label $(n+j\epsilon)$ by (j) and $A_\mu(n+j\epsilon)$ by $A_\mu(j)$. Under a gauge transformation (10.43) and (10.45), we have

$$e^{ig\epsilon A_\mu(j)} \rightarrow e^{ig\epsilon A'_\mu(j)}.$$

We consider the division of the interval between n and $n+\mu$ to be so fine that ϵ is infinitesimally small. We then have

$$e^{ig\epsilon A_\mu(j)} \approx 1 + ig\epsilon A_\mu(j),$$

and similarly

$$e^{ig\epsilon A'_\mu(j)} \approx 1 + ig\epsilon A'_\mu(j).$$

Under a gauge transformation, the gauge field transforms as Eq. (10.45) :

$$A_\mu(j) \rightarrow A'_\mu(j) = G(j)A_\mu(j)G^{-1}(j) - \frac{i}{g}[\partial_\mu G(j)]G^{-1}(j).$$

Therefore, we have

$$e^{ig\epsilon A'_\mu(j)} = 1 + ig\epsilon A'_\mu(j) + O(\epsilon^2)$$

$$= 1 + ig\epsilon\{G(j)A_\mu(j)G^{-1}(j) - \frac{i}{g}[\partial_\mu G(j)]G^{-1}(j)\} + O(\epsilon^2).$$

On the other hand, we have

$$G(j+1)e^{ig\epsilon A_\mu(j)}G^{-1}(j) = [G(j) + \epsilon\partial_\mu G(j)](1 + ig\epsilon A_\mu(j))G^{-1}(j) + O(\epsilon^2)$$

$$\sim e^{ig\epsilon A'_\mu(j)}.$$

Applying this relation to Eq. (1), we obtain the gauge transformation of the link variable

$$U_\mu(n) \rightarrow G(n+\mu)U_\mu(n)G^{-1}(n).$$]•

We consider a path-ordered product of the directed link variables along an elementary lattice square in Fig. 10.3

$$P_\square(n;\mu\nu) = U_{-\nu}(n+\nu)U_{-\mu}(n+\mu+\nu)U_\nu(n+\mu)U_\mu(n). \quad (10.57)$$

Here, one starts at the point n which is connected to the point $n+\mu$ by the link variable $U_\mu(n)$. The point at $n+\mu$ is in turn linked to the point $n+\mu+\nu$ by the link variable $U_\nu(n+\mu)$. The point $n+\mu+\nu$ is linked to the point $n+\nu$ by the link variable $U_{-\mu}(n+\mu+\nu)$. Finally, the lattice point $n+\mu+\nu$ is linked to the original point n by the link variable $U_{-\mu}(n+\nu)$. An elementary lattice square connected by four directed link variables is called a *plaquette* (Fig. 10.3). Associated with a plaquette is the product $P_\square(n,\mu\nu)$, the path-ordered product of four link variables along its four sides.

It is easy to see that under the gauge transformation (10.56) the product $P_\square(n;\mu\nu)$ in Eq. (10.57) along an elementary plaquette transforms into $P'_\square(n;\mu\nu)$ where

$$P'_\square(n;\mu\nu) = G(n)U_{-\nu}(n+\nu)G^{-1}(n+\nu)$$
$$\times G(n+\nu)U_{-\mu}(n+\mu+\nu)G^{-1}(n+\mu+\nu)$$
$$\times G(n+\mu+\nu)U_\nu(n+\mu)G^{-1}(n+\mu)G(n+\mu)U_\mu(n)G^{-1}(n)$$
$$= G(n)U_{-\nu}(n+\nu)U_{-\mu}(n+\mu+\nu)U_\nu(n+\mu)U_\mu(n)G^{-1}(n)$$
$$= G(n)P_\square(n;\mu\nu)G^{-1}(n).$$

Clearly, if we take the trace of $P'_\Box(n; \mu\nu)$ in the above equation and the trace $P_\Box(n; \mu\nu)$ in Eq. (10.57), we obtain

$$tr\{P'_\Box(n; \mu\nu)\} = tr\{P_\Box(n; \mu\nu)\}. \tag{10.58}$$

Thus, the trace of the product of the four link variables, $tr\{P_\Box(n; \mu\nu)\}$, is invariant under a gauge transformation. It can be used as the gauge field term for the lattice action [1]. Therefore, the part of the Euclidean action involving only the gauge field is given by

$$S_{EA} = -\frac{1}{2g^2} \sum_{n,\{\mu,\nu\}} tr\{U_{-\nu}(n+\nu)U_{-\mu}(n+\mu+\nu)U_\nu(n+\mu)U_\mu(n)\} + h.c.$$

$$\tag{10.59}$$

The summation is carried out over all lattice points and all combinations of $\{\mu, \nu\}$ directions. This summation is the same as the summation over all elementary plaquettes. It can easily be shown (in Exercise 10.5) that for classical, smooth fields, the above action, with the particular choice of the coefficient $1/2g^2$, reduces to the gauge field part of the action in the continuum description of a non-Abelian gauge field (Eq. (10.55)).

•⟦Exercise 10.5
By using the Taylor expansion for the link variable U about the point n, show that for a small value of the coupling constant and/or a small lattice spacing, the Euclidean action Eq. (10.59) leads to the same action as in the continuum theory Eq. (10.53).

◇Solution:
We study the plaquette variable (10.57) about the point n, with the directions μ and ν

$$P_\Box(n; \mu\nu) = U_{-\nu}(n + \nu)U_{-\mu}(n + \mu + \nu)U_\nu(n + \mu)U_\mu(n).$$

Using the definition of the link variable and the trapezoidal rule for integration in a small interval a

$$\int_n^{n+\mu} A_\mu(x)dx^\mu = \frac{a}{2}\{A_\mu(n) + A_\mu(n + \mu)\},$$

we can write down the link variable $U_\mu(n)$ as

$$\begin{aligned} U_\mu(n) &= P \ \exp(ig \int_n^{n+\mu} A_\mu(x)dx^\mu) \\ &= P \ \exp(ig\frac{a}{2}\{A_\mu(n) + A_\mu(n + \mu)\}) \\ &= \exp(ig\frac{a}{2}A_\mu(n + \mu)) \ \exp(ig\frac{a}{2}A_\mu(n)). \end{aligned}$$

10. Lattice Gauge Theory

Expanding the gauge field $A_\mu(n + \mu)$ about $A_\mu(n)$, we have

$$U_\mu(n) = \exp(ig\frac{a}{2}\{A_\mu(n) + a\partial_\mu A_\mu(n)\})\ \exp(ig\frac{a}{2}A_\mu(n))$$

$$= \exp(ig\frac{a}{2}\{2A_\mu(n) + a\partial_\mu A_\mu(n)\} + O(a^3)).$$

Here in the above equation and in this Exercise, as an exception to the summation convention, no summation is implied by repeated indices. The other link variables in the plaquette can be similarly expanded to give

$$U_\nu(n + \mu) = \exp(ig\frac{a}{2}\{A_\nu(n + \mu) + A_\nu(n + \mu + \nu)\} + O(a^3))$$

$$= \exp(ig\frac{a}{2}\{2A_\nu(n) + 2a\partial_\mu A_\nu(n) + a\partial_\nu A_\nu(n)\} + O(a^3)),$$

$$U_{-\mu}(n + \mu + \nu) = \exp(-ig\frac{a}{2}\{A_\mu(n + \nu) + A_\mu(n + \mu + \nu)\} + O(a^3))$$

$$= \exp(-ig\frac{a}{2}\{2A_\mu(n) + 2a\partial_\nu A_\mu(n) + a\partial_\mu A_\mu(n)\} + O(a^3)),$$

and

$$U_{-\nu}(n + \nu) = \exp(-ig\frac{a}{2}\{A_\nu(n) + A_\nu(n + \nu)\} + O(a^3))$$

$$= \exp(-ig\frac{a}{2}\{2A_\nu(n) + a\partial_\nu A_\nu(n) + O(a^3)).$$

Using the formula

$$e^x e^y = e^{x+y+\frac{1}{2}[x,y]+\cdots},$$

we obtain

$$U_\nu(n+\mu)U_\mu(n) = \exp(-ig\frac{a}{2}\{2A_\mu(n)+2A_\nu(n)+2a\partial_\mu A_\nu(n)+a\partial_\mu A_\mu(n)+a\partial_\nu A_\nu(n)\}$$

$$-\frac{a^2g^2}{2}[A_\nu(n), A_\mu(n)] + O(a^3))$$

and

$$U_{-\nu}(n+\nu)U_{-\mu}(n+\mu+\nu)=\exp(-ig\frac{a}{2}\{2A_\mu(n)+2A_\nu(n)+2a\partial_\nu A_\mu(n)+a\partial_\mu A_\mu(n)$$

$$+a\partial_\nu A_\nu(n)\} - \frac{a^2g^2}{2}[A_\nu(n), A_\mu(n)] + O(a^3)).$$

Combining the above two equations, we have

$$U_{-\nu}(n + \nu)U_{-\mu}(n + \mu + \nu)U_\nu(n + \mu)U_\mu(n)$$

$$= \exp(ia^2g\{\partial_\mu A_\nu - \partial_\nu A_\mu\} - a^2g^2[A_\nu, A_\mu] + O(a^3))$$

Using the results of Eq. (10.47b) for the gauge field strength tensor, we have

$$P_\square(n; \mu\nu) = U_{-\nu}(n + \nu)U_{-\mu}(n + \mu + \nu)U_\nu(n + \mu)U_\mu(n) = \exp(ia^2gF_{\mu\nu}).$$

Therefore, we can take the trace of the above equation. We find

$$tr\{P_\square(n;\mu\nu)\} = tr\{UUUU\} = tr\{\exp(ia^2 g F_{\mu\nu})\}.$$

In the continuum limit for which $a^2 g F_{\mu\nu}$ is small, one can expand the exponential factor on the right hand side of the above equation. We have

$$tr\{P_\square(n;\mu\nu)\} = tr\{1 + ia^2 g F_{\mu\nu} - \frac{a^4 g^2}{2} F_{\mu\nu} F^{\mu\nu}\} \qquad \text{(no sum over } \mu \text{ or } \nu). \qquad (1)$$

The trace of the unit matrix gives a constant value which is not important for the action. The trace of the term linear in $F_{\mu\nu} = (\lambda_i/2)F^i_{\mu\nu}$ gives a value of zero, because the trace of the generators λ_i of the $SU(3)$ group is zero. The trace of the term quadratic in $F_{\mu\nu}$ gives

$$tr\{F_{\mu\nu} F^{\mu\nu}\} = \frac{1}{2} F^i_{\mu\nu} F^{\mu\nu,i} \qquad \text{(no sum over } \mu \text{ or } \nu).$$

Therefore, we can write down the gauge field portion of the Euclidean action as

$$
\begin{aligned}
S_{EA} &= -\frac{1}{2g^2} \sum_{n,\mu,\nu} tr\{P_\square(n;\mu\nu)\} + h.c. \\
&= -\frac{1}{2g^2} \int \frac{d\tau d^3 x}{a^4} (-\frac{a^4 g^2}{2}) \frac{1}{2} \sum_{\mu,\nu} F^i_{\mu\nu} F^{\mu\nu,i} \times 2 \\
&= \frac{1}{4} \int d\tau d^3 x \sum_{\mu,\nu} F^i_{\mu\nu} F^{\mu\nu,i}.
\end{aligned}
$$

By the definition of Eq. (10.29), the gauge field part of the Euclidean Lagrangian obtained from the above Euclidean action is

$$\mathcal{L}_{EA} = \frac{1}{4} \sum_{\mu,\nu} F^i_{\mu\nu} F^{\mu\nu,i}. \qquad (2)$$

On the other hand, if one starts from the continuum theory in Minkowski space, the gauge field part of the Minkowski Lagrangian, as given by Eq. (10.55), is

$$\mathcal{L}_{MA} = -\frac{1}{4} \sum_{\mu,\nu} F^i_{\mu\nu} F^{\mu\nu,i}, \qquad (3)$$

where we have written out explicitly the summation over μ and ν. From this Minkowski Lagrangian in the continuum theory, we can get the corresponding Euclidean Lagrangian by using the relation (10.30),

$$\mathcal{L}_E = -\mathcal{L}_M \mid_{t=-i\tau}.$$

We find then that the Euclidean Lagrangian deduced from the continuum theory is

$$\mathcal{L}_{EA} = \frac{1}{4} \sum_{\mu,\nu} F^i_{\mu\nu} F^{\mu\nu,i},$$

which is identical to Eq. (2). Thus, for a small value of the coupling constant and/or a small lattice spacing, the Euclidean Lagrangian obtained from the continuum theory is identical to the Euclidean Lagrangian [Eq. (2)] obtained from the action (10.59) of the lattice gauge theory. One notes here that the factor g^{-2} in Eq. (10.59) arises from the fact that $tr\{UUUU\}$ is proportional to $g^2 F_{\mu\nu} F^{\mu\nu}$ and the factor g^{-2} is needed to cancel the g^2 factor in $g^2 F_{\mu\nu} F^{\mu\nu}$ in the expansion of $\exp(iag F_{\mu\nu})$ of Eq. (1).   ▮●

§10.7 Euclidean Space and Minkowski Space Compared and Contrasted

In lattice gauge theory with fermions, it is convenient to work with γ matrices whose anticommutation relations are related to the Euclidean metric tensor $\delta^{\mu\nu}$ with diagonal elements $(1,1,1,1)$. These γ matrices should be distinguished from the usual γ matrices whose anticommutation relations are related to the Minkowski metric tensor $g^{\mu\nu}$ with diagonal elements $(1,-1,-1,-1)$. To make the distinction explicit, we shall call these matrices the Euclidean γ_E matrices with a subscript E. They obey the anticommutation relation

$$\{\gamma^\mu_E, \gamma^\nu_E\}_+ = 2\delta^{\mu\nu}, \tag{10.60}$$

and they are related to the Minkowski γ matrices by

$$\gamma^0_E = \gamma^0, \tag{10.61a}$$

and

$$\gamma^i_E = -i\gamma^i, \quad \text{for} \quad i = 1, 2, 3. \tag{10.61b}$$

In terms of the Euclidean γ_E matrices, the operator $\gamma^\mu p_\mu$, which appears in the Minkowski Lagrangian, becomes

$$\gamma^\mu p_\mu = \not{p} = -\gamma^0_E \frac{\partial}{\partial \tau} - \gamma^1_E \partial_1 - \gamma^2_E \partial_2 - \gamma^3_E \partial_3.$$

Thus, in the Euclidean space of (τ, x, y, z), the Minkowski operator $\gamma^\mu p_\mu$ becomes

$$\gamma^\mu p_\mu = -\gamma^\mu_E \partial_\mu. \tag{10.62}$$

It is instructive to compare and contrast various quantities in Minkowski space and in Euclidean space in Table 10.1.

Table 10.1

Dynamical variables in Minkowski space and Euclidean space compared and contrasted.

	Minkowski space	Euclidean space	Relation
Coordinates (x^0, x^1, x^2, x^3)	(t, x, y, z)	(τ, x, y, z)	$t = -i\tau$
Volume element d^4x	$dt\, dx\, dy\, dz$	$d\tau\, dx\, dy\, dz$	$dt = -id\tau$
Line element ds^2	$g_{\mu\nu}dx^\mu dx^\nu$	$\delta_{\mu\nu}dx^\mu dx^\nu$	
Metric tensor	$g_{\mu\nu} = \begin{pmatrix} 1 & 0 & 0 & 0 \\ 0 & -1 & 0 & 0 \\ 0 & 0 & -1 & 0 \\ 0 & 0 & 0 & -1 \end{pmatrix}$	$\delta_{\mu\nu} = \begin{pmatrix} 1 & 0 & 0 & 0 \\ 0 & 1 & 0 & 0 \\ 0 & 0 & 1 & 0 \\ 0 & 0 & 0 & 1 \end{pmatrix}$	
Derivatives ∂_μ	$(\partial_t, \partial_x, \partial_y, \partial_z)$	$(\partial_\tau, \partial_x, \partial_y, \partial_z)$	$\partial_t = i\partial_\tau$
Gauge field (A_0, A_1, A_2, A_3)	(A_t, A_x, A_y, A_z)	(A_τ, A_x, A_y, A_z)	$A_t = iA_\tau$
Gamma matrices γ^μ	$\gamma^0, \gamma^1, \gamma^2, \gamma^3$ $\{\gamma^\mu, \gamma^\nu\} = 2g^{\mu\nu}$	$\gamma_E^0, \gamma_E^1, \gamma_E^2, \gamma_E^3$ $\{\gamma_E^\mu, \gamma_E^\nu\} = 2\delta^{\mu\nu}$	$\gamma^0 = \gamma_E^0$ $\gamma^j = i\gamma_E^j$ $j=1,2,3$
Action (in continuum limit)	$S_M = \int dtdx \mathcal{L}_M$	$S_E = \int d\tau dx \mathcal{L}_E$	$iS_M = -S_E$ $\mathcal{L}_M = -\mathcal{L}_E$
Lagrangian (scalar field)	$\mathcal{L}_M = \frac{1}{2}[(\partial_t\phi)^2 - |\nabla\phi|^2] - V(\phi)$	$\mathcal{L}_E = \frac{1}{2}[(\partial_\tau\phi)^2 + |\nabla\phi|^2] + V(\phi)$	$\mathcal{L}_M = -\mathcal{L}_E$

§10.8 Fermion Part of the Euclidean Action

The link variables describe the gauge field. The action (10.59) involving the link variables is therefore the action for the gauge field. What is the fermion part of the Euclidean action for the lattice gauge theory? Clearly, from the transformation rules Eqs. (10.42), (10.43), and (10.56), coupling terms of the form $\bar{\psi}(n)U_{-\mu}(n+\mu)\psi(n+\mu)$ and $\bar{\psi}(n)U_\mu(n-\mu)\psi(n-\mu)$, and the mass term $m\bar{\psi}(n)\psi(n)$ are invariant under a gauge transformation. We can use these terms to write down the fermion part of the Euclidean action:

$$S_{Eq} = \frac{a^3}{2} \sum_{n,\mu} \bar{\psi}(n)\gamma_E^\mu [U_{-\mu}(n+\mu)\psi(n+\mu) - U_\mu(n-\mu)\psi(n-\mu)]$$

$$+ma^4 \sum_n \bar{\psi}(n)\psi(n). \qquad (10.63)$$

Using the expansion of U for small a as in Exercise 10.5, it is easy to prove as shown in Exercise 10.6 that in the limit of very small lattice spacing, this action gives the correct continuum Euclidean action for the fermions:

$$\lim_{a\to 0} S_{Eq} = \int d\tau d^3x [\bar{\psi}\gamma_E^\mu(\partial_\mu - igA_\mu)\psi + m\bar{\psi}\psi]. \qquad (10.64)$$

The Euclidean action for lattice gauge theory is the sum of the gauge field part S_{EA} and the fermion part S_{Eq}. Explicitly, it is given by

$$S_E = -\frac{1}{2g^2} \sum_{n,\mu} tr\{U_{-\nu}(n+\nu)U_{-\mu}(n+\mu+\nu)U_\nu(n+\mu)U_\mu(n)\} + h.c.$$

$$+\frac{a^3}{2} \sum_{n,\mu} \bar{\psi}(n)\gamma_E^\mu [U_{-\mu}(n+\mu)\psi(n+\mu) - U_\mu(n-\mu)\psi(n-\mu)]$$

$$+ma^4 \sum_n \bar{\psi}(n)\psi(n). \qquad (10.65)$$

•[Exercise 10.6
Show that the Euclidean action of Eq. (10.63) gives the correct continuum limit.

◇Solution:
For a small value of the lattice spacing a, we can expand $U_{-\mu}(n + \mu)$ in terms of $A_\mu(n)$. Following similar steps as used in Exercise 10.5, we have

$$U_{-\mu}(n + \mu) = \exp(-ig\frac{a}{2}\{2A_\mu(n) + O(a)\}).$$

We can also expand $U_\mu(n - \mu)$ in terms of $A_\mu(n)$ as

$$U_\mu(n - \mu) = \exp(ig\frac{a}{2}\{2A_\mu(n) + O(a)\}).$$

In the limit $a \to 0$, the first term in Eq. (10.63) is then

$$\lim_{a\to 0} \frac{a^3}{2} \sum_{n,\mu} \bar\psi(n)\gamma^\mu_E\left[U_{-\mu}(n + \mu)\psi(n + \mu) - U_\mu(n - \mu)\psi(n - \mu)\right]$$

$$= \lim_{a\to 0} \frac{a^4}{2a} \sum_{n,\mu} \bar\psi(n)\gamma^\mu_E\left[\exp\{-igaA_\mu(n)\}\psi(n + \mu) - \exp\{ igaA_\mu(n)\}\psi(n - \mu)]\right.$$

$$= \lim_{a\to 0} \frac{a^4}{2a} \sum_{n,\mu} \bar\psi(n)\gamma^\mu_E\left[\{1 - igaA_\mu(n)\}\psi(n + \mu) - \{1 + igaA_\mu(n)\}\psi(n - \mu)\right]$$

$$= \lim_{a\to 0} a^4 \sum_{n,\mu} \bar\psi(n)\gamma^\mu_E\left[\partial_\mu\psi(n) - \frac{ig}{2}A_\mu(n)\{\psi(n + \mu) + \psi(n - \mu)\}\right]$$

$$= \lim_{a\to 0} a^4 \sum_{n,\mu} \bar\psi(n)\gamma^\mu_E\left[\partial_\mu\psi(n) - igA_\mu(n)\psi(n) + O(a)\right]$$

$$= \lim_{a\to 0} \int \frac{d\tau d^3x}{a^4} a^4\bar\psi(\tau,x)\gamma^\mu_E\left[\partial_\mu\psi(\tau,x) - igA_\mu(\tau,x)\psi(\tau,x)\right]$$

$$= \int d\tau d^3x \bar\psi(\tau,x)\gamma^\mu_E\left(\partial_\mu\psi(\tau,x) - igA_\mu(\tau,x)\psi(\tau,x)\right).$$

Thus, in the continuum limit, the fermion part of the action S_E of Eq. (10.65) becomes

$$\lim_{a\to 0} S_{E_q} = \int d\tau d^3x[\bar\psi\gamma^\mu_E(\partial_\mu - igA_\mu)\psi + m\bar\psi\psi]. \tag{1}$$

On the other hand, if we start from the continuum theory in Minkowski space, the Minkowski action from Eq. (10.63) is

$$S_{M_q} = \int dt d^3x[\bar\psi\gamma^\mu(i\partial_\mu + gA_\mu)\psi - m\bar\psi\psi],$$

then by making the change of variables $(t = -i\tau,\ A_t = iA_\tau,\ \gamma^j = i\gamma^j_E, ...)$ and using Eqs. (10.23), (10.38) and (10.63), we have

$$S_{Mq} = -i \int d\tau d^3x \{-\bar{\psi}\gamma^\mu_E\partial_\mu\psi + \bar{\psi}[\gamma^0_E(igA_\tau) + i\gamma^j_E gA_j]\psi - m\bar{\psi}\psi\},$$

$$= i \int d\tau d^3x \{\bar{\psi}\gamma^\mu_E(\partial_\mu - igA_\mu)\psi + m\bar{\psi}\psi\}.$$

Using the relation $S_E = -iS_M$ of Eq. (10.27), the above action, as deduced from the continuum theory, leads to the fermion part of the Euclidean action given by

$$S_{Eq} = \int d\tau d^3x [\bar{\psi}\gamma^\mu_E(\partial_\mu - igA_\mu)\psi + m\bar{\psi}\psi], \tag{2}$$

which is identical to Eq. (1). Therefore, in the limit of a small lattice spacing, the Euclidean action (10.63) leads to the same action [Eq. (2)] as obtained from the continuum theory in Minkowski space. ⟧•

§References for Chapter 10

1. K. G. Wilson, Phys. Rev. **D14**, 2455 (1974).
2. Excellent reviews of lattice gauge theory can be found in J. B. Kogut, Rev. Mod. Phys. 51, 659 (1979), J. B. Kogut, Rev. Mod. Phys. 55, 775 (1983), and M. Creutz, *Quarks, Gluons and Lattices*, Cambridge University Press, 1983.
3. An introduction to the renormalization theory can be found in Ta-Pei Cheng and Ling-Fong Li, *Gauge Theory of Elementary Particle Physics*, Clarendon Press, Oxford, 1984.
4. R. P. Feynman and A. R. Hibbs, *Quantum Mechanics and Path Integrals*, McGraw-Hill Book Company, New York, 1965.
5. L. D. Landau and E. M. Lifshitz, *Statistical Physics*, Pergamon Press, Oxford, 1980; Kerson Huang, *Statistical Mechanics*, John Wiley & Sons, N.Y., 1963.
6. C. Itzykson and J.-B. Zuber, *Quantum Field Theory*, McGraw-Hill Book Company, 1980.
7. See for example, J. I. Kapusta, *Finite-tempertaure Field Theory*, Cambridge University Press, 1989.
8. M. Gell-Mann, Phys. Rev. 125, 1067 (1962).
9. Ta-Pei Cheng and Ling-Fong Li, *Gauge Theory of Elementary Particle Physics*, Clarendon Press, Oxford, 1984.

11. Results from Lattice Gauge Theory

In the previous section, we introduced the basic ideas of the lattice gauge theory. We would like to discuss here some simple theoretical consequences of the theory. We shall examine the property of the confinement of a quark and an antiquark with the Wilson loop parameter, the deconfinement phase transition, the Monte Carlo calculations with the gauge field, and give a brief discussion of the lattice gauge theory with fermions.

§11.1 The Wilson Loop

The property of the interaction between a quark and an antiquark can be inferred from the expectation value of the trace of a product of link variables around a closed loop. The trace of the product of the link operators $U_\mu(n)$ around a closed loop c is called a *Wilson loop operator*,

$$(\text{Wilson loop operator}) = tr\{\prod_c U_\mu(n)\}, \qquad (11.1)$$

and the expectation value of the Wilson loop operator,

$$W(c) = < tr\{\prod_c U_\mu(n)\} >, \qquad (11.2)$$

is called a *Wilson loop parameter*. Here, the expectation value of an operator is defined according to Eq. (10.41). For brevity of notation, the term 'Wilson loop' is often used to represent both the Wilson loop operator and the Wilson loop parameter.

One can follow the same proof of the gauge invariance of $tr\{P_\square(n; \mu\nu)\}$ as in Eq. (10.58) to show that the Wilson loop operator is gauge invariant. Whether the Wilson loop parameter $W(c)$ depends on the area or the perimeter of the loop will reveal important characteristics of the confinement process.

We can relate the Wilson loop parameter to the interaction between a quark and an antiquark. We consider an antiquark at $x_q = 0$ and a quark $x_{\bar{q}} = R = (R, 0, 0)$. The quark and the antiquark are assumed to be so massive that their kinetic motion can be neglected. They appear as static external current source j^μ which can be expanded in terms of the generators of the $SU(3)_c$ group. For simplicity of notation and understanding, we shall focus our attention on one of the components in this expansion and shall not write out the color degrees of freedom explicitly.

We can study the action and the partition function arising from the presence of this $q\bar{q}$ pair from three different points of view.

In the simplest description of the $q\bar{q}$ pair, we treat the two-body system as a 'single-particle' system with a single relative coordinate $x_{\rm rel} = x_q - x_{\bar{q}} = R$. The Euclidean action for this 'single-particle' system is given by Eq. (10.13) in terms of the kinetic energy term involving $(dx/d\tau)^2$ and the effective $q\bar{q}$ potential $V(r)$. For the quark and the antiquark at fixed locations, the kinetic energy term gives no contribution. Staring from an initial (imaginary) time τ_a to a final (imaginary) time τ_b, the Euclidean action from the interaction of the $q\bar{q}$ pair is

$$S_{Ei} = \int_{\tau_a}^{\tau_b} d\tau V(R)$$
$$= T V(R), \tag{11.3}$$

which is just the product of the effective $q\bar{q}$ potential and the length of (imaginary) time $T(= \tau_b - \tau_a)$ over which the action is taken. The partition function, due to the presence of the quark-antiquark pair, is given by Eq. (10.14) as $e^{-S_{Ei}}$. It follows that the partition function arising from the interaction of the quark and the antiquark is

$$e^{-S_{Ei}} = e^{-V(R)T}. \tag{11.4}$$

In the second equivalent description of the $q\bar{q}$ pair, the antiquark at $x = 0$ generates a gauge field $A_\mu(x)$ at x which interacts with the quark at x, and the quark at x generates a gauge field $A_\mu(0)$ at $x = 0$ which interacts with the anitquark at $x = 0$. The 'external' sources j^μ produces a gauge fields A^μ which interact with the quark and the antiquark. In this description, the gauge field $A_\mu(x)$ arising from the presence of the quark and the antiquark is a classical field and is not a dynamical variable. That is, the gauge field $A_\mu(x)$ has a fixed value at a given space-time point x and is not allowed to have a distribution and a variation at that point. The time-independent external current can be represented by

$$j^0(x, \tau) = \delta^{(3)}(x - R) - \delta^{(3)}(x). \tag{11.5}$$

From Eq. (10.64), the Euclidean action which arises from the interaction of the quark current with the gauge field in the time interval from $\tau = 0$ to $\tau = T$, is

$$S_{Ei} = -ig \int j^\mu(x) A_\mu(x) d^3x d\tau.$$

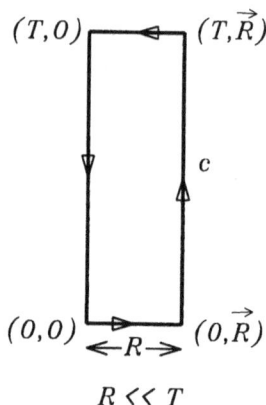

$$R \ll T$$

Fig. 11.1 The Wilson loop c used to study the interaction between a quark and an antiquark.

Upon substituting (11.5) into the above equation, the Euclidean action from the interaction is

$$S_{Ei} = -ig \int_0^T [A_\tau(\boldsymbol{R}) - A_\tau(\boldsymbol{0})]d\tau$$

$$= -ig \int_0^T A_\tau(\boldsymbol{R})d\tau + g \int_T^0 A_\tau(\boldsymbol{0})d\tau. \qquad (11.6)$$

We can construct a closed loop c as in Fig. 11.1, which follows the points $(0,0) \rightarrow (0,\boldsymbol{R}) \rightarrow (T,\boldsymbol{R}) \rightarrow (T,0) \rightarrow (0,0)$. The loop is so constructed that the dimension T in the time direction is much larger than the spatial dimension $R = |\boldsymbol{R}|$. Then in the limit of very large T the above Euclidean action (11.6) can be written as

$$S_{Ei} = -ig \oint_c A_\mu dx^\mu. \qquad (11.7)$$

The partition function arising from this action is

$$\lim_{T \to \infty} e^{-S_{Ei}} = e^{ig \oint_c A_\mu dx^\mu}, \qquad (11.8)$$

which is just the continuum limit of the Wilson loop operator for the loop c.

In the third description in lattice gauge theory, the gauge field is allowed to vary as a dynamical variable. The gauge field $A_\mu(\boldsymbol{x})$ at the space-time point $\boldsymbol{x}$ can take on all possible values, with the 'weighting factor' $e^{-S_{EA}}$ determined by the gauge field action S_{EA} (Eq. 10.59) which is a function of the gauge field at all space-time points. As we mentioned previously, in practical problems, instead of varying the gauge field A_μ itself, we use the link variable (10.37) which is a function of the gauge field. By evaluating the expectation value of the quantity in Eq. (11.8) with all possible variations of the link variables at different links, we can determine the partition function due to the interaction of the quark current with the gauge field. The partition function obtained in the gauge field theory can then be compared with the partition function obtained in the single-particle description (11.4) to infer the dependence of $V(R)$ as a function of R.

To take the expectation value of $\exp\{-S_{Ei}\}$, the Wilson loop on the righthand side of equation (11.8) is discretized in lattice gauge theory to become

$$tr\{U_{c_N c_1}(c_N)...U_{c_2 c_3}(c_2)U_{c_1 c_2}(c_1)\} = tr\{\prod_{i=1}^{N} U_{c_i c_{i+1}}(c_i)\}, \qquad (11.9)$$

where we have discretized the loop c into N ordered points $\{c_1, c_2, c_3, ..., c_N\}$ along the loop (Fig. 11.2), and $U_{c_i c_{i+1}}(c_i)$ is the link variable residing on the link between c_i and c_{i+1}. and directing from c_i to c_{i+1}. We shall study the expectation value of the Wilson loop operator in the case of strong coupling. From Eq. (10.41) and (10.40), we have

$$< tr\{\prod_{i=1}^{N} U_{c_i c_{i+1}}(c_i)\} > = \frac{\int \prod_{n,\mu} dU_\mu(n) e^{-S_E} tr\{\prod_{i=1}^{N} U_{c_i c_{i+1}}(c_i)\}}{\int \prod_{n,\mu} dU_\mu(n) e^{-S_E}}.$$

$$(11.10)$$

Because the operator $\prod_c U_\mu$ does not contain any fermion field, it is necessary to include only the link-variable part of the Euclidean action S_{EA} in Eq. (11.10). In the strong coupling limit $\beta = 1/g^2 \ll 1$, the exponentiation of the action S_{EA} in (10.59) can be written as a product over the plaquettes:

$$e^{-S_{EA}} = exp\left\{\frac{1}{g^2} \sum_{n,\mu} tr\{P_\square(n; \mu\nu)\}\right\}$$

$$= \prod_p exp\left\{\beta tr\{P_\square(p)\} + O(\beta^2)\right\}, \qquad (11.11)$$

where p runs through all the plaquettes. To the same order of approximation, the ordering of the different plaquettes can be interchanged. We can expand the exponential $e^{-S_{EA}}$ in powers of the products of the U matrices. Therefore, the expectation value of the Wilson loop becomes

$$< tr\{\prod_{i=1}^{N} U_{c_i c_{i+1}}(c_i)\} >$$

$$\propto \int \prod_{n,\mu} dU_\mu(n) tr\{\prod_{i=1}^{N} U_{c_i c_{i+1}}(c_i)\} \prod_{p}(1 - \beta tr\{UUUU\})_p. \quad (11.12)$$

We can carry out the integration over the space of the variables which define the U matrices. For this integration, we have (see Exercise 11.1)

$$\int dU_\mu(n) \, [U_\mu(n)]_{ij} = 0, \quad (11.13a)$$

and

$$\int dU_\mu(n) \, [U_\mu(n)]_{ij} [U_\mu(n)]_{kl}^\dagger = \frac{1}{N} \delta_{il} \delta_{jk}. \quad (11.13b)$$

The above equations mean that the integration of quantities with a single U matrix is zero, while the integration of the product of a U matrix and its Hermitian conjugate gives a Kronecker delta function. The U matrices obey an 'orthogonality' condition.

To carry out the evaluation of Eq. (11.12), we integrate over the variables which define the set of link variables U specified on all the links in the lattice. When we expand out the product $\prod_p(1 - \beta tr\{UUUU\})_p$ for all the plaquettes in the lattice, there are many terms. For each term, a plaquette can contribute the factor 1 or the factor $-\beta tr\{UUUU\}_p$. We consider one special term which arises from the expansion of $\prod_p(1 - \beta tr\{UUUU\})_p$ in the following way. We divide the set of plaquettes $\{p\}$ into two sets. The plaquettes in the set which we call $\{p_c\}$ are either completely inside the Wilson loop or border on the Wilson loop, and this set of plaquettes $\{p_c\}$ cover exactly the area inside the Wilson loop. For each one of these plaquettes in the set $\{p_c\}$, we take the factor $-\beta tr\{UUUU\}$ from the expansion $(1 - \beta tr\{UUUU\})_p$. The four U matrices are those associated with the four sides of the plaquette. For the set of plaquettes which lie outside the loop c we take the factor 1 in the expansion $(1 - \beta tr\{UUUU\})_p$. For this special term in the expansion of the integrand in Eq. (11.12), we have the product

$$tr\{\prod_{i=1}^{N} U_{c_i c_{i+1}}(c_i)\} \prod_{p_c}(-\beta tr\{UUUU\})_{p_c}. \quad (11.14)$$

In Fig. 11.2, we show, as an example, the link variables of the first product $tr\{\prod_{i=1}^{N} U_{c_i c_{i+1}}(c_i)\}$ of expression (11.14) as the outermost rectangle. We also show the set of plaquettes inside the Wilson loop

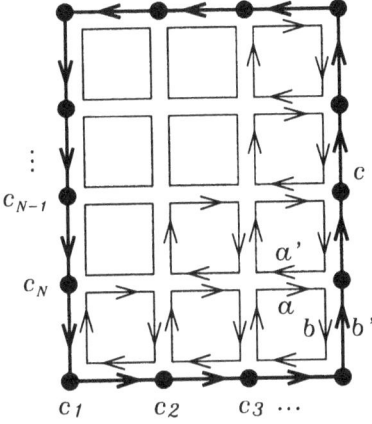

Fig. 11.2 An example of the link variables which form the product (11.14). The outermost rectangular loop c comes from the link variables of the Wilson loop $tr\{\prod_{i=1}^{N} U_{c_i c_{i+1}}(c_i)\}$. The squares inside the loop c represent the plaquettes and their link variables arising from the product $\prod_{p_c}(-\beta tr\{UUUU\})_{p_c}$, where $\{p_c\}$ is the set of plaquettes which lie inside the loop c or border on the loop c.

c and the link variables in each plaquette. The product (11.14) is obtained by multiplying all the link variables shown in Fig. 11.2. For each of those link variables which is in the interior of the Wilson loop c and belongs to a plaquette (such as link a in Fig. 11.2), there is another link variable along the same link but pointing in the opposite direction (such as link a' in Fig. 11.2). The integration of the product of these two link variables aa' will lead to a unit matrix because of Eq. (11.13b). For each of those link variables on the boundary of the loop coming from one of the plaquettes of this set $\{p_c\}$ (such as the link b in Fig. 11.2), there is another correspondingly link variable from the set $\prod_c U_\mu(n)$ which is along the same link but pointing in the opposite direction (such as the link b' in Fig. 11.2). The integration of the product of these two link variables bb' will also lead to a unit matrix. All other terms in the expansion of the integrand of Eq. (11.12) have configurations different from this one and will lead to zero because of Eqs. (11.13a-13b). In consequence, the result of integration is

$$\int \prod_{n,\mu} dU_\mu(n) tr\{\prod_{i=1}^{N} U_{c_i c_{i+1}}(c_i)\} \prod_{p_c}(\frac{1}{g^2}) tr\{UUUU\}_{p_c} \propto (\frac{1}{g^2})^{N_p}$$

where N_p is the number of the plaquettes in the set $\{p_c\}$ inside the loop c. We obtain the result

$$< tr\{\prod_{i=1}^{N} U_{c_i c_{i+1}}(c_i)\} > \propto (\frac{1}{g^2})^{N_p} = e^{-\ln(g^2)\mathcal{A}/a^2}, \qquad (11.15)$$

where $\mathcal{A} = RT = N_p a^2$ is the area enclosed by the Wilson loop. This is known as the area law.

What does the area law tell us about the interaction between the quark and the antiquark? A comparison of Eq. (11.15) with Eq. (11.4) shows that if we represent the quark and the antiquark system as a single-particle system with an effective interaction, then the effective interaction is given by

$$V(R)T = \frac{\ln(g^2)RT}{a^2}$$

or

$$V(R) = \sigma R, \qquad (11.16a)$$

where the string tension coefficient σ is

$$\sigma = \ln(g^2)/a^2. \qquad (11.16b)$$

The effective potential $V(R)$ is linearly proportional to the distance between a quark and an antiquark. The potential becomes infinitely large as a quark is separated from an antiquark. The isolation of a quark is therefore impossible. The potential (11.16) exhibits the property of quark confinement.

•⟦ Exercise 11.1
Prove Eqs. (11.13a) and (11.13b) for the case of $SU(2)$.

◇Solution:
An $SU(2)$ matrix can be written in general as

$$U = \begin{pmatrix} \alpha & -\beta \\ \beta^* & \alpha^* \end{pmatrix}$$

where

$$\alpha\alpha^* + \beta\beta^* = 1, \qquad (1)$$

such that U possesses the following two properties for *special* unitary matrices:

$$UU^\dagger = 1,$$

and

$$\det U = 1.$$

We can write α and β in terms of real coordinates x_1, x_2, x_3 and x_4 explicitly as

$$\alpha = x_1 + ix_2$$

and

$$\beta = x_3 + ix_4,$$

Eq. (1) can be written as

$$x_1^2 + x_2^2 + x_3^2 + x_4^2 = 1. \tag{1'}$$

The variation of the unitary matrix U can be represented in terms of the variation of x_1, x_2, x_3 and x_4 subject to the constraint of Eq. (1'). The integration over the variables which define U is explicitly given by

$$\int dU \rightarrow \int d^4x \; \delta(x_1^2 + x_2^2 + x_3^2 + x_4^2 - 1).$$

We normalize the integral to unity, subject to the constraint (1'):

$$\int dU = N \int d^4x \; \delta(x_1^2 + x_2^2 + x_3^2 + x_4^2 - 1) = 1.$$

We consider the integrals involving a first power of the matrix element of U. As the limits of x_i runs from $-\infty$ to $+\infty$, the integral of U_{11} is zero,

$$\int dU \; U_{11} = N \int d^4x \; \delta(x_1^2 + x_2^2 + x_3^2 + x_4^2 - 1)(x_1 + ix_2) = 0.$$

We have also

$$\int dU \; U_{12} = \int dU \; U_{21} = \int dU \; U_{22} = 0.$$

We consider next the integral involving a quadratic power of the matrix element of U. We have

$$\int dU \; U_{11}(U^\dagger)_{11} = N \int d^4x \; \delta(x_1^2 + x_2^2 + x_3^2 + x_4^2 - 1)\alpha\alpha^*$$

$$= N \int d^4x \; \delta(x_1^2 + x_2^2 + x_3^2 + x_4^2 - 1)(x_1^2 + x_2^2)$$

$$= \frac{1}{2} N \int d^4x \; \delta(x_1^2 + x_2^2 + x_3^2 + x_4^2 - 1)(x_1^2 + x_2^2 + x_3^2 + x_4^2)$$

$$= \frac{1}{2}.$$

One can show similarly that

$$\int dU \; U_{12}(U^\dagger)_{12} = \int dU \; (-\beta)\beta$$

$$= \int dU \; (x_3^2 - x_4^2 + 2ix_3x_4) = 0,$$

$$\int dU \ U_{12}(U^\dagger)_{21} = \frac{1}{2},$$

and

$$\int dU \ U_{22}(U^\dagger)_{22} = \frac{1}{2}.$$

Combining all the results, we have

$$\int dU \ U_{ij}(U^\dagger)_{j'i'} = \frac{1}{2}\delta_{ii'}\delta_{jj'}.$$

This result can be generalized to $SU(N)$

$$\int dU \ U_{ij}(U^\dagger)_{j'i'} = \frac{1}{N}\delta_{ii'}\delta_{jj'}. \qquad\qquad]\bullet$$

§11.2 The Order of a Phase Transition

The phase transition of matter can be classified [1] according to the discontinuity of the free energy F as a function of the temperature T. A phase transition is an *nth order phase transition* if the nth derivative of the free energy, $\partial^n F/\partial T^n$, is discontinuous while all its lower derivatives in T are continuous.

According to this definition, a *first-order phase transition* is characterised by a discontinuous $\partial F/\partial T$ and a continuous free energy F. On the other hand, $\partial F/\partial T$ is related to the energy density which is the total energy of the system E divided by the volume V:

$$\epsilon = \frac{E}{V} = \frac{F + T\frac{\partial F}{\partial T}}{V}. \qquad (11.17)$$

Therefore, in a first-order phase transition, the energy density ϵ is discontinuous as a function of temperature. The difference of the energy density ϵ at the discontinuity gives the value of the *latent heat*. As an example, the liquid-gas phase transition of water is a first-order phase transition. The energy density of water in the form of a liquid and the energy density of water in the form of a vapor is discontinuous at the critical temperature T_c.

In a *second-order phase transition*, the free energy F and $\partial F/\partial T$ are continuous while $\partial^2 F/\partial T^2$ is discontinuous in T. Because the specific heat at constant volume is related to $\partial E/\partial T$ or $\partial^2 F/\partial T^2$, a second-order phase transition is characterized by a continuous free energy and energy density but a discontinuous specific heat at constant volume. An example of a second-order phase transition is the transition which occurs when the symmetry of a crystal lattice changes discontinuously from a symmetric configuration to an non-symmetric configuration as the temperature changes [2], while the crystal lattice remains in the solid state.

We shall first examine the nature of the phase transition for the lattice gauge theory containing only the gauge field. The action S_E will then contain only link variables $U_\mu(n)$. Such a study is useful as a first step to pave the way for more extended calculations in which the quark degrees of freedom are also included.

To study the nature of the phase transition, one can calculate the energy density of the two phases of matter as a function of temperature. It is however more convenient to consider the behavior of the expectation value of the Wilson line operator as a function of temperature. The *Wilson line operator L* is defined as the trace of the product of the link operators along the time direction at a fixed spatial point (n_x, n_y, n_z):

$$L = tr\left\{ \prod_{n_\tau=1}^{n_\tau=N_\tau} U_\tau(n_x, n_y, n_z, n_\tau) \right\}. \tag{11.18}$$

A discontinuity in the energy density is accompanied by a discontinuity in the expectation value of the Wilson line [3]. Therefore, the expectation value of the Wilson line operator $< L >$ can be used as an *order parameter* for the investigation of the order of the phase transition. For simplicity of nomenclature, we shall use the term "Wilson line" to denote both the Wilson line operator and its expectation value, except in those cases which might lead to ambiguities.

The Wilson line links together the link variables in the lattice and describes a world line in space-time. This world line is fixed in spatial location but propagates forward in time. Physically, it describes a static quark located at the spatial position (n_x, n_y, n_z), propagating in time. As discussed previously in the analogous relation Eq. (10.24), the configuration of the gauge field system is periodic in the time variable τ. This requires that the link variable at $\tau = 0$ is equal to the link variable at $\tau = \beta_B = 1/T$,

$$U_\mu(n) = U_\mu(n + \beta_B) \tag{11.19}$$

for $n = (n_x, n_y, n_z, \tau = 0)$ and $n + \beta_B = (n_x, n_y, n_z, \tau = \beta_B)$. Thus, the end point of the Wilson line is identical to the beginning point of the line. The Wilson line (11.18) is in this sense a closed loop. Therefore, following arguments similar to the ones used to show the gauge-invariance of the plaquette operator in Eq. (10.58), it is easy to show that the Wilson line operator is gauge-invariant.

The expectation value of the Wilson line gives the quantity $e^{-\beta_B F}$ where F is the free energy of the system containing this static quark [4] at the spatial position (n_x, n_y, n_z) in a gauge field. For a quark to be confined, the interaction between the isolated static quark and the gauge field is infinite and the expectation value of the Wilson line should be zero. On the other hand, when a quark is deconfined, the interaction between the quark and the gauge field is finite and the expectation value of the Wilson line should be non-zero. Therefore, $< L >$ can be used as an order parameter for the investigation of the phase. A change of the Wilson line parameter from zero to a finite value corresponds to a change from the confined to the deconfined phase. A discontinuity in the value of the Wilson line will correspond to a first-order transition. It is expected that the discontinuity of the Wilson line will be accompanied by a discontinuity of other physical quantities such as the entropy and the energy density [3]. The system will show the usual characteristics of a first-order phase transition as described above.

We shall first study theoretically the symmetry associated with the confinement-deconfinement phase transition. Because the Euclidean action is given by the sum of the plaquette variables and each plaquette consists of the product of 4 link variables, the action is invariant under a more general periodic condition relating the link variables at $\tau = 0$ and at $\tau = \beta_B$. Specifically, for the $SU(N)$ gauge theory, one can consider a generalized periodic condition in which the usual boundary condition Eq. (11.19) is followed with an additional transformation for all the time-like link variables U_τ at the time sheet $\tau = 0$ represented by the set of points $(n_x, n_y, n_z, \tau = 0)$ [5]:

$$U_\tau(n_x, n_y, n_z, \tau = 0) \rightarrow (zI)\, U_\tau(n_x, n_y, n_z, \tau = 0), \qquad (11.20)$$

where (zI) is an element of the $Z(N)$ group with $z = \exp\{i2\pi\alpha/N\}$, $\alpha = 1, 2, .., N$, and I is a unit $N \times N$ matrix. In the transformation (11.20), zI has been judiciously chosen to be an element of the $Z(N)$ group which is the center of the $SU(N)$ group. That is, zI and its Hermitian conjugate commutes with all elements of $SU(N)$, which include $U_\tau(n_x, n_y, n_z, \tau = 0)$ of Eq. (11.20). This transformation is a global transformation for all spatial points at the time $\tau = 0$, and may be called the $Z(N)$ transformation. If an elementary plaquette contains a time-like link variable pointing from $\tau = 0$ to $\tau = a$, it will also contain another time-like link variable pointing in the opposite direction from $\tau = a$ to $\tau = 0$. The product of the four link variables in this elementary plaquette will be invariant under the $Z(N)$ transformation Eq. (11.20). Being a sum of the products of link variables in the plaquette, the gauge field action is invariant under the

$Z(N)$ transformation (11.20). In place of Eq. (11.19), a generalized periodic boundary condition for time-like links in the lattice gauge theory is

$$U_\tau(n) = (zI)\, U_\tau(n + \beta_B).\tag{11.21}$$

We show in Supplement 11.1 that the Euclidean action (10.59) is invariant under the $Z(N)$ transformation (11.20). On the other hand, from its definition, the Wilson line involves only a single time-like link variable at time $\tau = 0$. Under the $Z(N)$ transformation (11.20), the Wilson line transforms as

$$< L > \to z < L > .\tag{11.22}$$

Thus, the Wilson line is not invariant under a global $Z(N)$ transformation, except in the case when $< L >$ is zero.

⊕❲ **Supplement 11.1**
We would like to show here that the Euclidean action S_{EA} of Eq. (10.59) is invariant under the $Z(N)$ transformation (11.20).
The center of the $SU(N)$ group consists of all those elements of the group which commute with all the elements of the $SU(N)$ group. As the unit matrix commutes with all elements of the group, an element of the center of $SU(N)$ is proportional to the unit matrix I, which is an $N \times N$ matrix. We can write the element of the center as zI. Because the determinant of an element of the $SU(N)$ group is unity, we have $\det(zI) = z^N = 1$, with N solutions of z given by $z = e^{i2\pi\alpha/N}$, and $\alpha = 1, 2, ..., N$. It is easy to see that zI is an element of the $Z(N)$ group, which has the property that $(zI)^N = I$.
We consider the $Z(N)$ transformation for a link variable at $(n_x, n_y, n_z, \tau = 0)$ pointing in the time-like direction from $\tau = 0$ to the next lattice point at $\tau = a$:

$$U_\tau(n_x, n_y, n_z, \tau = 0) \to zI\, U_\tau(n_x, n_y, n_z, \tau = 0).\tag{1}$$

The link variable pointing in the opposite direction linking $\tau = a$ to $\tau = 0$ can be written as

$$U_{-\tau}(n_x, n_y, n_z, \tau = a) = [U_\tau(n_x, n_y, n_z, \tau = 0)]^{-1}.$$

Hence, under the $Z(N)$ transformation, the link variable pointing in the direction from $\tau = a$ to $\tau = 0$ transforms as

$$U_{-\tau}(n_x, n_y, n_z, \tau = a) \to [zIU_\tau(n_x, n_y, n_z, \tau = 0)]^{-1} = U_{-\tau}(n_x, n_y, n_z, \tau = a)z^{-1}I.\tag{2}$$

The Euclidean action involving the gauge field link variables is given by

$$S_A = -\frac{1}{2g^2} \sum_{n,\nu,\mu} tr\{U_\mu(n)U_\nu(n + \mu)U_{-\mu}(n + \mu + \nu)U_{-\nu}(n + \nu) + h.c.\}.$$

Under the $Z(N)$ transformation (11.20), either zero or two link variables in an elementary plaquette are affected by the transformation. A plaquette in which no link variables are affected will not change under the $Z(N)$ transformation. In a plaquette in which two of the links are affected by the transformation, the two link variables must be pointing in opposite directions on the time axis. Under the $Z(N)$ transformation, the plaquette transforms as

$$U_\mu(n)U_\tau(n+\mu)U_{-\mu}(n+\mu+a)U_{-\tau}(n+a) \rightarrow$$

$$U_\mu(n) \; zI \; U_\tau(n+\mu)U_{-\mu}(n+\mu+a)U_{-\tau}(n+a)z^{-1}I,$$

where we have used Eqs. (1) and (2). As zI and $z^{-1}I$ commute with U_μ and $zz^{-1} = 1$, the plaquette is invariant under the $Z(N)$ transformation. In consequence, the Euclidean action (10.59) is invariant under the $Z(N)$ transformation.

What type of $Z(N)$ symmetry does one expect as the temperature varies? At low temperatures, one expects quarks to be confined and the Wilson line to have the value zero. From Eq. (11.22) the system is expected to be invariant under $Z(N)$ and there will be a global $Z(N)$ symmetry. The value of the Wilson line will cluster around the origin in the complex plane. However, for high temperatures, when the Wilson line is not expected to vanish, the Wilson line will depend on z (Eq. (11.22)) and the global $Z(N)$ symmetry is broken. The value of the Wilson line will cluster around the three complex roots of $z^N = 1$. In the critical temperature region, the system tunnels between the confined phase and the deconfined phase, changing the value of the Wilson line $< L >$ from zero to $e^{-i2\pi\alpha/N}$ and vice versa. The global $Z(N)$ symmetry can be used to define the two different phases for a pure gauge field.

The $SU(N)$ group and the $Z(N)$ group have the same center and the phase transition in $SU(N)$ theory is accompanied by a change of the center $Z(N)$ symmetry. Because of the possibility of a universality of critical behavior of different groups having the same center symmetry, it has been suggested [5] that the type of phase transition of $SU(N)$ in $d+1$-dimensional space-time may be the same as that of the analogous $Z(N)$ spin system in d-dimensional configuration space.

According to this suggestion of "universality" [5], one expects that the critical behavior of an $SU(2)$ gauge system is analogous to the Ising model which has a global $Z(2)$ symmetry [6,7]. The Ising model in four dimensional space-time undergoes a continuous phase transition at the critical temperature with a power-law type singularity [6]; it has a second-order phase transition. By universality, the analogous $SU(2)$ gauge system in four dimensional space-time should also have a second-order phase transition. By a similar argument of universality, an $SU(3)$ gauge system in $(3+1)$ dimensional space-time should be analogous to a $Z(3)$ spin model in 3-dimensional space, which has a first-order phase transition [8,9]. One expects that $SU(3)$ gauge systems should have a first-order phase transition. Monte-Carlo calculations in $SU(2)$ and $SU(3)$ gauge theories bear out the predictions on the order of the phase transition [3,10-15].

§11.3 Mean-Field Approximation

To help us understand the dynamics of the phase transitions of $SU(N)$ systems in $(d+1)$-dimensional space-time, it is instructive to study the nature of the phase transitions of the analogous $Z(N)$ systems in d-dimensional configuration space, in a simple mean-field approximation [7-9]. This will provide an intuitive insight as to how different types of phase transition arise in the lowest order approximation.

In the *mean-field approximation*, the order of the phase transition depends on the functional form of the action. The action is different for the $Z(N)$ spin system [16] and for the $Z(N)$ lattice gauge system [7-9].

(a) $Z(N)$ Spin System

We discretize the d-dimensional space to contain N_v equally-spaced lattice vertices. For the $Z(N)$ spin system, the spin field at the lattice vertex site i is described by the spin variable $s(\alpha_i)$, which is an element of the $Z(N)$ group specified by an integer α_i. It is explicitly given by

$$s(\alpha_i) = e^{i2\pi\alpha_i/N}, \quad \text{(lattice vertices } i = 1 \text{ to } i_{\max}), \quad (11.23)$$

where $i_{\max}$ is the total number of spin variables on the lattice, and α_i takes on the values from 0 to $N-1$,

$$\alpha_i = 0, 1, 2, ..., \text{ or } N-1. \quad (11.24)$$

Because the spin variable $s(\alpha_i)$ is defined at each lattice vertex i, the total number of spin variables $i_{\max}$ is the same as the number of vertices N_v:

$$i_{\max} = N_v, \quad \text{for } Z(N) \text{ spin systems.} \quad (11.25)$$

The action for the spin system possessing a global $Z(N)$ symmetry can be written as [16]

$$S_E = -\beta_B J \sum_{\{1,2\}} \cos\{\frac{2\pi}{N}(\alpha_2 - \alpha_1)\},$$

$$= -\beta_B J \sum_{\{1,2\}} \mathcal{R}e\,\{s^*(\alpha_1)s(\alpha_2)\},$$

$$= -\beta_B J \sum_{\{1,2\}} \mathcal{R}e\,\{s(-\alpha_1)s(\alpha_2)\}, \tag{11.26}$$

where $\{1,2\}$ labels a nearest-neighboring pair and the summation $\sum_{\{1,2\}}$ is carried out over the whole set of nearest-neighboring pairs in the lattice. The coefficient β_B is the inverse temperature, $\beta_B = 1/T$, and $J > 0$ is the strength of the interaction of $s(\alpha_1)$ at lattice site 1 with $s(\alpha_2)$ at lattice site 2. With this action, the spin system will lower its energy when the spins of a nearest-neighbor pair have the same value, $\alpha_1 = \alpha_2$; that is, when the spins of a nearest-neighbor pair are aligned in the same direction. The action (11.26) is invariant under an arbitrary global phase rotation in which all $s(\alpha_i)$'s are transformed according to

$$s(\alpha_i) \to e^{i\theta}s(\alpha_i).$$

(b) $Z(N)$ Lattice Gauge System

On the other hand, for the $Z(N)$ lattice gauge system which possesses local gauge invariance, we specify the link variable $s(\alpha_i)$ at the directional link i between lattice vertices. The link variable is an element of the $Z(N)$ group and obeys the same equations as the spin variable of Eqs. (11.23) and (11.24),

$$s(\alpha_i) = e^{i2\pi\alpha_i/N}, \quad \text{(links } i = 1 \text{ to } i_{max}\text{)}, \tag{11.27}$$

where i_{max} is the total number of link variables in the lattice and α_i take on the values from 0 to $N - 1$,

$$\alpha_i = 0, 1, 2, ..., \text{ or } N - 1. \tag{11.28}$$

We shall also call the link variable in the $Z(N)$ gauge theory the spin variable.

For the $Z(N)$ gauge system, as there is a link variable specified at each link between two vertices, the total number of link variables i_{max} is the same as the number of links between vertices. There are N_v number of lattice vertices. Each vertex is connected to $2d$ links, but each link is shared by 2 vertices. The total number of link variables i_{max}, in the d-dimensional lattice, is

$$i_{max}(d) = N_v \times 2d \times \frac{1}{2} = N_v d \quad \text{for} \quad Z(N) \quad \text{gauge systems.} \quad (11.29)$$

The action of the $Z(N)$ gauge system is given by [7,8]

$$S_E = -\beta_B J \sum_{\{1,2,3,4\}} \mathcal{R}e \left\{s(\alpha_1)s(\alpha_2)s(\alpha_3)s(\alpha_4)\right\} \quad (11.30)$$

where 1, 2, 3, and 4 label the four links in a plaquette and the summation $\sum_{\{1,2,3,4\}}$ is carried out over all the plaquettes of the lattice. This action is similar to the Wilson action (10.59) and it satisfies local gauge invariance under the local gauge transformation

$$s(\alpha_i) \to e^{i\theta_{iL}} s(\alpha_i) e^{-i\theta_{iR}}, \quad (11.31)$$

where θ_{iL} and θ_{iR} are arbitrary phase rotations associated respectively with the left and the right lattice vertices forming the link i.

We can study the phase transition for both the $Z(N)$ spin system [Eq. (11.26)] and the $Z(N)$ gauge system [Eq. (11.30)] in a unified way by writing the actions (11.26) and (11.30) in the general form

$$S_E = -\beta_B J \sum_{\{p\}} \mathcal{R}e \left\{s(\pm\alpha_1)s(\alpha_2)...s(\alpha_r)\right\}. \quad (11.32)$$

The above equation is the action (11.26) for the $Z(N)$ spin system, when we set r equal to 2, take the first spin variable to be $s(-\alpha_1)$ with the negative argument, and consider the summation $\sum_{\{p\}}$ to stand for $\sum_{\{1,2\}}$, the sum over all pairs of nearest neighbors of the system. Eq. (11.32) equation is the action (11.30) for the $Z(N)$ gauge system, when we set r equal to 4, take the first spin variable to be $s(+\alpha_1)$ with the positive argument, and consider the summation $\sum_{\{p\}}$ to stand for $\sum_{\{1,2,3,4\}}$, the sum over all plaquettes of the system. It is important to note that for the $Z(N)$ spin system, the action is quadratic in s while for the $Z(N)$ gauge system, the action is quartic in s. We shall see that this difference is crucial for the determination of the different orders of phase transition in the mean-field approximation.

The partition function for the two $Z(N)$ systems under consideration is

$$Z = \sum_{\{\alpha\}} \exp\{\beta_B J \sum_{\{p\}} \mathcal{R}e\ \{s(\pm\alpha_1)s(\alpha_2)...s(\alpha_r)\}, \qquad (11.33)$$

where the summation $\sum_{\{\alpha\}}$ is carried out over all possible α_i values for the spin variable at all i's, with $i = 1, i_{max}$.

To obtain an approximate evaluation for Eq. (11.33), we introduce a variational parameter λ to rewrite the partition function in the form

$$Z = \sum_{\{\alpha\}} \exp\left\{\beta_B J \sum_{\{p\}} \mathcal{R}e\ \{s(\pm\alpha_1)s(\alpha_2)...s(\alpha_r)\} - \lambda \sum_{i=1}^{i_{max}} \mathcal{R}e\ s(\alpha_i)\right\}$$

$$\times \exp\{\lambda \sum_{i=1}^{i_{max}} \mathcal{R}e\ s(\alpha_i)\}. \qquad (11.34)$$

The meaning of the variational parameter λ will become clearer later (see Eqs. (11.43)-(11.45)). We define a statistical average of a quantity $O(\{\alpha\})$ by averaging it over all the spin fields at all lattice sites with a weighting factor $\exp\{\lambda \sum_{i=1}^{i_{max}} \mathcal{R}e\ s(\alpha_i)\}$:

$$< O(\{\alpha\}) > = \frac{\sum_{\{\alpha\}} O(\{\alpha\})\ \exp\{\lambda \sum_{i}^{i_{max}} \mathcal{R}e\ s(\alpha_i)\}}{\sum_{\{\alpha\}} \exp\{\lambda \sum_{i}^{i_{max}} \mathcal{R}e\ s(\alpha_i)\}}. \qquad (11.35)$$

Then, with respect to this averaging, Eq. (11.34) can be written as

$$Z = <\exp\left\{\beta_B J \sum_{\{p\}} \mathcal{R}e\ \{s(\pm\alpha)s(\alpha_2)...s(\alpha_r)\} - \lambda \sum_{i=1}^{i_{max}} \mathcal{R}e\ s(\alpha_i)\right\}>$$

$$\times \sum_{\{\alpha\}} \exp\{\lambda \sum_{i=1}^{i_{max}} \mathcal{R}e\ s(\alpha_i)\}. \qquad (11.36)$$

The free energy F is related to the partition function by

$$Z = e^{-\beta_B F}. \qquad (11.37)$$

To get an approximation for the partition function, we note that

$$< \exp\{f\} > = \exp\{< f >\} + \sum_{n=2}^{\infty} \frac{< f^n > - < f >^n >}{n!}. \qquad (11.38)$$

Applying this result to Eq. (11.36) by identifying f as

$$f = \beta_B J \sum_{\{p\}} \mathcal{R}e \; s(\pm\alpha_1)s(\alpha_2)...s(\alpha_r) - \lambda \sum_{i=1}^{i_{max}} \mathcal{R}e \; s(\alpha_i), \qquad (11.39)$$

then Eq. (11.36) can be written as

$$e^{-\beta_B F} = \left\{ e^{<f>} + \sum_{n=2}^{\infty} \frac{< f^n - < f >^n >}{n!} \right\} \sum_{\{\alpha\}} \exp\{\lambda \sum_{i=1}^{i_{max}} \mathcal{R}e \; s(\alpha_i)\}. \qquad (11.40)$$

Much intuitive insight into the dynamics of the phase transition can be obtained by neglecting the fluctuations represented by terms of the type $< f^n - < f >^n >$ in the curly bracket of the above equation. The free energy F obtained in this way will be called the mean-field free energy F_{MF} because it depends only on the mean value of f which varies with the mean value of the spin field, $< s(\alpha_i) >$. We would like to examine the behavior of the mean-field free energy F_{MF} as a function of the variational parameter λ. In this mean-field approximation, we have

$$e^{-\beta_B F_{MF}} = \exp\left\{ < \beta_B J \sum_{\{p\}} \mathcal{R}e \; \{s(\pm\alpha_1)s(\alpha_2)...s(\alpha_r)\} - \lambda \sum_{i=1}^{i_{max}} \mathcal{R}e \; s(\alpha_i) > \right\}$$

$$\times \sum_{\{\alpha\}} \exp\{\lambda \sum_{i=1}^{i_{max}} \mathcal{R}e \; s(\alpha_i)\}. \qquad (11.41)$$

This is equivalent to

$$\beta_B F_{MF} = - < \beta_B J \sum_{\{p\}} \mathcal{R}e \; \{s(\pm\alpha_1)s(\alpha_2)...s(\alpha_r)\} > + < \lambda \sum_{i=1}^{i_{max}} \mathcal{R}e \; s(\alpha_i) >$$

$$- \ln \sum_{\{\alpha\}} \exp\{\lambda \sum_{i=1}^{i_{max}} \mathcal{R}e \; s(\alpha_i)\}. \qquad (11.42)$$

We evaluate the above equation by using definition (11.35). We find (see Exercise 11.2)

$$< \sum_{\{p\}} \mathcal{R}e\{s(\pm\alpha_1)s(\alpha_2)...s(\alpha_r)\} > = B(d) \left[\frac{\sum_{\alpha_i} \mathcal{R}e \; s(\alpha_i) \exp\{\lambda \mathcal{R}e \; s(\alpha_i)\}}{\sum_{\alpha_i} \exp\{\lambda \mathcal{R}e \; s(\alpha_i)\}} \right]^r,$$

where $B(d)$ is the number of the combinations of $\{p\}$ in the lattice. It is equal to the total number of pairs of nearest neighbors in the $Z(N)$ spin system and to the total number of plaquettes in the $Z(N)$ gauge system. We introduce the function $t(\lambda)$ which is the mean value of the spin at a site (for a spin system) or a link (for a gauge system),

$$t(\lambda) \equiv < \mathcal{R}e\ s(\alpha_i) >= \frac{\displaystyle\sum_{\alpha_i} \mathcal{R}e\ s(\alpha_i) \exp\{\lambda \mathcal{R}e\ s(\alpha_i)\}}{\displaystyle\sum_{\alpha_i} \exp\{\lambda \mathcal{R}e\ s(\alpha_i)\}}. \tag{11.43}$$

Then, in terms of $t(\lambda)$, we have

$$< \sum_{\{p\}} \mathcal{R}e\{s(\pm\alpha_1)s(\alpha_2)...s(\alpha_r)\}>$$

$$= \begin{cases} N_v d\ [t(\lambda)]^2, & \text{for } Z(N) \text{ spin systems,} \\ N_v \frac{d(d-1)}{2}\ [t(\lambda)]^4, & \text{for } Z(N) \text{ gauge systems.} \end{cases} \tag{11.44}$$

We can evaluate the function $t(\lambda)$ explicitly for $Z(2)$ and $Z(3)$, and we obtain

$$t(\lambda) = \begin{cases} (e^\lambda - e^{-\lambda})/(e^\lambda + e^{-\lambda}) & \text{for } Z(2), \\ (e^\lambda - e^{-\lambda/2})/(e^\lambda + 2e^{-\lambda/2}) & \text{for } Z(3). \end{cases} \tag{11.45}$$

We are now in a position to understand the physical meaning of the variational parameter λ. Eqs. (11.43)-(11.45) indicate that the variational parameter λ depends on the expectation value of $\mathcal{R}e\ s(\alpha_i)$ which measures the degree of the alignment of the spins of the system. In particular, a zero value of λ is accompanied by the vanishing of $< \mathcal{R}e\ s(\alpha_i) >$ and a finite value of λ is accompanied by a non-vanishing value of $< \mathcal{R}e\ s(\alpha_i) >$. If $\lambda = 0$, which occurs when $< \mathcal{R}e\ s(\alpha_i) >$ is zero, there is a symmetry with respect to a random distribution of the different orientations of $s(\alpha)$ at different locations of the lattice. If $\lambda \neq 0$, which occurs when $< \mathcal{R}e\ s(\alpha_i) >$ is non-zero, there is a definite alignment of the spins at various locations of the lattice, and the symmetry of the system with respect to a random orientation of the spins is broken.

We evaluate the second term on the right-hand side of Eq. (11.42) (see Exercise 11.3). We have

$$< \sum_{i=1}^{i_{\max}} \mathcal{R}e \ s(\alpha_i) > = \frac{i_{\max} \sum_{\alpha_i} \mathcal{R}e \ s(\alpha_i) \exp\{\lambda \mathcal{R}e \ s(\alpha_i)\}}{\sum_{\alpha_i} \exp\{\lambda \mathcal{R}e \ s(\alpha_i)\}}$$

$$= i_{\max} t(\lambda), \tag{11.46}$$

$$= \begin{cases} N_v \ t(\lambda) & \text{for } Z(N) \text{ spin systems,} \\ N_v \ d \ t(\lambda) & \text{for } Z(N) \text{ gauge systems.} \end{cases}$$

Finally, we evaluate the third term on the right-hand side of Eq. (11.42). We note that the sum of all possible products can be written as the product of all possible sums. We find

$$\sum_{\{\alpha\}} \exp\{\lambda \sum_{i=1}^{i_{\max}} \mathcal{R}e \ s(\alpha_i)\} = \left[u(\lambda)\right]^{i_{\max}} \tag{11.47}$$

$$= \begin{cases} \left[u(\lambda)\right]^{N_v} & \text{for } Z(N) \text{ spin systems,} \\ \left[u(\lambda)\right]^{N_v d} & \text{for } Z(N) \text{ gauge systems,} \end{cases}$$

where $u(\lambda)$ is

$$u(\lambda) = \sum_{\alpha_i} \exp\{\lambda \ \mathcal{R}e \ s(\alpha_i)\}, \tag{11.48}$$

and $u(\lambda)$ and $t(\lambda)$ are related by

$$t(\lambda) = \frac{d \ln u(\lambda)}{d\lambda}. \tag{11.49}$$

The function $u(\lambda)$ for $Z(2)$ and $Z(3)$ is given explicitly by

$$u(\lambda) = \begin{cases} e^\lambda + e^{-\lambda} & \text{for } Z(2), \\ e^\lambda + 2e^{-\lambda/2} & \text{for } Z(3). \end{cases} \tag{11.50}$$

•⟦ Exercise 11.2
Prove Eqs. (11.43)-(11.45) for $Z(N)$ spin and gauge systems.

◇Solution:
We wish to evaluate $< \sum_{\{p\}} \mathcal{R}e \ \{s(\pm\alpha_1)s(\alpha_2)...s(\alpha_r)\} >$ which is defined by Eq. (11.35) as

$$< \sum_{\{p\}} \mathcal{R}e \left\{ s(\pm\alpha_1)s(\alpha_2)...s(\alpha_r) \right\} >$$

$$= \frac{\sum_{\{\alpha\}}\sum_{\{p\}} \mathcal{R}e \left\{ s(\pm\alpha_1)s(\alpha_2)...s(\alpha_r) \right\} \exp\{\lambda \sum_i^{i_{max}} \mathcal{R}e\, s(\alpha_i)\}}{\sum_{\{\alpha\}} \exp\{\lambda \sum_i^{i_{max}} \mathcal{R}e\, s(\alpha_i)\}} .$$

The summation $\sum_{\{\alpha\}} \sum_{\{p\}}$ is carried out over all possible α values for all sites (or links) and combinations of nearest neighbors $\{p\}$ in the lattice.

We first examine the $Z(N)$ spin system. The spin variables are defined on the sites of the lattice. From Eq. (11.32), we have $r = 2$. The summation over $\{p\}$ becomes a summation over all combinations $\{1, 2\}$ of nearest neighbors in the lattice. Each pair of nearest neighbors gives the same contribution. The total sum is the product of the number of nearest neighbor pairs times the contribution from one nearest neighbor pair. There are N_v number of lattice sites and each site has $2d$ number of neighbors, but each pair of nearest neighbors is shared by two sites. The total number of nearest neighbors in the lattice, $B(d)$, is

$$B(d) = N_v \times 2d \times \frac{1}{2}$$
$$= N_v\, d \qquad \text{for } Z(N) \text{ spin system} . \tag{1}$$

For a pair of nearest neighbors $\{1, 2\}$, the expectation value of $\mathcal{R}e\{s(-\alpha_1)s(\alpha_2)\}$ can be obtained by expanding the spin variables $s(-\alpha_1)$ and $s(\alpha_2)$ in terms of their real parts and imaginary parts. The contribution from $\mathcal{I}m\, s(\alpha_i)$ is zero when all possible values of α_i are summed over. Furthermore, $\mathcal{R}e\, s(-\alpha) = \mathcal{R}e\, s(\alpha)$. We obtain

$$< \mathcal{R}e\{s(-\alpha_1)s(\alpha_2) >$$

$$= \frac{\sum_{\{\alpha_1,\alpha_2,\alpha_3,...\}} \mathcal{R}e\{s(-\alpha_1)s(\alpha_2)\} \exp\{\lambda \sum_i^{i_{max}} \mathcal{R}e\, s(\alpha_i)\}}{\sum_{\{\alpha_1,\alpha_2,\alpha_3,...\}} \exp\{\lambda \sum_i^{i_{max}} \mathcal{R}e\, s(\alpha_i)\}}$$

$$= \frac{\sum_{\{\alpha_3,\alpha_4,...\}} \exp\{\lambda \sum_{i=3,4,...}^{i_{max}} \mathcal{R}e\, s(\alpha_i)\} \sum_{\alpha_1} \mathcal{R}e\, s(\alpha_1) \exp\{\lambda\mathcal{R}e\, s(\alpha_1)\} \sum_{\alpha_2} \mathcal{R}e\, s(\alpha_2) \exp\{\lambda\mathcal{R}e\, s(\alpha_2)\}}{\sum_{\{\alpha_3,\alpha_4,...\}} \exp\{\lambda \sum_{i=3,4,...}^{i_{max}} \mathcal{R}e\, s(\alpha_i)\} \sum_{\alpha_1} \exp\{\lambda\mathcal{R}e\, s(\alpha_1)\} \sum_{\alpha_2} \exp\{\lambda\mathcal{R}e\, s(\alpha_2)\}}$$

$$= \left\{ \frac{\sum_{\alpha_i} \mathcal{R}e\, s(\alpha_i) \exp\{\lambda\mathcal{R}e\, s(\alpha_i)\}}{\sum_{\alpha_i} \exp\{\lambda\mathcal{R}e\, s(\alpha_i)\}} \right\}^2 . \tag{2}$$

We therefore find, for $Z(N)$ spin systems,

$$< \sum_{\{p\}} \mathcal{R}e \ \{s(\pm\alpha_1)s(\alpha_2)...s(\alpha_r)\} > = N_v d \left[\frac{\sum_{\alpha_i} \mathcal{R}e \ s(\alpha_i) \exp\{\lambda \mathcal{R}e \ s(\alpha_i)\}}{\sum_{\alpha_i} \exp\{\lambda \mathcal{R}e \ s(\alpha_i)\}} \right]^2,$$

which is the result of Eq. (11.43).

We study next the $Z(N)$ gauge system. The spin variables are defined on the links between the lattice sites. From Eq. (11.32), we have $r = 4$. The summation over $\{p\}$ becomes a summation over all combinations $\{1, 2, 3, 4\}$ of plaquettes in the lattice. Each plaquette gives the same contribution. The total sum is the product of the number of plaquettes times the contribution from a plaquette. There are N_v number of lattice vertices. Consider one such vertex i. Associated with the vertex i, there are $d(d-1)/2$ lattice planes meeting at the vertex. On each plane, there are 4 plaquettes associated with the vertex i and each plaquette is shared by 4 vertices. The number of plaquettes in the lattice, $B(d)$, is

$$B(d) = N_v \times \frac{d(d-1)}{2} \times 4 \times \frac{1}{4}$$

$$= N_v \frac{d(d-1)}{2} \qquad \text{for } Z(N) \text{ gauge system}.$$

For each plaquette $\{1, 2, 3, 4\}$, the expectation value of $\mathcal{R}e \ \{s(\alpha_1)s(\alpha_2)s(\alpha_4)s(\alpha_4)\}$ can be obtained by expanding out the real and the imaginary parts of the spin variables. The imaginary parts give zero contribution when all possible values of α are summed over. Following the same steps as in deriving Eq. (2), we obtain

$$< \mathcal{R}e\{s(\alpha_1)s(\alpha_2)s(\alpha_3)s(\alpha_4)\} > = \left\{ \frac{\sum_{\alpha_i} \mathcal{R}e \ s(\alpha_i) \exp\{\lambda \mathcal{R}e \ s(\alpha_i)\}}{\sum_{\alpha_i} \exp\{\lambda \mathcal{R}e \ s(\alpha_i)\}} \right\}^4.$$

We therefore find for the $Z(N)$ gauge system,

$$< \sum_{\{p\}} \mathcal{R}e \ \{s(\pm\alpha_1)s(\alpha_2)...s(\alpha_r)\} > = N_v \frac{d(d-1)}{2} \left[\frac{\sum_{\alpha_i} \mathcal{R}e \ s(\alpha_i) \exp\{\lambda \mathcal{R}e \ s(\alpha_i)\}}{\sum_{\alpha_i} \exp\{\lambda \mathcal{R}e \ s(\alpha_i)\}} \right]^4,$$

which is the result of Eq. (11.43).

We can evaluate $t(\lambda)$ for $Z(2)$ and $Z(3)$ as follows:

a) In $Z(2)$, the spin variables $s(\alpha)$ take on the values $\{1, e^{i\pi}\}$, and $\mathcal{R}e \ s(\alpha)$ takes on the values $\{1, -1\}$. We have

$$t(\lambda) = \frac{\sum_{\alpha} \mathcal{R}e \ s(\alpha_i) \exp\{\lambda \mathcal{R}e \ s(\alpha_i)\}}{\sum_{\alpha} \exp\{\lambda \mathcal{R}e \ s(\alpha_i)\}}$$

$$= \frac{1 \times e^{\lambda \times 1} + (-1) \times e^{\lambda \times (-1)}}{e^{\lambda \times 1} + e^{\lambda \times (-1)}}$$

$$= \frac{e^{\lambda} - e^{-\lambda}}{e^{\lambda} + e^{-\lambda}} \qquad \text{for } Z(2).$$

b) In $Z(3)$, the spin variable $s(\alpha)$ takes on the values $\{1, e^{i2\pi/3}, e^{i4\pi/3}\}$, and $\mathcal{R}e \; s(\alpha)$ takes on the values $\{1, -\frac{1}{2}, -\frac{1}{2}\}$. We have

$$
\begin{aligned}
t(\lambda) &= \frac{\sum_\alpha \mathcal{R}e \; s(\alpha_i) \exp\{\lambda \mathcal{R}e \; s(\alpha_i)\}}{\sum_\alpha \exp\{\lambda \mathcal{R}e \; s(\alpha_i)\}} \\
&= \frac{1 \times e^{\lambda \times 1} + (-\frac{1}{2}) \times e^{\lambda \times (-\frac{1}{2})} + (-\frac{1}{2}) \times e^{\lambda \times (-\frac{1}{2})}}{e^{\lambda \times 1} + e^{\lambda \times (-\frac{1}{2})} + e^{\lambda \times (-\frac{1}{2})}} \\
&= \frac{e^\lambda - e^{-\lambda/2}}{e^\lambda + 2e^{-\lambda/2}} \quad \text{for } Z(3).
\end{aligned}
$$

]•

•[Exercise 11.3
Evaluate $< \sum_{i=1}^{i_{max}} \mathcal{R}e \; s(\alpha_i) >$ for $Z(N)$ spin and gauge systems.

◇Solution:
For the $Z(N)$ spin system, i refers to a vertex of the lattice. For the $Z(N)$ gauge system, i refers to the link between 2 vertices of the lattice. Because each ith element gives the same contribution, we have

$$
\begin{aligned}
< \sum_{i=1}^{i_{max}} \mathcal{R}e \; s(\alpha_i) > &= i_{max} < \mathcal{R}e \; s(\alpha_i) > \\
&= i_{max} t(\lambda).
\end{aligned}
$$

The quantity i_{max} is given by Eq. (11.25) for the $Z(N)$ spin systems, and by Eq. (11.29) for the $Z(N)$ gauge systems. Using these previous results, we have

$$
< \sum_{i=1}^{i_{max}} \mathcal{R}e \; s(\alpha_i) > = \begin{cases} N_v \, t(\lambda) & \text{for } Z(N) \text{ spin systems}, \\ N_v d \, t(\lambda) & \text{for } Z(N) \text{ gauge systems}. \end{cases}
$$

]•

From Eqs. (11.41-47), the free energy in the mean-field approximation $F_{MF}(\lambda)$ is given by

$$
F_{MF}(\lambda) = -J \, B(d) \, [t(\lambda)]^r + i_{max}(d) T \, \lambda t(\lambda) - i_{max}(d) T \, \ln u(\lambda).
$$
(11.51)

For the $Z(N)$ spin system, we have $r = 2$ and the free energy is quadratic in $t(\lambda)$. For the $Z(N)$ gauge system, we have $r = 4$ and the free energy is quartic in $t(\lambda)$. The mean-field free energy $F_{MF}(\lambda)$ has a minimum at

$$
\begin{aligned}
y(\lambda) &= r \, B(d) \, (J/T)[t(\lambda)]^{r-1} \\
&= i_{max}(d) \, \lambda.
\end{aligned}
$$
(11.52)

We shall specialize to the $Z(3)$ symmetry group and to the three-dimensional space with $d = 3$. We plot $F_{MF}(\lambda)/N_v T$ as a function of λ for different values of J/T for the $Z(3)$ spin system in Fig. 11.3 and for the $Z(3)$ gauge system in Fig. 11.4.

The first term of Eq. (11.51) for F_{MF} gives the effect of the interaction which we can call the interaction term. The second term represents the effect of the entropy due to thermal motion and can be called the entropy term. The third term is logarithmic and is slowly varying; it does not play an important role in the qualitative discussion of the dynamics of phase transition.

Z(3) Spin System

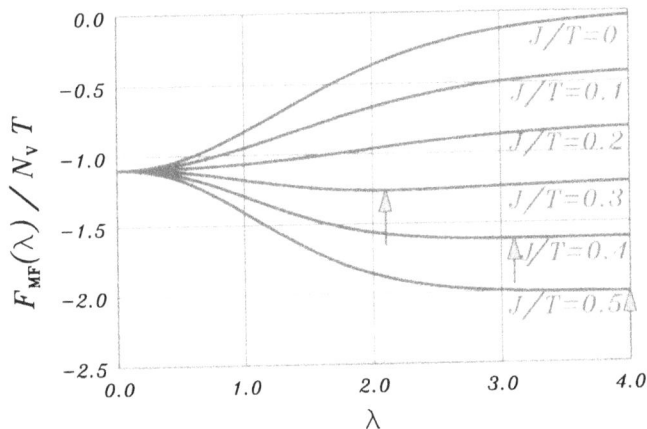

Fig. 11.3 Mean-field free energy F_{MF} for the $Z(3)$ spin system per lattice site, multiplied by β_B, as a function of λ for different values of $\beta_B J$ in three-dimensional space. The arrows indicate the locations of the minima of F_{MF}. This illustrates the occurrence of a second-order phase transition.

We study first the case when the temperature is large. Then, the interaction term is small compared to the entropy term. The free energy F_{MF} behaves as λ^2 around $\lambda \sim 0$. The free energy is a minimum at $\lambda = 0$ and the vacuum state is characterized by $\lambda = 0$. It has the property that $< s(\alpha_i) >$ is zero, with no breaking of symmetry.

What happens as the temperature decreases? In the mean-field approximation, the behavior of the phase transition as the temperature decreases is different for the $Z(3)$ spin system and the $Z(3)$ gauge system. The interaction term has a negative sign and is proportional to the strength of the interaction J. As one observes from Eq. (11.45), the function $t(\lambda)$ for small values of λ is linear in λ. Thus, in the region around $\lambda \sim 0$, the interaction term is quadratic in λ for the $Z(3)$ spin system and is quartic in λ for the $Z(3)$ gauge system. On the other hand, in the region around $\lambda \sim 0$, the entropy term, $\lambda t(\lambda)$, is positive and is quadratic in λ.

As T decreases, the interaction term becomes more important. In the $Z(3)$ spin system, the negative quadratic interaction term will eventually overwhelm the quadratic entropy term. The free energy behaves as $-\lambda^2$ and it has a maximum at $\lambda = 0$. The minimum of the free energy is shifted to a position at $\lambda \neq 0$. The vacuum of the system at low temperatures is characterized by $\lambda \neq 0$ and the symmetry is spontaneously broken. The transition from the symmetric vacuum to the vacuum with broken symmetry occurs in a continuous manner as shown in Fig. 11.3. The phase transition is a second-order phase transition.

Z(3) Gauge System

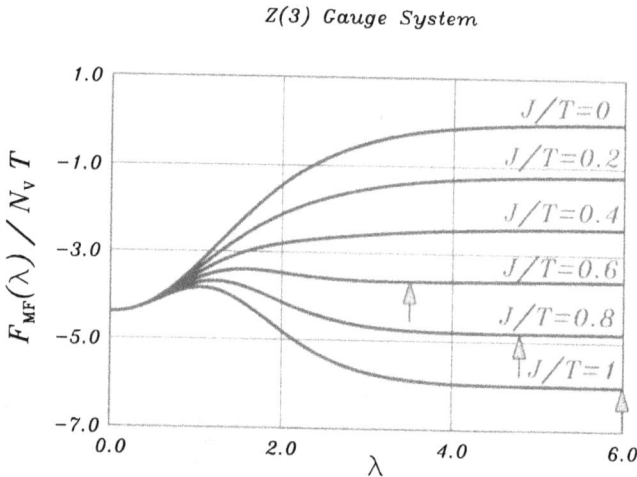

Fig. 11.4 Mean-field free energy F_{MF} for the $Z(3)$ gauge system per lattice site, multiplied by β_B, as a function of λ for different values of $\beta_B J$ in three-dimensional space. The arrows indicate the locations of the minima of F_{MF}. This illustrates the occurrence of a first-order phase transition.

The transition is different for the $Z(3)$ gauge system, as shown in Fig. 11.4. In the region around $\lambda \sim 0$, the interaction term is quartic in λ and goes as $-J\lambda^4$, while the entropy term is quadratic in λ and goes as $\lambda^2 T$. In this region of $\lambda \sim 0$, whatever the magnitude of the inverse temperature may be, the quadratic entropy terms dominates over the quartic interaction term and the free energy behaves as λ^2. It has a local minimum at $\lambda = 0$. As the temperature decreases, the interaction term becomes important for large values of λ and it overwhelms the entropy term. A minimum of the mean-field free energy occurs at $\lambda \neq 0$ (as indicated, for example, by the position of the arrow for $J/T = 0.6$ in Fig. 11.4). When the temperature reaches the critical temperature T_c, the minimum at $\lambda = 0$ and the minimum at $\lambda \neq 0$ lie on the same level of F_{MF}. At a lower temperature, the minimum at $\lambda \neq 0$ is lower than the minimum at $\lambda = 0$, and the vacuum, which is the state of the lowest free energy, switches from the minimum at $\lambda = 0$ to the minimum at $\lambda \neq 0$. The vacuum state is in the phase with a nonvanishing expectation value of $s(\alpha_i)$ and the symmetry with respect to a random distribution of the different orientations of s is broken. The transition at T_c is discontinuous and the phase transition is a first-order phase transition.

What one learns from this simple example is that the occurrence of a phase transition results from the interplay between the effects due to the interaction, and the effects due to the entropy. For low temperatures with a large J/T ratio, there is a tendency for the spins s at different sites of the system to line up to lower the energy. The extent to which the free energy is lowered depends on the the interaction strength J relative to the temperature T and the spatial dimension number d. (See the first term on the right-hand side of Eq. (11.51).) On the other hand, the effect of the entropy due to the thermal motion is to disrupt the alignment of the spins at different sites of the system. For a given dimension d, when the temperature is lower than a certain limit, the tendency to align overwhelms the tendency to disrupt the alignment, and the system acquires a non-zero average spin value. The symmetry of the system is then spontaneously broken.

As the temperature increases, the tendency to disrupt the spin alignment overwhelms the tendency to align, and the ground state of the system restores itself to a state with complete symmetry with respect to the different orientations of the spins at the sites in the lattice. The system resides in the state with full symmetry.

From the above analysis of the $Z(3)$ system, we have the following rough intuitive picture of the phase transition in the analogous $SU(3)$ gluon system. We describe the system in terms of the link variables at the lattice links which are elements of the color $SU(3)$ group. At low temperatures and a large interaction strength so that J/T is large, there is a tendency for the link variables at different sites of the system to correlate in space so as to lower the energy of the system. The extent to which the free energy is lowered depends on the dimension of the space d, and the ratio of the interaction strength J to the temperature T. On the other hand, the effect of the entropy due to thermal motion is to disrupt the correlation of the links at different sites of the system. When the temperature is lower than a certain limit, this tendency to correlate overwhelms the tendency to disrupt the correlation, and the system acquires a non-zero average value for the link variable. The symmetry of the system is then spontaneously broken. As the temperature increases, the tendency to disrupt the correlation overwhelms the tendency to align and the ground state of the system restores itself to a state with complete symmetry.

In the mean-field approximation, the manner in which the symmetry is broken depends on the functional form of the action. For an interaction which is quadratic in the spin variable, the phase transition is continuous while for an interaction that is quartic in the spin variable, the transition is discontinuous. Thus, one would predict in the mean-field approximation that all $Z(N)$ spin systems have second-order phase transition, and all $Z(N)$ gauge systems have first-order phase transitions.

It is necessary to point out that while the mean-field approximation provides an intuitively simple description of the dynamics leading to a phase transition, it may not be an accurate description for some cases. As is clear from Eqs. (11.38) and (11.40), the mean-field approximation neglects the effect of fluctuations which are large for systems with a small spatial dimension d and a small symmetry group order N. The phase transition for the $Z(N)$ gauge system is predicted to be a first-order transition for all N in the mean-field approximation. In numerical Monte-Carlo calculations, the phase transition for the $Z(3)$ gauge system is indeed first-order [10-15], as correctly predicted by the mean-field approximation. However, the phase transition for the $Z(2)$ gauge system is found numerically to be second-order [10-15]. As shown by Flyvbjerg et al. [17], the mean-field approximation using Eq. (11.51) represents the lowest order approximation and is not accurate for the $Z(2)$ gauge system. To obtain the correct behavior in the region around the critical temperature for the $Z(2)$ gauge system, it is necessary to perform a systematic expansion of the path integral around the mean-field solution. The addition of the correction terms alters the behavior of the free energy as a function of λ. When the higher-order corrections are included, one obtains the correct prediction of a second-order phase transition and an excellent agreement with numerical Monte-Carlo calculations [17].

In summary, the mean-field approximation is a good approximation only when $< f^n > - < f >^n$ is small. For the systems we study, the fluctuations may not be small. The numerical Monte-Carlo methods remain to be a more reliable way to study the order of phase transitions.

§11.4 Numerical Results from Lattice Gauge Theory

In a numerical lattice gauge theory calculation, the only input parameter is the strength of the coupling constant g, which is a dimensionless quantity. One needs to know the scale of the lattice spacing a (for example, in units of GeV^{-1}) for a given coupling constant g. For this purpose, one makes a calculation for some physical quantities such as the string tension, using the Wilson loop as discussed in section 11.1, for various values of the coupling constant g. The square root of the string tension is then expressed in units of the inverse of the lattice spacing $1/a$ (Eq. (11.16b)). By equating the string tension to the adopted value of 1 GeV/fm or 0.2 $GeV^2/\hbar c$, the scale of the lattice spacing a in units of GeV^{-1} can be obtained.

When the lattice spacing is small enough, one approaches the "scaling limit" in which the coupling constant and their corresponding length scale are related. The coupling constant $g(a)$ at one distance scale represented by a lattice spacing a, is related to the coupling constant $g_1(a_1)$ at another distance scale represented by a lattice spacing a_1, as follows:

$$g(a) = \frac{g_1(a_1)}{1 + g_1(a_1)^{\frac{(33-2n_f)}{6\pi}} \ln(\frac{a_1}{a})}, \tag{11.53}$$

where n_f is the number of flavors. With the string constant fixed to the adopted value, calculations with various values of the coupling constant will reveal whether the lattice spacing is fine enough such that the above "scaling" relation is satisfied. When the scaling limit is reached, one can be sure that the lattice spacing is fine enough to allow a quantitative estimate of various quantities of interest.

The next step of the calculation is to evaluate the expectation value of the Wilson line operator for various values of the coupling constant. The location g_c at which the Wilson line undergoes a rapid change is noted. The relation between g and a in Eq. (11.53) then allows one to obtain the value of a_c corresponding to that value of g_c. The critical temperature T_c is then given by $\beta T_c = [N_\tau a_c]^{-1}$ where N_τ is the number of lattice spacings in the τ direction. The shape of the transition of the Wilson line around g_c reveals the order of the phase

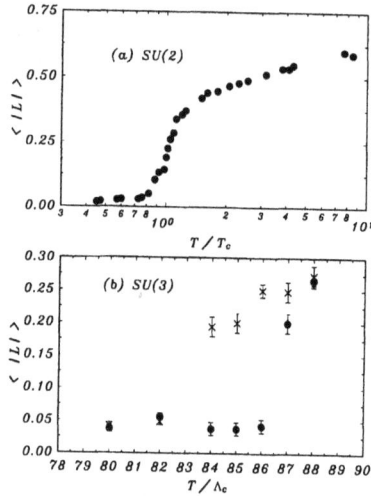

Fig. 11.5 The magnitude of the Wilson line parameter $|<L>|$ as a function of T/T_c or T/Λ_c for the $SU(2)$ and the $SU(3)$ lattice gauge systems (From Ref. 3).

transition. A discontinuity of the Wilson line signals a discontinuity in the energy density and a first-order phase transition [3].

Monte Carlo calculations for the pure gauge theory have been carried out by many groups [3,10-15]. The results for a pure gauge system show that there is a phase transition as the temperature changes. We show the results of Satz *et al.* in Figs. 11.5 and 11.6. At low temperatures, quarks are in a confined phase, as characterized by small expectation values of the Wilson line operator. At high temperatures the quarks are in a deconfined phase, as the absolute value of the Wilson line, $|<L>|$, is large. Quark gluon matter undergoes a transition from the confined phase at low temperatures to the unconfined phase at high temperatures. The order of the phase transition is different depending on the number of the color degrees of freedom. If there are two color degrees of freedom so that the symmetry group is $SU(2)$, the increase in $|<L>|$ as a function of the temperature is gradual (see Fig. 11.5a). The phase transition is a second-order transition. On the other hand, if there are three color degrees of freedom so that the symmetry group is $SU(3)$, the absolute value of the Wilson line increase discontinuously at the critical temperature and it exhibits the characteristics of a first-order phase transition (see Fig. 11.5b). The energy density is also a

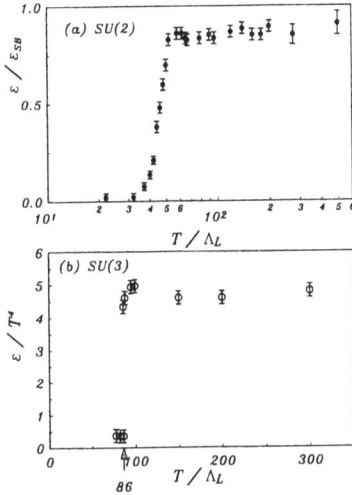

Fig. 11.6 The energy density ϵ as a function of the temperature for the $SU(2)$ and the $SU(3)$ system. [From Ref. 3].

discontinuous function of the temperature as shown in Fig. 11.6. Monte-Carlo calculations in $SU(2)$ and $SU(3)$ gauge theories bear out the predictions [5] of the universality relation between $SU(N)$ and $Z(N)$.

For the $SU(3)$ gauge system, the transition temperature has been found to be 210 MeV [3]. The discontinuity of the energy density at the critical temperature in Fig. 11.6 gives a difference of 0.9 GeV/fm^3. This is the amount of energy density which is needed to supply to the confined phase to make the transition to the deconfined plasma phase. The latent heat, which is the amount of energy per unit volume needed to make the transition, is therefore, 0.9 GeV/fm^3.

§11.5 Lattice Gauge Calculations with Fermions

The discussions in the last section deal with lattice gauge calculations for systems which possess only the gauge field degrees of freedom. The Euclidean action depends only of the link variables U which are functions of the color gauge fields. The quark field is not explicitly included in the dynamics. A quark and an antiquark appear only as a static external probe represented by the Wilson line, which is given in terms of a product of the link variables at a fixed spatial location along a time-like world-line. The expectation value of the Wilson line reveals how a static quark situated at that spatial location interacts with the gauge field. The examination of these gauge field systems provides valuable insight into the nature of quark confinement and the phase transition of quark gluon matter. However, there are additional effects which may be present when quarks are treated dynamically. The discussions of the hadron phase and its transition, chiral symmetry breaking and its restoration and the nature of the vacuum will require the consideration of quark fields as dynamical variables.

Quark fields are fermion fields. The inclusion of fermions explicitly in lattice gauge calculations as additional degrees of freedom is a difficult task. It remains a current subject of active research [18,19]. We shall outline the basic concepts to show how the fermion problem is treated and where the difficulties lie.

In a lattice gauge calculation with dynamical fermions, the basic variables are the gauge field A_μ, the fermion field ψ and its momentum conjugate $\bar{\psi}$. The gauge field A_μ can be represented by the link variable $U_\mu(n)$, which is a matrix in color space. A complete field configuration consists of specifying the link variables at the links and the fermion fields ψ and $\bar{\psi}$ at the lattice points, together with their associated configurations in color space. In principle, if we know the field configuration, the action S_E and the weight factor e^{-S_E} can be calculated. The partition function can then be obtained by summing the weight factor for all field configurations.

In color space, the fermion field $\psi(n)$ is a 3-dimensional complex column vector, and its momentum conjugate is a 3-dimensional row vector. For simplicity of notation and understanding, we shall temporarily suppress the color and flavor degrees of freedom for the discussions to be presented below. The inclusion of these degrees of freedom can be carried out by generalization.

The fermion fields in the path integral method are classical fields which must nevertheless satisfy the anticommuting properties of the corresponding quantum fields. The classical fermion fields $\psi(n)$ and $\bar{\psi}(n)$ which possess such a property are special types of variables known as Grassmann variables. They satisfies the following anticommutation relations:

$$\left\{\psi(n), \psi(m)\right\}_+ = 0, \tag{11.54a}$$

$$\left\{\bar{\psi}(n), \bar{\psi}(m)\right\}_+ = 0, \tag{11.54b}$$

and

$$\left\{\psi(n), \bar{\psi}(m)\right\}_+ = 0, \tag{11.54c}$$

where m can be equal to n. We note that the anticommutation relation (11.54c) for the case $n = m$ differs from the corresponding anticommutation relation for the quantum field case as the fields $\psi(n)$ and $\bar{\psi}(n)$ are classical fields. The anticommutation relation for the quantum field becomes the anticommutation relation for the classical field, when $\hbar$ is set to zero.

Because of these anticommutator relations, we have

$$\psi^2(n) = \bar{\psi}^2(n) = 0. \tag{11.55}$$

Lattice gauge theory with fermions can be simplified by integrating the fermion degrees of freedom analytically. This procedure will lead to an effective action for the remaining link variables. We shall discuss how this is accomplished.

$\oplus\lbrack$ Supplement 11.2

The properties of the Grassmann variables are given in standard textbooks [20,21,22] We shall follow Itzykson and Zuber [20], and Kaku [21] and outline some of the main concepts and results below.

Because of the properties (11.55), the most general polynomial function containing the variable $\psi(n)$ has at most two terms and is given in the form

$$g[\psi(n)] = g_0 + g_1\psi(n), \tag{1}$$

where the coefficients g_i are ordinary complex numbers. Similarly, the most general polynomial function containing the variables $\psi(n)$ and $\bar{\psi}(n)$ has at most four terms in the form

$$P[\psi(n), \bar{\psi}(n)] = P_0 + P_1\psi(n) + P_2\bar{\psi}(n) + P_3\psi(n)\bar{\psi}(n), \tag{2}$$

where the coefficients P_i are ordinary complex numbers.

We need the rules of differentiation and the rules of integration of functions of the Grassmann variables. The rules of integration should be such that the scalar product of two analytic functions of Grassmann variables behaves like the scalar product of two fermion (2-dimensional) spinors, as the spinor property is a distinct characteristic of fermions fields.

One can introduce the derivatives as

$$\frac{d}{d\psi(n)}g_0 = 0, \tag{3}$$

and

$$\frac{d}{d\psi(n)}\psi(n) = 1. \tag{4}$$

Then, the derivatives of the functions g of Eq. (1) and P of Eq. (2) with respect to $\psi(n)$ are given by

$$\frac{\partial}{\partial\psi(n)}g = g_1,$$

and

$$\partial P = \frac{\partial}{\partial \psi(n)} P = P_1 + P_3 \bar{\psi}(n).$$

To take the derivative of the function P with respect to $\bar{\psi}(n)$, we use the anticommutation property of $\psi(n)$ and $\bar{\psi}(n)$ to bring $\bar{\psi}(n)$ to the left in Eq. (2) before the differentiation. The result is

$$\bar{\partial} P = \frac{\partial}{\partial \bar{\psi}(n)} P = P_2 - P_3 \psi(n).$$

In consequence, we have

$$\partial^2 P = \bar{\partial}^2 P = 0, \tag{5}$$

and

$$\bar{\partial} \partial P = -\partial \bar{\partial} P = P_3. \tag{6}$$

Therefore, we obtain from Eqs. (5) and (6) that with respect to any polynomial function P, the differential operators satisfy anticommutator relations

$$\{\partial, \partial\}_+ = 0,$$

$$\{\bar{\partial}, \bar{\partial}\}_+ = 0,$$

and

$$\{\partial, \bar{\partial}\}_+ = 0.$$

We define an analytic function $u[\psi(n)]$ of the Grassmann variable $\psi(n)$ by $\partial u[\psi(n)]/\partial \psi(n) = 0$. The most general form of this function is

$$u[\psi(n)] = u_0 + u_1 \bar{\psi}(n).$$

In line with the scalar product of fermion spinors, we wish to define the scalar product of the analytic functions $u[\psi(n)]$ and $v[\psi(n)]$ as

$$(u, v) = \bar{u}_0 v_0 + \bar{u}_1 v_1, \tag{7}$$

where $\bar{u}_i$ is the complex conjugation of u_i. We shall introduce the rules of integration such that this scalar product of two functions of the Grassmann variable can be represented in terms of an integral over the Grassmann variables. It is desirable to bring this scalar product as an integral into the same form as in the case for bosons [20]. For the boson field, as shown in Eq. (9-38) of Reference [20], the scalar product for two coherent states represented by analytic functions $u(z)$ and $v(\bar{z})$ is proportional to

$$\int d\bar{z} \, dz e^{-\bar{z} z} \overline{u(z)} v(z). \tag{8}$$

The scalar product for the fermion field can be brought to this form by adopting integration rules of the Grassmann variable as

$$\int d\psi(n) = \int d\bar{\psi}(n) = 0, \tag{9}$$

$$\int d\psi(n)\psi(n) = \int d\bar\psi(n)\bar\psi(n) = 1, \tag{10}$$

and requiring that $d\bar\psi(n), \psi(n), d\psi(n)$ and $\bar\psi(n)$ anticommute with each other. Then, the integral (8) is equal to

$$\int d\bar\psi(n)\, d\psi(n) e^{-\bar\psi(n)\psi(n)} \overline{u(\psi(n))} v(\psi(n))$$

$$= \int d\bar\psi(n)\, d\psi(n)\{1 - \bar\psi(n)\psi(n)\}[\bar u_0 + \bar u_1\psi(n)][v_0 + v_1\bar\psi(n)]$$

$$= \bar u_0 v_0 + \bar u_1 v_1$$

$$= (u,v).$$

In the expansion of the exponential function of $\bar\psi(n)\psi(n)$ as a power series of $\bar\psi(n)\psi(n)$, powers higher than the first order do not contribute because of the Pauli exclusion principle represented by Eq. (11.55). Thus, the integration rules (9) and (10), together with the anticommuting properties of the Grassmann variables and their differential elements, gives the proper scalar product for functions of the Grassmann variables.

We can also understand the rules of integration (9) and (10) from a different viewpoint [22,23]. The fermion field variable $\psi(n)$ can assume values in the range from $-\infty$ to $+\infty$,

$$-\infty < \psi(n) < +\infty, \tag{11}$$

but they must also obey the condition (11.55) from the commutation relation,

$$\psi^2(n) = \bar\psi^2(n) = 0. \tag{12}$$

What kinds of integral rules can there be in order that $\psi(n)$ satisfies conditions (11) and (12)? Because of condition (12), the most general polynomial function containing the variable $\psi(n)$ has at most two terms and is given by Eq. (1). The integral of $g[\psi(n)]$ with respect to the whole range of $\psi(n)$ is

$$\int_{-\infty}^{+\infty} g[\psi(n)]d\psi(n) = g_0 \int_{-\infty}^{+\infty} d\psi(n) + g_1 \int_{-\infty}^{+\infty} \psi(n)d\psi(n). \tag{13}$$

Because of the translational invariance of the range of $\psi(n)$ as is evident from Eq. (11), the above integral is the same if $\psi(n)$ is displaced by a constant value c,

$$\int_{-\infty}^{+\infty} g[\psi(n)]d\psi(n) = \int_{-\infty}^{+\infty} g[\psi(n) + c]d\psi(n)$$

$$= (g_0 + c)\int_{-\infty}^{+\infty} d\psi(n) + g_1\int_{-\infty}^{+\infty} \psi(n)d\psi(n). \tag{14}$$

Equating the right-hand sides of Eqs. (13) and (14), we have

$$g_0 \int_{-\infty}^{+\infty} d\psi(n) = (g_0 + c)\int_{-\infty}^{+\infty} d\psi(n), \tag{15}$$

for any arbitrary value of c, and the other integral

$$\int_{-\infty}^{+\infty} \psi(n)d\psi(n) \tag{16}$$

can assume any arbitrary value. To satisfy Eq. (15) for any arbitrary value of c, we must have

$$\int_{-\infty}^{+\infty} d\psi(n) = 0,$$

which is the integration rule Eq. (9). Because the integral (16) can take on any value, we can normalize $\psi(n)$ such that

$$\int_{-\infty}^{+\infty} \psi(n)d\psi(n) = 1,$$

which is the other integration rule Eq. (10). The rules of integration (9) and (10) allow us to integrate general functions of Grassmann variables.

We can consider the case where there are only two independent fermion fields so that ψ is two-dimensional and has components ψ_1 and ψ_2. The conjugate momentum of $\bar\psi$ is also two-dimensional and has components $\bar\psi_1$ and $\bar\psi_2$. We are interested in the integral of the form $\int d\bar\psi_2 d\psi_2 d\bar\psi_1 d\psi_1 \, e^{-\bar\psi A\psi}$ where A is a 2×2 matrix. We have

$$\int d\bar\psi_2 d\psi_2 d\bar\psi_1 d\psi_1 \, e^{-\bar\psi A\psi}$$

$$= \int d\bar\psi_2 d\psi_2 d\bar\psi_1 d\psi_1 \, \frac{(-1)^2}{2}(\bar\psi_{i_1} A_{i_1 j_1}\psi_{j_1})(\bar\psi_{i_2} A_{i_2 j_2}\psi_{j_2})$$

$$= \frac{1}{2}\int d\psi_2 d\psi_1 d\bar\psi_2 d\bar\psi_1 \, \bar\psi_{i_1}\bar\psi_{i_2}\psi_{j_1}\psi_{j_2} A_{i_1 j_1} A_{i_2 j_2}$$

$$= \frac{1}{2}\int d\psi_2 d\psi_1 d\bar\psi_2 d\bar\psi_1 \, (\bar\psi_1\bar\psi_2\psi_1\psi_2 A_{11}A_{22} + \bar\psi_1\bar\psi_2\psi_2\psi_1 A_{12}A_{21}$$

$$+ \bar\psi_2\bar\psi_1\psi_1\psi_2 A_{21}A_{12} + \bar\psi_2\bar\psi_1\psi_2\psi_1 A_{22}A_{11})$$

$$= \frac{1}{2}(A_{11}A_{22} - A_{12}A_{21} - A_{21}A_{12} + A_{22}A_{11})$$

$$= \det A. \tag{17}$$

This result can be easily generalized. We can consider the case where there are N independent fermion fields, as for example when the fermion fields are specified at N lattice vertices. In this case, ψ is an N-dimensional array and has components ψ_i with $i = 1, ..., N$ and A is an $N \times N$ matrix. The integral of interest is

$$\int d\bar\psi_N d\psi_N ... d\bar\psi_2 d\psi_2 d\bar\psi_1 d\psi_1 \, e^{-\bar\psi A\psi}$$

$$= \int d\bar\psi_N d\psi_N ... d\bar\psi_2 d\psi_2 d\bar\psi_1 d\psi_1 \, \frac{(-1)^N}{N!}(\bar\psi_{i_1} A_{i_1 j_1}\psi_{j_1})(\bar\psi_{i_2} A_{i_2 j_2}\psi_{j_2})...(\bar\psi_{i_N} A_{i_N j_N}\psi_{j_N})$$

$$= \int d\psi_N d\bar{\psi}_N ... d\psi_2 d\bar{\psi}_2 d\psi_1 d\bar{\psi}_1 \frac{1}{N!} (\bar{\psi}_{i_1} A_{i_1 j_1} \psi_{j_1}) (\bar{\psi}_{i_2} A_{i_2 j_2} \psi_{j_2}) ... (\bar{\psi}_{i_N} A_{i_N j_N} \psi_{j_N})$$

$$= \frac{1}{N!} \int d\psi_N ... d\psi_2 d\psi_1 d\bar{\psi}_N d\bar{\psi}_2 d\psi_1 d\bar{\psi}_1 \ \bar{\psi}_{i_1} \bar{\psi}_{i_2} ... \bar{\psi}_{i_N} \psi_{j_1} \psi_{j_2} ... \psi_{j_N} A_{i_1 j_1} A_{i_2 j_2} ... A_{i_N j_N}$$

$$= \frac{1}{N!} \epsilon_{12...N} \epsilon_{i_1 i_2 ... i_N} \epsilon_{j_1 j_2 ... j_N} A_{i_1 j_1} A_{i_2 j_2} ... A_{i_N j_N}$$

$$= N! \frac{1}{N!} \epsilon_{12...N} \epsilon_{j_1 j_2 ... j_N} A_{1 j_1} A_{2 j_2} ... A_{N j_N}$$

$$= \det A. \tag{18}$$

This result can be used to integrate out the fermion variables. $\mathbb{1}\oplus$

The Euclidean action (10.65) for QCD on a lattice is of the form:

$$S_E = \sum_{n,n'}^{N} \bar{\psi}(n)(D_{n\,n'}(U) + m\delta_{n\,n'})\psi(n') + S_0(U), \tag{11.56}$$

where the first term is the Euclidean action for quarks including their kinetic energy, their rest mass m, and their coupling to the gauge field, which is represented by the link variables U. The second term S_0 is the action involving only the gauge field. The integer N is the number of independent fermion field variables $\psi(n)$. Each lattice site is associated with a fermion field variable $\psi(n)$; there are in addition the color, the flavor and the spinor degrees of freedom. Hence, N is a multiple of the number of lattice sites. The partition function (10.40) is

$$Z = \int \prod_n d\bar{\psi}(n) \ d\psi(n) \ [dU] \ e^{-S_E}$$

$$= \int \prod_n^N d\bar{\psi}(n) \ d\psi(n) \ [dU] \ e^{-\bar{\psi}(D(U)+m)\psi - S_0(U)},$$

where ψ represents the N-dimensional array with components $\psi(n)$ and $D(U)$ represents the $N \times N$ matrix with elements $D_{nn'}(U)$. From the result of Supplement 11.2, we can integrate out the fermion field variables and find

$$Z = \int \prod_n^N d\bar{\psi}(n) \ d\psi(n) \ [dU] \ \det(D(U) + m)e^{-S_0(U)}$$

$$= \int [dU] \ e^{-S_0(U) + \ln\{\det(D(U)+m)\}}.$$

$$= \int [dU] \ e^{-S_{\text{eff}}(U)}, \tag{11.57}$$

where

$$S_{\text{eff}}(U) = S_0(U) - \ln\{\det(D(U) + m)\}.$$

The *effective action* S_{eff} depends only on the gauge field link variables U as the fermion field variables have been integrated out. Unfortunately, the effective action S_{eff} is nonlocal and is difficult to deal with. The determinant of $(D(U)+m)$ is an $N \times N$ determinant. Because N is a multiple of the number of lattice sites, a direct evaluation of this $N \times N$ determinant is not practical. Furthermore, each term contributing to the determinant is associated with a sign factor 1 or (-1), which depends on its position in the determinant. In consequence, there are intricate cancellations from different terms in the determinant. The difficulty in finding an efficient way to deal with the determinant of $D(U) + m$ is one of the main obstacles associated with the lattice calculations with fermions.

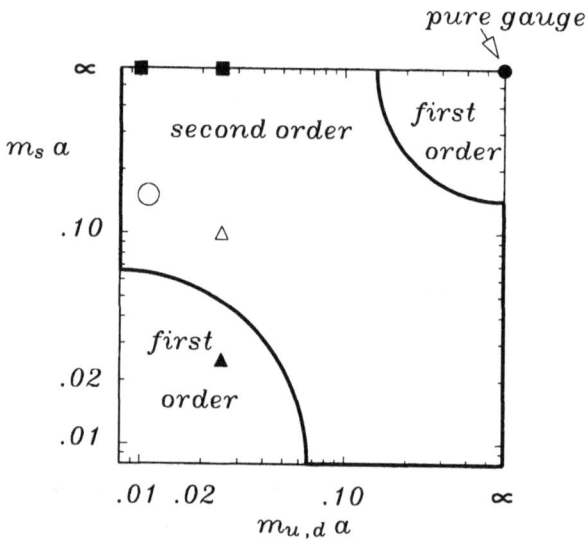

Fig. 11.7 The regions of phase transition as a function of the quark masses (From Ref. 19).

In one approximate treatment of the problem, the mass m is taken to be so large that the determinant of $D(U)+m$ is expanded in powers of $1/m$. This expansion is called the hopping parameter expansion, whose $m \to \infty$ limit is called the quenched approximation [24]. Other approximate methods have also been presented which make use of the sparse nature of the matrix, since the coupling exists only with nearest neighbor fermion fields in a plaquette. Discussions of these methods are beyond the scope of this book. Interested readers may consult Refs. [18], [19], and other references cited therein.

We shall discuss the results of recent lattice calculations with fermions from the Columbia group, performed on a $16^3 \times 4$ lattice using the algorithm of Gottlieb *et al.* [25]. The results are summarized in Fig. 11.7. The order of the phase transition appears to depend on the quark masses. In pure gauge theory, which is equivalent to the case when all the masses of the quarks are infinite, it has been shown that the phase transition is a first-order phase transition [3,10-15]. This is indicated as the solid circle in Fig. 11.7. For two degenerate flavors and quark masses $ma = 0.01$ and 0.025 (where a is the lattice spacing), the phase transition is found to be a second-order phase transition, as indicated by the square symbols in Fig. 11.7. For three degenerate flavors and a mass $ma = 0.025$, the phase transition is found to be a first-order transition, as indicated by the solid triangle in Fig. 11.7. For $m_{u,d}a = 0.025$ and $m_s a = 0.10$, the phase transition is found to be a second-order phase transition, as indicated by the open triangle in Fig. 11.7. The physical point, indicated by the open circle, lies in the region of second-order phase transition. However, because the number of lattice points in the τ direction is small, the scaling limit has not been reached and the results that the phase transition is second-order need to be confirmed by more refined calculations.

§**References for Chapter 11**

1. D. Gross and E. Witten, Phys. Rev. D21, 446 (1980).
2. L. D. Landau and E. M. Lifshitz, *Statistical Physics*, Pergamon Press, Oxford, 1980.
3. H. Satz, Nucl. Phys. A418, 447c (1984).
4. J. B. Kogut, Rev. Mod. Phys. 51, 659 (1979); J. B. Kogut, Rev. Mod. Phys. 55, 775 (1983).
5. B. Svetitsky and L. Yaffe, Nucl. Phys. B210 [FS6], 423 (1982); L. Yaffe and B. Svetitsky, Phys. Rev. D26, 963 (1982). B. Svetitsky Phys. Rep. 132, 1 (1986).
6. An excellent review of the three dimensional Ising gauge theory can be found in Section 5E of J. B. Kogut, Rev. Mod. Phys. 51, 659 (1979).
7. R. Balian, J.M. Drouffe, and C. Itzykson, Phys. Rev. D11, 2098 (1975).
8. C. P. Korthals Altes, Nucl. Phys. B142, 315 (1978).
9. A review of the mean-field approximation is given in M. Creutz, *Quarks, Gluons and Lattices*, Cambridge University Press, 1983, Chap. 14.
10. T. DeGrand, in *Quark Matter*, Springer-Verlag, Berlin, edited by K. Kajantie, 1984.
11. K. Kajantie, C. Montonen, and E. Pietarinen, Z. Phys. C9, 253 (1981).

12. J. Kogut *et al.*, Phys. Rev. Lett. **51**, 869 (1983).
13. T. Celik, J. Engels, and H. Satz, Phys. Lett. **125B**, 411 (1983).
14. F. Fucito and B. Svetitsky, Phys. Lett. **131B**, 165 (1983).
15. A. D. Kennedy, J. Kuti, S. Meyer, and B. Pendleton, Phys. Rev. Lett. **54**, 87 (1985).
16. R. Savit, Rev. Mod. Phys. **52**, 453 (1980).
17. H. Flyvbjerg, H. B. Lantrup, and J. B. Zuber, Phys. Lett. **110B**, 279 (1982); H. Flyvbjerg, P. Mansfield, and B. Soderberg, Nucl. Phys. **B240**, 171 (1984).
18. For a review of the status of lattice results of finite temperature QCD, see A. Ukawa, Nucl. Phys. **A498**, 227c (1989), B. Petersson, Nucl. Phys. **A525**, 237c (1991), T. Hatsuda, Nucl. Phys. **A544**, 27c (1992), N. Christ, Nucl. Phys. **A544**, 81c (1992), and references cited in these references.
19. F. R. Brown *et al.*, Phys. Rev. Lett. **65**, 2491 (1990).
20. C. Itzykson and J.-B. Zuber, *Quantum Field Theory*, McGraw-Hill Book Company, N.Y., 1980.
21. M. Kaku, *Quantum Field Theory, a Modern Introduction*, Oxford University Press, Oxford, 1993.
22. Ta-Pei Cheng and Ling-Fong Li, *Gauge Theory of Elementary Particle Physics*, Clarendon Press, Oxford, 1984.
23. J. I. Kapusta, *Finite-tempertaure Field Theory*, Cambridge University Press, 1989.
24. A. Hasenfratz and P. Hasenfratz, Phys. Lett. **104B**, 489 (1981); C. Lang and H. Nicolai, Nucl. Phys. **B200**, 135 (1982).
25. G. Gottlieb *et al.*, Phys. Rev. **D35**, 2531 (1987).

12. Nucleus-Nucleus Collisions

In the last few chapters, we discuss the static properties of a system of quarks and gluons at high temperatures and densities. Our interest is centered on the possible existence of a new phase of matter, the quark-gluon plasma, consisting of unconfined quarks and gluons at a high temperature or a high density. How is it possible to create matter with a high temperature or a high density favorable for the formation of matter of this kind? In the laboratory, nucleus-nucleus collisions at very high energies provide a promising way to produce high temperature or high density matter. They can be utilized to explore the possible existence of the quark-gluon plasma.

§12.1 Multiple Collisions and Nuclear Stopping

One of the many factors which lead to an optimistic assessment that matter at high temperatures and high densities may be produced with nucleus-nucleus collisions is the occurrence of *multiple collisions*. By multiple collisions, a nucleon of one nucleus may collide with many nucleons in the other nucleus and in the process deposit a large amount of energy in the collision region. Our understanding of QCD is not complete enough to describe this process from first principles. However, much of the evidence of the occurrence of multiple collisions comes from experimental investigations [1,2]. We may consider, for example, the head-on collision of one projectile nucleon onto a target nucleus. In the nucleon-nucleon center-of-mass system, the longitudinal inter-nucleon spacings between target nucleons are Lorentz contracted and can be smaller than 1 fm in high-energy collisions. On the other hand, as we learned from Chapter 6, particle production occurs only when the leading quark and antiquark are separated by a minimum distance of about 1 fm, in the nucleon-nucleon center-of-mass system. Therefore, when the projectile nucleon collides with many target nucleons, particle production arising from the first nucleon-nucleon collision is not finished before the collision of the projectile with another target nucleon begins. How is the second collision affected by the first collision? Although there are models [4-12] to describe how this will proceed, a fundamental theory to describe the process remains one of the unsolved challenges. Experimental data suggest that after the projectile nucleon makes a collision, the projectile-like object that emerges from the first collision appears to continue to collide with other nucleons in the target nucleus on its way through the target nucleus. In each collision, the object (or objects) that emerge along the projectile nucleon direction has a net baryon number of unity because of the conservation of baryon number. One can speak loosely of this object as the projectile or baryon-like object and can describe the multiple collision process in terms of the projectile nucleon making many collisions with the target nucleons, losing energy and momentum in the process, and emerging from the other side of the target nucleus with a much diminished energy and momentum. The number of collisions depend on the thickness of the target nucleus. Therefore, the greater the radius of the target nucleus through which the projectile nucleon traverses, the greater will be the loss of energy and momentum.

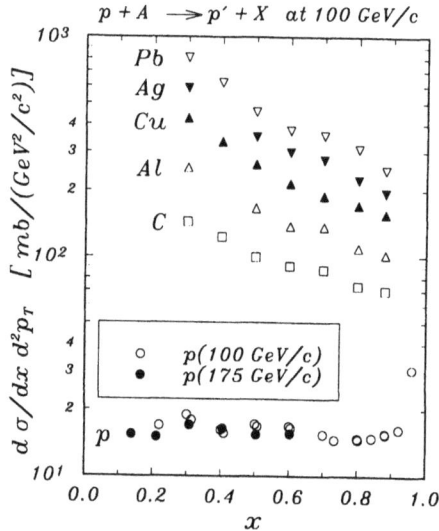

Fig. 12.1 The differential cross section $d\sigma/dx\,d^2p_T$ at $p_T = 0.3$ GeV/c for the reaction $p + A \rightarrow p' + X$ in the collision of a proton with a laboratory momentum of 100 GeV/c on various targets. For pp collisions, data at a laboratory momentum of 175 GeV/c are also included. The data are taken from Ref. [2] and Ref. [3].

Experimental evidence of the occurrence of the multiple collision process can be best illustrated with the data of $p + A \rightarrow p' + X$ reactions in the projectile fragmentation region. In Fig. 12.1 we plot the differential cross section $d\sigma/dx\,d^2p_T$ at $p_T = 0.3$ GeV/c as a function of x, for the collision of protons at 100 GeV/c on various targets. The value of x is a relativistically invariant measure of the fraction of the forward light-cone momentum which is retained by the incident particle after it emerges from the collision. (When no ambiguity arises, it is customary for simplicity of notation to use x to stand for x_+. See Eq. (2.10).) Except in the region near $x \sim 1$ which is the diffractive dissociation region, the distribution of x for the pp collisions is approximately flat, as noted earlier in Chapter 2. This implies that the probability of finding a proton, after an inelastic collision, is approximately a constant as a function of x, independent of the light-cone momentum fraction x. However, as the thickness of the target nucleus increases, the x-distribution of the proton emerging out of the target nucleus is shifted to smaller and smaller values of x. This means that the probability of finding the proton at smaller values of the forward momentum fraction increases as the thickness of the target increases. There is thus a slowing down of the incident proton as it traverses through the target nucleus. The degree of slowing down, or the *stopping power*, is quite large because the differential cross section at small values of x is substantially greater than the cross section near $x \sim 1$.

§12.2 Glauber Model of Nucleus-Nucleus Collision

To describe the dynamics of the nucleus-nucleus collision process, it is useful to consider the collision at the baryon level. After an incident projectile nucleon suffers a collision, the resultant energetic baryon-like object can be treated loosely as the projectile object which continue to make further collisions along the direction of the projectile. The *Glauber model* [13] of multiple-collision processes provides a quantitative consideration of the geometrical configuration of the nuclei when they collide. It is based on the concept of a mean-free path with the assumption of an elementary baryon-baryon cross section. Although a baryon of one nucleus may become excited and may, in principle, have a different cross section when it passes through the other nucleus, we can understand much of the geometrical concepts of the collision process if we take the basic baryon-baryon cross section to be the same throughout the passage of the baryon through the other nucleus.

We begin by defining $t(b)db$ as the probability for having a baryon-baryon collision within the transverse area element db when one baryon is situated at an impact parameter b relative to another baryon. The function $t(b)$ is called the baryon-baryon thickness function. Clearly, the total probability of a collision integrated over all impact parameters is unity and $t(b)$ is normalized according to

$$\int t(\boldsymbol{b})d\boldsymbol{b} = 1. \tag{12.1}$$

For collisions of baryons which are not polarized, the collision does not depend on the orientation of $\boldsymbol{b}$ and $t(\boldsymbol{b})$ depends only on the magnitude $|\boldsymbol{b}| = b$. We shall consider only this case of $t(\boldsymbol{b}) = t(b)$.

The nucleon-nucleon inelastic cross-section is approximately 32 mb and is relatively energy independent . About 6% of this can be attributed to diffractive dissociation for which the leading particle loses very little energy. Elastic or diffractive dissociation collisions lead to a small loss of the energy of the nucleon. For our discussion of particle production and stopping of baryons, two nucleons undergoing elastic or diffractive dissociation collisions can be considered as suffering essentially no collision at all. On this basis, we shall consider only non-diffractive inelastic collisions unless specified otherwise. For simplicity, by a nucleon-nucleon collision, we shall mean a non-diffractive inelastic nucleon-nucleon collision with a cross-section σ_{in} of about 30 mb (see Chapter 2).

When one baryon is situated at an impact parameter $\boldsymbol{b}$ relative to the other baryon, the probability of having a baryon-baryon (inelastic) collision is $t(\boldsymbol{b})\sigma_{in}$.

We consider next the collision of a beam nucleus B and a target nucleus A. We define the probability of finding a baryon in the volume element $d\boldsymbol{b}_B\, dz_B$ in nucleus B at the position $(\boldsymbol{b}_B, z_B)$ as $\rho_B(\boldsymbol{b}_B, z_B)d\boldsymbol{b}_B\, dz_B$ which is normalized according to

$$\int \rho_B(\boldsymbol{b}_B, z_B)d\boldsymbol{b}_B\, dz_B = 1. \tag{12.2a}$$

Note that ρ is equal to the usual number density function divided by the number of baryons in the nucleus. The probability function of finding a baryon in the nucleus A at the position $(\boldsymbol{b}_A, z_A)$ can be similarly defined as $\rho_A(\boldsymbol{b}_A, z_A)$ and normalized as

$$\int \rho_A(\boldsymbol{b}_A, z_A)d\boldsymbol{b}_A\, dz_A = 1. \tag{12.2b}$$

The probability element dP for the occurrence of a baryon-baryon collision when the nuclei B and A are situated at an impact parameter b relative to each other (Fig. 12.2), is the product of (i) the probability element $\rho_A(\boldsymbol{b}_A, z_A)d\boldsymbol{b}_A dz_A$ for finding a baryon in the volume element $d\boldsymbol{b}_B\, dz_B$ in nucleus B at the position $(\boldsymbol{b}_B, z_B)$, (ii) the probability element $\rho_B(\boldsymbol{b}_B, z_B)d\boldsymbol{b}_B dz_B$ for finding a baryon in the volume element

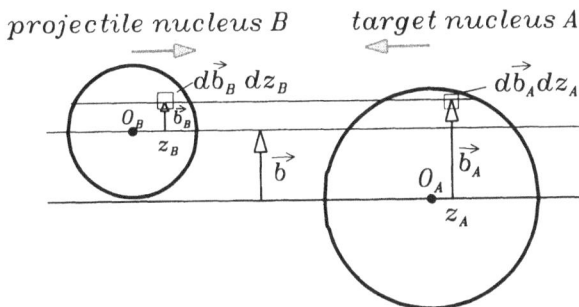

Fig. 12.2 The collision of the projectile nucleus B with the target nucleus A at an impact parameter $\boldsymbol{b}$.

$db_A \, dz_A$ in nucleus A at the position $(\boldsymbol{b}_A, z_A)$, and (iii) the probability $t(\boldsymbol{b} - \boldsymbol{b}_A - \boldsymbol{b}_B)\sigma_{in}$ for a baryon-baryon (inelastic) collision,

$$dP = \rho_A(\boldsymbol{b}_A, z_A)db_A dz_A \ \rho_B(\boldsymbol{b}_B, z_B)db_B dz_B \ t(\boldsymbol{b} - \boldsymbol{b}_A - \boldsymbol{b}_B)\sigma_{in}.$$

By integration, we can now write down the total probability for the occurrence of a baryon-baryon collision when the nuclei B and A are situated at an impact parameter $\boldsymbol{b}$ relative to each other. We call this probability $T(\boldsymbol{b})\sigma_{in}$:

$$T(\boldsymbol{b})\sigma_{in} = \int \rho_A(\boldsymbol{b}_A, z_A)db_A dz_A \rho_B(\boldsymbol{b}_B, z_B)db_B dz_B t(\boldsymbol{b} - \boldsymbol{b}_A - \boldsymbol{b}_B)\sigma_{in},$$

$$(12.3)$$

which defines the *thickness function* $T(\boldsymbol{b})$ for the collision of nucleus B on nucleus A:

$$T(\boldsymbol{b}) = \int \rho_A(\boldsymbol{b}_A, z_A)db_A dz_A \rho_B(\boldsymbol{b}_B, z_B)db_B dz_B t(\boldsymbol{b} - \boldsymbol{b}_A - \boldsymbol{b}_B). \quad (12.4)$$

From Eqs. (12.1)-(12.2), we can show that the thickness function $T(\boldsymbol{b})$ is normalized according to

$$\int T(\boldsymbol{b})d\boldsymbol{b} = 1. \qquad (12.5)$$

It is convenient to introduce the normalized thickness function $T_A(b_A)$ for the nucleus A as

$$T_A(b_A) = \int \rho_A(b_A, z_A) dz_A. \qquad (12.6)$$

Eq. (12.2b) gives the normalization condition for $T_A(b_A)$ as

$$\int T_A(b_A) db_A = 1.$$

The thickness function $T_B(b_B)$ for the other nucleus B can be similarly constructed as

$$T_B(b_B) = \int \rho_B(b_B, z_B) dz_B, \qquad (12.7)$$

and Eq. (12.2a) gives

$$\int T_B(b_B) db_B = 1.$$

In terms of T_A and T_B, Eq. (12.4) can be rewritten as

$$T(b) = \int db_A db_B T_A(b_A) T_B(b_B) t(b - b_A - b_B). \qquad (12.8)$$

For nuclei which are not deformed and oriented, $T_i(b_i)$ depends only on the magnitude of b_i. We shall consider for simplicity only this case of $T_i(b_i) = T_i(b_i)$ and $T(b) = T(b)$.

Having defined the thickness functions, we can write down the probability for various events. The probability for the occurrence of n inelastic baryon-baryon collisions at an impact parameter b is given by

$$P(n, b) = \binom{AB}{n} [T(b)\sigma_{in}]^n [1 - T(b)\sigma_{in}]^{AB-n}, \qquad (12.9)$$

where the first factor on the right-hand side represents the number of combinations for finding n collisions out of AB possible nucleon-nucleon encounters, the second factor gives the probability of exactly n collisions and the third factor gives the probability of having exactly $AB - n$ misses. The total probability for the occurrence of an inelastic event in the collision of A and B at an impact parameter b is the sum of Eq. (12.9) from $n = 1$ to $n = AB$:

$$\frac{d\sigma_{in}^{AB}}{db} = \sum_{n=1}^{AB} P(n, b) = 1 - [1 - T(b)\sigma_{in}]^{AB}. \qquad (12.10)$$

Therefore, from Eq. (12.10), the *total inelastic cross section* σ_{in}^{AB} for the collision of A and B is

$$\sigma_{in}^{AB} = \int d\mathbf{b}\{1 - [1 - T(b)\sigma_{in}]^{AB}\}. \tag{12.11}$$

With the probability of having n baryon-baryon collisions as given by Eq. (12.9), we can show in Supplement 12.1 that the average number of baryon-baryon collisions at an impact parameter b is

$$< n(b) >= \sum_{n=1}^{AB} nP(n,b) = ABT(b)\sigma_{in}. \tag{12.12}$$

We can show that for a nucleon-nucleus collision Eq. (12.9) for the probability having n collisions at an impact parameter b can be written in another simple form. For this case, we have $B = 1$ and

$$P(n,b) = \binom{A}{n}[T(b)\sigma_{in}]^n[1 - T(b)\sigma_{in}]^{A-n},$$

where $T(b)$ is the thickness function of the target nucleus A. Hence, we get

$$P(n,b) = \frac{A!}{n!\,(A-n)!}[T(b)\sigma_{in}]^n[1 - T(b)\sigma_{in}]^{A-n}$$

$$\approx \frac{A^n}{n!}[T(b)\sigma_{in}]^n e^{-T(b)\sigma_{in}(A-n)}$$

$$\approx \frac{[T(b)\sigma_{in}A]^n}{n!}e^{-T(b)\sigma_{in}A}.$$

The distribution of the number of collisions at an impact parameter b is a Poisson distribution with an average number $\bar{n}(b)$:

$$P(n,b) = \frac{[\bar{n}(b)]^n}{n!}e^{-\bar{n}(b)}, \tag{12.13a}$$

where

$$\bar{n}(b) = \sigma_{in}A\int \rho(b,z)dz. \tag{12.13b}$$

This type of probability distribution is also used in the Glauber model.

⊕[Supplement 12.1
We derive here the average number of collisions for a given impact parameter, Eq.
(12.12), and the standard deviation of the number of collisions.

The distribution function for the number of baryon-baryon collisions at a given
impact parameter b is given by Eq. (12.9):

$$P(n,b) = \binom{AB}{n}[T(b)\sigma_{in}]^n[1-T(b)\sigma_{in}]^{AB-n}.$$

The mean number of collisions after averaging over all possible numbers of collisions,
$<n(b)>$, is

$$<n(b)> = \sum_{n=0}^{AB} nP(n,b)$$

$$= \sum_{n=0}^{AB} n\binom{AB}{n}[T(b)\sigma_{in}]^n[1-T(b)\sigma_{in}]^{AB-n}$$

$$= \left[\tau\frac{\partial}{\partial\tau}\sum_{n=0}^{AB}\binom{AB}{n}[\tau\sigma_{in}]^n[1-T(b)\sigma_{in}]^{AB-n}\right]_{\tau=T(b)}$$

$$= \left[\tau\frac{\partial}{\partial\tau}\{1-T(b)\sigma_{in}+\tau\sigma_{in}\}^{AB}\right]_{\tau=T(b)}$$

$$= \left[\tau AB\sigma_{in}\{1-T(b)\sigma_{in}+\tau\sigma_{in}\}^{AB-1}\right]_{\tau=T(b)}$$

$$= ABT(b)\sigma_{in}.$$

We can calculate in a similar way $<n^2(b)>$. We have

$$<n^2(b)> = \sum_{n=0}^{AB} n^2P(n,b)$$

$$= \sum_{n=0}^{AB} n^2\binom{AB}{n}[T(b)\sigma_{in}]^n[1-T(b)\sigma_{in}]^{AB-n}$$

$$= \left[\tau\frac{\partial}{\partial\tau}\tau\frac{\partial}{\partial\tau}\sum_{n=0}^{AB}\binom{AB}{n}[\tau\sigma_{in}]^n[1-T(b)\sigma_{in}]^{AB-n}\right]_{\tau=T(b)}$$

$$= \left[\tau\frac{\partial}{\partial\tau}\tau\frac{\partial}{\partial\tau}\{1-T(b)\sigma_{in}+\tau\sigma_{in}\}^{AB}\right]_{\tau=T(b)}$$

$$= \left[\tau\frac{\partial}{\partial\tau}\tau AB\sigma_{in}\{1-T(b)\sigma_{in}+\tau\sigma_{in}\}^{AB-1}\right]_{\tau=T(b)}$$

$$= ABT(b)\sigma_{in} + AB(AB-1)\{T(b)\sigma_{in}\}^2.$$

Therefore, the standard deviation is

$$< n^2(b)- < n(b) >^2> = < n^2 >_n - < n >^2_n$$
$$= T(b)AB\sigma_{in} + AB(AB-1)[T(b)\sigma_{in}]^2 - (AB)^2[T(b)\sigma_{in}]^2$$
$$= ABT(b)\sigma_{in}\{1 - T(b)\sigma_{in}\}. \qquad \rceil \oplus$$

The average number of baryon-baryon collisions, under the additional condition of the occurrence of an inelastic collision, is

$$< n'(b) >= ABT(b)\sigma_{in}/\{1 - [1 - T(b)\sigma_{in}]^{AB}\}. \qquad (12.14)$$

When we further average $n'(b)$ over the impact parameter b with the weighing factor of the inelastic differential cross section (12.10), we get the average number of baryon-baryon collisions in an inelastic nucleus-nucleus collision:

$$< n' > = \frac{\int < n'(b) > \{d\sigma_{in}^{AB}/db\}db}{\int \{d\sigma_{in}^{AB}/db\}db}$$
$$= \frac{\int db < n'(b) > \{1 - [1 - T(b)\sigma_{in}]^{AB}\}}{\int db\{1 - [1 - T(b)\sigma_{in}]^{AB}\}}.$$

From Eqs. (12.11) and (12.14), we obtain the mean number of baryon-baryon collisions in an inelastic collision of nucleus A and nucleus B as

$$< n' >= AB\sigma_{in}/\sigma_{in}^{AB}. \qquad (12.15)$$

In an inelastic nucleus-nucleus collision without impact parameter selection, the number of baryon-baryon collisions n (for $n = 1$ to AB) has a probability distribution $\mathcal{P}(n)$. This is obtained by integrating $P(n, b)$ over all impact parameters:

$$\mathcal{P}(n) = \frac{\int db P(n, b)}{\sum_{n=1}^{AB} \int db P(n, b)},$$

where the denominator is to ensure that $\mathcal{P}(n)$ is properly normalized as

$$\sum_{n=1}^{AB} \mathcal{P}(n) = 1.$$

From Eqs. (12.9) and (12.11), we have

$$\mathcal{P}(n) = \int db \binom{AB}{n} [T(b)\sigma_{in}]^n [1 - T(b)\sigma_{in}]^{AB-n} \Big/ \sigma_{in}^{AB}. \qquad (12.16)$$

Nucleon-nucleus collisions are special cases of nucleus-nucleus collisions. For these case with $B = 1$, we can simplify the results by taking the nucleon thickness function $T_B(b_B) = \delta(b_B)$ and the baryon-baryon collision thickness function $t(b) = \delta(b)$. Then the thickness function for a nucleon colliding on a target nucleus of target mass number A is $T(b) = T_A(b)$. In an average collision with all possible impact parameters, the probability of having n collisions is

$$\mathcal{P}(n) = \int db \binom{A}{n} [T_A(b)\sigma_{in}]^n [1 - T_A(b)\sigma_{in}]^{A-n} \Big/ \sigma_{in}^{pA}, \qquad (12.17)$$

and the average number of collisions is given by

$$< n' > = \frac{A\sigma_{in}}{\sigma_{in}^{pA}}. \qquad (12.18)$$

The formalism developed above gives the total number of collisions for a given impact parameter and the total inelastic cross section. In some problems, it is also necessary to know the history of the collision process, as for example, in the dynamics involving the slowing-down of the baryons. It is convenient to adopt a row-on-row picture of one row of n nucleons from nucleus B colliding with another row of m nucleons in the other nucleus A.

The probability of finding a nucleon in a tube of cross section σ_{in} at the transverse coordinate b_B is $T_B(b_B)\sigma_{in}$. The probability of finding n nucleons in a tube of cross section σ_{in} at the transverse coordinate b_B in nucleus B is given by

$$\binom{B}{n} [T_B(b_B)\sigma_{in}]^n [1 - T_B(b_B)\sigma_{in}]^{B-n} \qquad (12.19)$$

which has an average value of

$$< n(b_B) > = B T_B(b_B)\sigma_{in}. \qquad (12.20)$$

Similarly, the probability of finding m nucleons in a tube of cross section σ_{in} at the transverse coordinate b_A in nucleus A is given by

$$\binom{A}{m} [T_A(b_A)\sigma_{in}]^m [1 - T_A(b_A)\sigma_{in}]^{A-m} \qquad (12.21)$$

which has an average value of

$$< m(b_A) > = A T_A(b_A)\sigma_{in}. \qquad (12.22)$$

The probability of having a nucleon-nucleon collision in the area element $d\boldsymbol{b}_B$ is $t(\boldsymbol{b} - \boldsymbol{b}_A + \boldsymbol{b}_B)db$ where $\boldsymbol{b}$ is the impact parameter of B relative to A. Putting all the factors together, we find the probability of having n nucleons of B in a tube of cross section σ_{in} colliding with m nucleons of A in a similar tube is

$$P(n, m, \boldsymbol{b}_B, \boldsymbol{b}) = \int t(\boldsymbol{b} - \boldsymbol{b}_A + \boldsymbol{b}_B) d\boldsymbol{b}_A \binom{B}{n} [T_B(b_B)\sigma_{in}]^n [1 - T_B(b_B)\sigma_{in}]^{B-n}$$

$$\times \binom{A}{m} [T_A(b_A)\sigma_{in}]^m [1 - T_A(b_A)\sigma_{in}]^{A-m} . \qquad (12.23)$$

The nucleon-nucleon thickness function t can be approximated by a delta function and the probability function P can then be simplified as the product

$$P(n, m, \boldsymbol{b}_B, \boldsymbol{b}) = \binom{B}{n} [T_B(b_B)\sigma_{in}]^n [1 - T_B(b_B)\sigma_{in}]^{B-n}$$

$$\times \binom{A}{m} [T_A(|\boldsymbol{b} - \boldsymbol{b}_B|)\sigma_{in}]^m [1 - T_A(|\boldsymbol{b} - \boldsymbol{b}_B|)\sigma_{in}]^{A-m}. \qquad (12.24)$$

For many problems, the dynamics is insensitive to the spatial transverse coordinates $\boldsymbol{b}_B$ of the tube but is more sensitive to the number of nucleon-nucleon collisions in the tube. One can then integrate the transverse coordinates $\boldsymbol{b}_B$ to obtain the probability for collisions of n nucleons in B with m nucleons in A at an impact parameter b :

$$P(n, m, b) = \int \frac{d\boldsymbol{b}_B}{\sigma_{in}^{AB}} \binom{B}{n} [T_B(b_B)\sigma_{in}]^n [1 - T_B(b_B)\sigma_{in}]^{B-n}$$

$$\times \binom{A}{m} [T_A(|\boldsymbol{b} - \boldsymbol{b}_B|)\sigma_{in}]^m [1 - T_A(|\boldsymbol{b} - \boldsymbol{b}_B|)\sigma_{in}]^{A-m} , \qquad (12.25)$$

where the total inelastic cross section σ_{in}^{AB} for this row-on-row method of decomposition is

$$\sigma_{in}^{AB} = \int d\boldsymbol{b} \sum_{n=1, m=1} P(n, m, b).$$

§12.3 Some Analytic Examples

The basic thickness function $t(b)$ can be well approximated by a Gaussian function with a standard deviation β_p. If the colliding nuclei are small, their density function ρ can also be taken to be a Gaussian function of the spatial coordinates. Consequently, the thickness function of Eq. (12.3) can be conveniently written as [14]

$$T(b) = \exp(-b^2/2\beta^2)/2\pi\beta^2 \qquad (12.26)$$

where

$$\beta^2 = \beta_A^2 + \beta_B^2 + \beta_p^2.$$

In terms of the standard root-mean-squared-radius parameter r_0' for nuclei A and B, the standard deviation β_A (or, similarly, β_B) is given by

$$\beta_A = r_0' A^{1/3}/\sqrt{3},$$

and β_p, the thickness function parameter for nucleon-nucleon collision, is 0.68 fm [14]. The Gaussian form of the thickness function is actually a more general shape than one may at first expect. It turns out that for the collision of two nuclei with a spatial density in the shape of a Fermi distribution (as in the case of a heavy nucleus), numerical calculations show that the thickness function can be well approximated by a Gaussian function, except that the effective parameter r_0' is slightly larger (r_0' is found to be ~ 1.05 fm in Ref. 14).

With a Gaussian thickness function, the total inelastic cross section is

$$\sigma_{in}^{AB} = -2\pi\beta^2 \sum_{n=1}^{AB} \binom{AB}{n} (-f)^n/n, \qquad (12.27)$$

where f is the dimensionless quantity

$$f = \sigma_{in}/2\pi\beta^2.$$

When there is no impact parameter selection, the probability distribution of the number of baryon-baryon collisions in an inelastic nucleus-nucleus collision is

$$\mathcal{P}(n) = \frac{2\pi\beta^2}{\sigma_{in}^{AB}} \binom{AB}{n} \sum_{m=0}^{AB-n} \binom{AB-n}{m} \frac{(-1)^m f^{m+n}}{m+n}. \qquad (12.28)$$

$\oplus$[Supplement 12.2
We derive here the results (12.27) and (12.28) for the Gaussian thickness function
(12.26). From Eq. (12.11), the inelastic nucleus-nucleus collision is

$$\sigma_{in}^{AB} = \int d\mathbf{b}\{1 - [1 - T(b)\sigma_{in}]^{AB}\}$$

$$= \int d\mathbf{b}\{1 - [1 - (\sigma_{in}/2\pi\beta^2)e^{-b^2/2\beta^2}]^{AB}\}$$

$$= \int d\mathbf{b}\{1 - [1 - fe^{-b^2/2\beta^2}]^{AB}\}$$

$$= -\int d\mathbf{b} \sum_{n=1}^{AB} \binom{AB}{n}(-f)^n e^{-nb^2/2\beta^2}$$

$$= -2\pi\beta^2 \sum_{n=1}^{AB} \binom{AB}{n}(-f)^n/n.$$

Another equivalent expression for the inelastic cross section can be obtained by
writing the inelastic cross section as

$$\sigma_{in}^{AB} = 2\pi\beta^2 I_{AB},$$

where

$$I_{AB} = \int_0^\infty dx\{1 - [1 - fe^{-x}]^{AB}\}.$$

We have

$$I_{AB} = \int_0^\infty dx\{1 - [1 - fe^{-x}]^{AB-1} + fe^{-x}[1 - fe^{-x}]^{AB-1}\},$$

$$= \int_0^\infty dx\{1 - [1 - fe^{-x}]^{AB-1}\} + \frac{[1 - fe^{-x}]^{AB}}{AB}\bigg|_0^\infty,$$

$$= I_{AB-1} + \frac{1 - (1 - f)^{AB}}{AB}.$$

The above recurrence relation gives

$$\sigma_{in}^{AB} = 2\pi\beta^2 \sum_{n=1}^{AB} \frac{1 - (1 - f)^n}{n},$$

which can also be written as

$$\sigma_{in}^{AB} = -2\pi\beta^2 \sum_{n=1}^{AB} \sum_{m=1}^{n} \binom{n}{m}(-f)^m/n.$$

If f is small, then

$$\sigma_{in}^{AB} = AB\sigma_{in}.$$

To obtain $\mathcal{P}(n)$, the distribution of the number of baryon-baryon collisions, we have from Eq. (12.16)

$$
\begin{aligned}
\mathcal{P}(n) &= \int \frac{d\boldsymbol{b}}{\sigma_{in}^{AB}} \binom{AB}{n} [T(b)\sigma_{in}]^n [1 - T(b)\sigma_{in}]^{AB-n}, \\
&= \int \frac{\pi db^2}{\sigma_{in}^{AB}} \binom{AB}{n} [f e^{-b^2/2\beta^2}]^n [1 - f e^{-b^2/2\beta^2}]^{AB-n}, \\
&= \frac{2\pi\beta^2}{\sigma_{in}^{AB}} \binom{AB}{n} \int_0^f dy\, y^{n-1} (1-y)^{AB-n} \\
&= \frac{2\pi\beta^2}{\sigma_{in}^{AB}} \binom{AB}{n} \sum_{m=0}^{AB-n} \binom{AB-n}{m} \frac{(-1)^m f^{m+n}}{m+n},
\end{aligned}
$$

which is Eq. (12.28).

We can consider another interesting limit in which one of the two colliding nuclei is much greater than the other. The thickness function can be approximated by using a sharp-cutoff density distribution. It has the form [14]

$$
T(b) = \frac{3}{2\pi R^3} \sqrt{R^2 - b^2}\ \theta(R - b) \tag{12.29}
$$

where R is the sum of the radii of the two colliding nuclei. We can easily show that the inelastic cross section is

$$
\sigma_{in}^{AB} = \pi R^2 \left\{ 1 + \frac{2}{F^2} \left[\frac{1 - (1-F)^{AB+2}}{AB+2} - \frac{1 - (1-F)^{AB+1}}{AB+1} \right] \right\} \tag{12.30}
$$

where F is a dimensionless ratio

$$
F = \frac{3\sigma_{in}}{2\pi R^2}.
$$

In an inelastic nucleus-nucleus collision, if there is no impact parameter selection the probability distribution of the number of baryon-baryon collisions is

$$
\mathcal{P}(n) = \frac{2\pi R^2}{\sigma_{in}^{AB}} \binom{AB}{n} \sum_{m=0}^{AB-n} \binom{AB-n}{m} \frac{(-1)^m F^{m+n}}{m+n+2}. \tag{12.31}
$$

⊕[Supplement 12.3

We derive here the results (12.30) and (12.31) for the sharp-cut-off thickness function (12.29). From Eq. (12.11), the inelastic nucleus-nucleus collision is

$$\sigma_{in}^{AB} = \int d\mathbf{b}\{1 - [1 - T(b)\sigma_{in}]^{AB}\}$$

$$= \int d\mathbf{b}\{1 - [1 - \frac{3\sigma_{in}}{2\pi R^2}(1 - b^2/R^2)^{1/2}\theta(R - b)]^{AB}\}$$

$$= \pi R^2 \int_0^1 dx\{1 - [1 - F(1 - x)^{1/2}]^{AB}\},$$

$$= \pi R^2 - \pi R^2 \int_1^0 (1 - Fy)^{AB}(-2y)dy$$

$$= \pi R^2 + \frac{2\pi R^2}{F} \int_1^0 [(1 - Fy)^{AB+1} - (1 - Fy)^{AB}]dy,$$

which leads to Eq. (12.30). To obtain $\mathcal{P}(n)$, the distribution of the number of baryon-baryon collisions, we have from Eq. (12.16)

$$\mathcal{P}(n) = \int \frac{d\mathbf{b}}{\sigma_{in}^{AB}} \binom{AB}{n} [T(b)\sigma_{in}]^n [1 - T(b)\sigma_{in}]^{AB-n},$$

$$= \int \frac{\pi db^2}{\sigma_{in}^{AB}} \binom{AB}{n} [F(1 - b^2/R^2)^{1/2}\theta(R - b)]^n [1 - F(1 - b^2/R^2)^{1/2}]^{AB-n},$$

$$= \frac{2\pi R^2}{\sigma_{in}^{AB}} \binom{AB}{n} \int_0^1 dy[F(1 - x)^{1/2}]^n [1 - F(1 - x)^{1/2}]^{AB-n}$$

$$= \frac{2\pi R^2}{\sigma_{in}^{AB} F^2} \binom{AB}{n} \int_0^F dy\, y^{n+1}(1 - y)^{AB-n}$$

$$= \frac{2\pi R^2}{\sigma_{in}^{AB}} \binom{AB}{n} \sum_{m=0}^{AB-n} \binom{AB - n}{m} \frac{(-1)^m F^{m+n}}{m + n + 2},$$

which is Eq. (12.31).

1⊕

§References for Chapter 12

1. W. Busza, and A. Goldhaber, Phys. Lett. 139B, 235 (1984).
2. A. Barton *et al.*, Phys. Rev. D27, 2580 (1983).
3. A. Brenner, *et al.*, Phys. Rev. D26, 1497 (1982).
4. B. Andersson, G. Gustavson, and B. Nilsson-Almqvist, Nucl. Phys. B281, 289 (1987).
5. B. Nilsson-Almqvist and E. Stenlund, Comp. Phys. Comm. 43, 387 (1987).
6. A. Capella, U. Sukhatame, C-I. Tan and J. Tran Thanh Van, Phys. Lett. 81B, 68 (1979); A. Capella, U. Sukhatame, and J. Tran Thanh Van, Zeit. Phys. C3, 329 (1980); A. J. A. Casado, C. Pajares, A. V. Ramello, and J. Tran Thanh

Van, Zeit. Phys.C33, 541 (1987); C. Pajares, A. V. Ramello, Phys. Rev. D31, 2800 (1985).

7. K. Werner, Phys. Rev. D39, 780 (1989).

8. K. Kinoshita, A. Minaka, and H. Sumiyoshi, Prog. Theo. Phys. 63, 1268 (1980).

9. C. Y. Wong, Phys. Rev. Lett. 52, 1393 (1984); C. Y. Wong, Phys. Rev. D30, 961 (1984); C.Y. Wong, Phys. Rev. D30, 972 (1984); C.Y. Wong D32, 94 (1985); C. Y. Wong and Z. D. Lu, Phys. Rev. D39, 2606 (1989).

10. H. Sorge, H.Stöcker, and W. Greiner, Nucl. Phys. A498, 567c (1989); H. Sorge, A. von Keitz, R. Mattiello, H. Stöcker, and W. Greiner, Phys. Lett. B243, 7 (1990); R. Matiello et al, Nucl. Phys. B24, 221 (1991).

11. X. N. Wang and M. Gyulassy, Phys. Rev. D44, 3501 (1991); M. Gyulassy and X. N. Wang, LBL Report LBL-32682 (1993).

12. K. Geiger and B. Müller, Nucl. Phys. A544, 467c (1992); K. Geiger, Phys. Rev. D46, 4965 (1992); K. Geiger, Phys. Rev. D47, 133 (1993).

13. R. J. Glauber, in Lectures in Theoretical Physics, edited by W. E. Brittin and L. G. Dunham (Interscience, N.Y., 1959), Vol. 1, p. 315.

14. C. Y. Wong, Phys. Rev. D30, 961 (1984).

13. High-Energy Heavy-Ion Collisions and Quark-Gluon Plasma

The nuclear stopping power, as revealed by nucleon-nucleus and nucleus-nucleus collisions, indicates that the colliding nuclear matter loses a substantial fraction of its energy in the collision process. Because the energy lost by the colliding nuclear matter is deposited in the vicinity of the center of mass with the production of hadrons, high-energy nucleus-nucleus collisions provide an excellent tool to produce regions of very high energy densities. As estimated by Bjorken [1], the energy density can be so high that these reactions might be utilized to explore the existence of the quark-gluon plasma. We shall discuss the basis for this suggestion and the space-time scenario put forth by Bjorken [1]. We shall examine the hydrodynamical evolution of the quark gluon plasma to see how its energy density and its temperature depend on the proper time.

§13.1 Nuclear Stopping Power and the Baryon Content

The term *"nuclear stopping power"* was introduced in high-energy nucleus-nucleus collisions by Busza and Goldhaber [2] to refer to the degree of stopping an incident nucleon suffers when it impinges on the nuclear matter of another nucleus. Besides depicting an important aspect of the reaction mechanism, it is related to the question of quark-gluon plasma formation [1]. The loss of the incident nuclear matter kinetic energy is accompanied by the production of a large number of particles (mostly pions). Therefore, in high-energy nucleus-nucleus central collisions, a large fraction of the longitudinal energy is converted into the energy of the hadronic matter produced in the vicinity of the center of mass of the colliding system. The degree of stopping will reveal whether the energy density attained is high enough to allow a phase transition, leading to the formation of a quark-gluon plasma.

Qualitatively speaking, the collision of high-energy heavy ions can be divided into two different energy regions: the " baryon-free quark-gluon plasma" region (or the " pure quark-gluon-plasma" region) with $\sqrt{s} > 100$ GeV per nucleon, and the "baryon-rich quark-gluon plasma" region (or the " stopping" region) with $\sqrt{s} \sim 5 - 10$ GeV per nucleon, which corresponds to about many tens of GeV per projectile nucleon in the laboratory system. In the baryon-free quark-gluon-plasma region, we need to know the nuclear stopping power to determine whether the beam baryons and the target baryons will recede away from the center of mass without being completely stopped, leaving behind quark-gluon plasma with very little baryon content. In the baryon-rich region or the stopping region, the nuclear stopping power determines whether the colliding baryons will be stopped in the center-of-mass system and pile up to form a quark-gluon plasma with a large baryon density.

⊕⟦ Supplement 13.1

We can get an idea of baryon stopping from some simple considerations by following the light-cone momentum fraction, or the rapidity variable, of a baryon as it traverses through another nucleus, making successive collisions on its way [3,4].

From the phenomenon of Feynman scaling (Chapter 3), we learn that the momentum distribution of nucleons after an inelastic collision can be characterized by the forward light-cone momentum fractions, and is rather insensitive to the total energy. We denote the forward light-cone momentum fraction of a baryon after the nth collision, relative to its initial forward light-cone momentum before any collision, by x_n. We can study a simple stopping law in which the probability distribution of finding the baryon with a light-cone momentum fraction x_1 after a collision is related to the initial momentum fraction $x_0 = 1$ by [3,4]

$$w(x_0, x_1) = \alpha x_1^{\alpha-1}. \tag{1}$$

In subsequent collisions, we shall assume that the same stopping law is applicable. (The case can be easily generalized to a different stopping law.) The probability distribution of finding the baryon with a light-cone momentum fraction x_n after the nth collision is related to the forward light-cone momentum fraction x_{n-1} by

$$w(x_{n-1}, x_n) = \alpha \left(\frac{x_n}{x_{n-1}} \right)^{\alpha-1} \frac{1}{x_{n-1}}. \tag{2}$$

As x_n cannot exceed x_{n-1}, the distribution $w(x_{n-1}, x_n)$ in Eq. (2) has been normalized according to

$$\int_0^{x_{n-1}} dx_n\, w(x_{n-1}, x_n) = 1.$$

The stopping law (2) means that if the momentum distribution of the baryon after the $(n-1)$th collision is $D^{(n-1)}(x_{n-1})$, then the momentum distribution of the baryon after the nth collision $D^{(n)}(x_n)$ is, by folding,

$$D^{(n)}(x_n) = \int_{x_n}^1 dx_{n-1}\, D^{(n-1)}(x_{n-1})\, w(x_{n-1}, x_n), \tag{3}$$

where the momentum distribution of the baryon is normalized according to

$$\int dx_n\, D^{(n)}(x_n) = 1.$$

We start with a nucleon with an initial momentum distribution

$$D^{(0)}(x_0) = \delta(x_0 - 1).$$

We obtain from Eq. (3)

$$D^{(1)}(x_1) = \int_{x_1}^{1} dx_0 \delta(x_0 - 1)\alpha\left(\frac{x_1}{x_0}\right)^{\alpha-1}\frac{1}{x_0} = \alpha x_1^{\alpha-1},$$

$$D^{(2)}(x_2) = \int_{x_2}^{1} dx_1 \alpha x_1^{\alpha-1}\alpha\left(\frac{x_2}{x_1}\right)^{\alpha-1}\frac{1}{x_1} = \alpha x_2^{\alpha-1}(-\alpha \ln x_2),$$

$$D^{(3)}(x_3) = \int_{x_3}^{1} dx_2 \alpha x_2^{\alpha-1}(-\alpha \ln x_2)\alpha\left(\frac{x_3}{x_2}\right)^{\alpha-1}\frac{1}{x_2} = \alpha x_3^{\alpha-1}\frac{(-\alpha \ln x_3)^2}{2},$$

and

$$D^{(n)}(x_n) = \alpha x_n^{\alpha-1}\frac{(-\alpha \ln x_n)^{n-1}}{(n-1)!}. \tag{4}$$

From these distributions, we find that the average value of the light-cone variable x_n is given by

$$< x_n >= \left(\frac{\alpha}{\alpha+1}\right)^n. \tag{5}$$

The result is that for the stopping law (2), the average momentum fraction decreases by a factor of $\alpha/(\alpha+1)$ after each collision,

$$< x_n >= \frac{\alpha}{\alpha+1} < x_{n-1} > . \tag{6}$$

From the light-cone momentum distribution, we can get the distribution of the rapidity y_n of the baryon after n collisions by making the change of variables as

$$x_n = \frac{m_T}{m}e^{y_n - y_B},$$

where m_T is the transverse mass of the baryon and y_B is the beam rapidity. Taking the transverse mass to be a constant for simplicity, the distribution of the rapidity variable y_n after n collisions is

$$D^{(n)}(y_n) = \alpha\left(\frac{m_T}{m}\right)^\alpha e^{\alpha(y_n - y_B)}\frac{[-\alpha\{\ln(m_T/m) + y_n - y_B\}]^{n-1}}{(n-1)!}. \tag{7}$$

From this rapidity distribution, we find that the average value of the rapidity y_n is given by the relation

$$y_B - < y_n > - \ln\left(\frac{m_T}{m}\right) = \frac{n}{\alpha}, \tag{8}$$

and the rapidity loss in the nth collision is

$$< y_{n-1} > - < y_n >= \frac{1}{\alpha}. \tag{9}$$

The result is that for the stopping law (2), the baryon loses $1/\alpha$ units of rapidity after each collision.

Experimentally, for a single nucleon-nucleon collision, the value of α is 1 (see Chapter 3). This means that the average forward light-cone momentum after a collision is half of the forward light-cone momentum of the incident beam nucleon and the average rapidity loss in a collision is one unity of rapidity. For a collision of a baryon with a succession of nucleons, various estimates places $\alpha = 1$ for all collisions [4], or $\alpha = 1$ for one of the collisions and $\alpha = 3$ for the other collisions [3]. The average rapidity loss in a succession of n collisions is in the range of n and $1 + (n - 1)/3$.

On the other hand, the mean-free-path of a baryon is of the order of 2.5 fm while the radius of a nucleus such as Au is 7 fm. For a nucleon on a Au target nucleus, a baryon would suffer an average of about 6 collisions if it travels through the diameter of a Au nucleus and when we average over all impact parameters the number of collisions in an inelastic event is about 4. The average rapidity loss for a baryon colliding on a Au nucleus is roughly from 2 to 4 units of rapidity.

For a collision at an energy of a few GeV per nucleon in the center-of-mass system (which corresponds to a projectile energy of many tens of GeV per nucleon on a fixed target in the laboratory), the separation between the beam rapidity and the target rapidity is about 3-4 units. This is of the same order as the average rapidity loss of the baryons in a central collision involving two heavy nuclei. Furthermore, in this center-of-mass system where the baryons may be stopped, the Lorentz contraction also places the baryon matter in a closer longitudinal proximity before the collisions take place. As a consequence, as suggested by Busza and Goldhaber [2], the baryon density of the matter created can be very high. The baryon density may be so high that the ground state of the baryon matter with this baryon density may be in the deconfined quark-gluon plasma phase rather than in the hadron phase, as discussed in Chapter 9. A transition from hadron matter to the deconfined phase of a quark-gluon plasma may take place. High-energy heavy-ion collisions in this "stopping" region may lead to the formation of a quark-gluon plasma with a high baryon content.

Indirect evidence for the occurrence of the stopping of baryon matter in heavy-ion collisions comes from many measurements. [5,6,7]. We show results from the WA80 Collaboration for collisions of ^{16}O on various targets in Fig. 13.1. In the experimental measurement, one sets up a "zero-degree-calorimeter" which collects the particles emitted within 0.3° of the incident beam ($\eta > 6$). It measures the amount of energy which is transmitted into the forward direction during the collision. If the stopping of the incident projectile is not effective, as in the case when the target nucleus is small, the probability for the incident nucleus to lose a large fraction of its

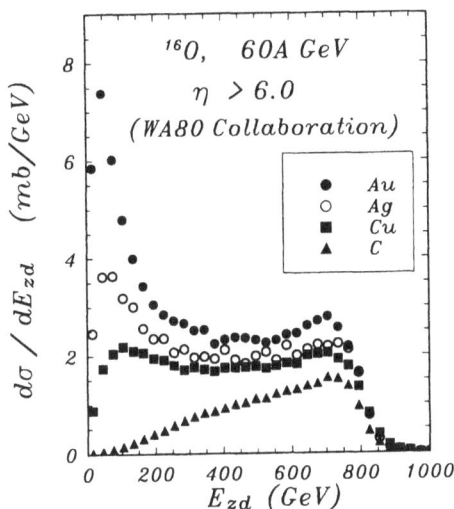

Fig. 13.1   The differential cross section as a function of the energy deposited in the zero-degree calorimeter for the collision of ^{16}O on various target nuclei at an energy of 60A GeV. The data are from the WA80 Collaboration [5,6].

incident energy will be small. On the other hand, when the target is thick enough, and the impact parameter is small, the transmission of the forward energy can be much diminished, as is evident from Fig. 13.1. As the target nucleus varies from from C to Au, the nuclear radius becomes larger and larger, and the highest peak of the distribution of the forward energy of the inelastically scattered proton moves from an energy close to the incident energy to an energy which is only a small fraction of the incident energy.

Another piece of experimental data which provides information on nuclear stopping in heavy-ion collisions is the measurement of the proton rapidity distribution in the collision of Si on Al at 14.6 GeV per projectile nucleon in the laboratory system [8]. The protons shown in Fig. 13.2 are predominantly from the baryons which have slowed down by the collision, and not from the production of baryon-antibaryon pairs in the collision process. A measurement of the difference between the Λ and $\bar{\Lambda}$ production at a higher energy shows a contribution of the pair production contribution to the central rapidity region that is much smaller than the rapidity density shown in Fig. 13.2 [9]. For the collision of Si on Al at 14.6A GeV shown in Fig. 13.2, the beam rapidity is 3.35. The rapidity distribution has a broad plateau shape which appears to be nearly constant in the central rapidity region. On the average, the rapidity of a baryon is shifted by about 1.5 units of rapidity, with a broad distribution in rapidity. This large shift in rapidity for a projectile and target as small as Si and Al indicates that the stopping of baryons when heavy-nuclei are used, will be very substantial. Experiments with Au projectile at about $15A$ GeV has just begun at Brookhaven National Laboratory. The use of Au beams on Au targets at an energy in the range of 15 GeV per projectile nucleon will be of great interest to explore the possibility of quark-gluon plasma formation with a high baryon content.

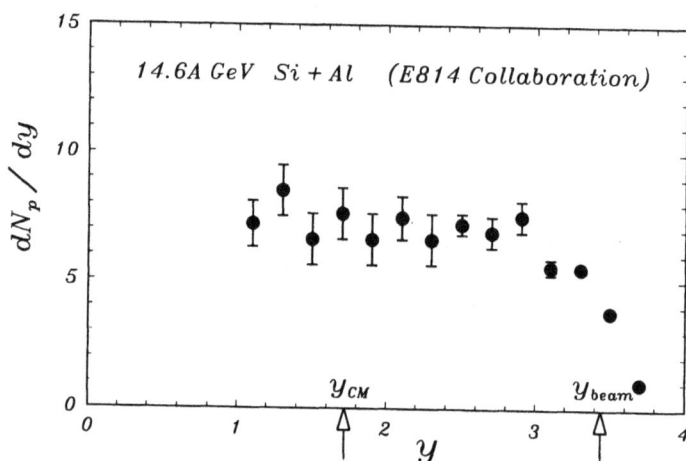

Fig. 13.2 The rapidity distribution of protons in the collision of Si on Al at an energy of 14.6A GeV. The data are from the E814 Collaboration [8].

In high-energy central nucleus-nucleus collisions involving heavy nuclei, the baryon matter will be slowed down and will lose a few units of rapidity. If the collision is such that the separation between the beam rapidity and the target rapidity is much greater than a few units, then after a central collision, the rapidities of the beam baryons and target baryons will remain substantially far away from the central rapidity region. When this happens, the net baryon content in the central rapidity region will be very small. For nucleus-nucleus collisions at an energy of 100 GeV per nucleon in the center-of-mass system, the separation between the projectile rapidity and the target rapidity is 10.7 units. On the other hand, from the analysis of Supplement 13.1, the average rapidity loss of the baryons in a central collision of Au on Au is roughly 2 to 4 units. Thus, the separation between the initial beam and target rapidities may be large enough to give a central rapidity region that is low in baryon content. Reactions at these energies will be useful to explore the "baryon-free quark-gluon plasma" or "pure quark-gluon plasma" region.

We shall discuss the "pure quark-gluon-plasma" region in the next two sections.

§13.2 Bjorken's Estimate of the Initial Energy Density in High-Energy Nucleus-Nucleus Collisions

For simplicity, we consider the head-on collision of two equal nuclei in the center-of-mass frame. There is a substantial Lorentz contraction in the longitudinal direction. We can represent the two colliding nuclei by two thin disks. The picture is much simplified if we consider the extremely high energy case so that we can neglect the longitudinal thickness of the nuclei, and the longitudinal coordinates of the nucleons of the same nucleus can be approximated to be the same.

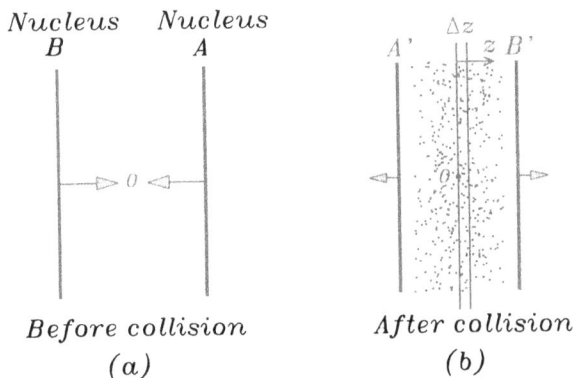

Before collision   After collision
(a)   (b)

Fig. 13.3   (a) The configuration of two colliding nuclei A and B before collision. (b) The configuration after collision with energy deposited in the region around $z \sim 0$.

In Fig. 13.3a, we show the configuration of the two nuclei before collision, in the center-of-mass system. The projectile nucleus B comes from $z = -\infty$ with a velocity close to the speed of light and meets the target nucleus A which comes from $z = +\infty$, also with a speed close to the speed of light. They meet at $z = 0$ and $t = 0$. At the point $(z, t) = (0, 0)$, collisions of the nucleons of the projectile nucleus with the nucleons of the target nucleus take place.

The dynamics can be viewed from a different perspective in the space-time diagram with the longitudinal coordinate z and the time coordinate t, as shown in Fig. 13.4. The trajectories of the colliding projectile nucleus and target nucleus are shown as thick lines.

As we saw from the multiple collision model in the last section, the number of nucleon-nucleon collisions in a nucleus-nucleus collision can be very large. For example, in the head-on collision of a uranium nucleus on another uranium nucleus, the number of inelastic nucleon-nucleon collisions is about 800 [10]. From the experimental data of nuclear collisions, we know that each inelastic nucleon-nucleon collision is accompanied by a large loss of the energy of the colliding baryons (see Chapter 2). As the baryons lose energy and momentum, they slow down after the collision. At very high energies (of the order of 100 GeV/nucleon and above in the center-of-mass system), the slowed-down baryons after the collision can still have enough momentum to proceed forward, and move away from the region of collision, as shown schematically in Fig. 13.3(b), where the projectile baryon matter after the collision is denoted by B' and the target baryon matter by A'. The energy lost by the baryons is deposited in the collision region around $z = 0$. This energy deposition is approximately additive in nature. In consequence, as the colliding nuclear matter B' and A' recede from each other after the collision (if they are not stopped as a result of the collision), a large amount of energy is deposited in a small region of space in a short duration of time [Fig. 13.3(b)]. The matter created in the collision region has a very high energy density, but a small net baryon content. As the net baryon content of the early universe is very small, the type of quark-gluon plasma which may be produced is of special astrophysical interest [1,11].

The quanta which carry the energy deposited in the collision region around $z \sim 0$ can be in the form of quarks, gluons, or hadrons. It is as yet an unresolved question in what form the quanta appear in the first instant after the collision. Whatever the form of the material, the energy density in the region around $z \sim 0$ is very high. This led to Bjorken's suggestion [1] of the space-time scenario for a high-energy nucleus-nucleus collision illustrated in Fig. 13.4. Soon after the collision of the two nuclei at $(z, t) = (0, 0)$, the energy density may be sufficiently high to make it likely that a system of quark-gluon plasma may be formed in the central rapidity region, since the ground state of matter with such an energy density is in the quark-gluon plasma

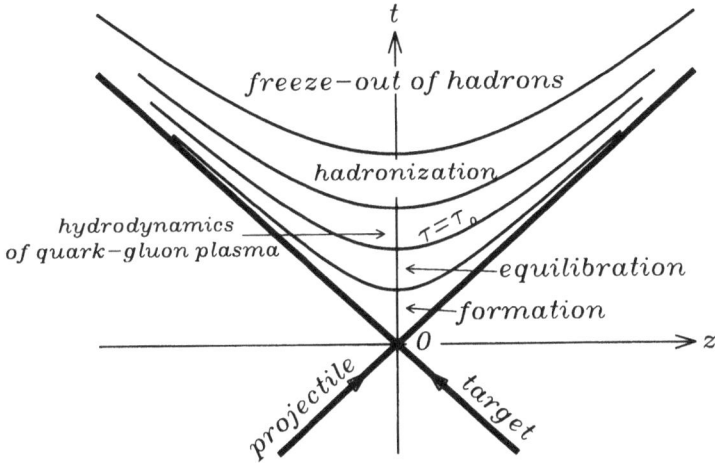

Fig. 13.4 The space-time picture of a nucleus-nucleus collision.

phase and not in the hadron phase. The plasma initially may not be in thermal equilibrium, but subsequent equilibrium may bring it to local equilibrium at the proper time τ_0, and the plasma may then evolve according to the laws of hydrodynamics thereafter. As the plasma expands, its temperature drops down and the hadronization of the plasma will take place at a later proper time. The hadrons will stream out of the collision region when the temperature falls below the freeze-out temperature.

To estimate the initial energy density ϵ_0 before the hydrodynamical evolution, we need to find out the content of energy deposited in the collision region and the relevant volume. Because the energy deposited in the collision region manifests itself eventually in the form of produced hadrons streaming out from the collision region, we can estimate the initial energy density from the particles which come out from this region after the collision. This way of estimating the initial energy density amounts to reconstructing the trajectories of these particles to bring them backward close to their space-time point of origin, finding out their energy content and the volume they occupy, in order to get the initial energy density. As the particles stream out from the collision point $(z, t) = (0, 0)$, the volume they occupy depends on time. As a consequence, the initial energy density of the matter created by the collision depends on the proper time. We are interested in the energy density at $z = 0$ and at the proper time τ_0 when a quark-gluon plasma may be formed as a result of the collision.

Accordingly, we shall use the produced particles to estimate the initial energy density. The particles which eventually stream out consist mostly of pions which have a transverse momentum of about 0.35 GeV/c (see Chapter 3). The transverse mass of the particles is about $m_T \sim 0.38$ GeV/c. These particles are characterized by their *rapidity distribution* or *rapidity density*, dN/dy, as a function of the rapidity variable y. To reconstruct their initial spatial distribution, we need to relate the space-time positions of these particles to their rapidity variables.

From the definition of the rapidity variable in Chapter 2, the longitudinal momentum p_z and the energy p_0 of a particle with a rapidity y are

$$p_z = m_T \sinh y, \qquad (2.16)$$

and

$$p_0 = m_T \cosh y. \qquad (2.17)$$

The velocity of the particle in the longitudinal direction is therefore

$$v_z = \frac{p_z}{p_0} = \tanh y.$$

For a particle streaming out from the origin $(0,0)$, its coordinates z and t are related by

$$\frac{z}{t} = v_z = \tanh y. \qquad (13.1)$$

From these results, one can relate the space-time position of a particle with the rapidity y of the particle by

$$z = \tau \sinh y, \qquad (13.2a)$$

and

$$t = \tau \cosh y, \qquad (13.2b)$$

where τ is the proper time variable defined by

$$\tau = \sqrt{t^2 - z^2}. \qquad (13.3)$$

Conversely, one can express the rapidity variable y in terms of t and z by

$$y = \frac{1}{2} \ln \frac{t+z}{t-z}. \qquad (13.4)$$

In the center-of-mass system, the region of small rapidity is called the *central rapidity region*. The relation (13.2*a*) indicates that for a given proper time τ, a small value of rapidity is associate with a small value of z. Hence, the central rapidity region is associated with the central spatial region around $z \sim 0$, where the nucleon-nucleon collisions have taken place. Because of this association, the term "central rapidity region" is often used to refer to the associated central spatial region around $z \sim 0$.

With the expression (13.2*a*) relating the rapidity variable with the spatial coordinate z, the rapidity distribution dN/dy can be transcribed as a spatial distribution from which the initial energy density can be inferred.

The initial energy density of a fluid element is defined in the frame in which the fluid element is at rest. In the center-of-mass frame, matter is at rest at $z = 0$. We therefore focus our attention to $z = 0$ and consider a longitudinal length of thickness Δz at $z = 0$ as in Fig. 13.3(*b*). We denote the transverse overlapping area in the collision of the two nuclei by $\mathcal{A}$. The volume formed by Δz and $\mathcal{A}$ is $\mathcal{A}\Delta z$. We fix our consideration to the proper time τ_0, at which a quark-gluon plasma may have been formed and equilibrated. The number density in this volume at $z = 0$ and at $\tau = \tau_0$ is

$$\frac{\Delta N}{\mathcal{A}\Delta z} = \frac{1}{\mathcal{A}}\frac{dN}{dy}\frac{dy}{dz}\bigg|_{y=0}$$

$$= \frac{1}{\mathcal{A}}\frac{dN}{dy}\frac{1}{\tau_0 \cosh y}\bigg|_{y=0}.$$

The energy of a particle with a rapidity y is $m_T \cosh y$. Therefore, the initial energy density is

$$\epsilon_0 = m_T \cosh y \frac{\Delta N}{\mathcal{A}\Delta z}.$$

The initial energy density averaged over the transverse area $\mathcal{A}$ at the proper time τ_0 is therefore

$$\epsilon_0 = \frac{m_T}{\tau_0 \mathcal{A}}\frac{dN}{dy}\bigg|_{y=0}. \tag{13.5}$$

This relation which connects the initial energy density to the rapidity density, is an important result first derived by Bjorken [1].

The quantity τ_0, the proper time at which a plasma is produced, is an unknown quantity. Bjorken estimated it to be 1 fm/c [1]. A proper determination of this time scale requires a knowledge of the dynamics of the formation and the equilibration of the quark-gluon plasma. The formation process and the equilibration process remain to be subjects of current research [12]. It is suggested that this quantity can be estimated from the time needed for the production of particles. It can be argued that as the baryons of the colliding nuclei separate after the collision, the energy is initially stored up in the field between these separating particles and cannot manifest itself in other forms such as free quarks and free gluons. Only after the lapse of a finite proper time when the field is stretched beyond a certain distance can this stored energy appear in other forms, such as in the form of hadrons, and perhaps even in the form of free quarks and free gluons, if the energy density is high enough. The plasma formation time should be comparable to the particle production time, at which particles begin to be produced out of the field spontaneously. Accordingly, one can estimate the initial time τ_0 from the particle production time τ_{pro}. Various estimates of particle production time place τ_{pro} at about $0.4 - 1.2$ fm/c. The initial time τ_0 is probably in this time range.

⊕⟦ Supplement 13.2

We give here various estimates of the particle production time τ_{pro}. From our discussions of the Schwinger particle production mechanism in Chapter 5, we learn from Eq. (5.14) that particle production is possible only if the two parallel plates are separated by a distance exceeding the minimum distance

$$L_{\min} = 2m_T/\kappa.$$

If one takes the transverse mass m_T of a constituent quark to be 0.400 GeV, and κ to be 1 GeV/fm, then the separation between the two sources of the field should be greater than 0.8 fm in order for a quark-antiquark pair to be produced. The time it takes for the two sources to travel to this separation is 0.4 fm/c. Thus, this estimate gives a particle production time of $\tau_{\text{pro}} \sim 0.4$ fm/c.

Another estimate of the particle production time may be obtained from the model of QED$_2$ as discussed in Chapter 6. In this picture, particles are produced as the two sources move apart, perturbing the field between the sources. The excitation of the field leads to the production of bosons which are stable quanta of the system. The particle production time is given by Eq. (6.46),

$$\tau_{\text{pro}} = 2.405\hbar/mc^2. \tag{6.46}$$

To apply the model of QED$_2$ to pion production, we identify the mass m of the produced boson as the transverse mass of the pions, which is about 0.4 GeV. The above equation then gives a production time $\tau_{\text{pro}} \sim 1.19$ fm/c.

There is another estimate of the particle production time, which may be obtained from dN/dy. It is shown in Exercise 7.1 and Ref. [13] that if all the vertices fall on the curve of the proper time $\tau = \tau_{\text{pro}}$, the rapidity distribution of the produced particles is a constant given by

$$\frac{dN}{dy} = \frac{\kappa \tau_{\text{pro}}}{m_T}.$$

Using a rapidity density of $dN/dy \sim 3$ (including the production of neutral particles) for $\sqrt{s} = 62.8$ GeV, and assuming that after a nucleon-nucleon collision two quark-diquark strings are formed, we find $\tau_{\text{pro}} \sim 0.6$ fm/c.

Finally, as discussed in Chapter 7, particle production time can also be estimated by the classical string model as the average proper time at which the vertices begin to emerge. Because of the detail phenomenological analysis of the Lund Model, this way of estimating the particle production time from the locations of the vertices is probably the most reliable among the different methods. From the form of the splitting function in the Symmetric Lund model in Eq. (7.36),

$$\rho(\Gamma) = C\Gamma^a e^{-b\Gamma}, \tag{7.36}$$

the average proper time at which the vertices occur is

$$< \tau_{\text{pro}} >= \frac{\Gamma(a+1+\frac{1}{2})}{\kappa\sqrt{b}\Gamma(a+1)}. \tag{7.37}$$

In the Lund Model, the parameters have been determined to be $a = 0.5$ and $b = 0.9$ GeV^{-2} [14], and the average proper time is $\tau_{\text{pro}} \sim 1.19$ fm/c. A more careful analysis using a splitting function which depends explicitly on the proper time gives $a = 0.4$, and $b = 1.2$ GeV^{-2}, leading to $\tau_{\text{pro}} \sim 0.99$ fm/c [13]. ⬛⊕

Experimental investigations of the collision process have shown that in nucleus-nucleus collisions the rapidity density of charged particles, dN_{ch}/dy, is quite large. For example, in central collisions of ^{32}S or ^{16}O beams on a Au target nucleus at an incident energy of 200A GeV, the peak pseudorapidity density (which is approximately the peak rapidity density) is about 160 for ^{32}S, and 110 for ^{16}O [15,16] (Fig. 13.5). In this measurement, central collisions are defined as those for which the remaining forward energy deposited in the zero-degree calorimeter with an acceptance angle of $0.3°$ is less than 20% of the beam energy for ^{32}S, and is less than 9.4% of the beam energy for ^{16}O. If one uses Eq. (13.5) with an initial time $\tau_0 = 1$ fm/c, $\mathcal{A} = \pi(1.2)^2 A^{2/3}$ fm^2, and $dN/dy = (3/2)dN_{ch}/dy$, then the average initial energy density for a central collision is about 1 GeV/fm^3 for collisions of ^{32}S on Al, about 2.1 GeV/fm^3 for ^{32}S on Au, and about 2.3 GeV/fm^3 for ^{16}O on Au. The distribution of peak values $(dN_{ch}/d\eta)_{\text{max}}$ as a function of the transverse energy, (which is correlated with the impact parameter), is quite wide. For example, for the collision of ^{16}O on Au at 200A GeV, while the average peak value of $(dN_{ch}/d\eta)_{max}$ is 110, there are events, presumably for collisions at small impact parameters, for which $(dN_{ch}/d\eta)_{max}$ is as large as 160. Those events with $dN_{ch}/d\eta \sim 160$ would give an initial energy density $\epsilon_0 \sim 3.3$ GeV/fm^3. The collision of ^{32}S or ^{16}O on Au at 200 GeV per nucleon may not conform to the idealized collision conditions as discussed in the derivation of Eq. (13.5), nevertheless the experimental results indicate that the effect of multiple collisions on the rapidity density dN/dy (and consequently the energy density) is approximately cumulative and a large initial energy density may be produced in relativistic heavy-ion collisions.

^{32}S or ^{16}O + Nucleus, 200A GeV

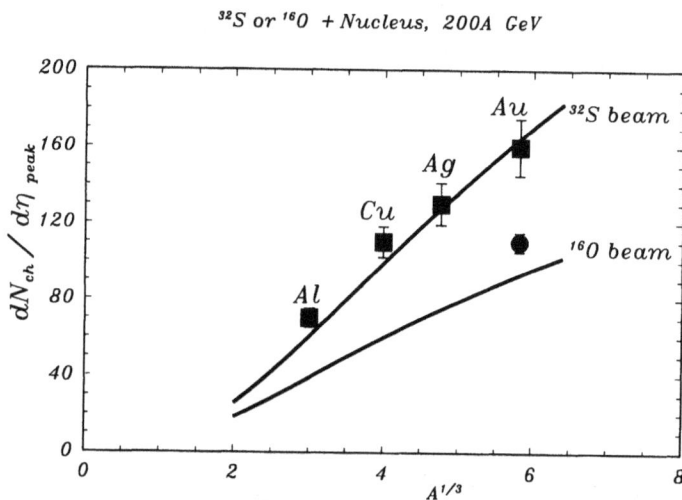

Fig. 13.5 The peak pseudorapidity density of charge particles as a function of $A^{1/3}$, in central collisions of ^{16}O or ^{32}S on various target nuclei A at an incident energy of 200 GeV per projectile nucleon. The data are from the WA80 Collaboration (Refs. [15] and [16]). The solid curves are estimates using Eq. (2) of Supplement 13.3.

It is worth noting that the estimate (13.5) is based on an average over the transverse overlapping area $\mathcal{A}$. The thickness of the projectile nucleus and of the target nucleus are not transversely uniform. There are large variations. The target nucleus and the projectile nucleus are obviously thicker at the center of their transverse axes rather than at other transverse locations. One expects that at those transverse locations where the target nucleus and the projectile nucleus are thickest, the initial energy densities are much greater than the estimate of Eq. (13.5).

The Relativistic Heavy-Ion Collider (RHIC), which is being constructed at Brookhaven National Laboratory, is designed to accelerate very heavy nuclei up to an energy of about 100 GeV per nucleon. It is expected that matter with an initial energy density of many GeV/fm^3 will be produced using such a collider. On the other hand, from the results of the lattice gauge theory as discussed in Chapters 10 and 11, the energy density at which the transition from the hadron phase to the quark-gluon plasma phase is expected to occur is a few GeV/fm^3. The initial energy density of the matter produced in such high-energy nucleus-nucleus collisions may be sufficiently high to make it possible to form a quark-gluon plasma in the central rapidity region.

From the experimental data of ^{16}O-induced reactions at 60A GeV and 200A GeV, very useful semi-empirical relations, between the transverse energy E_T and the number of participant nucleons or the number of binary nucleon-nucleon collisions, have been found by the WA80 Collaboration [17]. Besides measuring the energy E_{zd} deposited at the zero-degree calorimeter, the WA80 Collaboration also measured the energy of particles deposited in the mid-range calorimeter between 0.5^o and 10°. This range of angles corresponds to the pseudorapidity range $2.4 < \eta < 5.5$. From the angle θ and the energy deposited at that angle, the transverse energy $E \sin \theta$ at θ is obtained, and the total transverse energy E_T is the sum of all the transverse energies within the range of pseudorapidity $2.4 < \eta < 5.5$. Sorensen *el al.* first studied the relation between the energy deposited at the zero-degree calorimeter E_{zd} as a function of impact parameters, using the theoretical model of FRITIOF. The number of participant nucleons and the number of binary collisions are then determined as a function of E_{zd}. From the empirical measurements of E_T and E_{zd}, one obtains the transverse energy per participant nucleon and the transverse energy per nucleon-nucleon collision, as a function of E_{zd}. The experimental results indicate that the transverse energy per participant nucleon is nearly independent of E_{zd}, which is a measure of the impact parameters. The transverse energy per nucleon-nucleon collision is also nearly independent of the impact parameter, but there is a slight decrease as the impact parameter decreases. One expects that there are greater degradation of the energies of the colliding nucleons in central collisions as compared to peripheral collisions. The small decrease of the transverse energy per collision as the impact parameter decreases is an indication of the effect of energy degradation when nucleons of one nucleus traverse through the other nucleus.

Since the transverse energy E_T is related to the rapidity distribution dN/dy, the semi-empirical results of Ref. [17] gives support that in a crude approximation, dN/dy is proportional to the number of participant nucleons or to the number of nucleon-nucleon collisions, subject to energy-degradation corrections.

⊕[Supplement 13.3

We can give here a quick and approximate estimate of dN/dy in the central rapidity region for nucleus-nucleus collisions. Consider a nucleus-nucleus collision at an impact parameter b for which there are $n'(b)$ number of inelastic nucleon-nucleon collisions. Each collision contributes to the production of particles. If all the collisions contribute in the same way, then the rapidity distribution dN/dy for nucleus-nucleus collisions would just be $n'(b)$ times the rapidity distribution dN_{pp}/dy for nucleon-nucleon collisions.

However, the contributions from all of the collisions are not the same. One can envisage that in a nucleus-nucleus collision, those nucleon-nucleon collisions which occur later in the process contribute less than those collisions from earlier collisions, because of the stopping of the baryons of one nucleus as they go through the other nucleus. The degree of stopping depends on the thickness of the projectile and target nuclei. The resultant effect of energy degradation should lead to a reduction factor of the form $1/[1 + a(A^{1/3} + B^{1/3})]$, when one uses a reduction factor to the first power in the thickness of the projectile and of the target nucleus. For the purpose of making a simple estimate, it is useful to use this reduction factor to relate $dN/dy(b)$ for a nucleus-nucleus collision at an impact parameter b with the rapidity distribution dN_{pp}/dy for nucleon-nucleon collision:

$$\frac{dN}{dy}(b) \approx \frac{n'(b)}{1 + a(A^{1/3} + B^{1/3})} \frac{dN_{pp}}{dy}, \tag{1}$$

where a is a parameter. In a nucleus-nucleus collision at an impact parameter b, the number of nucleon-nucleon collisions, as given in the multiple collision model from Eq. (12.14), is

$$n'(b) = ABT(b)\sigma_{in}/\{1 - [1 - T(b)\sigma_{in}]^{AB}\}.$$

The thickness function can be approximated by a Gaussian distribution (12.26)

$$T(b) = \frac{1}{2\pi\beta^2}e^{-b^2/2\beta^2},$$

where

$$\beta^2 = \beta_A^2 + \beta_B^2 + \beta_p^2,$$

$$\beta_A = r_0'A^{1/3}/\sqrt{3},$$

$r_0' = 1.05$ fm, and β_p, the thickness function parameter for nucleon-nucleon collision, is 0.68 fm [10]. For central collisions, $T(b)$ does not vanish and the denominator in (12.14) can be approximated to be unity. Furthermore, one can neglect β_p which is small in comparison with β_A or β_B. We have therefore,

$$\frac{dN}{dy}(b) \approx \frac{3\sigma_{in}}{2\pi(r_0')^2} \frac{AB}{A^{2/3} + B^{2/3}} \frac{1}{1 + a(A^{1/3} + B^{1/3})}e^{-b^2/2\beta^2}\frac{dN_{pp}}{dy}.$$

For two equal nuclei, the estimate is then

$$\frac{dN}{dy}(b) \approx 0.64 \, A^{4/3} \frac{1}{1 + a(A^{1/3} + B^{1/3})} e^{-b^2/2\beta^2} \frac{dN_{pp}}{dy}.$$

For the case when $A > B$, nucleus A is thicker than nucleus B. The baryons in the nucleus B will lose more energy than the baryons in the nucleus A, as they make successive collisions. Therefore, the peak position of dN/dy is shifted toward the rapidity of A. If one considers Eq. (1) to remain applicable to the peak rapidity densities, as an estimate, we then have

$$\left.\frac{dN}{dy}(b)\right|_{peak} \approx 1.28 \frac{AB}{A^{2/3} + B^{2/3}} \frac{1}{1 + a(A^{1/3} + B^{1/3})} e^{-b^2/2\beta^2} \left.\frac{dN_{pp}}{dy}\right|_{peak}. \quad (2)$$

We shall use the above result to analyze the data in Fig. 13.5. The data in Fig. 13.5 represent a range of impact parameters in central collisions, corresponding to 20% of the beam energy remaining in the zero-degree calorimeter for the ^{32}S beam, and 9.4% of the beam energy for the ^{16}O beam. The spectra of the energy deposited in the zero-degree calorimeter depend on the mass of the target nucleus. For this group of central collision events considered, we estimate that the number of collisions is about 0.9 times the number of collisions at $b = 0$, with an uncertainty of perhaps about 10%. A value of $a = 0.09$ gives the two solid curves shown in Fig. 13.5. The experimental data suggest that the peak value of $dN/dy(b)$ scales approximately as $[AB/(A^{2/3} + B^{2/3})]/[1 + a(A^{1/3} + B^{1/3})]$. I⊕

§13.3 Hydrodynamics of the Quark-Gluon Plasma

To follow the dynamics of the quark-gluon plasma, we envisage the space-time scenario in Fig. 13.4 in which a quark-gluon plasma is formed and is in local thermal equilibrium at the proper time τ_0. The initial energy density at τ_0 is given by Eq. (13.5) and is inversely proportional to τ_0. Thereafter, the quark-gluon plasma will evolve according to the laws of hydrodynamics [1]. During this stage of hydrodynamical evolution, how do the energy density and other thermodynamical variables depend on the proper time?

In the hydrodynamical description of the quark-gluon plasma, the complete dynamics of the system is described by the energy density field ϵ, the pressure field p, the temperature field T, and the 4-velocity field $u^\mu = dx^\mu/d\tau$, at different space-time points as the system evolves. The energy density ϵ and the pressure p at a space-time point are measured in the frame in which the velocity of a fluid element at that point is zero, corresponding to the energy density in the frame F^* in which the velocity at that point, $(u^{*0}, u^{*1}, u^{*2}, u^{*3})$, is $(1, 0, 0, 0)$. The energy density ϵ, the pressure p and the temperature T are related by the equation of state $\epsilon = \epsilon(p, T)$.

The *energy-momentum tensor* $T^{\mu\nu}$ is defined as the momentum in the μ direction per unit 3-surface area perpendicular to the ν-diretcion. Thus, T^{00} is the momentum in the 0th-direction (the energy) per unit 3-surface area perpendicular to the 0th-direction. The 3-surface area perpendicular to the 0th-direction is just the spatial volume. Therefore, in the frame F^* in which the fluid is at rest, we have

$$T^{*\,00} = \frac{\Delta E}{\Delta x^1 \, \Delta x^2 \, \Delta x^3} = \epsilon. \qquad (13.6)$$

Similarly, T^{ij} for $i, j = 1, 2$, and 3 is the momentum in the ith-direction per unit 3-surface area in the jth-direction. The 3-surface area element in the 1st-direction is just $\Delta t \, \Delta x^2 \, \Delta x^3$. The energy momentum tensor $T^{*\,11}$ in the frame F^* is therefore

$$T^{*\,11} = \frac{\Delta p^1}{\Delta t \, \Delta x^2 \, \Delta x^3}.$$

We note that $\Delta p^1 / \Delta t$ is the force acting on a unit mass in the 1st-direction. The force acting on a unit mass in the 1st-direction, per unit area in the $\{2, 3\}$ direction, is the pressure in the 1st-direction on a $\{2, 3\}$ surface. Therefore, for a fluid for which the pressure is isotropic, we have

$$T^{*\,ij} = p\delta^{ij}. \qquad (13.7)$$

Equations (13.6) and (13.7) give the the energy momentum tensor in terms of energy density and pressure, when the fluid element is at rest. In any other frame F, in which the the fluid element under consideration is moving with a 4-velocity u^μ, the energy-momentum tensor $T^{\mu\nu}$ can be obtained from the energy momentum tensor $T^{*\,\mu\nu}$ by the law of transformation of tensors. After carrying out the transformation [18], one obtains

$$T^{\mu\nu} = (\epsilon + p)u^\mu u^\nu - g^{\mu\nu} p. \qquad (13.8)$$

To make the picture simple, we consider the collision of two equal nuclei at an energy $\sqrt{s} > 100$ GeV per nucleon For these high energies, the rapidity distribution for a nucleon-nucleon collision has a plateau structure in the central rapidity region (see Fig. 3.2), which implies that the rapidity density is approximately constant over a large range of the rapidity variable. Because of the cumulative nature of a nucleus-nucleus collision, the rapidity density in a nucleus-nucleus collision should have similar features. That is, it should be approximately constant over a range of the rapidity variable,

$$\left.\frac{dN}{dy}\right|_{\text{any } y \text{ in } y_L \leq y \leq y_R} = \left.\frac{dN}{dy}\right|_{y=0}. \qquad (13.9)$$

This constancy of the rapidity density leads to a state of Lorentz invariance when one makes a Lorentz transformation in the longitudinal direction, over a finite range of Lorentz frames (see Supplement 13.4 below).

⊕[Supplement 13.4

If one makes a Lorentz transformation from the center-of-mass frame F to a frame F' which moves with the speed β in the longitudinal direction relative to the frame F, the rapidity y' of a particle in the new frame F' is related to the rapidity y of the particle in the center-of-mass frame F by [Eq. (2.18)]:

$$y' = y - y_\beta, \tag{1}$$

where y_β is the rapidity of the moving frame given in Eq. (2.19) as

$$y_\beta = \frac{1}{2} \ln \frac{(1 + \beta)}{(1 - \beta)}.$$

From Eq. (1), the rapidity density in the new frame F', dN/dy', is then

$$\frac{dN}{dy'}\bigg|_{y'} = \frac{dN}{dy}\bigg|_{y=y'+y_\beta}. \tag{2}$$

The initial energy density of matter, $\epsilon_0'(F')$, in the new frame F' is given by Eq. (13.5) as

$$\epsilon_0'(F') = \frac{m_T}{\tau_0 A} \frac{dN}{dy'}\bigg|_{y'=0}.$$

Using Eq. (2), the above equation becomes

$$\epsilon_0'(F') = \frac{m_T}{\tau_0 A} \frac{dN}{dy}\bigg|_{y=y_\beta}. \tag{3}$$

From Eqs. (3) and Eq. (13.9), we find that if

$$y_L \leq y_\beta \leq y_R, \tag{4}$$

then, we have

$$\epsilon_0'(F') = \frac{m_T}{\tau_0 A} \frac{dN}{dy}\bigg|_{y=0}. \tag{5}$$

On the other hand, according to Eq. (13.5) the energy density ϵ_0 in the frame F is,

$$\epsilon_0(F) = \frac{m_T}{\tau_0 A} \frac{dN}{dy}\bigg|_{y=0}.$$

Therefore, we have

$$\epsilon_0'(F') = \epsilon_0(F).$$ (6)

Equations (4) and (6) state that under the range of Lorentz transformations specified by y_β of Eq. (4), the initial energy densities in these frames are the same. The range of y_β in Eq. (4) increases with the energy of collision. The range can be so large in high-energy collision that one can speak (approximately) of Lorentz invariance of the system: the initial configuration of the system does not appear different when one looks at the system in one Lorentz frame or another, within the limits imposed by Eq. (4).

The approximate Lorentz invariance of the initial configuration over a large range of Lorentz transformations makes it simple to study the hydrodynamics of the quark-gluon plasma. In Bjorken's hydrodynamical model [1], the system is approximated as an idealized continuum with a longitudinal translational symmetry so that a Lorentz transformation along the longitudinal direction (within a limited range) leads to the same initial condition and subsequently the same dynamics of the system. In any one of these translationally-invariant frames of reference, the complete dynamics of the system can be just specified by the thermodynamical variables as a function only of Lorentz invariant quantities such as the proper time coordinate, but independent of the rapidity variable:

$$\epsilon = \epsilon(\tau),$$ (13.10a)

$$p = p(\tau),$$ (13.10b)

and

$$T = T(\tau).$$ (13.10c)

The equation of motion for the quark-gluon plasma is governed by the hydrodynamical equation [1,18]

$$\frac{\partial T^{\mu\nu}}{\partial x^\mu} = 0,$$ (13.11)

which comes from energy and momentum conservation. In the set of four equations for $\nu = 0, 1, 2, 3$ in (13.11), the independent fields are ϵ, p and (u^0, u^1, u^2, u^3). However, p is related to ϵ and T by the equation of state, and the velocity fields u^ν are related by

$$\begin{aligned} u_\nu u^\nu &= g_{\nu\beta} u^\beta u^\nu \\ &= g_{\nu\beta} \frac{dx^\beta}{d\tau} \frac{dx^\nu}{d\tau} \\ &= 1. \end{aligned}$$ (13.12)

Thus, there are only four independent variables for the set of four equations (13.11).

The longitudinal expansion proceeds at a greater magnitude as compared to the transverse motion. For simplicity and clarity of insight, it is useful to study the hydrodynamics of the plasma in two-dimensional space-time containing only the longitudinal coordinate and the time coordinate. We shall solve Eq. (13.11) subject to the Lorentz-invariance conditions (13.10) that the thermodynamical variables depend only on τ. The independent variables are ϵ, u^0 and u^1 subject to the constraint (13.12), with the initial energy density ϵ_0 given by Eq. (13.5) and the initial longitudinal velocity field specified by Eq. (13.1).

It is convenient to form the contraction of u_ν with Eq. (13.11) which can be written as

$$
\begin{aligned}
u_\nu \frac{\partial T^{\mu\nu}}{\partial x^\mu} &= g_{\nu\beta}u^\beta \frac{\partial}{\partial x^\mu}[(\epsilon + p)u^\mu u^\nu - g^{\mu\nu}p] \\
&= g_{\nu\beta}u^\beta \frac{\partial(\epsilon + p)}{\partial \tau}\frac{\partial \tau}{\partial x^\mu}u^\mu u^\nu + g_{\nu\beta}u^\beta(\epsilon + p)\frac{\partial u^\mu}{\partial x^\mu}u^\nu \\
&\quad + g_{\nu\beta}u^\beta(\epsilon + p)u^\mu \frac{\partial u^\nu}{\partial x^\mu} - g_{\nu\beta}u^\beta \frac{\partial p}{\partial \tau}\frac{\partial \tau}{\partial x^\mu}g^{\mu\nu} \\
&= 0 .
\end{aligned}
\tag{13.13}
$$

The above equation can be simplified by noting that

$$
\frac{\partial \tau}{\partial x^\mu} = \frac{g_{\mu\alpha}x^\alpha}{\tau} ,
$$

$$
\begin{aligned}
\frac{\partial \tau}{\partial x^\mu}u^\mu &= \frac{g_{\mu\alpha}x^\alpha}{\tau}\frac{\partial x^\mu}{\partial \tau} \\
&= \frac{\partial \tau^2}{2\tau\partial\tau} = 1 .
\end{aligned}
$$

and

$$
g_{\nu\beta}u^\beta \frac{\partial u^\nu}{\partial x^\mu} = \frac{1}{2}\frac{\partial g_{\nu\beta}u^\beta u^\nu}{\partial x^\mu} = 0 .
$$

Eq. (13.13) becomes

$$
g_{\nu\beta}u^\beta \frac{\partial(\epsilon + p)}{\partial \tau}u^\nu + g_{\nu\beta}u^\beta(\epsilon + p)\frac{\partial u^\mu}{\partial x^\mu}u^\nu - \frac{\partial p}{\partial \tau} = 0 .
$$

Using Eq. (13.12), we obtain

$$
\frac{\partial \epsilon}{\partial \tau} + (\epsilon + p)\frac{\partial u^\mu}{\partial x^\mu} = 0 .
$$

In this equation, because ϵ and p depend only on τ, the velocity field in the above equation must satisfy

$$\frac{\partial u^{\mu}}{\partial x^{\mu}} = f(\tau),$$

where the function $f(\tau)$ must match the function $\partial u^{\mu}/\partial x^{\mu}$ at $\tau = \tau_0$ as determined by the initial velocity field at τ_0. With the initial velocity field given by Eq. (13.1), we have initially at $\tau = \tau_0$

$$u^{\mu} = \frac{x^{\mu}}{\tau}$$

and

$$\begin{aligned}
\frac{\partial u^{\mu}}{\partial x^{\mu}} &= \frac{\delta^{\mu}_{\mu}}{\tau} - \frac{x^{\mu}}{\tau^2}\frac{\partial \tau}{\partial x^{\mu}} \\
&= \frac{2}{\tau} - \frac{x^{\mu}}{\tau^2}\frac{1}{\tau}g_{\mu\alpha}x^{\alpha} \\
&= \frac{1}{\tau},
\end{aligned}$$

where we have used the result $\delta^{\mu}_{\mu} = 2$ for the two-dimensional space-time under consideration. The function $f(\tau)$, which is consistent with the initial velocity (13.1) at $\tau = \tau_0$ and depends only on τ, is

$$f(\tau) = \frac{1}{\tau}.$$

The equation for the thermodynamic variables ϵ and p is then

$$\frac{\partial \epsilon}{\partial \tau} + \frac{(\epsilon + p)}{\tau} = 0. \tag{13.14}$$

We can proceed to solve the above equation for a simple equation of state. In the case of an ideal gas of massless quarks and gluons, the quark-gluon plasma has the equation of state of a relativistic gas for which the energy density and the pressure are related by

$$p = \epsilon/3. \tag{13.15}$$

For this equation of state, equation (13.14) becomes

$$\frac{d\epsilon}{d\tau} = -\frac{4}{3}\frac{\epsilon}{\tau}, \tag{13.16}$$

which has the solution

$$\frac{\epsilon(\tau)}{\epsilon(\tau_0)} = \frac{\epsilon(\tau)}{\epsilon_0} = \left(\frac{\tau_0}{\tau}\right)^{4/3}, \tag{13.17a}$$

and the pressure decrease with proper time as

$$\frac{p(\tau)}{p(\tau_0)} = \left(\frac{\tau_0}{\tau}\right)^{4/3}. \qquad (13.17b)$$

For an ideal relativistic gas, the energy density and the pressure are proportional to T^4 as given by Eq. (9.2); the temperature of the plasma depends on the proper time according to

$$\frac{T(\tau)}{T(\tau_0)} = \left(\frac{\tau_0}{\tau}\right)^{1/3}. \qquad (13.18)$$

From Eqs. (5) and (9) of Supplement 9.2, we get

$$\frac{n_q(\tau)}{n_q(\tau_0)} = \frac{n_{\bar{q}}(\tau)}{n_{\bar{q}}(\tau_0)} = \frac{n_g(\tau)}{n_g(\tau_0)} = \frac{\tau_0}{\tau}. \qquad (13.19)$$

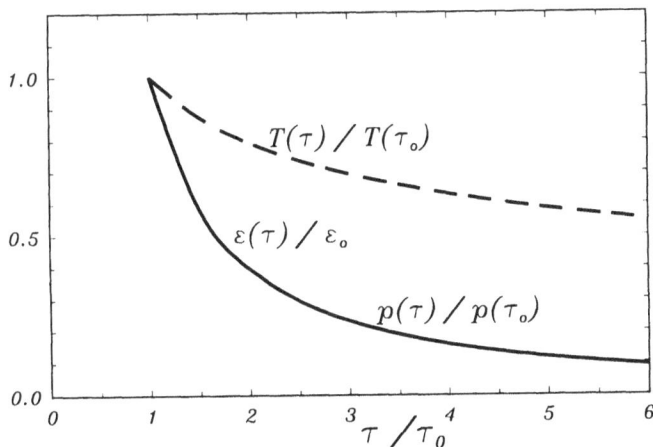

Fig. 13.6   The solid curve gives the ratio $\epsilon(\tau)/\epsilon_0$ and $p(\tau)/p(\tau_0)$, and the dashed curve gives $T(\tau)/T(\tau_0)$, as a function of τ/τ_0, in Bjorken's hydrodynamical model.

We can summarize the dynamics of the plasma in the hydrodynamical phase of the evolution, with the idealized description of the plasma as a relativistic gas in Bjorken's hydrodynamical model. Initially at $\tau = \tau_0$, local thermal equilibrium is reached and the initial energy density is ϵ_0 and the initial temperature $T(\tau_0)$ is proportional to $\epsilon_0^{1/4}$.

Thereafter, the energy density and the pressure decrease with proper time as $\tau^{-4/3}$, while the temperature drops down as $\tau^{-1/3}$. The decrease in temperature is a slower function of the proper time, as compared to the decrease of the energy density and the pressure. Fig. 13.6 shows how the various thermodynamical quantities vary with the proper time τ.

From the thermodynamic quantities we have already found, we can obtain other thermodynamic quantities such as the entropy and the entropy density. Under a constant temperature and pressure, the variation of the energy is related to the variation of volume and entropy by (Eq. (12.3) of Ref. 19)

$$dE = -pdV + TdS.$$

This can be rewritten to give the relation between entropy density $s = dS/dV$ in terms of ϵ and p as

$$s \equiv \frac{dS}{dV} = \frac{\epsilon + p}{T}. \tag{13.20}$$

Therefore, the entropy density depends on the proper time as

$$\frac{s(\tau)}{s(\tau_0)} = \frac{\epsilon(\tau) + p(\tau)}{\epsilon(\tau_0) + p(\tau_0)} \frac{T(\tau_0)}{T(\tau)}$$

$$= \left(\frac{\tau_0}{\tau}\right)^{4/3} \left(\frac{\tau}{\tau_0}\right)^{1/3}$$

$$= \frac{\tau_0}{\tau}.$$

Thus, the entropy density is inversely proportional to the proper time [1], and we have

$$s(\tau)\tau = s(\tau_0)\tau_0. \tag{13.21}$$

As the volume element dV is given by $d\mathbf{x}_\perp \tau dy$, the above equality implies that

$$\frac{dS}{d\mathbf{x}_\perp dy} = s\tau = (\text{constant of proper time }).$$

Therefore, we have

$$\frac{d}{d\tau}\left(\frac{dS}{dy}\right) = 0.$$

The hydrodynamical motion of the fluid is characterized by a constant entropy per unit of rapidity [1].

We see from the above discussion that as the quark-gluon plasma evolves, its temperature decreases as $\tau^{-1/3}$ in accordance with Eq.

(13.18). The temperature of the plasma drops down to T_c, at the proper time τ_c given by

$$\tau_c = \left(\frac{T(\tau_0)}{T_c}\right)^3 \tau_0 . \qquad (13.22)$$

Thereafter, the transition from the quark-gluon matter to the hadron matter will take place. During the transition, both the quark-gluon matter and the hadron matter are present and the system is in a mixed-phase state. The dynamics of the system at that point are characterized by a temperature maintained at the critical temperature T_c, while the entropy density of the matter decreases by converting the quark-gluon plasma into hadronic matter. The fraction of matter in the quark-gluon plasma phase f is a function of the proper time. The entropy density of the matter in the mixed phase is

$$s(\tau) = f(\tau)s_{qg}(T_c) + [1 - f(\tau)]s_h(T_c) , \qquad (13.23)$$

where $s_{qg}(T_c)$ is the entropy density of the quark-gluon plasma and $s_h(T_c)$ is the entropy density of the hadron matter, evaluated at the temperature T_c. We can evaluate this function $f(\tau)$ by noting that in the mixed phase, Eqs. (13.16) and (13.17) remain valid. Thus, we have

$$\frac{\epsilon(\tau) + p(\tau)}{\epsilon(T_c) + p(T_c)} = \left(\frac{T_c}{\tau}\right)^{4/3} .$$

In the mixed phase, because the temperature is maintained at the transition temperature $T = T_c$, the ratio of the entropy density is

$$\frac{s(\tau)}{s(\tau_c)} = \frac{\epsilon(\tau) + p(\tau)}{T_c} \frac{T_c}{\epsilon(\tau_c) + p(\tau_c)}$$

$$= \left(\frac{T_c}{\tau}\right)^{4/3} ,$$

or

$$s(\tau)\tau^{4/3} = s(\tau_c)\tau_c^{4/3} .$$

We have therefore,

$$\{f(\tau)s_{qg}(T_c) + [1 - f(\tau)]s_h(T_c)\}\tau^{4/3}$$
$$= \{f(\tau_c)s_{qg}(T_c) + [1 - f(\tau_c)]s_h(T_c)\}\tau_c^{4/3} .$$

We get

$$f(\tau) = \frac{1}{s_{qg}(T_c) - s_h(T_c)} \left(\{f(\tau_c)s_{qg}(T_c) + [1 - f(\tau_c)]s_h(T_c)\}\frac{\tau_c^{4/3}}{\tau^{4/3}} - s_h(T_c)\right) .$$

Using Eqs. (13.20) and the approximate relativistic energy densities in Supplements 9.2 and 9.3, we have

$$f(\tau) = \frac{1}{g_{qg} - g_h}\left(\{f(\tau_c)g_{qg} + [1 - f(\tau_c)]g_h\}\frac{\tau_c^{4/3}}{\tau^{4/3}} - g_h\right). \quad (13.24)$$

where the quark-gluon plasma degeneracy number $g_{qg} \sim 37$ and the hadronic matter degeneracy number $g_h \sim 3$. The system will be in the mixed phase in the proper time interval $\tau_c \leq \tau \leq \tau_h$ where τ_h is determined by

$$\tau_h = \left[\frac{g_{qg}}{g_h}f(\tau_c) + 1 - f(\tau_c)\right]^{3/4}\tau_c. \quad (13.25)$$

If one starts with a quark-gluon matter with $f(\tau_c) = 1$, the system will stay in the mixed phase for a proper time interval of about $6.16\tau_c$.

After the proper time τ_h, the system will be in the hadron phase. The dynamics of the hadron system can still be described by the equations of hydrodynamics, with the appropriate hadron equation of state. If one assumes that the hadronic matter is also a relativistic gas and there is an (approximate) Lorentz invariance, then the results relating the energy density, the pressure, and the temperature, as given by Eqs. (13.16)-(13.18) remain valid for the hadron phase, except that the quantity τ_0 in these equations must be replaced by τ_h, the proper time at which the system first emerges as hadronic matter. The temperature of the hadronic matter will decrease until it reaches the freeze-out temperature below which the hadron will not maintain thermal contact.

§References for Chapter 13

1. J. D. Bjorken, Phys. Rev. D27, 140 (1983).
2. W. Busza and A. Goldhaber, Phys. Lett. 139B, 235 (1984).
3. K. Kinoshita, A. Minaka, and H. Sumiyoshi, Prog. Theo. Phys. 63, 1268 (1980); S. Daté, M. Gyulassy, and H. Sumiyoshi, Phys. Rev. D32, 619 (1985).
4. C. Y. Wong, Phys. Rev. Lett. 52, 1393 (1985); C. Y. Wong, Phys. Rev. D32, 94 (1985); C. Y. Wong, Phys. Rev. C33, 1340 (1986); C. Y. Wong and Z. D. Lu, Phys. Rev. D39, 2606 (1989).
5. For an excellent review of the experimental status, see J. Stachel and G. R. Young, Ann. Rev. Nucl. Part. Phys. 42, 537 (1992).
6. WA80 Collaboration, R. Albrecht et al, Phys. Lett. B199, 297 (1987); WA80 Collaboration, R. Albrecht et al, Phys. Lett. B201, 390 (1987); WA80 Collaboration, H. Löhner et al, Zeit. Phys. C38, 97 (1988); WA80 Collaboration, S. Sorensen et al, Zeit. Phys. C38, 3 (1988).

7. NA35 Collaboration, A. Bamberger *et al.*, Phys. Lett. B184 271 (1987); NA35 Collaboration, A. Sandoval *et al.*, Nucl. Phys. A461, 465 (1987); NA35 Collaboration, H. Ströbele *et al.*, Zeit. Phys. C38, 89 (1988); NA35 Collaboration, T. Humanic *et al.*, Zeit. Phys. C38, 79 (1988); NA35 Collaboration, G. Vesztergombi *et al.*, Zeit. Phys. C38, 129 (1988).

8. P. Braun-Munzinger *et al.* (E814 Collaboration), Nucl. Phys. A544, 137c (1992).

9. J. Bartke *et al.* (NA35 Collaboration), Zeit. Phys. C48, 191 (1990).

10. C. Y. Wong, Phys. Rev. D30, 961 (1984).

11. S. Weinberg, *The First Three Minutes*, Basic Books,N.Y.,1977.

12. G. Baym, Quark Matter '84, Springer-Verlag, Berlin, 1985, p. 39; H. Heiselberg, G. Baym, C. J. Pethick, and J. Popp, Nucl. Phys. A544, 569c (1992); U. Heinz, Nucl. Phys. A461, 49c (1987).

13. C. Y. Wong, and R. C. Wang, Phys. Rev. D44, 679 (1991).

14. B. Andersson, G. Gustafson and B. Söderberg, Z. Phys. C20, 317 (1983); T. Sjöstrand, Comp. Phys. Comm. 39, 347 (1986); T. Sjöstrand, and M. Bengtsson, Comp. Phys. Comm. 43, 367 (1987).

15. WA80 Collaboration, I. Lund *et al.*, Zeit. Phys. C38, 52 (1988).

16. R. Albrecht *et al.*, WA80 Collaboration, Zeit. Phys. C55, 539 (1992).

17. S. Sorensen *et al*, WA80 Collaboration, Zeit. Phys. C38, 3 (1988).

18. See for example, S. Weinberg, *Gravitation and Cosmology*, John Wiley and Sons, New York, 1972, p. 47.

19. L. D. Landau and E. M. Lifshitz, *Statistical Physics*, Pergamon Press, Oxford, Third Edition, 1980.

14. Signatures for the Quark-Gluon Plasma (I)

In the previous chapter, we discussed the evolution of the matter produced after the collision of two heavy nuclei at high energies. The produced matter may make an excursion from the hadron phase into the quark-gluon plasma phase. Subsequent cooling allows the matter to return to the hadron phase and to appear as hadrons. During the time when the matter is in the quark-gluon plasma phase, particles which arise from the interactions between the constituents of the plasma will provide information concerning the state of the plasma. The detection of the products of their interactions will be useful as a plasma diagnostic tool.

It is generally recognized that there is no single unique signal which allows an unequivocal identification of the quark-gluon plasma phase. What can be achieved may be an accumulative set of data which taken together may indicate the presence of the deconfined phase. Chapter 14 to Chapter 19 will discuss different signatures used to search for the quark-gluon plasma.

§14.1 Dilepton Production in the Quark-Gluon Plasma

In the quark-gluon plasma, a quark can interact with an antiquark to form a virtual photon γ^* and the virtual photon subsequently decays into a lepton l^- and an antilepton l^+. The diagram which describes the reaction $q + \bar{q} \rightarrow l^+ + l^-$ is shown in Fig. 14.1.

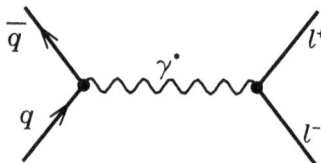

Fig. 14.1   The diagram for the reaction $q + \bar{q} \rightarrow l^+ + l^-$.

The system of the produced lepton-antilepton pair is called a dilepton. It is also called an l^+l^- pair, or simply a lepton pair. The dilepton is characterized by a dilepton invariant mass squared $M^2 = (l^+ + l^-)^2$, a dilepton four-momentum $P = (l^+ + l^-)$ and a dilepton transverse momentum $P_T = l_T^+ + l_T^-$.

After the lepton l^- and its antiparticle partner l^+ are produced, they must pass through the collision region to reach the detectors, in order to be observed. Since the leptons interact with the particles in the collision region only through the electromagnetic interaction, their interaction is not strong. The lepton-(charge particle) cross section is of the order $(\alpha/\sqrt{s})^2$ where $\alpha = 1/137$ is the fine structure constant and $\sqrt{s}$ is the lepton-(charge particle) center-of-mass energy. Consequently, the mean-free path of the leptons are expected to be quite large and the leptons are not likely to suffer further collisions after they are produced. On the other hand, the production rate and the momentum distribution of the produced l^+l^- pairs depend on the momentum distributions of quarks and antiquarks in the plasma, which are governed by the thermodynamic condition of the plasma. Therefore, l^+l^- pairs carry information on the thermodynamical state of the medium at the moment of their production [1-11].

We consider for simplicity a plasma where the net baryon density is zero so that the quark distribution $f(E)$ can be taken to be the same as the antiquark distribution. The case with unequal quark and antiquark distributions can be easily generalized (see Section 18.2 for an example how this can be carried out). The number of quarks with momentum p_1 in the spatial volume element d^3x, and in the momentum element d^3p_1 is

$$dN_q = g_q \frac{d^3x\, d^3p_1}{(2\pi)^3} f(E_1),$$

where g_q is the degeneracy of the quarks (or antiquarks) as given by Eq. (9.5),

$$g_q = (\text{ no. of colors } N_c) \times (\text{ no. of flavors } N_f) \times (\text{ no. of spins } N_s),$$

and E_1 is the energy of the quark with a rest mass m_q,

$$E_1 = \sqrt{p_1^2 + m_q^2}.$$

The integration of the phase space density with respect to the momentum coordinates gives the quark spatial density n_q,

$$\frac{dN_q}{d^3x} = n_q = g_q \int \frac{d^3p_1}{(2\pi)^3} f(E_1).$$

We shall show in Exercise 14.2 that the number of l^+l^- pairs produced per unit spatial volume per unit time is

$$\frac{dN_{l+l-}}{dt\,d^3x} = N_c N_s^2 \sum_{f=1}^{N_f} \left(\frac{e_f}{e}\right)^2 \int \frac{d^3p_1 d^3p_2}{(2\pi)^6} f(E_1)f(E_2)\sigma(M)v_{12}\,, \quad (14.1)$$

and the number of l^+l^- pairs produced per unit dilepton invariant mass squared M^2, per unit four-volume, is given by [11]

$$\frac{dN_{l+l-}}{dM^2 d^4x} = N_c N_s^2 \sum_{f=1}^{N_f} \left(\frac{e_f}{e}\right)^2 \frac{\sigma(M)}{2(2\pi)^4} M^2 \left(1 - \frac{4m_q^2}{M^2}\right)^{\frac{1}{2}} f(\epsilon)F\left(\frac{M^2}{4\epsilon}\right)\sqrt{\frac{2\pi}{w(\epsilon)}}.$$
$$(14.2)$$

where e_f is the electric charge of a quark with flavor f, $\sigma(M)$ is the $q\bar{q} \to l^+l^-$ cross section

$$\sigma(M) = \frac{4\pi}{3}\frac{\alpha^2}{M^2}\left(1 - \frac{4m_q^2}{M^2}\right)^{-\frac{1}{2}}\sqrt{1 - \frac{4m_l^2}{M^2}}\left(1 + 2\frac{m_q^2 + m_l^2}{M^2} + 4\frac{m_q^2 m_l^2}{M^4}\right),$$
$$(14.3)$$

(see Exercise 14.1), and m_l is the rest mass of the lepton. In Eq. (14.2), the function $F(E)$ is the indefinite integral of $[-f(E)]$,

$$F(E) = -\int_\infty^E f(E')dE'\,, \quad (14.4)$$

and $\epsilon = \epsilon(M)$ is the location of the extremum of the function $g(E)$,

$$g(E) = \ln f(E) + \ln F\left(\frac{M^2}{4E}\right). \quad (14.5a)$$

In other words, the location ϵ is the root of the extremum condition

$$\left\{\frac{d}{dE}\left[\ln f(E) + \ln F\left(\frac{M^2}{4E}\right)\right]\right\}_{E=\epsilon} = 0\,. \quad (14.5b)$$

The quantity $-w(\epsilon)$ is the second derivative of $g(E)$ with respect to E at the extremum, or,

$$w(\epsilon) = -\left\{\frac{d^2}{dE^2}\left[\ln f(E) + \ln F\left(\frac{M^2}{4E}\right)\right]\right\}_{E=\epsilon}. \quad (14.5c)$$

The distribution of l^+l^- pairs, with respect to dilepton invariant mass squared M^2 and transverse mass squared $M_T^2(=M^2+P_T^2)$, per unit four-volume, is

$$\frac{dN_{l+l-}}{dM^2dM_T^2d^4x} = N_cN_s^2\sum_{f=1}^{N_f}\left(\frac{e_f}{e}\right)^2\frac{\sigma(M)}{2(2\pi)^4}M^2\left(1-\frac{4m_q^2}{M^2}\right)^{\frac{1}{2}}$$

$$\times\left\{-\frac{d}{dM_T^2}\left(\left[f(\epsilon)F\left(\frac{M_T^2}{4\epsilon}\right)\sqrt{\frac{2\pi}{w(\epsilon)}}\right]_{\epsilon=\epsilon(M_T)}\right)\right\}. \quad (14.6)$$

The above results are valid for a general distribution $f(E)$ that is a function of E only. It shows that the dilepton distribution depends on the quark distribution $f(E)$ and its integral $F(E)$. If we can extract the dilepton distribution coming from the quark-gluon plasma by experimental measurements, then we can determine the characteristics of the quark distribution in the quark-gluon plasma.

We can specialize to a quark-gluon plasma in which the quark and the antiquark distribution $f(E)$ is given by $e^{-E/T}$ characterized by a temperature T. We shall show in Exercise 14.3 that the M^2 distribution of the l^+l^- pairs produced in the plasma per unit four-volume, is [2]

$$\frac{dN_{l+l-}}{dM^2d^4x} \sim N_cN_s^2\sum_{f=1}^{N_f}\left(\frac{e_f}{e}\right)^2\frac{\sigma(M)}{2(2\pi)^4}M^2\left(1-\frac{4m_q^2}{M^2}\right)^{\frac{1}{2}}TMK_1\left(\frac{M}{T}\right),$$
$$(14.7)$$

where K_ν is the modified Bessel function of order ν [12]. The distribution in dilepton invariant mass squared M^2 and transverse mass squared M_T^2, per unit space-time volume, is [2]

$$\frac{dN_{l+l-}}{dM^2dM_T^2d^4x} \sim N_cN_s^2\sum_{f=1}^{N_f}\left(\frac{e_f}{e}\right)^2\frac{\sigma(M)}{4(2\pi)^4}M^2\left(1-\frac{4m_q^2}{M^2}\right)^{\frac{1}{2}}K_0\left(\frac{M_T}{T}\right).$$
$$(14.8)$$

The above results are for a static plasma in which the temperature of the plasma is held fixed. We learned in the last chapter that the temperature of a plasma will decrease as a function of the proper time. In Bjorken's hydrodynamical model for the evolution of the plasma, the temperature T as a function of the proper time τ is given by Eq.

(13.18),

$$T(\tau) = T_0 \left(\frac{\tau_0}{\tau}\right)^{1/3},$$

where T_0 is the initial temperature of the plasma formed in heavy-ion collisions at the initial proper time τ_0. When the temperature drops below the transition temperature T_c at the proper time τ_c, the system will be in the mixed phase and no longer in the quark-gluon plasma phase. The l^+l^- pairs produced during the quark-gluon plasma phase can be obtained by integrating the production rate from the proper time τ_0 to τ_c.

We shall show in Exercise 14.4 that after we integrate over the contributions from the proper time τ_0 to τ_c in which the system is in the quark-gluon plasma phase, the distribution of l^+l^- pairs in dilepton invariant mass squared M^2 and dilepton rapidity y is [2]

$$\frac{dN_{l^+l^-}}{dM^2 dy} \sim \pi R_A^2 N_c N_s^2 \sum_{f=1}^{N_f} \left(\frac{e_f}{e}\right)^2 \frac{\sigma(M)}{2(2\pi)^4} \left(1 - \frac{4m_q^2}{M^2}\right)^{\frac{1}{2}} \frac{3\tau_0^2 T_0^6}{M^2}$$

$$\times \left[H\left(\frac{M}{T_0}\right) - H\left(\frac{M}{T_c}\right)\right], \qquad (14.9)$$

where R_A is the radius of the colliding nuclei (which are taken to be equal), and

$$H(z) = z^2(8 + z^2)K_0(z) + 4z(4 + z^2)K_1(z).$$

In a similar way, we can integrate the contributions from the proper time τ_0 to τ_c, to obtain the distribution of the l^+l^- pairs in M^2, M_T^2, and rapidity, given by [2]

$$\frac{dN_{l^+l^-}}{dM^2 dM_T^2 dy} \sim \pi R_A^2 N_c N_s^2 \sum_{f=1}^{N_f} \left(\frac{e_f}{e}\right)^2 \frac{\sigma(M)}{4(2\pi)^4} M^2 \left(1 - \frac{4m_q^2}{M^2}\right)^{\frac{1}{2}} \frac{3\tau_0^2 T_0^6}{M_T^6}$$

$$\times \left[G\left(\frac{M_T}{T_0}\right) - G\left(\frac{M_T}{T_c}\right)\right], \qquad (14.10)$$

where

$$G(z) = z^3(8 + z^2)K_3(z).$$

Eqs. (14.9) and (14.10) give the distribution of the l^+l^- pairs as a function of the initial temperature T_0 and the transition temperature

T_c of the plasma, after we take into account the hydrodynamical evolution of the system. Thus, the measurement of the l^+l^- pairs, if they can be identified as arising from this phase of matter, will reveal the thermodynamical state of the plasma.

It is instructive to obtain some intuitive insight into the dilepton distribution if the dileptons originate from $q\bar{q}$ annihilations in the quark-gluon plasma. For definiteness, we consider a quark-gluon plasma with flavors u and d whose quark and antiquark momentum distribution is described by a Boltzmann distribution $\exp\{-e/T\}$ with temperature T. For $M \gg T_0 \gg T_c$, one can use an exponential function to approximate the modified Bessel function, as given according to Eq. (9.7.2) of Ref. [12] by

$$K_1(z) \sim \sqrt{\frac{\pi}{2z}} e^{-z}.$$

For simplicity, we can neglect the quark mass and the lepton mass. It is easy to show that upon substituting Eq. (14.3) into Eq. (14.9), the dilepton distribution in invariant mass and rapidity is approximately given by

$$\frac{dN_{l+l-}}{dM\,dy} \sim \frac{5}{3\pi^2}\sqrt{\frac{\pi}{2}}\alpha^2\tau^2 R_A^2 T_0^3 \left(\frac{M}{T_0}\right)^{\frac{1}{2}} e^{-M/T_0} f\left(\frac{M}{T_0}\right)$$
$$\times \left\{1 - \left(\frac{T_0}{T_c}\right)^{\frac{7}{2}}\frac{f(M/T_c)}{f(M/T_0)}e^{-(\frac{M}{T_c}-\frac{M}{T_0})}\right\}, \quad (14.11)$$

where
$$f(z) = 1 + \frac{4}{z} + \frac{8}{z^2} + \frac{16}{z^3}.$$

The dominant invariant mass dependence comes from the factor e^{-M/T_0}. If we parametrize the dilepton invariant mass distribution in terms of a dilepton temperature in the form $M^{1/2}\exp\{-M/T_{\text{dilepton}}\}$, then the dilepton 'temperature' T_{dilepton} is approximately the same as the quark initial temperature T_0. Therefore, the dilepton invariant mass distribution varies approximately as the initial quark distribution. If one can extract the dilepton spectrum coming from the quark-gluon plasma, then one may determine the plasma initial temperature T_0.

•⟦ Exercise 14.1

Prove that the cross section for the process $q + \bar{q} \to l^+ + l^-$ depicted in Fig. 14.1 for unpolarized quark and antiquark is given by

$$\sigma(q\bar{q} \to l^+ l^-) = \left(\frac{e_q}{e}\right)^2 \sigma(M),$$ (1a)

where

$$\sigma(M) = \frac{4\pi}{3} \frac{\alpha^2}{M^2} \left(1 - \frac{4m_q^2}{M^2}\right)^{-1/2} \sqrt{1 - \frac{4m_l^2}{M^2}} \left(1 + 2\frac{m_q^2 + m_l^2}{M^2} + 4\frac{m_q^2 m_l^2}{M^4}\right),$$ (1b)

e_q is the charge of the quark, m_q is the mass of the quark, m_l is the rest mass of the lepton, and $M^2 = (q + \bar{q})^2$ is the square of the invariant mass of the $q\bar{q}$ system.

◇Solution:

This is a standard problem in perturbative QED to which Feynman diagram techniques can be applied [13-16]. For the limiting case with a vanishing mass of q and $\bar{q}$, the result can be found on page 329 of Ref. [16] and in the literature [2]. In order to assist those readers who may not be familiar with the Feynman diagram techniques, we shall present the derivation in some detail. To carry out the mathematical manipulations, it is useful to consult the Appendices of the books of Itzykson and Zuber [13], Bjorken and Drell [14], and Cheng and Li [15], where many useful formulas are listed.

Following the notations of Chapters 2 and 3, we use the same letter to represent a particle and its four-momentum. We denote l^- by l and its antiparticle l^+ by $\bar{l}$. The differential cross section for the process $q + \bar{q} \to l + \bar{l}$ is

$$d\sigma = \frac{2m_q \, 2m_{\bar{q}}}{4[(q \cdot \bar{q})^2 - m_q^2 m_{\bar{q}}^2]^{1/2}} |T_{fi}|^2 \frac{d^3l}{(2\pi)^3} \frac{m_l}{l_0} \frac{d^3\bar{l}}{(2\pi)^3} \frac{m_{\bar{l}}}{\bar{l}_0} (2\pi)^4 \delta^4(q + \bar{q} - l - \bar{l}).$$ (2)

Here, in accordance with Eq. (A-43) of the Appendix of Itzykson and Zuber [13], the factor $2m_q 2m_{\bar{q}}$ comes from the two incident massive fermions, the denominator comes from the relative velocity factor, $E_q E_{\bar{q}} |v_q - v_{\bar{q}}|$, (see Eq. (12) of Supplement 4.1), T_{fi} is the matrix element for a transition from the initial state i to the final state f, the differential elements $(d^3l/(2\pi)^3)(m_l/l_0)$ and $(d^3\bar{l}/(2\pi)^3)(m_{\bar{l}}/\bar{l}_0)$ come from the densities of states of the final particles l and $\bar{l}$, and the four-dimensional δ function originates from energy-momentum conservation.

To obtain the total cross section for unpolarized q and $\bar{q}$, we need to average over the initial states and sum over the final states. This means that we must sum over the momenta d^3l and $d^3\bar{l}$ and the polarizations of l and $\bar{l}$, and we must average over the initial polarizations of q and $\bar{q}$. Before we perform these integrations and summations, we introduce the Mandelstam variables s, t and u as

$$s = (q + \bar{q})^2,$$ (3a)

$$t = (q - l)^2,$$ (3b)

and

$$u = (q - \bar{l})^2 . \tag{3c}$$

Because of the energy-momentum conservation condition, $q + \bar{q} = l + \bar{l}$, s is also equal to

$$s = (l + \bar{l})^2 .$$

For convenience, we shall work in the $q\bar{q}$ center-of-mass frame. In this frame, we have $\boldsymbol{q} = -\bar{\boldsymbol{q}}$ and

$$s = (q_0 + \bar{q}_0)^2 = (l_0 + \bar{l}_0)^2 .$$

In consequence, we have

$$q_0 = \bar{q}_0 = l_0 = \bar{l}_0 = \frac{1}{2}\sqrt{s} , \tag{4a}$$

$$|\boldsymbol{q}| = |\bar{\boldsymbol{q}}| = \frac{1}{2}\sqrt{s - 4m_q^2} , \tag{4b}$$

and

$$|\boldsymbol{l}| = |\bar{\boldsymbol{l}}| = \frac{1}{2}\sqrt{s - 4m_l^2} . \tag{4c}$$

The quantity $\sqrt{s}$ is the invariant mass M of the $q\bar{q}$ system and is a constant of motion in this problem. The variable t measures the angle between $\boldsymbol{q}$ and $\boldsymbol{l}$, and u measures the angle between $\boldsymbol{q}$ and $\bar{\boldsymbol{l}}$.

We first evaluate the factor in the denominator of Eq. (2). We have

$$(q \cdot \bar{q})^2 - m_q^2 m_{\bar{q}}^2 = \frac{1}{4}(s - 2m_q^2)^2 - m_q^4$$

$$= \frac{1}{4}s(s - 4m_q^2) .$$

So, the denominator in the first factor of Eq. (2) is

$$4[(q \cdot \bar{q})^2 - m_q^2 m_{\bar{q}}^2]^{1/2} = 2\sqrt{s}\sqrt{s - 4m_q^2}$$

We next integrate out $d\bar{l}$ and dl_0 in Eq. (2), using the four-dimensional δ function. We have

$$\int \frac{d^3l}{(2\pi)^3}\frac{m_l}{l_0}\frac{d^3\bar{l}}{(2\pi)^3}\frac{m_{\bar{l}}}{\bar{l}_0}(2\pi)^4\delta^4(q + \bar{q} - l - \bar{l})\mathcal{F}$$

$$= \frac{1}{(2\pi)^2}\int \frac{|\boldsymbol{l}|^2 d|\boldsymbol{l}|\ d\cos\theta_l\ d\phi_l\ m_l}{l_0}\frac{d\bar{l}\ m_{\bar{l}}}{\bar{l}_0}\delta(\sqrt{s} - l_0 - \bar{l}_0)\delta(\boldsymbol{q} + \bar{\boldsymbol{q}} - \boldsymbol{l} - \bar{\boldsymbol{l}})\mathcal{F} ,$$

where

$$\mathcal{F} = \frac{2m_q\ 2m_{\bar{q}}}{4[(q \cdot \bar{q})^2 - m_q^2 m_{\bar{q}}^2]^{1/2}}|T_{fi}|^2 .$$

The integration over $d\bar{l}$ gives $l = -\bar{l}$, $\bar{l}_0 = l_0$, and the above integral becomes

$$\frac{1}{(2\pi)^2} \int \frac{|l| \, l_0 \, dl_0 \, d\cos\theta_l \, d\phi_l \, m_l^2}{l_0^2} \delta(\sqrt{s} - l_0 - \bar{l}_0)\bigg|_{\bar{l}_0 = l_0} \mathcal{F} \, .$$

After the integration over dl_0, this integral becomes

$$\frac{1}{2(2\pi)^2} \int \frac{|l| \, d\cos\theta_l \, d\phi_l \, m_l^2}{l_0} \mathcal{F} \, .$$

It is convenient to change the variable from $\cos\theta_l$ to t. Without loss of generality, we can choose q to lie along the z-axis. The angle θ_l is thus the angle between l and q, in our $q\bar{q}$ center-of-mass system. We have

$$
\begin{aligned}
t &= (q - l)^2 \\
&= q^2 - 2q \cdot l + l^2 \\
&= m_q^2 + m_l^2 - 2q_0 l_0 + 2\,|q|\,|l|\cos\theta_l \\
&= m_q^2 + m_l^2 - \frac{1}{2}s + \frac{1}{2}\sqrt{s - 4m_q^2}\sqrt{s - 4m_l^2}\cos\theta_l \, .
\end{aligned}
\tag{5}
$$

After changing the variable from $\cos\theta_l$ to t, we get

$$d\cos\theta_l = \frac{dt}{2\,|q|\,|l|} \, .$$

The integral in $d\cos\theta_l$ and $d\phi_l$ becomes

$$
\begin{aligned}
\frac{1}{2(2\pi)^2} \int \frac{|l| \, d\cos\theta_l \, d\phi_l \, m_l^2}{l_0} \mathcal{F} &= \frac{1}{2(2\pi)^2} \int \frac{|l| \, dt \, d\phi_l \, m_l^2}{l_0 \, 2\,|q|\,|l|} \mathcal{F} \\
&= \frac{1}{2(2\pi)^2} \int \frac{dt \, d\phi_l \, m_l^2}{l_0 \, 2\,|q|} \mathcal{F} \, .
\end{aligned}
$$

Using Eq. (4) to writing l_0 and $|q|$ in terms of s, the above integral becomes

$$\frac{1}{(2\pi)^2} \int \frac{dt \, d\phi_l \, m_l^2}{\sqrt{s}\sqrt{s - 4m_q^2}} \mathcal{F} \, .$$

Thus, after the integration over $d\bar{l}$ and dl_0, the differential cross section (2) becomes

$$d\sigma = \frac{2 \, m_q^2 \, m_l^2}{(2\pi)^2} \frac{dt \, d\phi_l}{s(s - 4m_q^2)} |T_{fi}|^2 \, . \tag{6}$$

Using the rules of Feynman diagrams, one can write down the explicit form of the matrix element T_{fi} by associating various parts of the diagram in Fig. 14.2 with different functions and matrices. We begin by specifying the momenta of the lines in the diagram. In Fig. 14.2, the letters q, $\bar{q}$, l, $\bar{l}$ label the corresponding particles

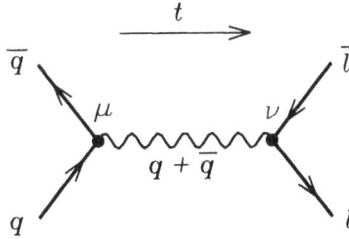

Fig. 14.2 The Feynman diagram for the reaction $q + \bar{q} \to l + \bar{l}$, in which the vertices and the momentum of the lines are labelled.

and their momenta. The momentum of the photon line (the wavy line) is $q + \bar{q}$. We also need to label the interaction vertices μ and ν and indicate the direction of the time axis.

The external fermion lines are associated with various spinor functions. We can illustrate this association by considering the quark field operator $\psi(x)$ and its conjugate $\bar{\psi}(x)$ written in terms of the quark and the antiquark creation and annihilation operators as given by Eq. (3-157) of Itzykson and Zuber [13],

$$\psi(x) = \sum_{\epsilon=1}^{2} \int \frac{d^3 k}{(2\pi)^3} \frac{m}{k^0} [q(k,\epsilon) u(k,\epsilon) e^{-ik \cdot x} + \bar{q}^\dagger(k,\epsilon) v(k,\epsilon) e^{ik \cdot x}], \tag{7}$$

and

$$\bar{\psi}(x) = \sum_{\epsilon=1}^{2} \int \frac{d^3 k}{(2\pi)^3} \frac{m}{k^0} [q^\dagger(k,\epsilon) \bar{u}(k,\epsilon) e^{ik \cdot x} + \bar{q}(k,\epsilon) \bar{v}(k,\epsilon) e^{-ik \cdot x}], \tag{8}$$

where the operators $q(k,\epsilon)$ and $\bar{q}(k,\epsilon)$ are respectively the quark annihilation operator and the antiquark annihilation operator for a quark or an antiquark with 4-momentum k, polarization ϵ, and mass m. The operators $q^\dagger(k,\epsilon)$ and $\bar{q}^\dagger(k,\epsilon)$ are the corresponding creation operators. In Eqs. (7) and (8), the spinors u and v are respectively the spinor function of the positive-energy solution and the negative-energy solution for the Dirac equation of a free particle. They are given explicitly by Eqs. (2-36), and (2-37) of Itzykson and Zuber [2] as

$$u(k,\epsilon) = \frac{\not{k} + m}{\sqrt{2m(m+k^0)}} \begin{pmatrix} \delta_{\epsilon 1} \\ \delta_{\epsilon 2} \\ 0 \\ 0 \end{pmatrix}, \tag{9}$$

and

$$v(k,\epsilon) = \frac{-\not{k} + m}{\sqrt{2m(m+k^0)}} \begin{pmatrix} 0 \\ 0 \\ \delta_{\epsilon 1} \\ \delta_{\epsilon 2} \end{pmatrix}, \tag{10}$$

with $k^0 = (m_q^2 + \mathbf{k}^2)^{1/2} \geq 0$.

The Feynman rules [13,14] for an external fermion line with a particle momentum k and polarization ϵ are as follows: First, we determine whether the fermion line represents a particle or an antiparticle. A fermion line with its arrow pointing along the time direction represents a particle and is associated with the spinors $u(k, \epsilon)$ or $\bar{u}(k, \epsilon)$. A fermion line with its arrow pointing opposite to the time direction represents an antiparticle and is associated with the spinors $v(k, \epsilon)$ or $\bar{v}(k, \epsilon)$. After this step, we need to pick a spinor out of the set of two possible spinors by determining whether the line represents an initial state or a final state. For a particle in an initial state, we take the spinor $u(k, \epsilon)$, and for a particle in a final state, we take the spinor $\bar{u}(k, \epsilon)$. To get the corresponding rules for the antiparticle, we note that the time direction of an antiparticle is opposite to that of a particle. Consequently, the roles of the initial state and the final state of an antiparticle are the reverse of those of a particle. Thus, the Feynman rules for the external lines of an antiparticle are: for an antiparticle in a final state, we take the spinor $v(k, \epsilon)$, and for an antiparticle in an initial state, we take the spinor $\bar{v}(k, \epsilon)$.

We can illustrate these considerations by examining Fig. 14.2. The quark line q enters into the interaction vertex μ. It represents the annihilation of a quark q at the vertex μ. From the direction of time, the quark line q is a particle line and represents an initial state. Hence, the quark line q is associated with the spinor function $u(q, \epsilon_q)$.

The vertex μ is associated with $-ie_q\gamma^\mu$, the product of $(-i)$ times the charge of the quark e_q and the gamma matrix γ^μ.

The line $\bar{q}$ leaves the vertex μ and the direction of its arrow indicates that it describes a quark travelling backward in time. This is equivalent to depicting the particle as an antiquark travelling forward in time. Therefore, in the physical world where particles travel forward in time, the line $\bar{q}$ can be treated as an antiquark. From the time direction, we see that it represents an initial state. Hence, the antiquark line $\bar{q}$ of the diagram in Fig. 14.2 is associated with the spinor function $\bar{v}(\bar{q}, \epsilon_{\bar{q}})$.

The dashed photon line in Fig. 14.2, has momentum $q + \bar{q}$. It represents the photon propagator and is associated with the factor $[-ig_{\mu\nu}/(q + \bar{q})^2]$.

The lepton line l leaves the vertex ν and represents the creation of a lepton particle with momentum l at the vertex ν after the interaction. It appears in the final state. So, the lepton line l is associated with the spinor function $\bar{u}(l, \epsilon_l)$.

The vertex ν is associated $-ie\gamma^\nu$, the product of $(-i)$ times the lepton electric charge e and the gamma matrix γ^ν.

The antilepton line $\bar{l}$ describes a lepton which travels backward in time. It is an antiparticle which appears in the final state. Here, the antiquark line $\bar{l}$ is associated with the spinor function $v(\bar{l}, \epsilon_{\bar{l}})$.

By putting all factors together, the diagram in Fig. 14.2 gives the transition matrix element

$$\mathcal{T}_{fi} = \bar{u}(l, \epsilon_l)(-ie\gamma^\nu)v(\bar{l}, \epsilon_{\bar{l}})\frac{-ig_{\mu\nu}}{(q + \bar{q})^2}\bar{v}(\bar{q}, \epsilon_{\bar{q}})(-ie_q\gamma^\mu)u(q, \epsilon_q). \tag{11}$$

The square of the matrix element is

$$
|T_{fi}|^2 = \frac{e^2 e_q^2}{s^2} [\bar{u}(l,\epsilon_l)\gamma_\mu v(\bar{l},\epsilon_{\bar{l}})\bar{v}(\bar{q},\epsilon_{\bar{q}})\gamma^\mu u(q,\epsilon_q)]^\dagger \, [\bar{u}(l,\epsilon_l)\gamma_\mu v(\bar{l},\epsilon_{\bar{l}}) \, \bar{v}(\bar{q},\epsilon_{\bar{q}})\gamma^\mu u(q,\epsilon_q)]
$$

$$
= \frac{e^2 e_q^2}{s^2} [\bar{u}(q,\epsilon_q)\gamma^0\gamma^{\nu\dagger}\gamma^0 v(\bar{q},\epsilon_{\bar{q}}) \, \bar{v}(\bar{l},\epsilon_{\bar{l}})\gamma^0\gamma_\nu^\dagger\gamma^0 u(l,\epsilon_l)]
$$

$$
\times [\bar{u}(l,\epsilon_l)\gamma_\mu v(\bar{l},\epsilon_{\bar{l}}) \, \bar{v}(\bar{q},\epsilon_{\bar{q}})\gamma^\mu u(q,\epsilon_q)]
$$

$$
= \frac{e^2 e_q^2}{s^2} [\bar{u}(\bar{q},\epsilon_q)\gamma^\nu v(\bar{q},\epsilon_{\bar{q}}) \, \bar{v}(\bar{l},\epsilon_{\bar{l}})\gamma_\nu u(l,\epsilon_l)] \, [\bar{u}(l,\epsilon_l)\gamma_\mu v(\bar{l},\epsilon_{\bar{l}}) \, \bar{v}(\bar{q},\epsilon_{\bar{q}})\gamma^\mu u(q,\epsilon_q)] \,,
$$

where we have used the relation $\gamma^0\gamma^{\nu\dagger}\gamma^0 = \gamma^\nu$.

It is useful to write the above expression into a density matrix form. To accomplish this, we write out the components of the spinor functions and the gamma matrices. We have

$$
|T_{fi}|^2 = \frac{e^2 e_q^2}{s^2} [\bar{u}_\alpha(q,\epsilon_q)(\gamma^\nu)_{\alpha\beta} v_\beta(\bar{q},\epsilon_{\bar{q}})\bar{v}_a(\bar{l},\epsilon_{\bar{l}})(\gamma_\nu)_{ab} u_b(l,\epsilon_l)]
$$

$$
\times [\bar{u}_c(l,\epsilon_l)(\gamma_\mu)_{cd} v_d(\bar{l},\epsilon_{\bar{l}})\bar{v}_\gamma(\bar{q},\epsilon_{\bar{q}})(\gamma^\mu)_{\gamma\delta} u_\delta(q,\epsilon_q)]
$$

$$
= \frac{e^2 e_q^2}{s^2} [u_\delta(q,\epsilon_q)\bar{u}_\alpha(q,\epsilon_q)(\gamma^\nu)_{\alpha\beta} v_\beta(\bar{q},\epsilon_{\bar{q}})\bar{v}_\gamma(\bar{q},\epsilon_{\bar{q}})(\gamma^\mu)_{\gamma\delta}]
$$

$$
\times [v_d(\bar{l},\epsilon_{\bar{l}})\bar{v}_a(\bar{l},\epsilon_{\bar{l}})(\gamma_\nu)_{ab} u_b(l,\epsilon_l)\bar{u}_c(l,\epsilon_l)(\gamma_\mu)_{cd}] \,.
$$

We need to sum over the final polarizations ϵ_l and $\epsilon_{\bar{l}}$ and average over the initial polarizations ϵ_q and $\epsilon_{\bar{q}}$. We have

$$
\frac{1}{4} \sum_{\epsilon_q \epsilon_{\bar{q}} \epsilon_l \epsilon_{\bar{l}}}^2 |T_{fi}|^2 = \frac{e^2 e_q^2}{4s^2} \sum_{\epsilon_q \epsilon_{\bar{q}} \epsilon_l \epsilon_{\bar{l}}}^2 [u_\delta(q,\epsilon_q)\bar{u}_\alpha(q,\epsilon_q)(\gamma^\nu)_{\alpha\beta} v_\beta(\bar{q},\epsilon_{\bar{q}})\bar{v}_\gamma(\bar{q},\epsilon_{\bar{q}})(\gamma^\mu)_{\gamma\delta}]
$$

$$
\times [v_d(\bar{l},\epsilon_{\bar{l}})\bar{v}_a(\bar{l},\epsilon_{\bar{l}})(\gamma_\nu)_{ab} u_b(l,\epsilon_l)\bar{u}_c(l,\epsilon_l)(\gamma_\mu)_{cd}]
$$

$$
= \frac{e^2 e_q^2}{4s^2} [\, \rho_{\delta\alpha}(q) \, (\gamma^\nu)_{\alpha\beta} \, \rho_{\beta\gamma}(\bar{q}) \, (\gamma^\mu)_{\gamma\delta}][\, \rho_{da}(\bar{l}) \, (\gamma_\nu)_{ab} \, \rho_{bc}(l) \, (\gamma_\mu)_{cd}], \quad (12)
$$

where we have introduced density matrices $\rho(q)$ and $\rho(\bar{q})$ defined as [16]

$$
\rho_{\delta\alpha}(q) = \sum_{\epsilon_q=1}^2 u_\delta(q,\epsilon_q)\bar{u}_\alpha(q,\epsilon_q) \,, \quad (13)
$$

and

$$
\rho_{\beta\gamma}(\bar{q}) = \sum_{\epsilon_{\bar{q}}=1}^2 v_\beta(\bar{q},\epsilon_{\bar{q}})\bar{v}_\gamma(\bar{q},\epsilon_{\bar{q}}) \,. \quad (14)
$$

The density matrices for the lepton l and the antilepton $\bar{l}$ are similarly defined. In matrix notation, Eq. (12) can be rewritten as the trace of a product of matrices as

$$
\frac{1}{4} \sum_{\epsilon_q \epsilon_{\bar{q}} \epsilon_l \epsilon_{\bar{l}}}^2 |T_{fi}|^2 = \frac{e^2 e_q^2}{4s^2} \, tr\{ \, \rho(q) \, \gamma^\nu \, \rho(\bar{q}) \, \gamma^\mu \, \} \, tr\{ \, \rho(\bar{l}) \, \gamma_\nu \, \rho(l) \, \gamma_\mu) \, \} \,. \quad (15)
$$

From the definitions (13) and (14), the density matrices can be evaluated by using the explicit expressions of the spinor functions (9) and (10). We find

$$\rho(q) = \sum_{\epsilon=1}^{2} u(q, \epsilon)\bar{u}(q, \epsilon)$$

$$= \frac{1}{2m_q(m_q + q^0)}(\not{q} + m_q)\sum_{\epsilon=1}^{2}\begin{pmatrix} \delta_{\epsilon 1} \\ \delta_{\epsilon 2} \\ 0 \\ 0 \end{pmatrix}(\begin{array}{cccc} \delta_{\epsilon 1} & \delta_{\epsilon 2} & 0 & 0 \end{array})(\not{q}^\dagger + m_q)\gamma^0, \quad (16)$$

where $\not{q}^\dagger = \gamma^{\mu\dagger}q_\mu = \gamma^0\gamma^\mu\gamma^0 q_\mu$. Direct matrix multiplication yields

$$\sum_{\epsilon=1}^{2}\begin{pmatrix} \delta_{\epsilon 1} \\ \delta_{\epsilon 2} \\ 0 \\ 0 \end{pmatrix}(\begin{array}{cccc} \delta_{\epsilon 1} & \delta_{\epsilon 2} & 0 & 0 \end{array}) = \begin{pmatrix} 1 & 0 & 0 & 0 \\ 0 & 1 & 0 & 0 \\ 0 & 0 & 0 & 0 \\ 0 & 0 & 0 & 0 \end{pmatrix}$$

$$= \frac{1 + \gamma^0}{2}. \quad (17)$$

Therefore, from Eqs. (16) and (17), we have

$$\rho(q) = \frac{1}{2m_q(m_q + q^0)}(\not{q} + m_q)\frac{1 + \gamma^0}{2}(\not{q}^\dagger + m_q)\gamma^0$$

$$= \frac{1}{4m_q(m_q + q^0)}(\not{q} + m_q)\{(\not{q}^\dagger + m_q)\gamma^0 + \gamma^0(\not{q}^\dagger + m_q)\gamma^0\}$$

$$= \frac{1}{4m_q(m_q + q^0)}(\not{q} + m_q)\{(\gamma^0\gamma^\mu\gamma^0 q_\mu + m_q)\gamma^0 + \gamma^0(\gamma^0\gamma^\mu\gamma^0 q_\mu + m_q)\gamma^0\}.$$

Using the anticommutation relation, $\gamma^\alpha\gamma^\beta + \gamma^\beta\gamma^\alpha = 2g^{\alpha\beta}$, we get

$$\rho(q) = \frac{1}{4m_q(m_q + q^0)}(\not{q} + m_q)\{[(2g^{0\mu} - \gamma^\mu\gamma^0)\gamma^0 q_\mu + m_q]\gamma^0 + (\gamma^\mu q_\mu + m_q)\}$$

$$= \frac{1}{4m_q(m_q + q^0)}\{(\not{q} + m_q)[2q^0\gamma^0 - \gamma^\mu q_\mu + m_q]\gamma^0 + (\not{q} + m_q)(\not{q} + m_q)\}$$

$$= \frac{1}{4m_q(m_q + q^0)}\{[(\not{q} + m_q)2q^0\gamma^0 + (\not{q} + m_q)(-\not{q} + m_q)]\gamma^0 + m_q^2 + 2m_q\not{q} + m_q^2\}$$

$$= \frac{\not{q} + m_q}{2m_q}. \quad (18)$$

Similarly, we find the density matrix $\rho(\bar{q})$ as

$$\rho(\bar{q}) = \sum_{\epsilon=1}^{2} v(\bar{q},\epsilon)\bar{v}(\bar{q},\epsilon)$$

$$= \frac{1}{2m_{\bar{q}}(m_{\bar{q}}+\bar{q}^0)}(-\bar{\not{q}}+m_q)\sum_{\epsilon=1}^{2}\begin{pmatrix}0\\0\\\delta_{\epsilon 1}\\\delta_{\epsilon 2}\end{pmatrix}(\,0\ \ 0\ \ \delta_{\epsilon 1}\ \ \delta_{\epsilon 2}\,)(-\bar{\not{q}}^\dagger+m_q)\gamma^0$$

$$= \frac{1}{2m_q(m_q+\bar{q}^0)}(-\bar{\not{q}}+m_q)\begin{pmatrix}0&0&0&0\\0&0&0&0\\0&0&1&0\\0&0&0&1\end{pmatrix}(-\bar{\not{q}}^\dagger+m_q)\gamma^0$$

$$= \frac{1}{2m_q(m_q+\bar{q}^0)}(-\bar{\not{q}}+m_q)\frac{1-\gamma^0}{2}(-\bar{\not{q}}^\dagger+m_q)\gamma^0$$

$$= \frac{1}{4m_q(m_q+\bar{q}^0)}(-\bar{\not{q}}+m_q)\{(-\bar{\not{q}}^\dagger+m_q)\gamma^0-\gamma^0(-\bar{\not{q}}^\dagger+m_q)\gamma^0\}\,.$$

Expanding out $\bar{\not{q}}$ and using the anticommutation relation for the gamma matrices, we obtain

$$\rho(\bar{q}) = \frac{1}{4m_q(m_q+\bar{q}^0)}(-\bar{\not{q}}+m_q)\{(-\gamma^0\gamma^\mu\gamma^0\bar{q}_\mu+m_q)\gamma^0-\gamma^0(-\gamma^0\gamma^\mu\gamma^0\bar{q}_\mu+m_q)\gamma^0\}$$

$$= \frac{1}{4m_q(m_q+\bar{q}^0)}(-\bar{\not{q}}+m_q)\{[-(2g^{0\mu}-\gamma^\mu\gamma^0)\gamma^0\bar{q}_\mu+m]\gamma^0-(-\gamma^\mu\bar{q}_\mu+m_q)\}$$

$$= \frac{1}{4m_q(m_q+\bar{q}^0)}\{(-\bar{\not{q}}+m_q)[-2\bar{q}^0\gamma^0+\gamma^\mu\bar{q}_\mu+m]\gamma^0-(-\bar{\not{q}}+m_q)(-\bar{\not{q}}+m_q)\}$$

$$= \frac{1}{4m_q(m_q+\bar{q}^0)}\{[-(-\bar{\not{q}}+m_q)2\bar{q}^0\gamma^0+(-\bar{\not{q}}+m_q)(\bar{\not{q}}+m_q)]\gamma^0-m_q^2+2m_q\bar{\not{q}}-m_q^2\}$$

$$= -\frac{-\bar{\not{q}}+m_q}{2m_q}\,, \tag{19}$$

which is just Eq. (29.17) of Ref. [16]. Using Eqs. (18) and (19), the summation (15) becomes

$$\frac{1}{4}\sum_{\epsilon_q\epsilon_{\bar{q}}\epsilon_l\epsilon_{\bar{l}}}|T_{fi}|^2 = \frac{e^2e_q^2}{4s^2}tr\{\frac{\not{q}+m_q}{2m_q}\gamma^\nu\frac{-\bar{\not{q}}+m_q}{2m_q}\gamma^\mu\}tr\{\frac{-\bar{\not{l}}+m_l}{2m_l}\gamma_\nu\frac{\not{l}+m_l}{2m_l}\gamma_\mu\}$$

$$= \frac{e^2e_q^2}{4m_q^2m_l^2s^2}\frac{1}{4}tr\{(\gamma^\alpha q_\alpha+m_q)\gamma^\nu(-\gamma^\beta\bar{q}_\beta+m_q)\gamma^\mu\}\frac{1}{4}tr\{(-\gamma^a\bar{l}_a+m_l)\gamma_\nu(\gamma^b l_b+m_l)\gamma_\mu\}$$

$$= \frac{e^2e_q^2}{4m_q^2\,m_l^2s^2}\frac{1}{4}tr\{m_q^2\gamma^\nu\gamma^\mu-\gamma^\alpha\gamma^\nu\gamma^\beta\gamma^\mu q_\alpha\bar{q}_\beta\}\frac{1}{4}tr\{m_l^2\gamma_\nu\gamma_\mu-\gamma^a\gamma_\nu\gamma^b\gamma_\mu\bar{l}_a l_b\}\,, \tag{20}$$

where we have used the property that the trace of the product of an odd number of γ matrices is zero. From the anticommutation relation of the γ matrices and the cyclic property of trace, we get

$$tr\{\gamma^\nu\gamma^\mu\} = tr\{2g^{\nu\mu}\} - tr\{\gamma^\mu\gamma^\nu\}$$
$$= tr\{2g^{\nu\mu}\} - tr\{\gamma^\nu\gamma^\mu\}$$
$$= tr\{g^{\nu\mu}\}$$
$$= 4g^{\nu\mu}.$$

We also have

$$tr\{\gamma^\alpha\gamma^\nu\gamma^\beta\gamma^\mu\} = tr\{2g^{\alpha\nu}\gamma^\beta\gamma^\mu\} - tr\{\gamma^\nu\gamma^\alpha\gamma^\beta\gamma^\mu\}$$
$$= 8g^{\alpha\nu}g^{\beta\mu} - tr\{\gamma^\nu 2g^{\alpha\beta}\gamma^\mu\} + tr\{\gamma^\nu\gamma^\beta\gamma^\alpha\gamma^\mu\}$$
$$= 8g^{\alpha\nu}g^{\beta\mu} - 8g^{\nu\mu}g^{\alpha\beta} + tr\{\gamma^\nu\gamma^\beta 2g^{\alpha\mu}\} - tr\{\gamma^\nu\gamma^\beta\gamma^\mu\gamma^\alpha\}$$
$$= 4g^{\alpha\nu}g^{\beta\mu} - 4g^{\nu\mu}g^{\alpha\beta} + 4g^{\nu\beta}g^{\alpha\mu}.$$

With these results, we obtain

$$\frac{1}{4}tr\{(\gamma^\alpha q_\alpha + m_q)\gamma^\nu(-\gamma^\beta\bar{q}_\beta + m_q)\gamma^\mu\} = \frac{1}{4}tr\{m_q^2\gamma^\nu\gamma^\mu - \gamma^\alpha\gamma^\nu\gamma^\beta\gamma^\mu q_\alpha\bar{q}_\beta\}$$
$$= m_q^2 g^{\nu\mu} - (g^{\alpha\nu}g^{\beta\mu} - g^{\nu\mu}g^{\alpha\beta} + g^{\nu\beta}g^{\alpha\mu})q_\alpha\bar{q}_\beta$$
$$= m_q^2 g^{\nu\mu} - q^\nu\bar{q}^\mu + g^{\nu\mu}q\cdot\bar{q} - q^\mu\bar{q}^\nu$$
$$= g^{\nu\mu}(m_q^2 + q\cdot\bar{q}) - q^\nu\bar{q}^\mu - q^\mu\bar{q}^\nu.$$

The summation (20) becomes

$$\frac{1}{4}\sum_{\epsilon_q\epsilon_{\bar{q}}\epsilon_l\epsilon_{\bar{l}}}^{2}|T_{fi}|^2 = \frac{e^2 e_q^2}{4m_q^2 m_l^2 s^2}[g^{\nu\mu}(m_q^2+q\cdot\bar{q})-q^\nu\bar{q}^\mu-q^\mu\bar{q}^\nu][g_{\nu\mu}(m_l^2+l\cdot\bar{l})-l_\nu\bar{l}_\mu-l_\mu\bar{l}_\nu]$$

$$= \frac{e^2 e_q^2}{4m_q^2 m_l^2 s^2}[g^{\nu\mu}g_{\nu\mu}(m_q^2+q\cdot\bar{q})(m_l^2+l\cdot\bar{l}) - (m_q^2+q\cdot\bar{q})2l\cdot\bar{l} - (m_l^2+l\cdot\bar{l})2q\cdot\bar{q}$$
$$+ (q\cdot l)(\bar{q}\cdot\bar{l}) + (q\cdot\bar{l})(\bar{q}\cdot l) + (q\cdot\bar{l})(\bar{q}\cdot l) + (q\cdot l)(\bar{q}\cdot\bar{l})]. \quad (21)$$

We note that

$$q\cdot l = -\frac{1}{2}[(q-l)^2 - q^2 - l^2] = -\frac{1}{2}[t - m_q^2 - m_l^2] \quad (22)$$
$$= -\frac{1}{2}[(\bar{q}-\bar{l})^2 - q^2 - l^2] = \bar{q}\cdot\bar{l},$$

and

$$q\cdot\bar{l} = -\frac{1}{2}[(q-\bar{l})^2 - q^2 - l^2] = -\frac{1}{2}[u - m_q^2 - m_l^2] \quad (23)$$
$$= -\frac{1}{2}[(\bar{q}-l)^2 - q^2 - l^2] = \bar{q}\cdot l.$$

Using the relations of Eqs. (22) and (23), Eq. (21) becomes

$$\frac{1}{4} \sum_{\epsilon_q \epsilon_{\bar{q}} \epsilon_l \epsilon_{\bar{l}}}^{2} |T_{fi}|^2 = \frac{e^2 e_q^2}{4m_q^2 m_l^2 s^2} [2(m_q^2 + q \cdot \bar{q})(m_l^2 + l \cdot \bar{l}) - (m_q^2 + q \cdot \bar{q})2l \cdot \bar{l}$$

$$+ 2(m_q^2 + q \cdot \bar{q})(m_l^2 + l \cdot \bar{l}) - (m_l^2 + l \cdot \bar{l})2q \cdot \bar{q} + 2(q \cdot l)^2 + 2(q \cdot \bar{l})^2]$$

$$= \frac{e^2 e_q^2}{4m_q^2 m_l^2 s^2} [2(m_q^2 + q \cdot \bar{q})m_l^2 + 2(m_l^2 + l \cdot \bar{l})m_q^2 + 2(q \cdot l)^2 + 2(q \cdot \bar{l})^2]. \quad (24)$$

Noting that
$$2(m_q^2 + q \cdot \bar{q}) = 2(m_l^2 + l \cdot \bar{l}) = s,$$

and substituting Eqs. (22) and (23) into (24), we obtain

$$\frac{1}{4} \sum_{\epsilon_q \epsilon_{\bar{q}} \epsilon_l \epsilon_{\bar{l}}}^{2} |T_{fi}|^2 = \frac{e^2 e_q^2}{4m_q^2 m_l^2 s^2} [sm_l^2 + sm_q^2 + \frac{1}{2}(t - m_q^2 - m_l^2)^2 + \frac{1}{2}(u - m_q^2 - m_l^2)^2]. \quad (25)$$

Having obtained the above result, we can go back to Eq. (6) to get the differential cross section for unpolarized q and $\bar{q}$ as an average over the initial polarizations and a sum over the final polarizations,

$$d\sigma = \frac{2m_q^2 m_l^2}{(2\pi)^2} \frac{dt \, d\phi_l}{s(s - 4m_q^2)} \frac{1}{4} \sum_{\epsilon_q \epsilon_{\bar{q}} \epsilon_l \epsilon_{\bar{l}}}^{2} |T_{fi}|^2$$

$$= \frac{2m_q^2 m_l^2}{(2\pi)^2} \frac{dt \, d\phi_l}{s(s - 4m_q^2)} \frac{e^2 e_q^2}{4m_q^2 m_l^2 s^2} [s(m_l^2 + m_q^2) + \frac{(t - m_q^2 - m_l^2)^2}{2} + \frac{(u - m_q^2 - m_l^2)^2}{2}].$$

Integration over ϕ yields

$$d\sigma = \frac{e^2 e_q^2}{8\pi} \frac{dt}{s^3(s - 4m_q^2)} [2s(m_l^2 + m_q^2) + (t - m_q^2 - m_l^2)^2 + (u - m_q^2 - m_l^2)^2]. \quad (26)$$

We shall now integrate Eq. (26) with respect to the variable t, keeping s as a constant. We need to determine the integration limits of t. From Eq. (5), the upper limit t_+ occurs when $\theta_l = 0$ and the lower limits t_- occurs when $\theta_l = \pi$. Therefore, the limits are given by

$$t_{\pm} = m_q^2 + m_l^2 - \frac{1}{2}s \pm \frac{1}{2}\sqrt{s - 4m_q^2}\sqrt{s - 4m_l^2}. \quad (27)$$

and we have
$$t_+ - t_- = \sqrt{s - 4m_q^2}\sqrt{s - 4m_l^2}. \quad (28)$$

To integrate Eq. (26) with respect to t, we note that the variable u is related to t by the relation
$$s + t + u = 2m_q^2 + 2m_l^2.$$

Therefore, we have

$$\int_{t_-}^{t_+} dt \, (u - m_q^2 - m_l^2)^2 = -\int_{t=t_-}^{t=t_+} du \, (u - m_q^2 - m_l^2)^2$$

$$= -\frac{1}{3}(u - c)^3 \Big|_{u(t=t_-)}^{u(t=t_+)}.$$

From the definition of u, we find

$$u = (q - \bar{l})^2$$
$$= m_q^2 + m_l^2 - 2q_0\bar{l}_0 + 2\mathbf{q} \cdot \bar{\mathbf{l}}$$
$$= m_q^2 + m_l^2 - 2q_0\bar{l}_0 - 2\mathbf{q} \cdot \mathbf{l} \qquad\qquad (29)$$
$$= m_q^2 + m_l^2 - \frac{1}{2}s - \frac{1}{2}\sqrt{s - 4m_q^2}\sqrt{s - 4m_l^2}\cos\theta_l \, .$$

Hence, by comparing Eq. (29) with Eqs. (5) and (27), we find that when t is at either limits, the quantity u is given by

$$u\big|_{t=t_+} = t_- \, ,$$

and

$$u\big|_{t=t_-} = t_+ \, .$$

Therefore, the total cross section is

$$\sigma = \frac{e^2 e_q^2}{8\pi s^3(s - 4m_q^2)}\{2s(m_l^2 + m_q^2)(t_+ - t_-) + \frac{1}{3}[(t_+ - m_q^2 - m_l^2)^3 - (t_- - m_q^2 - m_l^2)^3]$$

$$- \frac{1}{3}[(t_- - m_q^2 - m_l^2)^3 - (t_+ - m_q^2 - m_l^2)^3]\}$$

$$= \frac{e^2 e_q^2}{8\pi s^3(s - 4m_q^2)}\{2s(m_l^2 + m_q^2)(t_+ - t_-) + \frac{2}{3}[(t_+ - m_q^2 - m_l^2)^3 - (t_- - m_q^2 - m_l^2)^3]\}.$$

Using the results of Eqs. (27) and (28), we can simplify the above expression to

$$\sigma = \frac{e^2 e_q^2 \sqrt{s - 4m_q^2}\sqrt{s - 4m_l^2}}{8\pi s^3(s - 4m_q^2)}\frac{2}{3}s[s + 2(m_q^2 + m_l^2) + \frac{4m_q^2 m_l^2}{s}].$$

We relate the electric charge to the fine structure constant α by [13]

$$e^2 = 4\pi\alpha \, .$$

Then the total cross section for the process $q + \bar{q} \to l + \bar{l}$ is

$$\sigma = \frac{4\pi}{3}\left(\frac{e_q}{e}\right)^2\frac{\alpha^2}{s}\left(1 - \frac{4m_q^2}{s}\right)^{-1/2}\sqrt{1 - \frac{4m_l^2}{s}}\left(1 + 2\frac{m_q^2 + m_l^2}{s} + 4\frac{m_q^2 m_l^2}{s^2}\right), \qquad (30)$$

which is the result that had to be proved.   ]•

•[Exercise 14.2
Derive Eq. (14.2) for the rate of production and the distribution of l^+l^- pairs in a quark-gluon plasma where the quark momentum distribution is the same as the antiquark momentum distribution and is a general function of the energy E.

◇Solution:
The number of quarks with flavor f and momentum p_1, in the volume element d^3x and the momentum element d^3p_1, is

$$dN_{qf} = \frac{d^3x\, d^3p_1}{(2\pi)^3}\, N_c N_s\, f(E_1). \tag{1}$$

Not all of the reactions between a quark and an antiquark lead to the production of l^+l^- pairs. To produce a dilepton, the quark and the antiquark must annihilate each other; the antiquark must be the antiparticle of the quark, having the anticolor and the same flavor as the quark.

We use the symbol v_{12} to denote the relative velocity between the quark and the antiquark, and we use the symbol $\sigma_f(M)$ to denote the cross section for a quark q_f and an antiquark $\bar{q}_f$ of flavor f (and of the same color) to produce a dilepton of invariant mass M. The volume swept by the cross section $\sigma_f(M)$ due to the relative motion of the quark and the antiquark per unit time is $\sigma_f(M)v_{12}$. As a consequence, the number of antiquarks with momentum p_2 and energy $E_2 = (p_2^2 + m_q^2)^{1/2}$, in the momentum element d^3p_2, with which a quark can interact in a unit time (to produce l^+l^- pairs), is

$$\sigma_f(M)v_{12}\, N_s\, f(E_2)\frac{d^3p_2}{(2\pi)^3}. \tag{2}$$

From Exercise 14.1, the cross section $\sigma_f(M)$ for the process $q_f + \bar{q}_f \to l^+ + l^-$ is

$$\sigma_f(M) = (e_f/e)^2 \sigma(M), \tag{3}$$

where e_f is the charge of a quark with flavor f ($e_f = (2/3)|e|$ for an up quark and $e_f = -(1/3)|e|$ for a down quark) and $\sigma(M)$ is given explicitly in Exercise 14.1. When the mass of the quark is neglected, it has the form [2]

$$\sigma(M) = \frac{4\pi}{3}\frac{\alpha^2}{M^2}\left(1 + \frac{2m_l^2}{M^2}\right)\sqrt{1 - \frac{4m_l^2}{M^2}}. \tag{4}$$

The number of l^+l^- pairs produced per unit time is the integral of the product of Eqs. (1) and (2), summed over all flavors f:

$$\frac{dN_{l+l-}}{dt} = N_c N_s^2 \sum_{f=1}^{N_f}\left(\frac{e_f}{e}\right)^2 \int \frac{d^3x\, d^3p_1\, d^3p_2}{(2\pi)^6} f(E_1)f(E_2)\sigma(M)v_{12}. \tag{5}$$

Therefore, we have

$$\frac{dN_{l+l-}}{dt\, d^3x} = N_c N_s^2 \sum_{f=1}^{N_f}\left(\frac{e_f}{e}\right)^2 \int \frac{d^3p_1 d^3p_2}{(2\pi)^6} f(E_1)f(E_2)\sigma(M)v_{12}, \tag{6}$$

which is Eq. (14.1)

For a momentum distribution $f(E)$ that depends only on E, as in the case when there is thermal equilibrium, there is symmetry with respect to the orientation of the vector $\boldsymbol{p}_1$. Without any loss of generality, we can choose $\boldsymbol{p}_1$ to be along the z-axis. The angle θ_{12} between $\boldsymbol{p}_1$ and $\boldsymbol{p}_2$ becomes just θ_2, which is the angle between $\boldsymbol{p}_2$ and the z-axis. Equation (6) becomes

$$\frac{dN_{l^+l^-}}{d^4x} = N_c N_s^2 \sum_{f=1}^{N_f} \left(\frac{e_f}{e}\right)^2 \int \frac{4\pi \boldsymbol{p}_1^2 d|\boldsymbol{p}_1|\, 2\pi \boldsymbol{p}_2^2 d|\boldsymbol{p}_2|\, d(\cos\theta_2)}{(2\pi)^6} f(E_1) f(E_2) \sigma(M) v_{12}.$$

(7)

The six-fold integration of $d^3p_1 d^3p_2$ in Eq. (6) is now reduced to a three-fold integration.

The invariant mass M of the q-$\bar{q}$ pair or the l^+l^- pair is given by

$$M^2 = (p_1 + p_2)^2$$
$$= 2m_q^2 + 2E_1 E_2 - 2|\boldsymbol{p}_1|\,|\boldsymbol{p}_2|\cos\theta_2.$$

(8)

The relative velocity v_{12} between the quark and the antiquark is given by Eq. (12) of Supplement 4.1 as

$$v_{12} = \frac{[(p_1 \cdot p_2)^2 - m_q^4]^{1/2}}{E_1 E_2}.$$

We note that

$$p_1 \cdot p_2 = \frac{(p_1 + p_2)^2 - 2m_q^2}{2},$$
$$= \frac{M^2 - 2m_q^2}{2}.$$

Thus, we can write v_{12} in terms of M, E_1 and E_2 as

$$v_{12} = \frac{M\sqrt{M^2 - 4m_q^2}}{2E_1 E_2}.$$

(9)

With Eqs. (3) and (9), the integrand of Eq. (7) can be written as a function of E_1, E_2 and M. Therefore, it is convenient to transform the variables $|\boldsymbol{p}_1|$, $|\boldsymbol{p}_2|$ and $\cos\theta_2$ in terms of the variables E_1, E_2, and M^2. When E_1 and E_2 are held fixed, Eq. (8) gives

$$dM^2 = 2|\boldsymbol{p}_1|\,|\boldsymbol{p}_2|\, d(\cos\theta_2).$$

Equation (7) becomes

$$\frac{dN_{l^+l^-}}{d^4x} = N_c N_s^2 \sum_{f=1}^{N_f} \left(\frac{e_f}{e}\right)^2 \frac{2}{(2\pi)^4} \int |\boldsymbol{p}_1| E_1 dE_1 |\boldsymbol{p}_2| E_2 dE_2 \frac{dM^2}{2|\boldsymbol{p}_1|\,|\boldsymbol{p}_2|} f(E_1) f(E_2)$$

$$\times \sigma(M) \frac{M\sqrt{M^2 - 4m_q^2}}{2E_1 E_2}$$

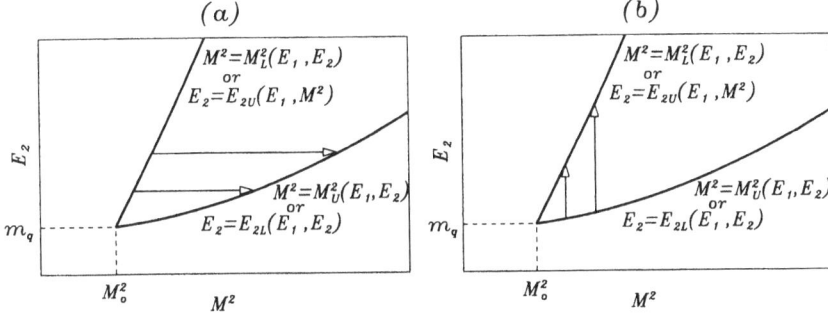

Fig. 14.3 (a) Integration is first carried out in the M^2 direction and then in the E_2 direction. (b) The order of integration is interchanged by carrying out the integration first along the E_2 direction and then along the M^2 direction.

$$= N_c N_s^2 \sum_{f=1}^{N_f} \left(\frac{e_f}{e}\right)^2 \frac{1}{2(2\pi)^4} \int dE_1 dE_2 dM^2 f(E_1) f(E_2) \sigma(M) M \sqrt{M^2 - 4m_q^2}$$

$$= N_c N_s^2 \sum_{f=1}^{N_f} \left(\frac{e_f}{e}\right)^2 \frac{1}{2(2\pi)^4} \int dE_1 dE_2 dM^2 f(E_1) f(E_2) \Sigma(M), \tag{10}$$

where

$$\Sigma(M) = \sigma(M) M \sqrt{M^2 - 4m_q^2}. \tag{11}$$

In Eq. (10), the integration is first carried out over M^2 from the lower limit $M_L^2(E_1, E_2)$ to the upper limit $M_U^2(E_1, E_2)$, as depicted in Fig. 14.3a. These limits are functions of E_1 and E_2. From Eq. (8), the lower limit of M_L^2 is given by

$$M^2 = M_L^2(E_1, E_2) = 2m_q^2 + 2E_1 E_2 - 2|\boldsymbol{p}_1| |\boldsymbol{p}_2|, \tag{12a}$$

and the upper limit M_U^2 is given by

$$M^2 = M_U^2(E_1, E_2) = 2m_q^2 + 2E_1 E_2 + 2|\boldsymbol{p}_1| |\boldsymbol{p}_2|. \tag{12b}$$

After the integration over M^2, the integration is then carried out over the variable E_2 from m_q to ∞.

It is useful to change the order of integration by first carrying out the integration in E_2 from the lower limit E_{2L} to the upper limit E_{2U}, and then the integration is carried out over M^2 from $M_0^2 = 2m_q^2 + 2E_1 m_q$ to ∞, as depicted in Fig. 14.3b.

As one can see from Fig. 14.3, the upper limit E_{2U} of E_2 follows the same curve as described by $M^2 = M_L^2(E_1, E_2)$. Thus, the function E_{2U} can be obtained from Eq. (12a) by inverting $M^2 = M_L^2(E_1, E_2)$ to get E_2 as a function of M^2 and E_1:

$$E_2 = E_{2U}(E_1, M^2). \tag{13a}$$

Similarly, the lower limit E_{2L} of E_2 follows the same curve as described by $M^2 = M_U^2(E_1, E_2)$ (see Fig. 14.3). Thus, the function E_{2L} can be obtained from Eq. (12b) by inverting $M^2 = M_U^2(E_1, E_2)$ to get E_2 as a function of M^2 and E_1:

$$E_2 = E_{2L}(E_1, M^2). \tag{13b}$$

After interchanging the order of integration, we have

$$\int_{m_q}^{\infty} dE_2 \int_{M_L^2(E_1,E_2)}^{M_U^2(E_1,E_2)} dM^2 f(E_2)\Sigma(M^2)$$
$$= \int_{M_0^2}^{\infty} dM^2 \int_{E_{2L}(E_1,M^2)}^{E_{2U}(E_1,M^2)} dE_2 f(E_2)\Sigma(M^2)$$
$$= \int_{M_0^2}^{\infty} dM^2 \Sigma(M) \int_{E_{2L}(E_1,M^2)}^{E_{2U}(E_1,M^2)} [-dF(E_2)]$$
$$= \int_{M_0^2}^{\infty} dM^2 \Sigma(M)[F(E_{2L}(E_1, M^2)) - F(E_{2U}(E_1, M^2))]. \tag{14}$$

Here, the function $F(E)$ is defined as the indefinite integral of $[-f(E)]$,

$$F(E) = -\int_{\infty}^{E} f(E')dE'.$$

In the approximation when the quark mass is taken to be very small compared to M, E_{2U} is large and $F(E_{2U})$ in Eq. (14) approaches zero as E_{2U} becomes large. (For example, for a thermal distribution $f(E) = e^{-E/T}$, the function $F(E) = Te^{-E/T}$ and $F(E) \to 0$, as $E \to \infty$). After neglecting $F(E_{2U})$ in Eq. (14) and substituting Eq. (14) into Eq. (10), we obtain

$$\frac{dN_{l+l-}}{d^4x} = N_c N_s^2 \sum_{f=1}^{N_f} \left(\frac{e_f}{e}\right)^2 \frac{1}{2(2\pi)^4} \int_{m_q}^{\infty} dE_1 f(E_1) \int_{M_0^2}^{\infty} dM^2 \Sigma(M)F(E_{2L}(E_1, M^2)). \tag{15}$$

We consider the case of very small quark masses so that Eq. (12b) and Eq. (13b) lead to the lower limit of E_2 given by

$$E_2 = E_{2L}(E_1, M^2)$$
$$\sim M^2/4E_1.$$

Substituting this into Eq. (15) and interchanging the order of integration, we have

$$\frac{dN_{l+l-}}{d^4x} = N_c N_s^2 \sum_{f=1}^{N_f} \left(\frac{e_f}{e}\right)^2 \frac{1}{2(2\pi)^4} \int_{4m_q^2}^{\infty} dM^2 \Sigma(M) \int_{m_q}^{(M^2-2m_q^2)/2m_q} dE_1 f(E_1) F\left(\frac{M^2}{4E_1}\right). \tag{16}$$

We shall carry out the integration over E_1 by the saddle-point method. We write the integrand in E_1 as

$$f(E_1)F\left(\frac{M^2}{4E_1}\right) = e^{\ln f(E_1) + \ln F(M^2/4E_1)} = e^{g(E_1)},$$

where

$$g(E_1) = \ln f(E_1) + \ln F\left(\frac{M^2}{4E_1}\right). \tag{17}$$

We expand the function $g(E_1)$ about its extremum at $E_1 = \epsilon(M)$, which is determined as the solution of the extremum condition

$$g'(\epsilon) = \left\{\frac{d}{dE_1}\left[\ln f(E_1) + \ln F\left(\frac{M^2}{4E_1}\right)\right]\right\}_{E_1 = \epsilon} = 0. \tag{18}$$

Then, by expanding $g(E_1)$ about $E_1 = \epsilon$, the function $g(E_1)$ can be written as

$$g(E_1) = g(\epsilon) - \frac{w}{2}(E_1 - \epsilon)^2,$$

where $-w(\epsilon)$ is the second derivative of $g(E_1)$ with respect to E_1 at the extremum,

$$w(\epsilon) = -\left\{\frac{d^2}{dE_1^2}\left[\ln f(E_1) + \ln F\left(\frac{M^2}{4E_1}\right)\right]\right\}_{E_1 = \epsilon}. \tag{19}$$

Equation (16) then becomes

$$\frac{dN_{l^+ l^-}}{d^4 x} = N_c N_s^2 \sum_{f=1}^{N_f}\left(\frac{e_f}{e}\right)^2 \frac{1}{2(2\pi)^4}\int_{4m_q^2}^{\infty} dM^2 \Sigma(M) f(\epsilon(M)) F\left(\frac{M^2}{4\epsilon(M)}\right)$$
$$\times \int_{m_q}^{(M^2 - 2m_q^2)/2m_q} dE_1 e^{-\frac{w}{2}(E_1 - \epsilon)^2}. \tag{20}$$

For large values of M, the quantity ϵ is approximately $M/2$ (see Exercise 14.3), which is much greater than m_q. Approximating this integral over E_1 as having the limits from $-\infty$ to $+\infty$, we obtain

$$\frac{dN_{l^+ l^-}}{d^4 x} = N_c N_s^2 \sum_{f=1}^{N_f}\left(\frac{e_f}{e}\right)^2 \frac{1}{2(2\pi)^4}\int_{4m_q^2}^{\infty} dM^2 \Sigma(M) f(\epsilon(M)) F\left(\frac{M^2}{4\epsilon(M)}\right)\sqrt{\frac{2\pi}{w(\epsilon)}}. \tag{21}$$

From this equation, we find

$$\frac{dN_{l^+ l^-}}{dM^2 d^4 x} = N_c N_s^2 \sum_{f=1}^{N_f}\left(\frac{e_f}{e}\right)^2 \frac{\sigma(M) M^2}{2(2\pi)^4}\left(1 - \frac{4m_q^2}{M^2}\right)^{\frac{1}{2}} f(\epsilon(M)) F\left(\frac{M^2}{4\epsilon(M)}\right)\sqrt{\frac{2\pi}{w(\epsilon)}}, \tag{22}$$

which is Eq. (14.2). ∎

•⟦ Exercise 14.3

Obtain the rate of production and the distribution of l^+l^- pairs in a quark gluon plasma, in which the momentum distribution for the quarks and the antiquarks is $f(E) = e^{-E/T}$.

◇Solution:

For the quark and the antiquark momentum distribution $f(E)$ of the form

$$f(E) = e^{-E/T}, \tag{1}$$

which is the thermal distribution for quarks and antiquarks in a quark-gluon plasma at high temperatures, the integral of $[-f(E)]$ yields

$$F(E) = -\int_\infty^E f(E')dE' \\ = Te^{-E/T}. \tag{2}$$

Knowing $f(E)$ and $F(E)$, we can determine the quantity ϵ of Eq. (14.2), which is the location of the extremum of the function $g(E) = \ln f(E) + \ln F(M^2/4E)$. It is given as the solution of the extremum condition

$$g'(\epsilon) = \left\{ \frac{d}{dE}\left[\ln f(E) + \ln F\left(\frac{M^2}{4E}\right)\right]\right\}_{E=\epsilon} = 0.$$

For our case this equation is

$$-\frac{1}{T} + \frac{M^2}{4\epsilon^2 T} = 0, \tag{3}$$

so the solution is

$$\epsilon = \frac{M}{2}. \tag{4}$$

The quantity w, the second derivative of the function $g(E)$ at the extremum multiplied by (-1), is

$$w(\epsilon) = -\left\{ \frac{d^2}{dE^2}\left[\ln f(E) + \ln F\left(\frac{M^2}{4E}\right)\right]\right\}_{E=\epsilon} \\ = \frac{4}{MT}. \tag{5}$$

Substituting all of these quantities into Eq. (14.2), we obtain

$$\frac{dN_{l+l-}}{dM^2 d^4 x} = N_c N_s^2 \sum_{f=1}^{N_f}\left(\frac{e_f}{e}\right)^2 \frac{\sigma(M)}{2(2\pi)^4}M^2\left(1 - \frac{4m_q^2}{M^2}\right)^{\frac{1}{2}} T\sqrt{\frac{\pi MT}{2}}e^{-M/T}. \tag{6}$$

We note from the asymptotic expansion of the Bessel function $K_\nu(z)$ (Eq. (9.7.2) of Ref. [12]) that for large values of z

$$K_1(z) \sim \sqrt{\frac{\pi}{2z}}e^{-z}. \tag{7}$$

Consequently, Eq. (6) can be rewritten as [2]

$$\frac{dN_{l+l-}}{dM^2 d^4x} \sim N_c N_s^2 \sum_{f=1}^{N_f} \left(\frac{e_f}{e}\right)^2 \frac{\sigma(M)}{2(2\pi)^4} M^2 \left(1 - \frac{4m_q^2}{M^2}\right)^{\frac{1}{2}} TM K_1\left(\frac{M}{T}\right). \qquad (8)$$

which is Eq. (14.7).

The distribution over M_T^2 cannot actually be obtained from the distribution in d/dM^2. However, a comparison with the M_T distribution obtained previously in Ref. [2] indicates that the distribution in M_T is related to the above as

$$\frac{dN_{l+l-}}{dM^2 dM_T^2 d^4x} \sim N_c N_s^2 \sum_{f=1}^{N_f} \left(\frac{e_f}{e}\right)^2 \frac{\sigma(M)}{2(2\pi)^4} M^2 \left(1 - \frac{4m_q^2}{M^2}\right)^{\frac{1}{2}} T\left\{\frac{-d}{dM^2} MK_1\left(\frac{M}{T}\right)\right\}_{M^2 = M_T^2}.$$

Noting from Eq. (9.6.28) of Ref. [12] that

$$\frac{d}{dz^2}[zK_1(z)] = -\frac{K_0(z)}{2},$$

we get [2]

$$\frac{dN_{l+l-}}{dM^2 dM_T^2 d^4x} \sim N_c N_s^2 \sum_{f=1}^{N_f} \left(\frac{e_f}{e}\right)^2 \frac{\sigma(M)}{4(2\pi)^4} M^2 \left(1 - \frac{4m_q^2}{M^2}\right)^{\frac{1}{2}} K_0\left(\frac{M_T}{T}\right), \qquad (9)$$

which is Eq. (14.8).   ]•

•[Exercise 14.4

Obtain the distribution functions (14.9) and (14.10) for dileptons produced in a quark-gluon plasma, which undergoes a hydrodynamical evolution in the interval from the proper time τ_0 to τ_c. The momentum distribution of the quarks and the antiquarks is assumed to be $f(E) = e^{-E/T}$.

◇Solution:

We learn from Eq. (13.18) of the last chapter that for a quark-gluon plasma which undergoes a hydrodynamical evolution, the temperature T of the plasma decreases with the proper time τ as $T \propto \tau^{-1/3}$. As a consequence, the rate of dilepton production also varies with the proper time.

With the momentum distribution $f(E)$ of the quarks and antiquarks given by $e^{-E/T}$, the rate of dilepton production is related to the temperature T by Eq. (14.7)

$$\frac{dN_{l+l-}}{dM^2 d^4x} \sim N_c N_s^2 \sum_{f=1}^{N_f} \left(\frac{e_f}{e}\right)^2 \frac{\sigma(M)}{2(2\pi)^4} M^2 \left(1 - \frac{4m_q^2}{M^2}\right)^{\frac{1}{2}} TM K_1\left(\frac{M}{T}\right).$$

To obtain all contributions from the system while it is in the quark-gluon phase, we need to integrate over the proper time interval from τ_0 to τ_c, during which the system is in the quark-gluon plasma phase. We also need to integrate over the transverse coordinates $\mathbf{x}_\perp$. For these purposes, we rewrite the space-time 4-volume element as

$$d^4x = d\mathbf{x}_\perp \tau d\tau dy. \qquad (1)$$

The integration over $dx_\perp$ gives a factor πR_A^2, which is the transverse area of the plasma. We have therefore

$$\frac{dN_{l+l-}}{dM^2 dy} \sim \pi R_A^2 N_c N_s^2 \sum_{f=1}^{N_f} \left(\frac{e_f}{e}\right)^2 \frac{\sigma(M)}{2(2\pi)^4} M^2 \left(1 - \frac{4m_q^2}{M^2}\right)^{\frac{1}{2}} M \int_{\tau_0}^{\tau_c} \tau d\tau \, T K_1\left(\frac{M}{T}\right).$$

(2)

It is convenient to change the variable from τ to T. Using Eq. (13.18), we have

$$\tau = \tau_0 \left(\frac{T_0}{T}\right)^3.$$

(3)

Substituting Eq. (3) into (2), we have therefore

$$\int_{\tau_0}^{\tau_c} \tau d\tau \, T K_1\left(\frac{M}{T}\right) = \tau_0^2 T_0^6 \int_{T_0}^{T_c} \frac{(-3)}{T^7} dT \, T K_1\left(\frac{M}{T}\right)$$

$$= \frac{3\tau_0^2 T_0^6}{M^5} \int_{M/T_0}^{M/T_c} z^4 K_1(z) \, dz.$$

(4)

Using the relation (9.6.27) of Ref. [12]

$$\frac{d}{dz} K_0(z) = -K_1(z),$$

(5)

and Eq. (9.6.28) of Ref. [12]

$$\left(\frac{1}{z}\frac{d}{dz}\right)\{z K_1(z)\} = -K_0(z),$$

(6)

the integral in Eq. (4) can be evaluated by integration by parts. We finally obtain

$$\int z^4 K_1(z) = -z^2 (8 + z^2) K_0(z) - 4z(4 + z^2) K_1(z)$$

$$\equiv -H(z).$$

(7)

Substituting Eqs. (4) and (7) into Eq. (2) leads to [2]

$$\frac{dN_{l+l-}}{dM^2 dy} \sim \pi R_A^2 N_c N_s^2 \sum_{f=1}^{N_f} \left(\frac{e_f}{e}\right)^2 \frac{\sigma(M)}{2(2\pi)^4} \left(1 - \frac{4m_q^2}{M^2}\right)^{\frac{1}{2}} \frac{3\tau_0^2 T_0^6}{M^2} \left[H\left(\frac{M}{T_0}\right) - H\left(\frac{M}{T_c}\right)\right],$$

which is Eq. (14.9).

To get the distribution of dileptons with respect to M^2 and M_T^2, we use Eq. (14.8) and integrate over dx_T and $d\tau$. We get

$$\frac{dN_{l+l-}}{dM^2 dM_T^2 dy} \sim \pi R_A^2 N_c N_s^2 \sum_{f=1}^{N_f} \left(\frac{e_f}{e}\right)^2 \frac{\sigma(M)}{4(2\pi)^4} M^2 \left(1 - \frac{4m_q^2}{M^2}\right)^{\frac{1}{2}} \int_{\tau_0}^{\tau_c} \tau d\tau \, K_0\left(\frac{M_T}{T}\right).$$

A change of the variable from τ to T leads to

$$\int_{\tau_0}^{\tau_c} \tau d\tau \, K_0\left(\frac{M_T}{T}\right) = \tau_0^2 T_0^6 \int_{T_0}^{T_c} (-1)\frac{3}{T^7} K_0\left(\frac{M_T}{T}\right) dT$$

$$= \frac{3\tau^2 T_0^6}{M_T^6} \int_{M_T/T_0}^{M_T/T_c} z^5 K_0(z)\, dz \,.$$

Again, using Eqs. (5) and (6), we find,

$$\int z^5 K_0(z) = -z(z^2+8)^2 K_1(z) - 4z^2(z^2+8)K_0(z)$$

$$= -z^3(8+z^2)K_3(z)$$

$$\equiv -G(z)\,.$$

Therefore, the distribution of l^+l^- pairs in M^2, M_T^2 and y is given by [2]

$$\frac{dN_{l^+l^-}}{dM^2 dM_T^2 dy}$$

$$\sim \pi R_A^2 N_c N_s^2 \sum_{f=1}^{N_f} \left(\frac{e_f}{e}\right)^2 \frac{\sigma(M)M^2}{4(2\pi)^4}\left(1-\frac{4m_q^2}{M^2}\right)^{\frac{1}{2}} \frac{3\tau_0^2 T_0^6}{M_T^6}\left[G\left(\frac{M_T}{T_0}\right) - G\left(\frac{M_T}{T_c}\right)\right],$$

which is Eq. (14.10). $]\bullet$

§14.2 Dilepton Production from Other Processes

In a high-energy nucleus-nucleus collision, the possible formation of the quark-gluon plasma is not the only source of l^+l^- pairs. There are other processes which also contribute to dilepton production. To separate out the portion due to the quark-gluon plasma, it is necessary to analyze the contributions from the other sources.

§14.2.1 Drell-Yan Process

An important contribution to dilepton production comes from the Drell-Yan process [17-20], which is specially important for large values of the invariant mass of the l^+l^- pair. In the Drell-Yan process in a nucleus-nucleus collision, we envisage that a valence quark of a nucleon of one of the nuclei can interact with a sea antiquark of a nucleon of the other nucleus. They annihilate to form a virtual photon which subsequently decays into an l^+l^- pair. In this process, the effects of the correlations of the nucleons within a nucleus are not important and

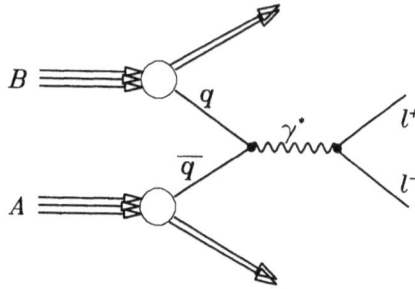

Fig. 14.4   The diagram for the Drell-Yan process leading to the production of a dilepton pair.

the nucleons can be considered to be independent. The production of l^+l^- pairs under the Drell-Yan process in heavy-ion collisions can be considered as arising from a collection of independent nucleon-nucleon collisions.

The diagram for the Drell-Yan process in a nucleon-nucleon collision is illustrated in Fig. 14.4 where A and B represent nucleons. The valence quark q (or sea antiquark $\bar{q}$) of nucleon B interact with the sea antiquark $\bar{q}$ (or the valence quark q) of A to form a virtual photon γ^* which decays into a pair of l^+ and l^-. The l^+l^- pair is characterized by a momentum C which is the sum of the momentum of l^+ and l^-,

$$C = l^+ + l^-, \qquad (14.12a)$$

and the l^+l^- pair has an invariant mass squared M^2 given by

$$M^2 = C^2 = (l^+ + l^-)^2. \qquad (14.12b)$$

It is useful to introduce the Feynman scaling variable x_F for the l^+l^- pair,

$$x_F = \frac{C_z}{\sqrt{s}/2}, \qquad (14.13)$$

where $C_z = l_z^+ + l_z^-$ is the longitudinal momentum of the l^+l^- pair in the nucleon-nucleon center-of-mass system and $\sqrt{s}$ is the nucleon-nucleon center-of-mass energy.

It is easy to show (see Exercise 14.5) that the differential cross section in M^2 and x_F for the process $NN \to l^+l^-X$ in a nucleon-nucleon collision is [18]-[20]

$$\frac{d^2\sigma}{dM^2 dx_F} = \frac{1}{sN_c}\sigma(M)\sum_f^{N_f}\left(\frac{e_f}{e}\right)^2\frac{q_f^B(x_1)\bar{q}_f^A(x_2) + \bar{q}_f^B(x_1)q_f^A(x_2)}{\sqrt{x_F^2 + 4M^2/s}} \qquad (14.14)$$

where $q_f^{(B,A)}(x)$ and $\bar{q}_f^{(B,A)}(x)$ are the probability of finding respectively a quark and an antiquark of flavor f, with a light-cone variable x in the beam nucleon B or in the target nucleon A. The quantities $x_{1,2}$ are related to x_F and M^2 by

$$x_{1,2} = \frac{1}{2}\left(\sqrt{x_F^2 + \frac{4M^2}{s}} \pm x_F \right), \qquad (14.15a)$$

$$x_1 - x_2 = x_F, \qquad (14.15b)$$

and

$$x_1 x_2 s = M^2. \qquad (14.15c)$$

From Eqs. (14.13) and (14.15b), we can understand the kinematic configurations in the production of the l^+l^- pair. Working in the center-of-mass system of the colliding nucleons and using $\sqrt{s}/2$ as the unit of momentum, we define x_F as the longitudinal momentum of the l^+l^- pair with Eq. (14.13). The longitudinal momentum x_F of the dilepton comes from the parton with the longitudinal momentum x_1 and another parton with the longitudinal momentum $-x_2$, in units of $\sqrt{s}/2$. Hence, the sum of the parton longitudinal momentum $x_1 + (-x_2)$ gives the dilepton longitudinal momentum x_F, as in Eq. (14.15b).

The square of the momentum of a parton gives the invariant mass of the parton. At very high energies, the invariant mass of the parton can be neglected and the energy of the parton is approximately the same as the magnitude of its longitudinal momentum. Consequently, $x_1\sqrt{s}$ is the forward light-cone momentum of one parton and $x_2\sqrt{s}$ is the backward light-cone momentum of the other parton. Hence, the product of $x_1\sqrt{s}$ and $x_2\sqrt{s}$ gives the invariant mass squared of the parton pair, which is the same as the invariant mass squared of the l^+l^- pair. Therefore, the invariant mass squared of the l^+l^- pair is $x_1 x_1 s$, as in Eq. (14.15c).

Eq. (14.15a) reveals that given an l^+l^- pair which is characterized by the Feynman scaling variable x_F and an invariant mass squared M^2, one can determine the light-cone momentum fractions x_1 and x_2 from which the l^+l^- pair originated. Thus, another way to present the result of Eq. (14.14) is to rewrite it as a differential cross section in x_1 and x_2

$$\frac{d\sigma}{dx_1 dx_2} = \frac{1}{N_c}\sigma(M)\sum_f^{N_f}\left(\frac{e_f}{e}\right)^2 \{q_f^B(x_1)\bar{q}_f^A(x_2) + \bar{q}_f^B(x_1)q_f^A(x_2)\}. \qquad (14.16)$$

In terms of the variables M and y, the differential cross section is

$$\frac{d\sigma}{dMdy} = \frac{8\pi\alpha^2}{3sN_cM}\sum_f^{N_f}\left(\frac{e_f}{e}\right)^2[q_f^B(xe^y)\bar{q}_f^A(xe^{-y}) + \bar{q}_f^A(xe^y)q_f^B(xe^{-y})], \tag{14.17}$$

where $x = M/\sqrt{s}$. The differential cross section in M and y at $y = 0$ assumes the form given by

$$\left.\frac{d\sigma}{dMdy}\right|_{y=0} = \frac{8\pi\alpha^2}{3N_cM^3}\sum_f^{N_f}\left(\frac{e_f}{e}\right)^2[xq_f^B(x)x\bar{q}_f^A(x) + x\bar{q}_f^B(x)xq_f^A(x)]. \tag{14.18}$$

When one plots the left-hand side quantity of Eq. (14.18) as a function of $M/\sqrt{s}$, the data points from different measurements at different energies should fall on the same curve (see Fig. 1.7 of Ref. 19). This is confirmed by experimental dilepton production data which can be well represented by the relation

$$\left.M^3\frac{d^2\sigma}{dMdy}\right|_{y=0} = 3\times10^{-32}e^{-15M/\sqrt{s}} \quad (\text{cm}^2\text{GeV}^2). \tag{14.19}$$

This 'scaling behavior' and the results of Eqs. (14.14) and (14.16) allow one to extract the quark and antiquark distribution functions in a nucleon.

Eqs. (14.14)-(14.18) for the Drell-Yan process are obtained by using the leading order diagram (Fig. 14.1 and Fig. 14.4) in which there are no initial-state or final-state interactions. The main result is that the cross section consists of three factors: a distribution function which pertains to the constituent in the projectile nucleon, another distribution function for another constituent in the target nucleon, and the basic elementary constituent-constituent cross section $\sigma(M)$. This property of the Drell-Yan cross section is called the *factorization property*. It turns out that the factorization property has validity beyond the leading order because of a cancellation of amplitudes [21]. Because of this factorization property, the parton distribution functions obtained in the Drell-Yan process should be the same as the parton distribution functions (also known as the parton structure functions) obtained from lepton-nucleon deep-inelastic collisions, in which there is a large momentum transfer Q from the lepton to a constituent of the nucleon.

Upon using the quark and antiquark structure functions deduced from deep-inelastic lepton-nucleon measurements to analyze the dilepton data, one finds that the experimental dilepton data will be consistent with the theoretical distributions if the lowest order theoretical results using the lowest order diagram of Fig. 14.1 are multiplied by an overall factor, the K factor [18,20]. Experimentally, this K factor is found to range from 1.6 to 2.8 [24].

If one considers only the lowest order diagram of Fig. 14.4, the Drell-Yan dilepton cross section will be proportional to $\alpha^2 \alpha_s^0$, which is second order in the electromagnetic coupling constant α but zeroth order in the strong coupling constant α_s. Higher-order QCD corrections to the Drell-Yan cross section, up to $O(\alpha_s)$ [22] and $O(\alpha_s^2)$ [23], have been worked out. These investigations show that the K factor can be accounted for by including high-order QCD corrections [22,23]. The most important contribution to the dilepton cross section due to the α_s-order (the next-to-leading-order) diagrams is the vertex correction at the $q\bar{q}\gamma$ vertex, which leads to a factor equal to $(1 + 2\pi\alpha_s/3)$ [18,20]. For a coupling constant of $\alpha_s = 0.3$, the vertex correction gives a factor of about 1.7. Additional contributions from the α_s-order Compton diagrams bring the K factor within the observed range of 1.6 to 2.8 [22,23].

The large magnitude of the α_s-order corrections raises questions concerning the convergence of the perturbation series. From the higher-order α_s^2-order corrections, Hamberg *et al.* [23] find that for the case of very high energies and very large dilepton masses, the total contribution from the α_s^2-order corrections is small compared to that from the α_s-order corrections.

Using the parton model and perturbative QCD, with a K factor either taken to be a constant [26] or determined by the α_s-order QCD corrections [27-28], a large set of experimental data, including dilepton and deep-inelastic data, have been analyzed globally. Consistent sets of parameters for the parton distributions have been obtained by many workers [25-31]. Most groups use a functional representation of the parton distributions of the form

$$xq^a(x, Q) = A_0^a x^{A_1^a}(1 - x)^{A_2^a} P^a(x), \qquad (14.20)$$

where the superscript a is a flavor label (including the gluon and the antiquark label), and $P^a(x)$ is a smooth function. The choice of the function $P^a(x)$ varies considerably. For example, Ref. [26] uses a polynomial in x while Ref. [27] uses a logarithmic function of $(1+1/x)$. The constants A_i^a are functions of Q^2.

The parton distribution function depends on the momentum scale in which the parton distribution is probed. In Eq. (14.20), such a dependence is expressed by having xq^a as a function of Q^2. In a deep-inelastic lepton-nucleon scattering, the quantity Q is the momentum transfer from the lepton to the constituent and is equal to the four-momentum of the probing intermediate space-like virtual photon. (A virtual photon is space-like or time-like if the square of the photon four-momentum Q^2 is negative or positive.) In the Drell-Yan process, the quantity Q is the four-momentum of the intermediate time-like virtual photon and Q^2 is the square of the invariant mass of the photon, which is the same as the square of the invariant mass of the l^+l^- pair.

It is instructive to gain an intuitive insight into the main features of the dilepton distribution arising from the Drell-Yan process. From Eq. (14.18), we infer that the Drell-Yan dilepton differential cross section at $y = 0$ is given by the product of the quark distribution $xq(x)$ and the antiquark distribution $x\bar{q}(x)$. Using the quark and the antiquark distributions of the form of Eq. (14.20), the dilepton differential cross section at $y = 0$ is given from Eq. (14.18) by

$$\left.\frac{d\sigma}{dMdy}\right|_{y=0} \propto \frac{1}{M^3} x^{A_1^q + A_1^{\bar{q}}} (1 - x)^{A_2^q + A_2^{\bar{q}}} P^q(x) P^{\bar{q}}(x).$$

Because the functions $P^q(x)$ and $P^{\bar{q}}(x)$ are smooth functions of x, the dilepton differential cross section is approximately

$$\left.\frac{d\sigma}{dMdy}\right|_{y=0} \sim \frac{\text{constant}}{M^3} \left(\frac{M}{\sqrt{s}}\right)^{A_1^q + A_1^{\bar{q}}} \left(1 - \frac{M}{\sqrt{s}}\right)^{A_2^q + A_2^{\bar{q}}}$$

$$\sim \frac{\text{constant}}{M^3} \left(\frac{M}{\sqrt{s}}\right)^{A_1^q + A_1^{\bar{q}}} e^{-M(A_2^q + A_2^{\bar{q}})/\sqrt{s}}. \quad (14.21)$$

The above differential cross section, with the exponential dependence $\exp\{-M(A_2^q + A_2^{\bar{q}})/\sqrt{s}\}$, is in approximate agreement with the experimental data represented by Eq. (14.19), with $(A_2^q + A_2^{\bar{q}}) \sim 15$.

We can introduce a parameter T_{DY} to express the Drell-Yan dilepton differential cross section (14.21) in the form

$$\left.\frac{d\sigma}{dMdy}\right|_{y=0} \sim \frac{1}{M^3} \left(\frac{M}{\sqrt{s}}\right)^{A_1^q + A_1^{\bar{q}}} e^{-M/T_{DY}}. \quad (14.22)$$

The Drell-Yan dilepton differential cross section at $y = 0$ behaves as if there is an effective 'temperature' T_{DY} arising from the intrinsic motion of the quarks and antiquarks in the nucleon given by

$$T_{DY} \sim \frac{\sqrt{s}}{A_2^q + A_2^{\bar{q}}}. \quad (14.23)$$

As the parameters A_2^q and $A_2^{\bar{q}}$ depend on Q^2, the effective Drell-Yan 'temperature' T_{DY} also depends on Q^2.

To provide a better insight to the dilepton spectrum, we consider Set I of the structure functions of Duke and Owens [26]. For the valence quark distribution (averaged over up quarks and down quarks), the parameters are

$$A_1^q = 0.419 + 0.004k - 0.007k^2, \tag{14.24a}$$

and

$$A_2^q = 3.46 + 0.724k - 0.066k^2, \tag{14.24b}$$

and for the antiquark distribution, the parameters are

$$A_1^{\bar{q}} = -0.327k - 0.029k^2, \tag{14.24c}$$

and

$$A_2^{\bar{q}} = 8.05 + 1.59k - 0.153k^2, \tag{14.24d}$$

where k is related to Q^2 by

$$k = \ln[(\ln Q^2/\Lambda^2)/(\ln Q_0^2/\Lambda^2)], \tag{14.24e}$$

with $Q_0^2 = 4$ GeV^2 and $\Lambda = 0.2$ GeV.

Upon using the parameters Set I of Duke and Owens [26], the dilepton differential cross section at $y = 0$ is characterized approximately by a temperature

$$T_{DY} \sim \frac{\sqrt{s}}{11.51 + 2.31k - 0.219k^2}, \tag{14.25}$$

which gives $T_{DY} \sim \sqrt{s}/11.51$ for a dilepton with an invariant mass $M = 2$ GeV.

•⟦ Exercise 14.5
Obtain the dilepton production differential cross section (14.14) for the Drell-Yan process in nucleon-nucleon collisions as depicted in Fig. 14.4. Show that the differential cross section can be rewritten in the forms of Eqs. (14.16) and (14.17).

◇Solution:
Following the hard-scattering model described in Chapter 4, we can write down the differential cross section for the production of a dilepton with a total momentum $C = l^+ + l^-$ as given by

$$\frac{d^4\sigma}{dC^4}\bigg|_{BA \to l^+ l^- X} = \sum_{ba} \int dx_b d\vec{b}_T dx_a d\vec{a}_T G_{b/B}(x_b, \vec{b}_T) G_{a/A}(x_a, \vec{a}_T)$$

$$\times r(s, s', x_b, x_a) \frac{d^3\sigma}{dC^4}(ba \to l^+ l^-). \tag{1}$$

Here, $r(s, s', x_b, x_a)$ is a kinematic factor defined by Eq. (4.17). For high-energy collisions, the kinematic factor r is approximately 1 (see page 45). We shall henceforth take r to be unity. The structure function $G_{b/B}(x_b, \vec{b}_T)$ is the probability of finding a quark (or an antiquark) of type b in the particle B with a momentum fraction x_b and a transverse momentum $\vec{b}_T$. The other structure function $G_{a/A}(x_a, \vec{a}_T)$ is similarly defined as the probability of finding a constituent a in the particle A with a momentum fraction x_a and a transverse momentum $\vec{a}_T$. The quantity

$d^4\sigma/dC^4(ba \to l^+l^-)$ is the invariant cross section for the basic process $b + a \to l^+ + l^-$. The summation over b and a is carried out over all possible constituents with different colors, flavors, spins and types (quark or antiquark).

(Note that for the present dilepton production process, the momentum C of the l^+l^- pair is not on the mass shell, as the momentum components C^0 and $\vec{C}$ do not satisfy the condition that $C_0^2 - \vec{C}^2$ is always a fixed constant. This is in contrast to the case of Eq. (4.16) which describes the case for the detected particle C on the mass shell. This is the origin of the difference between Eq. (1) above and Eq. (4.16).)

In order to produce an l^+l^- pair, the constituent a must be the antiparticle of b. The cross section for the process $a + b \to l^+ + l^-$ is

$$\left.\frac{d^4\sigma}{dC^4}\right|_{ba \to l^+l^-} = \frac{|e_a e_b|}{e^2}\sigma(M)\,\delta(C - a - b),$$

where e_b is the charge of the constituent b which depends on its flavor and the cross section $\sigma(M)$ is given in Exercise 14.1. We note that

$$dC^0 d\vec{C}\,\delta(C^0 - (a^0 + b^0))\,\delta(\vec{C} - (\vec{a}+\vec{b}))$$
$$= \frac{dC^2}{2C^0}d\vec{C}\,\delta(\sqrt{C^2 + |\vec{C}|^2} - (a^0+b^0))\,\delta(\vec{C}-(\vec{a}+\vec{b}))$$
$$= \frac{dC^2}{2C^0}d\vec{C}\frac{\delta(C^2 - (a+b)^2)}{|\frac{d}{dC^2}\{\sqrt{C^2+|\vec{C}|^2} - (a^0+b^0)\}|}\,\delta(\vec{C}-(\vec{a}+\vec{b}))$$
$$= dC^2 d\vec{C}\,\delta(C^2 - (a+b)^2)\,\delta(\vec{C}-(\vec{a}+\vec{b}))$$
$$= dM^2 d\vec{C}\,\delta(M^2 - (a+b)^2)\,\delta(\vec{C}-(\vec{a}+\vec{b})).$$

Therefore, Eq. (1) becomes

$$\left.\frac{d^4\sigma}{dM^2 d\vec{C}}\right|_{BA \to l^+l^-X} = \sum_{ba}\int dx_b d\vec{b}_T dx_a d\vec{a}_T G_{b/B}(x_b,\vec{b}_T)G_{a/A}(x_a,\vec{a}_T)\frac{|e_a e_b|}{e^2}\sigma(M)$$
$$\times\,\delta(M^2 - (a+b)^2)\,\delta(\vec{C}-(\vec{a}+\vec{b})),\qquad(2)$$

where particle a is constrained to be the antiparticle of b.

We shall work in the center-of-mass frame of the colliding nucleon B and nucleon A. The average transverse momentum of the constituents a and b is of the order of a few hundred MeV/c, which is small compared to the longitudinal momenta and the energies of a and b we wish to consider. For this purpose, we can neglect $\vec{b}_T$ and $\vec{a}_T$ and the integrations over $\vec{b}_T$, $\vec{a}_T$ and $\vec{C}_T$ can be independently carried out. We define

$$G_b(x_b) = \int d\vec{b}_T G_{b/B}(x_b,\vec{b}_T),$$
$$G_a(x_a) = \int d\vec{a}_T G_{a/A}(x_a,\vec{a}_T),$$

then

$$\frac{d\sigma}{dM^2dC_z}\bigg|_{BA\to l^+l^-X} = \sum_{ba} \int dx_b dx_a G_b(x_b) G_a(x_a) \frac{|e_a e_b|}{e^2} \sigma(M)$$

$$\times \delta(M^2 - (a+b)^2)\, \delta(C_z - a_z - b_z)\,.$$

In the infinite momentum frame, the momentum of a and b can be written as in Eqs. (4.10) and (4.11) in the form:

$$b = \left(x_b P_1 + \frac{b^2 + b_T^2}{4x_b P_1},\ \ \vec{b}_T,\ \ x_b P_1 - \frac{b^2 + b_T^2}{4x_b P_1}\right)$$

$$a = \left(x_a P_2 + \frac{a^2 + a_T^2}{4x_a P_2},\ \ \vec{a}_T,\ \ -x_a P_2 + \frac{a^2 + a_T^2}{4x_a P_2}\right).$$

In high-energy collisions with $s \gg A^2, B^2$, and in the center-of-mass frame of A and B, the quantities P_1 and P_2 are given by Eq. (4.5) and Eq. (4.6) as

$$P_1 \sim P_2 \sim \sqrt{s}/2\,.$$

Hence, a and b can be approximated by

$$b \sim \left(\frac{x_b \sqrt{s}}{2},\ \ \vec{b}_T,\ \ \frac{x_b \sqrt{s}}{2}\right)$$

$$a \sim \left(\frac{x_a \sqrt{s}}{2},\ \ \vec{a}_T,\ \ -\frac{x_a \sqrt{s}}{2}\right)$$

Neglecting the transverse momentum, we have

$$(a+b)^2 \sim x_b x_a s\,,$$

and

$$a_z + b_z \sim (x_b - x_a)\frac{\sqrt{s}}{2}\,.$$

We find

$$\frac{d\sigma}{dM^2dC_z}\bigg|_{BA\to l^+l^-X} = \sum_{ba} \int dx_b dx_a G_b(x_b) G_a(x_a) \frac{|e_a e_b|}{e^2} \sigma(M)$$

$$\times\ \delta(x_b x_a s - M^2)\, \delta(C_z - (x_b - x_a)\sqrt{s}/2)\,.$$

We introduce the Feynman scaling variable for the l^+l^- pair

$$x_F = \frac{C_z}{\sqrt{s}/2}\,,$$

then we have

$$\frac{d\sigma}{dM^2dC_z}\bigg|_{BA\to l^+l^-X} = \frac{2}{s^{3/2}} \sum_{ba} \int dx_a dx_{\bar{a}} G_b(x_b) G_a(x_a) \frac{|e_a e_b|}{e^2} \sigma(M)$$

$$\times\ \delta(x_b x_a - M^2/s)\, \delta(x_F - (x_b - x_a))\,.$$

By solving a quadratic equation, we can write the product of the delta functions as

$$\delta(x_b x_a - M^2/s)\, \delta(x_F - (x_b - x_a)) = \delta((x_b - x_1)(x_b + x_2))\, \delta((x_b - x_1) - (x_a - x_2)),$$

where $x_{1,2}$ are given by

$$x_{1,2} = \frac{1}{2}\left(\sqrt{x_F^2 + \frac{4M^2}{s}} \pm x_F\right).$$

Integration over the two delta functions leads to

$$
\begin{aligned}
\left.\frac{d\sigma}{dM^2 dx_F}\right|_{BA \to l^+ l^- X} &= \frac{1}{s}\sum_{ba}\frac{|e_a e_b|}{e^2}\sigma(M)\frac{G_b(x_1)G_a(x_2)}{x_1 + x_2} \\
&= \frac{1}{s}\sum_{ba}\frac{|e_a e_b|}{e^2}\sigma(M)\frac{G_b(x_1)G_a(x_2)}{\sqrt{x_F^2 + 4M^2/s}}.
\end{aligned}
\tag{3}
$$

We define the quark structure function $q_f^B(x)$ for a quark with flavor f in the nucleon B as the probability of finding a quark of flavor f with any color and spin. It is given as the sum of $G_b(x)$ over b where b denotes a quark of flavor f with any color and spin,

$$q_f^B(x) = \left.\sum_{b=\text{all color, spin}} G_b(x)\right|_{b=\text{quark of flavor } f}.$$

We similarly define the antiquark structure function $\bar{q}_f^B(x)$ for an antiquark of fixed flavor f in nucleon B as the probability of finding an antiquark of flavor f with any color and spin in nucleon B. It is given by

$$\bar{q}_f^B(x) = \left.\sum_{b=\text{all color, spin}} G_b(x)\right|_{b=\text{antiquark of flavor } f}.$$

We are now ready to carry out the summation over b and a in Eq. (3). There is the implicit constraint that a must be an antiparticle of b. The summation over ba must be carried out with this constraint condition. Then Eq. (3) becomes

$$\left.\frac{d^2\sigma}{dM^2 dx_F}\right|_{BA \to l^+ l^- X} = \frac{1}{sN_c}\sigma(M)\sum_f^{N_f}\left(\frac{e_f}{e}\right)^2\frac{q_f^B(x_1)\bar{q}_f^A(x_2) + \bar{q}_f^B(x_1)q_f^A(x_2)}{\sqrt{x_F^2 + 4M^2/s}},$$

which is Eq. (14.14).

We can rewrite the above differential cross section in terms of x_1 and x_2 by transforming the variables $\{M^2, x_F\}$ to $\{x_1, x_2\}$, using Eq. (14.15). We note that

$$x_1 - x_2 = x_F, \tag{4}$$

and

$$x_1 x_2 = \frac{M^2}{s}. \tag{5}$$

Hence, we have

$$dx_1 - dx_2 = dx_F ,$$

$$x_2\,dx_1 + x_1\,dx_2 = \frac{dM^2}{s} ,$$

and taking the cross product of the above equations, we get

$$dM^2\,dx_F = s(x_1 + x_2)\,dx_1\,dx_2 = s\sqrt{x_F^2 + 4M^2/s}\,dx_1\,dx_2 .$$

Using this relation, we can write Eq. (14.14) as

$$\frac{d\sigma}{dx_1 dx_2} = \frac{1}{N_c}\sigma(M)\sum_f^{N_f}\left(\frac{e_f}{e}\right)^2\{q_f^B(x_1)\bar{q}_f^A(x_2) + \bar{q}_f^A(x_1)q_f^B(x_2)\} , \qquad (6)$$

which is Eq. (14.16). Finally, we change the variable from x_F to the variables M and the rapidity variable y by

$$C_z = x_F\sqrt{s}/2 = M\sinh y. \qquad (7)$$

then the variables x_1 and x_2 can be written in terms of M and y as

$$x_1 = \frac{M}{\sqrt{s}}e^y , \qquad (8)$$

and

$$x_2 = \frac{M}{\sqrt{s}}e^{-y} , \qquad (9)$$

which clearly satisfies Eqs. (4), (5) and (7). From Eqs. (8) and (9), we have

$$dx_1 = \frac{dM}{\sqrt{s}}e^y + \frac{M}{\sqrt{s}}e^y dy ,$$

and

$$dx_2 = \frac{dM}{\sqrt{s}}e^{-y} - \frac{M}{\sqrt{s}}e^{-y} dy .$$

Consequently, we have

$$|dx_1 dx_2| = \frac{2M}{s}dM dy .$$

Eq. (6) can be transformed into

$$\frac{d\sigma}{dM dy} = \frac{8\pi\alpha^2}{3N_c M s}\sum_f^{N_f}\left(\frac{e_f}{e}\right)^2[q_f^B(\frac{M}{\sqrt{s}}e^y)\bar{q}_f^A(\frac{M}{\sqrt{s}}e^{-y}) + \bar{q}_f^A(\frac{M}{\sqrt{s}}e^y)q_f^B(\frac{M}{\sqrt{s}}e^{-y})] ,$$

where we have written out $\sigma(M)$ explicitly, neglecting terms of the order of $(m_l/M)^2$. The above equation is Eq. (14.17a).

For the events with $y = 0$, we have $C_z = 0$, $x_F = 0$ and $x_1 = x_2 = x = M/\sqrt{s}$. Hence, we have

$$M^3 \frac{d\sigma}{dMdy}\bigg|_{y=0} = \frac{8\pi\alpha^2}{3N_c} \sum_f^{N_f} \left(\frac{e_f}{e}\right)^2 [xq_f^B(x)x\bar{q}_f^A(x) + xq_f^A(x)x\bar{q}_f^B(x)],$$

which is Eq. (14.17b).

$\rule{0.5em}{0.5em}$

§14.2.2 Drell-Yan Process in Nucleus-Nucleus Collisions

In the collision of a beam nucleus B and a target nucleus A, how does the probability for a Drell-Yan process depend on the mass numbers of the colliding nuclei? Following arguments similar to those leading to Eq. (12.3), the total probability for the occurrence of a Drell-Yan process in a nucleus-nucleus collision when the nuclei B and A are situated at an impact parameter b relative to each other, is the sum of the products of three factors: the probability element $\rho_A(b_A, z_A)db_A dz_A$ for finding a nucleon in the volume element $db_A dz_A$ in nucleus A at the position (b_A, z_A), (ii) the probability element $\rho_B(b_B, z_B)db_B dz_B$ for finding a nucleon in the volume element $db_B dz_B$ in nucleus B at the position (b_B, z_B), and (iii) the probability $t(b - b_A - b_B)\sigma_{DY}^{NN}$ for a nucleon-nucleon Drell-Yan process. We call this probability $T(b)\sigma_{DY}^{NN}$:

$$T(b)\sigma_{DY}^{NN} = \int \rho_A(b_A, z_A)db_A dz_A \rho_B(b_B, z_B)db_B dz_B t(b - b_A - b_B)\sigma_{DY}^{NN},$$

(14.26)

where ρ is the density function (12.2), $t(b)$ is the nucleon-nucleon thickness function, and σ_{DY}^{NN} is the nucleon-nucleon Drell-Yan cross section. Eq. (14.26) leads to the thickness function for nucleus-nucleus collision as given by

$$T(b) = \int db_A db_B T_A(b_A)T_B(b_B)t(b - b_A - b_B),$$

which is just Eq. (12.8). The total probability for the occurrence of a Drell-Yan process in the collision of A and B at an impact parameter b is the sum

$$P_{DY}^{AB}(b) = \sum_{n=1}^{AB} \binom{AB}{n} [T(b)\sigma_{DY}^{NN}]^n [1 - T(b)\sigma_{DY}^{NN}]^{AB-n}.$$

(14.27)

The nucleon-nucleon Drell-Yan cross section being proportional to α^2, the summation is dominated by the first term with $n = 1$. Terms with $[T(b)\sigma_{DY}^{NN}]^n$ and $n > 1$ represent multiple Drell-Yan collisions and the term $[1 - T(b)\sigma_{DY}^{NN}]^{AB-n}$ the shadowing corrections. As an approximation, these corrections can be neglected as the cross section σ_{DY}^{NN} is small and the probability for a Drell-Yan process in a nucleus-nucleus collision is

$$P_{DY}^{AB}(b) = AB[T(b)\sigma_{DY}^{NN}]. \tag{14.28}$$

Consequently, the differential probability for finding an l^+l^- pair with an invariant mass M and a rapidity y in a nucleus-nucleus collision is given by

$$\frac{dP_{DY}^{AB}}{dMdy}(b) = AB\, T(b)\frac{d\sigma_{DY}^{NN}}{dMdy}. \tag{14.29}$$

Each Drell-Yan event produces one pair of dilepton. Consequently, the differential number of l^+l^- pairs produced with a rapidity y and an invariant mass M is given by

$$\frac{dN_{l^+l^-}}{dMdy}(b) = AB\, T(b)\frac{d\sigma_{DY}^{NN}}{dMdy}. \tag{14.30}$$

Therefore, when we integrate over the transverse area, we have

$$\frac{d\sigma_{DY}^{AB}}{dMdy} = \int d\mathbf{b}\, \frac{dN_{l^+l^-}}{dMdy}(b)$$

$$= AB\frac{d\sigma_{DY}^{NN}}{dMdy}.$$

How does one parametrize the nucleus-nucleus thickness function $T(\mathbf{b})$? We found earlier that for nucleus-nucleus collisions, the thickness function can be approximated by Eq. (12.26),

$$T(b) = \exp(-b^2/2\beta^2)/2\pi\beta^2\,,$$

where

$$\beta^2 = \beta_A^2 + \beta_B^2 + \beta_p^2\,,$$

$$\beta_A = r_0' A^{1/3}/\sqrt{3},$$

$r'_0 \sim 1.05$ fm, and $\beta_p = 0.68$ fm [32]. The quantity β_p is small in comparison with the radii of heavy nuclei and can be neglected in qualitative estimates.

Accordingly, the differential number of l^+l^- pairs depends on the impact parameter b between the two nuclei as given by

$$\frac{dN_{l+l^-}}{dMdy}(b) = \frac{3}{2\pi(r'_0)^2} \frac{AB}{A^{2/3} + B^{2/3}} e^{-\frac{b^2}{2\beta^2}} \frac{d\sigma^{NN}_{DY}}{dMdy} \qquad (14.31)$$

When we consider the collision of two equal nuclei with mass number A, we have

$$\frac{dN_{l+l^-}}{dMdy}(b) = \frac{3}{4\pi(r'_0)^2} A^{4/3} e^{\frac{-b^2}{2\beta^2}} \frac{d\sigma^{NN}_{DY}}{dMdy}$$

In particular, for a head-on collision, the number of l^+l^- pairs from the Drell-Yan process is given by

$$\left.\frac{dN_{l+l^-}}{dMdy}\right|_{b=0} = \frac{3}{4\pi(r'_0)^2} A^{4/3} \frac{d\sigma^{NN}_{DY}}{dMdy}. \qquad (14.32)$$

Thus, the number of l^+l^- pairs from the Drell-Yan process for the head-on collision of two equal nuclei scales as $A^{4/3}$.

§14.2.3   Dilepton Production from Hadrons and Resonances

Dilepton pairs can be produced from the interaction of charged hadrons and their antiparticles by processes such as $\pi^+ + \pi^- \to l^+ + l^-$. Dilepton pairs can also come from the decay of hadron resonances such as the ρ, ω, ϕ, and J/ψ. Therefore, hadron collisions and hadron resonance decays are additional sources of l^+l^- pairs.

Hadrons and resonances are produced in the initial nucleus-nucleus collision. If the quark-gluon plasma is produced, hadronic matter will also be present when the quark-gluon plasma cools down below the transition temperature. Dileptons from hadron sources must be identified separately in order to look for dilepton production from the quark-gluon plasma.

One can estimate the contribution of the l^+l^- pairs which come from the hadronic matter. As the dominant constituents of the hadronic matter are pions, we shall for simplicity consider the

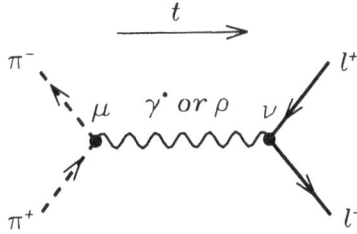

Fig. 14.5   The diagram for the reaction $\pi^+ + \pi^- \to l^+ + l^-$.

hadronic matter to consist only of pions. The collision of a π^+ with a π^- leads to an l^+l^- pair through the diagram depicted in Fig. 14.5. The wavy line in Fig. 14.5 represents a virtual photon γ^* in scalar electrodynamics. In the vector dominance model, it represents a ρ meson intermediate state.

For the production of l^+l^- pairs in hadronic matter, the evaluation of the production rate follows the same considerations as those from quark-gluon matter discussed previously in Section 14.1. The only differences are the different degeneracies of the quarks and pions, and the different basic annihilation cross sections due to the difference in the coupling and the formation of a ρ resonance intermediate state. The hadron matter starts to interact at the initial temperature T_i and stops interacting at the freeze-out temperature T_f. As a consequence, all of the results of Eqs. (14.1)-(14.10) for l^+l^- pair production in the quark-gluon plasma can be used for l^+l^- pair production in the hadronic matter, with the following replacements:

$$N_c \to 1$$
$$N_f \to 1$$
$$N_s \to 1$$
$$m_q \to m_\pi$$
$$e_f \to e$$
$$T_0 \to T_i$$
$$T_c \to T_f$$

and

$$\sigma(M) \to \tilde{\sigma}(M),$$

where m_π is the mass of the pion and $\tilde{\sigma}(M)$ is the cross section for the process $\pi^+\pi^- \rightarrow l^+l^-$ given by (see Exercise 14.6),

$$\tilde{\sigma}(M) = \frac{4\pi}{3}\frac{\alpha^2}{M^2}\left(1 - \frac{4m_\pi^2}{M^2}\right)^{1/2}\left(1 - \frac{4m_l^2}{M^2}\right)^{1/2}\left(1 + \frac{2m_l^2}{M^2}\right)|F_\pi(m_\rho)|^2.$$

(14.33)

The square of the absolute value of the form factor $F_\pi(m_\rho)$ is

$$|F_\pi(m_\rho)|^2 = \frac{m_\rho^4}{(M^2 - m_\rho^2)^2 + \Gamma^2 m_\rho^2},$$

(14.34)

where m_ρ and Γ are the mass and the width of the ρ meson. With these changes, Eqs. (14.7)-(14.10) can be used to estimate the rate of l^+l^- pair production from the hadron phase.

The decays of the hadron resonances will show up as sharp peaks in the invariant mass spectrum of l^+l^- pairs, with a width reflecting the mean lifetime of the resonance and a magnitude depending on the abundance of the resonance. Hadron resonances such as the ρ, ω, and ϕ may arise from the initial nucleus-nucleus collision before thermalization. They may also come from collisions of pions in dense pion gas during the thermalization of the hadron gas. The decay of the J/ψ resonance will give a peak at a dilepton invariant mass of about 3.1 GeV. The large mass of the J/ψ resonance makes it unlikely that it can be produced in soft processes or in the thermalization of the hadronic matter. Thus, the J/ψ production comes mainly from hard-scattering processes. We shall consider J/ψ production and its suppression in more detail in the next chapter.

●⟦ Exercise 14.6
Prove that the cross section for the process $\pi^+ + \pi^- \rightarrow l^+ + l^-$ depicted in Fig. 14.5 is given by Eqs. (14.33) and (14.34).

◇ Solution:
We shall first evaluate the cross section for the case when the wavy line in Fig. 14.5 represents a virtual photon. The Feynman rules for boson particles π and $\bar{\pi}$ are slightly different from those for fermions. Following the rules as given by the Appendix of the book of Itzykson and Zuber [13], the differential cross section for the process $\pi^+ + \pi^- \rightarrow l^- + l^+$ is

$$d\sigma = \frac{1}{4[(\pi \cdot \bar{\pi})^2 - m_\pi^2 m_{\bar{\pi}}^2]^{1/2}}|T_{fi}|^2\frac{d^3l}{(2\pi)^3}\frac{m_l}{l_0}\frac{d^3\bar{l}}{(2\pi)^3}\frac{m_{\bar{l}}}{\bar{l}_0}(2\pi)^4\delta^4(\pi + \bar{\pi} - l - \bar{l}), \quad (1)$$

where we have used the notations π for π^+, $\bar{\pi}$ for π^-, l for l^- and $\bar{l}$ for l^+. We introduce the Mandelstam variables s, t, and u as

$$s = (\pi + \bar{\pi})^2,$$

(2a)

$$t = (\pi - l)^2, \tag{2b}$$

and

$$u = (\pi - \bar{l})^2. \tag{2c}$$

Following Exercise 14.1, we again integrate Eq. (1) with respect to $d\bar{l}$ and dl_0, using the four-dimensional δ function. After these integrations, the differential cross section (1) is

$$d\sigma = \frac{m_l^2}{2(2\pi)^2} \frac{dt \, d\phi_l}{s(s - 4m_\pi^2)} |T_{fi}|^2. \tag{3}$$

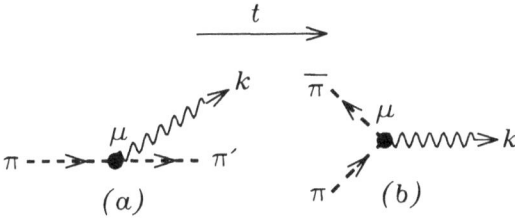

Fig. 14.6   Two types of vertices in scalar electrodynamics.

As before, by using the rules of Feynman diagrams, one can write down the explicit form of the matrix element T_{fi}. The vertex μ involves the coupling of the charged pion field with the photon field. The boson-photon coupling at the vertex μ in scalar electrodynamics is different from the fermion-photon coupling in spinor electrodynamics studied previously in Exercise 14.1. This difference leads to different Feynman rules for a vertex of the type μ. If the boson lines is in the form of Fig. 14.6a, the vertex will be associated with the integral (see the Appendix of Itzykson and Zuber [13])

$$\int \frac{d^4k}{(2\pi)^4} (-ie)(\pi_\mu + \pi_\mu')(2\pi)^4 \delta^4(\pi - \pi' - k).$$

For our case of interest where we consider the annihilation of a $\pi\bar{\pi}$ pair as depicted in Fig. 14.6b, the destruction of a $\bar{\pi}$ travelling backward in time with momentum $\bar{\pi}$ at the vertex μ is equivalent to the creation of a π with momentum $-\bar{\pi}$ travelling forward in time at the vertex μ. The function associated with vertex 14.6b is

$$\int \frac{d^4k}{(2\pi)^4} (-ie)(\pi_\mu - \bar{\pi}_\mu)(2\pi)^4 \delta^4(\pi + \bar{\pi} - k).$$

The boson lines at the vertex μ do not bring in additional functions or factors, in contrast to the case with fermion lines which comes with additional spinor functions.

The lepton spinor functions which are associated with the lepton lines at the vertex ν are the same as in Exercise 14.1 and need not be discussed again. By putting all factors together, the diagram 14.5 gives the transition matrix element

$$T_{fi} = \bar{u}(l, \epsilon_l)(-ie\gamma^\nu)v(\bar{l}, \epsilon_{\bar{l}}) \frac{-ig_{\mu\nu}}{(\pi + \bar{\pi})^2} [-ie(\pi^\mu - \bar{\pi}^\mu)]. \tag{4}$$

The square of the matrix element is

$$
\begin{aligned}
|T_{fi}|^2 &= \frac{e^4}{s^2}[(\pi^\nu - \bar\pi^\nu)\,\bar{v}(\bar{l},\epsilon_{\bar{l}})\gamma^0\gamma_\nu{}^\dagger\gamma^0 u(l,\epsilon_l)][\bar{u}(l,\epsilon_l)\gamma_\mu v(\bar{l},\epsilon_{\bar{l}})(\pi^\mu - \bar\pi^\mu)] \\
&= \frac{e^4}{s^2}\,\bar{v}(\bar{l},\epsilon_{\bar{l}})\gamma_\nu u(l,\epsilon_l)\,\bar{u}(l,\epsilon_l)\gamma_\mu v(\bar{l},\epsilon_{\bar{l}})(\pi^\nu - \bar\pi^\nu)(\pi^\mu - \bar\pi^\mu) \\
&= \frac{e^4}{s^2}[v_d(\bar{l},\epsilon_{\bar{l}})\bar{v}_a(\bar{l},\epsilon_{\bar{l}})(\gamma_\nu)_{ab}u_b(l,\epsilon_l)\bar{u}_c(l,\epsilon_l)(\gamma_\mu)_{cd}](\pi^\nu - \bar\pi^\nu)(\pi^\mu - \bar\pi^\mu).
\end{aligned}
$$

We need to sum over the final polarizations ϵ_l and $\epsilon_{\bar{l}}$. We have

$$
\begin{aligned}
\sum_{\epsilon_l\epsilon_{\bar{l}}}^{2}|T_{fi}|^2 &= \frac{e^4}{s^2}\sum_{\epsilon_l\epsilon_{\bar{l}}}^{2} v_d(\bar{l},\epsilon_{\bar{l}})\bar{v}_a(\bar{l},\epsilon_{\bar{l}})(\gamma_\nu)_{ab}u_b(l,\epsilon_l)\bar{u}_c(l,\epsilon_l)(\gamma_\mu)_{cd}\,(\pi^\nu - \bar\pi^\nu)(\pi^\mu - \bar\pi^\mu) \\
&= \frac{e^4}{s^2}[\,\rho_{da}(\bar{l})\,(\gamma_\nu)_{ab}\,\rho_{bc}(l)\,(\gamma_\mu)_{cd}](\pi^\nu - \bar\pi^\nu)(\pi^\mu - \bar\pi^\mu),
\end{aligned}
\tag{5}
$$

where the density matrices [16] have been defined as before in Exercise 14.1,

$$
\rho_{bc}(l) = \sum_{\epsilon_l=1}^{2} u_b(l,\epsilon_l)\bar{u}_c(l,\epsilon_l),
\tag{6}
$$

and

$$
\rho_{da}(\bar{l}) = \sum_{\epsilon_{\bar{l}}=1}^{2} v_d(\bar{l},\epsilon_{\bar{l}})\bar{v}_a(\bar{l},\epsilon_{\bar{l}}).
\tag{7}
$$

In matrix notation, Eq. (5) can be written as the trace of a product of matrices,

$$
\sum_{\epsilon_q\epsilon_{\bar{q}}\epsilon_l\epsilon_{\bar{l}}}^{2}|T_{fi}|^2 = \frac{e^4}{s^2}\,tr\{\,\rho(\bar{l})\,\gamma_\nu\,\rho(l)\,\gamma_\mu\,\}(\pi^\nu - \bar\pi^\nu)(\pi^\mu - \bar\pi^\mu)\}.
\tag{8}
$$

From Exercise 14.1, we have

$$
\rho(l) = \frac{\slashed{l} + m_l}{2m_l}
$$

and

$$
\rho(\bar{l}) = -\frac{-\slashed{l} + m_l}{2m_l}.
$$

The trace in Eq. (8) can be evaluated. Following Exercise 14.1, we find

$$
-\frac{1}{4}tr\{(-\slashed{\bar{l}} + m_l)\,\gamma_\nu\,(\slashed{l} + m_l)\,\gamma_\mu\} = -[g_{\nu\mu}(m_l^2 + l\cdot\bar{l}) - l_\nu\bar{l}_\mu - l_\mu\bar{l}_\nu].
$$

The summation (8) becomes

$$\sum_{\epsilon_l \epsilon_{\bar{l}}}^{2} |T_{fi}|^2 = -\frac{e^4}{m_l^2 s^2} [g_{\nu\mu}(m_l^2 + l \cdot \bar{l}) - l_\nu \bar{l}_\mu - l_\mu \bar{l}_\nu](\pi^\nu - \bar{\pi}^\nu)(\pi^\mu - \bar{\pi}^\mu)$$

$$= -\frac{e^4}{m_l^2 s^2} \{(m_l^2 + l \cdot \bar{l})(\pi - \bar{\pi})^2 - 2[(\pi - \bar{\pi}) \cdot l] [(\pi - \bar{\pi}) \cdot \bar{l}]\}.$$

This expression can be simplified by noting that

$$(\pi - \bar{\pi}) \cdot l = \pi \cdot (l - \bar{l})$$

and

$$(\pi - \bar{\pi}) \cdot \bar{l} = \pi \cdot (\bar{l} - l)$$

We have then

$$\sum_{\epsilon_l \epsilon_{\bar{l}}}^{2} |T_{fi}|^2 = -\frac{e^4}{m_l^2 s^2} \{(m_l^2 + l \cdot \bar{l})(\pi - \bar{\pi})^2 + 2[\pi \cdot (l - \bar{l})]^2\}. \tag{9}$$

Using Eq. (2), we can write the scalar products in terms of the Mandelstam variables. We have

$$m_l^2 + l \cdot \bar{l} = \frac{s}{2}$$

$$(\pi - \bar{\pi})^2 = -s + 4m_\pi^2$$

and

$$\pi \cdot (l - \bar{l}) = -\frac{1}{2}(t - u).$$

Using these relations, we can rewrite the sum Eq. (9) as

$$\sum_{\epsilon_l \epsilon_{\bar{l}}}^{2} |T_{fi}|^2 = -\frac{e^4}{m_l^2 s^2} [-\frac{s}{2}(s - 4m_\pi^2) + \frac{1}{2}(t - u)^2]. \tag{10}$$

After summing over the polarizations of the final particles l and $\bar{l}$, the differential cross section (3) becomes

$$d\sigma = \frac{m_l^2}{2(2\pi)^2} \frac{dt \, d\phi_l}{s(s - 4m_\pi^2)} \sum_{\epsilon_l \epsilon_{\bar{l}}}^{2} |T_{fi}|^2$$

$$= \frac{m_l^2}{2(2\pi)^2} \frac{dt \, d\phi_l}{s(s - 4m_\pi^2)} \frac{e^4}{m_l^2 s^2} \left[\frac{s}{2}(s - 4m_\pi^2) - \frac{1}{2}(t - u)^2\right].$$

Integration over ϕ_l yields

$$d\sigma = \frac{e^4}{4\pi} \frac{dt}{s^3(s - 4m_\pi^2)} \left[\frac{s}{2}(s - 4m_\pi^2) - \frac{1}{2}(t - u)^2\right]. \tag{11}$$

To perform the integration over t, we note that

$$\int (t-u)^2 \, dt = \int dt (t - 2m_l^2 - 2m_\pi^2 + s + t)^2$$
$$= \frac{1}{6}(t-u)^3 .$$

As in Exercise 14.1, the upper and lower limit of integration of t are t_+ and t_- given by

$$t_\pm = m_\pi^2 + m_l^2 - \frac{1}{2}s \pm \frac{1}{2}\sqrt{s - 4m_\pi^2}\sqrt{s - 4m_l^2} ,$$

and we have

$$t_+ - t_- = \sqrt{s - 4m_\pi^2}\sqrt{s - 4m_l^2}. \qquad (12)$$

At these limits of t, the variable u assumes the values

$$u\big|_{t=t_+} = t_- ,$$

and

$$u\big|_{t=t_-} = t_+ .$$

Therefore, after integrating Eq. (11) with respect to t, the total cross section is

$$\sigma = \frac{e^4}{8\pi s^3(s - 4m_\pi^2)}\left[s(s - 4m_\pi^2)(t_+ - t_-) - \frac{1}{6}\{[t_+ - u(t = t_+)]^3 - [t_- - u(t = t_-)]^3\}\right]$$
$$= \frac{e^4}{8\pi s^3(s - 4m_\pi^2)}\left[s(s - 4m_\pi^2)(t_+ - t_-) - \frac{1}{6}\{(t_+ - t_-)^3 - (t_- - t_+)^3\}\right]$$
$$= \frac{e^4}{8\pi s^3(s - 4m_\pi^2)}(t_+ - t_-)\left[s(s - 4m_\pi^2) - \frac{1}{3}(t_+ - t_-)^2\right]$$

Using Eq. (12), and the fine structure constant $\alpha = e^2/4\pi$, we have

$$\sigma = \frac{2\pi\alpha^2}{3s^3(s - 4m_\pi^2)}\sqrt{s - 4m_\pi^2}\sqrt{s - 4m_l^2}\left[3s(s - 4m_\pi^2) - (s - 4m_\pi^2)(s - 4m_l^2)\right]$$
$$= \frac{4\pi\,\alpha^2}{3\,s}\left(1 - \frac{4m_\pi^2}{s}\right)^{1/2}\left(1 - \frac{4m_l^2}{s}\right)^{1/2}\left(1 + 2\frac{m_l^2}{s}\right), \qquad (13)$$

The above result was obtained for the annihilation of π^+ and π^- through an intermediate virtual photon state. However, the meson dominance model suggests that the dominant process for l^+l^- pair production goes through the intermediate state of the ρ resonance. In this model, the dashed line in diagram Fig. 14.5 represents the ρ-meson propagator

$$\frac{1}{(\pi + \bar\pi)^2 - (m_\rho + i\Gamma/2)^2}. \qquad (14)$$

The result of Eq. (13) is obtained with the photon propagator

$$\frac{1}{(\pi + \bar{\pi})^2} = \frac{1}{s}. \tag{15}$$

Therefore, the cross section for the case of an intermediate ρ meson can be obtained by multiplying Eq. (13) with a factor given by the ratio of the square of the absolute values of (14) and (15),

$$\begin{aligned}
|F_\pi(m_\pi)|^2 &= \left| \frac{1}{(\pi + \bar{\pi})^2 - (m_\rho + i\Gamma/2)^2} \right|^2 \div \frac{1}{s^2} \\
&= \frac{s^2}{(s - m_\rho^2 + (\Gamma^2/4))^2 + \Gamma^2 m_\rho^2} \\
&\simeq \frac{m_\rho^4}{(M^2 - m_\rho^2)^2 + \Gamma^2 m_\rho^2}.
\end{aligned}$$

Therefore, in the vector dominance model, the cross section for the process $\pi^+ + \pi^- \to l^+ + l^-$ is given by Eqs. (14.33) and (14.34), as had to be proved.]•

§14.2.4 Dileptons from the Decay of Charm Particles

In nucleon-nucleon hard-scattering processes, charm mesons such as D^+ and D^- are produced by the interaction of a constituent of one nucleon with a constituent of the other colliding nucleon. In particular, in the lowest order QCD theory, the quark of one nucleon interacts with the antiquark of the other nucleon to form a virtual gluon which decays into a $c\bar{c}$ pair,

$$q + \bar{q} \to g^* \to c + \bar{c}, \tag{14.35}$$

as illustrated in Fig. 14.7a. A $c\bar{c}$ pair can also be produced by the interaction of a gluon of one nucleon with a gluon of the other colliding nucleon,

$$g + g \to c + \bar{c}, \tag{14.36}$$

as illustrated in Fig. 14.7b, 14.7c, and 14.7d.

A charm meson D^+ is a composite particle which consists of a charm quark and a $\bar{u}$, $\bar{d}$, or $\bar{s}$ antiquark. The corresponding antiparticle D^- meson is the composite particle which consists of a charm antiquark and a u, d or s quark. Following the production of the $c\bar{c}$ pair in a nucleon-nucleon hard-scattering collision by the processes (14.35) and (14.36), the fragmentation of the c quark into a D^+ meson and the fragmentation of the $\bar{c}$ antiquark into a D^- meson result in the production of a $D^+ D^-$ pair. The subsequent decay of the D^+

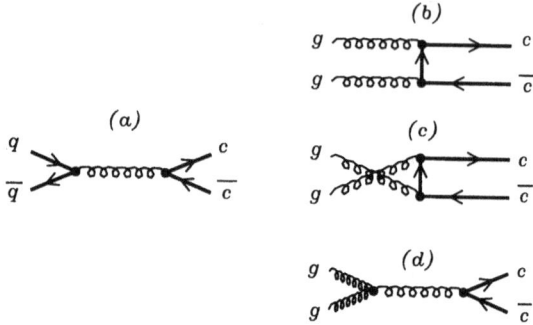

Fig. 14.7   The Feynman diagrams for $c\bar{c}$ pair production.

into $l^{+}X$ (for example, through the decay of $D^{+} \rightarrow l^{+}\overline{K}^{0}\nu_{l}$) and the decay of and D^{-} into $l^{-}X$ (for example, through the decay of $D^{-} \rightarrow l^{-}K^{0}\bar{\nu}_{l}$) give rise to an $l^{+}l^{-}$ pair.

Dilepton production by nucleon-nucleon collisions through the intermediate $D^{+}D^{-}$ production provide an additional contribution to the dilepton spectrum. This contribution must be subtracted in order to extract the dileptons arising from the transient quark-gluon plasma.

There has been much work to investigate the mechanism for the production of $D^{+}D^{-}$ pairs [33-40]. Lowest order QCD results to study heavy quark production using the diagrams in Fig. 14.7 have been presented in Refs. [33]-[35]. It is generally recognized that charm production cannot be fully described by lowest order perturbative QCD diagrams in Fig. 14.7 because of the small value of the charm quark mass [36]-[41]. An additional phenomenological factor, similar to the K factor in the Drell-Yan process (see Section §14.2.1), is needed to bring the calculated cross section to agree with experimental data [35].

Much work remains to be done to understand the spectrum of dileptons by charm production in nucleus-nucleus collisions. Higher order QCD effects can be included order by order in terms of the summation of amplitudes using Feynman diagrams, as is done in Ref. [36]-[39]. It can also be carried out using the phenomenological parton shower program PYTHIA where a parton splits into other partons with a probability distribution and the resultant strings are fragmented according to the Lund model of string fragmentation [40,41].

When we consider the production of $\mu^{+}\mu^{-}$ pairs through the intermediate stage of $D^{+}D^{-}$ production, we note that because of the large masses of the charm quark and $D^{\pm}$ mesons, the partons which participate in charm production need to have large momenta, which occur with a small probability. Furthermore, the $l^\pm$ particle is the product of a multi-particle decay of the $D^\pm$ meson. The invariant mass of the l^+l^- pair is only a fraction of the invariant mass of the originating D^+D^- or the $c\bar{c}$ pair. In addition, the branching ratio for the the $D^\pm$ meson to decay into an $l^\pm$ is of the order of 10 percent. As a consequence of these factors, the estimates from Ref. [41] using the PYTHIA program show that the number of $\mu^+\mu^-$ pair with $M > 1.5$ GeV arising from charm production is less than the number of dileptons from the Drell-Yan process, in nucleon-nucleon collision at 200 GeV on fixed targets. The dileptons produced via charm production have an approximately exponential invariant mass distribution with a slope parameter corresponding to an effective 'temperature' much smaller than the corresponding Drell-Yan process.

§14.3 Spectrum of Dileptons

It is of interest to inquire whether the dilepton yield arising from the produced matter in the quark-gluon plasma phase can be strong enough to make it observable. In order to be observable, the dilepton yield from the transient quark-gluon plasma must be greater than or comparable to the dilepton yields from non-quark-gluon-plasma sources.

In the region below an invariant mass of 1 GeV, the decays from ρ, ω, and ϕ dominate over the production of low-mass l^+l^- pairs arising from the possible formation of the quark-gluon plasma [42]. Furthermore, hadron collisions and charm meson production are dilepton sources with low temperatures and they contribute dileptons in the low invariant mass region. In the region below about 1 GeV, it may be difficult to separate out the quark-gluon plasma contribution to the dilepton spectrum.

If we examine dileptons with an invariant mass greater than 1.5 GeV and away from the resonance peaks, dileptons from hadron-hadron interactions and charm meson production may not be important. The dominant non-quark-gluon plasma contribution comes from the Drell-Yan process. It is of interest to compare dilepton production from the Drell-Yan process and from the quark-gluon plasma.

Equation (14.11) in Section 14.1 shows that dileptons produced by the transient quark-gluon plasma has an apparent temperature about the same as the initial temperature T_0 of the quarks in the plasma. Since the magnitude of the transition temperature is about 200 MeV (see Chapter 9), the magnitude of the initial temperature T_0 is of the order of a few hundred MeV. On the other hand, dileptons from the Drell-Yan process has a distribution with an effective temperature T_{DY} approximately $\sqrt{s}/(12 \text{ to } 15)$, as given by Eq. (14.25) or (14.19). For the RHIC collider with an energy of 100 GeV per nucleon, $\sqrt{s}$ is 200 GeV. The dileptons from the Drell-Yan process have an effective temperature T_{DY} of about 13 to 17 GeV, which is much greater than the effective temperature of the dileptons from the quark-gluon plasma. The large value of the effective temperature implies that the dilepton yield from the Drell-Yan process will be greater than the dilepton yield from the quark-gluon plasma at large dilepton invariant masses. At what value of the dilepton invariant mass will the dilepton yields from these two different sources be comparable and the role of the dominance of one signal over the other begins to reverse?

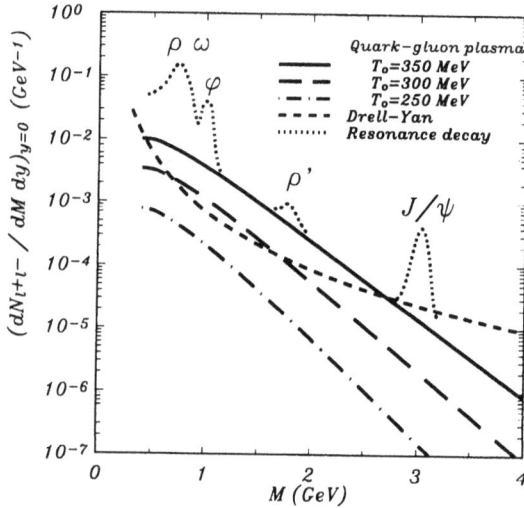

Fig. 14.8   The distribution $dN_{l^+l^-}/dMdy$ of l^+l^- pairs at $y = 0$, as a function of the dilepton invariant mass M.

Clearly, the location where the dilepton yield from the quark-gluon plasma exceeds the dilepton yield from the Drell-Yan process depends on the quark-gluon plasma initial temperature, which is related to the initial dynamics of the collision process. The initial temperature can be obtained from the energy density, for which various estimates have been made [43]-[47]. Since the initial dynamics of the collision process remains a subject of current research, we shall examine the dilepton yield as a function of the initial temperature. We shall discuss $\mu^+\mu^-$

results; the e^+e^- distributions behave in a similar way. Using Eq. (14.9) and assuming a transition temperature $T_c = 200$ MeV and an initial time $\tau_0 = 1$ fm/c, we show in Fig. 14.8 dilepton yields for a quark-gluon plasma with initial temperatures $T_0 = 250$, 300 and 350 MeV, in a head-on collision of Pb on Pb. In Fig. 14.8, we show the dilepton distribution for the Drell-Yan process estimated by using Eq. (14.32) and (14.19), at $\sqrt{s} = 200$ GeV. Schematic dilepton yields from resonances are also shown in Fig. 14.8, to indicate the locations where they may be important. Fig. 14.8 indicates that if the quark-gluon plasma initial temperature is greater than 350 MeV, the dilepton yield from the quark-gluon plasma will be much greater than the dilepton yield from the Drell-Yan process in the region 1 GeV $< M <$ 2.8 GeV and the observation of dileptons from the quark-gluon plasma may be possible. If the quark-gluon plasma temperature is about 300 MeV, the dilepton yield from the quark-gluon plasma will be greater than the dilepton yield from Drell-Yan processes only around 1 to 1.6 GeV. Observation of the dileptons from the quark-gluon plasma is possible but not as clean as at higher temperatures. On the other hand, at a still lower temperature, $T_0 = 250$ MeV, the dilepton yield from the Drell-Yan process is greater than the dilepton yield from the quark-gluon plasma and the signal from the quark-gluon plasma may be masked. The observation of dileptons from the quark-gluon plasma will depend on the initial temperature of the plasma. Because the energy density is approximately proportional to the fourth power of the temperature, an initial temperature of 300 MeV corresponds to an energy density of the quark-gluon plasma about 5 times the transition energy density. Future experiments will reveal whether high-energy heavy-ion collisions will go through the transient state of the quark-gluon plasma with such an energy density or not.

§References for Chapter 14

1. Excellent reviews of dilepton production in high-energy heavy-ion collisions can be found in P. V. Ruuskanen, Nucl. Phys. A522, 255c (1991), P. V. Ruuskanen, Nucl. Phys. A544, 169c (1992), and Ref. 2.
2. K. Kajantie, J. Kapusta, L. McLerran, and A. Mekjian, Phys. Rev. D34, 2746 (1986).
3. E. L. Feinberg, Nuovo Cim. 34A, 391 (1976); E. V. Shuryak, Phys. Lett. 78B, 150 (1978).
4. G. Demokos and J. I. Goldman, Phys. Rev. D23, 203 (1981); G. Domokos Phys. Rev. D23, 203 (1981).
5. S. Chin, Phys. Lett. 119B, 51 (1982).
6. R. C. Hwa and K. Kajantie Phys. Rev. D32, 1109 (1985).
7. L. D. McLerran and T. Tiomela, Phys. Rev. D31, 545 (1985).
8. J. Cleymans and J. Fingberg, Phys. Lett. 168B, 405 (1986); J. Cleymans, J.

Fingberg, and K. Redlich, Phys. Rev. D35, 2153 (1987).

9. S. Raha and B. Sinha, Int. J. Mod. Phys. A6, 517 (1991).

10. L. H. Xia, C. M. Ko, and C. T. Li, Phys. Rev. C41, 572 (1990).

11. C. Y. Wong, Phys. Rev. C48, 902 (1993).

12. M. Abramowitz and I. A. Stegun, *Handbook of Mathematical Tables*, Dover Publications, New York, 1965.

13. C. Itzykson and J.-B. Zuber, *Quantum Field Theory*, McGraw-Hill Book Company, N.Y., 1980.

14. J. D. Bjorken and S. D. Drell, *Relativistic Quantum Mechanics*, (McGraw-Hill Book Company, N.Y. 1964) and J. D. Bjorken and S. D. Drell, *Relativistic Quantum Fields*, (McGraw-Hill Book Company, N.Y. 1965).

15. Ta-Pei Cheng and Lin-Fong Li, *Gauge Theory of Elementary Particle Physics*, Clarendon Press, Oxford, 1984.

16. V. B. Berestetskii, E. M. Lifshitz and L. I. Pitaevskii, *Quantum Electrodynamics*, Pergamon Press, Oxford, 1982.

17. S. D. Drell and T. M. Yan, Phys. Rev. Lett. 25, 316 (1970).

18. Excellent reviews of dilepton production in hadron collisions can be found in Ref. 19, Ref. 20, and C. Grosso-Pilcher and M. J. Shochet, Ann. Rev. Nucl. Part. Sci. 36, 1 (1986).

19. N. S. Craigie, Phys. Rep. 47, 1 (1978).

20. R. Brock *et al.*, *Handbook of Perturbative QCD*, (CTEQ Collaboration), editor George Sterman, Fermilab Report Fermilab-pub-93/094, 1993.

21. J. C. Collins, D. E. Soper, G. Sterman, Phys. Lett. 134B, 263 (1984).

22. J. Kubar, M. Le Bellac, J. L. Meunier, and G. Plaut, Nucl. Phys. B175, 251 (1980).

23. R. Hamberg, W. L. van Neerven, and T. Matsuura, Nulc. Phys. B359, 343 (1991).

24. See Table 6.2 of Ref. 20.

25. For a review of the parton distribution function of hadrons, see for example J. F. Owens and Wu-Ki Tung, Ann. Rev. Nucl. Part. Sci. 42, 291 (1992).

26. D. W. Duke and J. F. Owens, Phys. Rev. D30, 49 (1984).

27. J. G. Morfin and W. K. Tung, Zeit. Phys. C52, 13 (1990).

28. J. Botts *et al.*, CTEQ Collaboration, Phys. Lett. B304, 159 (1993).

29. M. Glück, E. E. Reya and A. Vogt, Zeit. Phys. C48, 471 (1990).

30. P. N. Harriman, A. D. Martin, W. J. Stirling, and R. G. Roberts, Phys. Rev. D42, 798 (1990).

31. A. D. Martin, W. J. Stirling, and R. G. Roberts, Phys. Rev. D43, 3648 (1991).

32. C. Y. Wong, Phys. Rev. D30, 961 (1984).

33. M. Glück and E. Reya, Phys. Lett. 79B, 453 (1978).

34. B. L. Combridge, Nucl. Phys. B151, 429 (1979).

35. R. Vogt, S. J. Brodsky, and P. Hoyer, Nucl. Phys. B383, 643 (1992).

36. P. Nason, S. Dawson, and R. K. Ellis, Nucl. Phys. B303, 607 (1989); P. Nason, S. Dawson, and R. K. Ellis, Nucl. Phys. B327, 49 (1989).

37. J. C. Collins and R. K. Ellis, Nulc. Phys. B360, 3 (1991).

38. M. L. Mangan, P. Nason, G. Ridolfi, Nucl. Phys. B405 507 (1993).

39. W. Beenakker, W. L. van Neerven, R. Meng, G. A. Schuler, and J. Smith, Nucl. Phys. B351 (1991).

40. J. Sjostrand, *PYTHIA 5.6 and JETSET 7.3, Physics and Manual*, CERN Report CERN-TH-6488-92, 1992.

41. M. C. Abreu *et al*, NA38 Collaboration, Nucl. Phys. A566, 77c (1994).

42. T. Altherr and P. V. Rauskanen, Nucl. Phys. B380, 377 (1992).

43. J. D. Bjorken, Phys. Rev. D27, 140 (1983).
44. C. Y. Wong, Phys. Rev. D30, 961 (1984).
45. H. Satz, Nucl. Phys. A544, 371c (1992).
46. E. Shurayak, Phys. Rev. Lett. 22, 3270 (1992); E. Shurayak and L. Xiong, Phys. Rev. Lett. 70, 2241 (1993).
47. K. Geiger, Phys. Rev. D46, 4965 (1992); K. Geiger, Phys. Rev. D47, 133 (1993).

15. Signatures for the Quark-Gluon Plasma (II)

§15.1 Debye Screening in the Quark-Gluon Plasma

In a quark-gluon plasma, the color charge of a quark is subject to screening due to the presence of quarks, antiquarks, and gluons in the plasma. This phenomenon is called the *Debye screening*, in analogy to the familiar Debye screening of an electric charge in QED. If we place a J/ψ particle, which is the bound state of a charm quark c and a charm antiquark $\bar{c}$, in the plasma, the Debye screening will weaken the interaction between c and $\bar{c}$. Furthermore, in the quark-gluon plasma, quarks and gluons are deconfined and the string tension between c and $\bar{c}$ vanishes. Because of these two combined effects, a J/ψ particle placed in the quark-gluon plasma at high temperatures will become dissociated, leading to the suppression of its production in high-energy nucleus-nucleus collisions, as first suggested by Matsui and Satz [1].

How do we understand the influence of the quark-gluon plasma on a J/ψ particle? Before we discuss this topic, let us consider the J/ψ particle as a two-body system of a charm quark interacting with a charm antiquark, without the quark-gluon plasma. We place the c quark with a color charge $q > 0$ at the origin and the $\bar{c}$ antiquark with a color charge $(-q)$ at r. The color potential from the c quark as seen by the antiquark $\bar{c}$ at the point r can be represented phenomenologically by the Coulomb potential

$$V_0(r) = \frac{q}{4\pi r}. \tag{15.1}$$

There is also the confining linear potential between c and $\bar{c}$ which increases with their separation,

$$V_{\text{linear}}(r) = \kappa r, \tag{15.2}$$

where κ is the string tension coefficient. The potential energy for the $c\bar{c}$ system is [2]

$$H_I = (-q)\frac{q}{4\pi r} + \kappa r, \tag{15.3}$$

The Hamiltonian for the $c\bar{c}$ system is [2]

$$H = \frac{p^2}{2\mu} - \frac{\alpha_{\text{eff}}}{r} + \kappa r, \tag{15.4}$$

where $\mu (= m_c/2)$ is the reduced mass of the $c\bar{c}$ system and $\alpha_{\text{eff}} = q^2/4\pi$. This simple Hamiltonian (15.4) provides a good quantitative description of the observed spectroscopy of $c\bar{c}$ systems. The charmonium states are well described by Eq. (15.4) with the set of parameters: $\alpha_{\text{eff}} = 0.52$, $\kappa = 0.926$ GeV/fm, and $m_c = 1.84$ GeV, (or by the alternative set: $\alpha_{\text{eff}} = 0.30$, $\kappa = 1.18$ GeV/fm, and $m_c = 1.65$ GeV) [2].

●〚Exercise 15.1
What is the relationship between the QCD coupling constant g and the effective color charge q which appears in Eqs. (15.1) and (15.3)?

◇Solution:
In perturbative QCD, the gluon propagator in the Coulomb gauge is [3]

$$\frac{-i\delta_{ab}g_{\mu\nu}}{k^2},$$

where a and b are color indices and $\boldsymbol{k}$ is the gluon momentum. The quark-gluon vertex is associated with the operator

$$ig\gamma^\mu \frac{\lambda^a}{2},$$

where $\{\lambda^a, a = 1, ..., 8\}$ are the generators of the color-$SU(3)$ group (see e.g. Section §10.5 or Appendix B of Ref. [3]). Thus, the interaction energy due to a gluon exchange between the quark c and the antiquark $\bar{c}$ is given in the Coulomb gauge by

$$H_I = \frac{\lambda_c}{2} \cdot \frac{\lambda_{\bar{c}}}{2} \frac{g^2}{4\pi r}, \tag{1}$$

where $\lambda_c = \{\lambda_c^1, \lambda_c^2, ..., \lambda_c^8\}$ is the set of eight generators of the color-$SU(3)$ group, acting on the color wave function of the quark c, and $\lambda_{\bar{c}}$ is the analogous set of eight generators for the antiquark $\bar{c}$. The product $\lambda_c \cdot \lambda_{\bar{c}}$ is

$$\lambda_c \cdot \lambda_{\bar{c}} = \frac{1}{2}[(\lambda_c + \lambda_{\bar{c}})^2 - \lambda_c^2 - \lambda_{\bar{c}}^2]. \tag{2}$$

From the definition of the $SU(3)$ generators in Eq. (10.34), explicit matrix multiplication gives

$$\lambda_c^2 = (\lambda_c^1)^2 + (\lambda_c^2)^2 + ... + (\lambda_c^8)^2$$
$$= \frac{16}{3} \begin{pmatrix} 1 & 0 & 0 \\ 0 & 1 & 0 \\ 0 & 0 & 1 \end{pmatrix}$$
$$= \lambda_{\bar{c}}^2.$$

Therefore, from Eq. (2), we have

$$\lambda_c \cdot \lambda_{\bar{c}} = \frac{1}{2}(\lambda_c + \lambda_{\bar{c}})^2 - \frac{16}{3}.$$

Of all the color states that can be formed by c and $\bar{c}$, we are only interested in the color singlet states of the $c\bar{c}$ system, as all observed isolated particles are in color singlet states. Thus, for all possible coupling of λ_c and $\lambda_{\bar{c}}$, we shall only deal with the case when $(\lambda_c + \lambda_{\bar{c}})^2 = 0$. For this case, we have $\lambda_c \cdot \lambda_{\bar{c}} = -16/3$ and

$$H_I = -\frac{4}{3}\frac{g^2}{4\pi r}.\tag{3}$$

A comparison of the above equation and the gluon-exchange interaction term in Eq. (15.3) shows that

$$q^2 = \frac{4}{3}g^2 = \frac{4}{3}4\pi\alpha_s,$$

where α_s is the strong coupling constant. It is convenient to introduce the effective coupling constant α_{eff} defined by

$$\alpha_{eff} = \frac{4}{3}\alpha_s.\tag{4}$$

The interaction energy (3) due to the exchange of a gluon between c and $\bar{c}$ becomes

$$H_I = -\frac{\alpha_{eff}}{r}.\tag{5}] \bullet$$

After considering a $c\bar{c}$ system by itself, we now place the $c\bar{c}$ system in the quark-gluon plasma, with the heavy quark c at $r = 0$ and the antiquark at r. The presence of the quarks, antiquarks and the gluons of the plasma affect the $c\bar{c}$ system in two important ways [1]. First, since the string tension depends on temperature, the quark matter at a finite temperature alters the string tension coefficient κ between c and $\bar{c}$ [4-12]. Second, the presence of quark matter leads to the rearrangement of the densities of quarks, antiquarks, and gluons around c and $\bar{c}$. The rearrangement leads to the screening of the color charge of c from $\bar{c}$ and vice versa. As a consequence, the interaction between c and $\bar{c}$ is modified from a Coulomb interaction Eq. (15.1) into a Yukawa-type short-range interaction. We shall discuss these in detail below.

The string tension κ depends on the temperature. Confinement occurs when the string tension does not vanish. Conversely, deconfinement of quarks and gluons is accompanied by the vanishing of the string tension κ.

As discussed in Chapters 10 and 11, the study of confinement-deconfinement transition of quark matter at finite temperature has been carried out with lattice-gauge theory [4]-[9]. The occurrence of a deconfinement transition is a general lattice gauge theory result, although the order of the transition may depend on the dynamics and the quark masses, as discussed in Section 11.5. Model calculations based on the flux-sheet model [10] and the flux-tube model [12] give an indication of the critical behavior of the the string tension κ as the temperature approaches the critical temperature. To search for signals associated with deconfinement, we shall examine the case when the string tension vanishes in the deconfined quark-gluon plasma.

The absence of the string tension does not automatically mean that a charm quark and a charm antiquark cannot form a bound state. They remain interacting with each other with the Coulomb interaction $-\alpha_{\text{eff}}/r$. This mutual Coulomb interaction is however modified because of Debye screening. The screening may be so severe that c and $\bar{c}$ cannot form a bound state and the $c\bar{c}$ system will dissociate into a separate c quark and another $\bar{c}$ antiquark without forming a bound state.

How do we understand qualitatively the phenomenon of Debye screening for the $c\bar{c}$ system in the quark-gluon plasma? The essential concepts of this Debye screening can be best illustrated by treating the color interaction in the Abelian approximation. (The treatment of the Debye screening in perturbative non-Abelian gauge theory can be found in Ref. [13].) In this overly-simplified Abelian approximation, the screening arises from the spatial re-distribution of quarks and antiquarks in the plasma.

We envisage initially a quark-gluon plasma in thermal equilibrium at a temperature T. The number density of quarks n_q and the number density of antiquarks $n_{\bar{q}}$ are the same, with a zero chemical potential μ. The number densities and chemical potential depend only on the temperature and are independent of spatial location. The chemical potential, $\mu = 0$, is also independent of spatial location. The chemical composition of quarks with different flavors and masses m_q depends on the Boltzmann factor $e^{-m_q/T}$. For a temperature in the range of a few hundred MeV, the fraction of charm quarks is small and the plasma can be considered to consist essentially only of light and strange quarks and antiquarks. In the plasma, the energy density of the quarks ϵ_q and the energy density of the antiquarks $\epsilon_{\bar{q}}$ are equal and spatially constant.

In the environment of the quark-gluon plasma, there is no linear confining interaction between a quark and an antiquark. There remains only the $(\pm q^2/r)$-type interaction arising from the exchange of a gluon. We consider the charm quark c at $r = 0$. With the gluon-exchange interaction, the presence of the charm quark c will pull the antiquarks in the plasma (represented schematically by solid circles in Fig. 15.1) towards c but will push the quarks (represented by the open circles in Fig. 15.1) away from c. In consequence, the static equilibrium must be readjusted anew, resulting in a modification of the density distribution $n_q(\mu)$ of the quarks and the density $n_{\bar{q}}(\mu)$

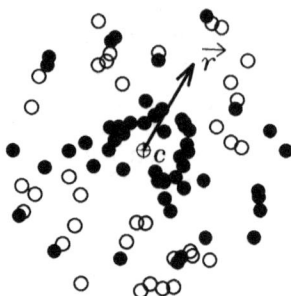

Fig. 15.1   A quark c is placed at $r = 0$ in the quark-gluon plasma. It polarizes the surrounding medium as it attracts antiquarks (represented by solid circles) and repels quarks (represented by open circles).

of the antiquarks, which are now characterized by a spatially-varying chemical potential $\mu(r)$. There will be more antiquarks than quarks around c. The presence of this cloud of excess antiquarks surrounding the quark c will lessen the magnitude of the net color charge seen by a test quark and will modify the potential.

We shall show in Supplement 15.1 that, for the idealized case of the plasma as a massless quark and antiquark gas, the potential $V(r)$ as seen by a test quark at r will be modified from the long-range Coulomb potential (Eq. (15.1)) to the short-range Yukawa potential

$$V(r) = \frac{q}{4\pi} \frac{e^{-r/\lambda_D}}{r}, \qquad (15.5)$$

where the Debye screening length λ_D is given, in the Abelian case, by

$$\lambda_D(\text{Abelian}) = \sqrt{\frac{6}{g_q\, q^2}\, \frac{1}{T}}$$

$$= \sqrt{\frac{9 \times 1.202\, T}{\pi^2 q^2 (n_q + n_{\bar{q}})}}, \qquad (15.6)$$

and $g_q (= 6N_f)$ is the degeneracy of the quark gas (see Eq. (9.5)). It is convenient to introduce the Debye screening mass m_D as the inverse of the Debye screening length,

$$m_D = \frac{1}{\lambda_D}. \qquad (15.7)$$

The results of Eqs. (15.5) and (15.6) are obtained in the Abelian approximation in which the gauge bosons do not carry color charges and there is no interaction between gauge bosons. In the proper description of QCD as a non-Abelian gauge theory, the screening phenomenon involves not only the polarization of quarks and antiquarks in the vicinity of the quark c, but also the polarization of gluons which also carry color charge. The polarization of these gluons leads to a change of the gluon density and additional screening of the quark c. Within the one-loop calculations in perturbative QCD (PQCD), the Debye screening mass m_D and the screening length have been calculated. They are found to be given by

$$\lambda_D(\text{PQCD}) = \frac{1}{m_D(\text{PQCD})} = \frac{1}{\sqrt{\left(\frac{N_c}{3} + \frac{N_f}{6}\right)g^2}\,T}, \qquad (15.8)$$

(see Appendix C of Ref. [13]). Thus, the Debye screening length is a function of the plasma temperature and the coupling constant.

The Debye screening length has been calculated using the lattice gauge theory [8,9, 14-19]. It is found that the Debye screening length obtained in the lattice gauge theory, λ_D(lattice gauge), is about a factor of 2 shorter than the screening length calculated from the lowest-order perturbative QCD theory (see Fig. 7 of Ref. [19]).

⊕[Supplement 15.1

Here, we would like to find within the Abelian approximation the screened color potential $V(\boldsymbol{r})$ arising from a quark c placed at the origin and show that it is given by the Yukawa potential Eq. (15.5) with λ_D of Eq. (15.6).

In the Abelian case, the screened color potential $V(\boldsymbol{r})$ at $\boldsymbol{r}$ arises from three contributions: (1) the potential $V_0(\boldsymbol{r})$ due the quark c placed at the origin $\boldsymbol{r} = 0$, (2) the potential $V_q(\boldsymbol{r})$ due to the quarks of the plasma, and (3) the potential $V_{\bar{q}}(\boldsymbol{r})$ due to the antiquarks of the plasma,

$$V(\boldsymbol{r}) = V_0(\boldsymbol{r}) + V_q(\boldsymbol{r}) + V_{\bar{q}}(\boldsymbol{r})\,. \qquad (1)$$

A quark at $\boldsymbol{r}$ will experience a force $-q\nabla V(\boldsymbol{r})$ and an antiquark a force $-(-q)\nabla V(\boldsymbol{r})$.

We consider a fluid element at $\boldsymbol{r}$. The fluid element contains quarks of density n_q and antiquarks of density $n_{\bar{q}}$, and is subject to color forces arising from the potential $V(\boldsymbol{r})$ at $\boldsymbol{r}$. The fluid element will experience a force per unit volume $-qn_q\nabla V(\boldsymbol{r})$ acting on its quarks and a force per unit volume $-(-q)n_{\bar{q}}\nabla V(\boldsymbol{r})$ acting on its antiquarks. In addition, this fluid element will be subject to a force per unit volume ∇p arising from the spatial re-distribution of the quarks and antiquarks due to the presence of the quark c.

The condition for the fluid element at $\boldsymbol{r}$ to be in equilibrium is that the total force (per unit volume) acting on this fluid element is zero,

$$\nabla p(n_q(\mu), n_{\bar{q}}(\mu)) - qn_q(\mu)\nabla V - (-q)n_{\bar{q}}(\mu)\nabla V = 0\,. \qquad (2)$$

For simplicity, we consider quarks to be highly relativistic so that the rest mass of quarks and antiquarks can be neglected. Then, the pressure is related to the energy density ϵ by [20]

$$p(\mu) = \frac{1}{3}\epsilon(\mu) = \frac{1}{3}\left[\epsilon_q(\mu) + \epsilon_{\bar{q}}(\mu)\right].$$ (3)

We shall use the equilibrium condition (2) to obtain a relation between the chemical potential $\mu(\mathbf{r})$ and the potential $V(\mathbf{r})$. In order to do that, it is useful to rewrite the number densities and the energy densities of the quarks and the antiquarks of the plasma in terms of chemical potential μ and temperature T explicitly. (For brevity of notation, we shall not exhibit the dependence on T explicitly.) Because the equilibrium densities are characterized by $\mu = 0$, we expand the density quantity up to the first power in μ. We have

$$
\begin{aligned}
n_q(\mu) &= \frac{g_q}{2\pi^2}\int_0^\infty \frac{p_0^2 dp_0}{1 + e^{(p_0-\mu)/T}} \\
&= \frac{g_q T^3}{2\pi^2}\int_0^\infty \frac{z^2 dz}{1 + e^{z-(\mu/T)}} \\
&= \frac{g_q T^3}{2\pi^2}\int_0^\infty z^2 dz\left[\frac{1}{1+e^z} - \frac{\mu}{T}\frac{d}{dz}\frac{1}{1+e^z}\right] + \cdots \\
&= \frac{g_q T^3}{2\pi^2}\int_0^\infty dz\left[\frac{z^2}{1+e^z} + \frac{\mu}{T}\frac{2z}{1+e^z}\right] + \cdots .
\end{aligned}
$$ (4)

We can prove the result that (see [20] and Section 9.3.1)

$$\int_0^\infty dz\,\frac{z^{x-1}}{1+e^z} = (1 - 2^{1-x})\Gamma(x)\zeta(x)\,,$$

where $\zeta(x)$ is the Riemann zeta function. We shall need the values of $\zeta(2)$, $\zeta(3)$, and $\zeta(4)$ which are given by [21]

$$\zeta(2) = \frac{\pi^2}{6}\,,$$

$$\zeta(3) = 1.202\,,$$

and

$$\zeta(4) = \frac{\pi^4}{90}\,.$$

From these results, we have the number density of quarks in the plasma

$$n_q(\mu) = \frac{g_q T^3}{2\pi^2}\left[\frac{3}{2}\zeta(3) + \frac{\mu}{T}\frac{\pi^2}{6}\right].$$ (5)

We can follow a similar procedure to obtain the energy density of the quarks. We have

$$\epsilon_q(\mu) = \frac{g_q}{2\pi^2} \int_0^\infty \frac{p_0^3 dp_0}{1 + e^{(p_0 - \mu)/T}}$$

$$= \frac{g_q T^4}{2\pi^2} \int_0^\infty \frac{z^3 dz}{1 + e^{z - (\mu/T)}}$$

$$= \frac{g_q T^4}{2\pi^2} \int_0^\infty z^3 dz \left[\frac{1}{1 + e^z} - \frac{\mu}{T} \frac{d}{dz} \frac{1}{1 + e^z} + \frac{1}{2} \left(\frac{\mu}{T} \right)^2 \frac{d^2}{dz^2} \frac{1}{1 + e^z} \right] + \dots$$

$$= \frac{g_q T^4}{2\pi^2} \int_0^\infty dz \left[\frac{z^3}{1 + e^z} + \frac{\mu}{T} \frac{3z^2}{1 + e^z} + \frac{1}{2} \left(\frac{\mu}{T} \right)^2 \frac{6z}{1 + e^z} \right] + \dots .$$

The energy density of the quarks in the plasma, up to second order in the chemical potential, is

$$\epsilon_q(\mu) = \frac{g_q T^4}{2\pi^2} \left[\frac{7}{4} \frac{\pi^4}{30} + \frac{\mu}{T} \frac{9}{2} \zeta(3) + \frac{1}{2} \left(\frac{\mu}{T} \right)^2 \frac{\pi^2}{2} \right]. \tag{6}$$

Given the chemical potential, we can obtain the density of the antiquarks. The presence of antiquarks corresponds to the absence of quarks in negative energy states. The number density of antiquarks is therefore

$$n_{\bar{q}}(\mu) = \frac{g_q}{2\pi^2} \int_{-\infty}^0 p_0^2 dp_0 \left[1 - \frac{1}{1 + e^{(p_0 - \mu)/T}} \right]$$

$$= \frac{g_q}{2\pi^2} \int_{-\infty}^0 p_0^2 dp_0 \frac{e^{(p_0 - \mu)/T}}{1 + e^{(p_0 - \mu)/T}}$$

$$= \frac{g_q}{2\pi^2} \int_{-\infty}^0 p_0^2 dp_0 \frac{1}{1 + e^{-(p_0 - \mu)/T}} .$$

By making a transformation $p_0 = -\bar{p}_0$, with $\bar{p}_0 \geq 0$, we have

$$n_{\bar{q}}(\mu) = \frac{g_q}{2\pi^2} \int_0^\infty \bar{p}_0^2 d\bar{p}_0 \frac{1}{1 + e^{(\bar{p}_0 + \mu)/T}} . \tag{7}$$

Comparing Eq. (7) with Eqs. (4) and (5), we get the number density of the antiquarks given by

$$n_{\bar{q}}(\mu) = \frac{g_q T^3}{2\pi^2} \left[\frac{3}{2} \zeta(3) - \frac{\mu}{T} \frac{\pi^2}{6} \right]. \tag{8}$$

Similarly, the energy density of the antiquarks can be found to be

$$\epsilon_{\bar{q}}(\mu) = \frac{g_q}{2\pi^2} \int_0^\infty \frac{\bar{p}_0^3 d\bar{p}_0}{1 + e^{(\bar{p}_0 + \mu)/T}}$$

$$= \frac{g_q T^4}{2\pi^2} \left[\frac{7}{4} \frac{\pi^4}{30} - \frac{\mu}{T} \frac{9}{2} \zeta(3) + \frac{1}{2} \left(\frac{\mu}{T} \right)^2 \frac{\pi^2}{2} \right]. \tag{9}$$

Having obtained all the relevant quantities explicitly as functions of the chemical potential μ, we can find the equation for $\mu(r)$. We expand equation (3) with respect to $\mu = 0$, the term linear in μ vanishes and we have

$$p(\mu) = \frac{1}{3} \left[\epsilon(\mu = 0) + \frac{1}{2} \mu^2 \frac{\partial^2 \epsilon}{\partial \mu^2} \right],$$

where $\partial^2\epsilon/\partial\mu^2$ is evaluated at $\mu = 0$ and is given from Eqs. (6) and (9) by

$$\frac{\partial^2\epsilon}{\partial\mu^2} = \frac{g_qT^4}{2\pi^2}\frac{\pi^2}{T^2}.$$

(10)

Therefore, Eq. (2) becomes

$$\nabla\frac{1}{6}\mu^2\frac{\partial^2\epsilon}{\partial\mu^2} - q\left[n_q(\mu) - n_{\bar{q}}(\mu)\right]\nabla V = 0.$$

(11)

The difference $n_q(\mu) - n_{\bar{q}}(\mu)$ can be expressed as a function of the chemical potential μ. We have, from Eqs. (5) and (8),

$$n_q(\mu) - n_{\bar{q}}(\mu) = n_q(\mu = 0) + \mu\frac{\partial n_q}{\partial\mu} - n_{\bar{q}}(\mu = 0) - \mu\frac{\partial n_{\bar{q}}}{\partial\mu}$$

$$= 2\mu\frac{\partial n_q}{\partial\mu},$$

(12)

where $\partial n_q/\partial\mu$ is evaluated at $\mu = 0$ and is given from Eq. (5) by

$$\frac{\partial n_q}{\partial\mu} = \frac{g_qT^3}{2\pi^2}\frac{\pi^2}{6T}.$$

(13)

In Eq. (12), we have used the relations $n_q = n_{\bar{q}}$ and $\partial n_{\bar{q}}/\partial\mu = -\partial n_q/\partial\mu$ at $\mu = 0$. The equilibrium condition (11) becomes

$$\frac{1}{3}\frac{\partial^2\epsilon}{\partial\mu^2}\mu\nabla\mu - 2q\mu\frac{\partial n_q}{\partial\mu}\nabla V = 0.$$

This condition is satisfied if

$$\mu\frac{\partial^2\epsilon}{\partial\mu^2} - 6q\frac{\partial n_q}{\partial\mu}V = (\text{ a constant of } r).$$

(14)

At $r \to \infty$, μ and V approaches zero. The constant in Eq. (14) is zero.

Eq. (14) is a relation between the chemical potential μ and the potential V. The chemical potential is related to the number densities which are connected to the potentials V_q and $V_{\bar{q}}$ through the Poisson equation. Therefore, we can rewrite Eq. (14) as an equation for the potential V only. From Eq. (12), we express μ as a function of the densities,

$$\mu = \frac{n_q(\mu) - n_{\bar{q}}(\mu)}{2\frac{\partial n_q}{\partial\mu}},$$

From the Poisson equation, the density $n_q(\mu)$ of quarks and the potential V_q generated by these quarks are related by

$$\nabla^2V_q = -qn_q(\mu).$$

The density $n_{\bar{q}}(\mu)$ of antiquarks and the potential $V_{\bar{q}}$ generated by the antiquarks are related by
$$\nabla^2 V_{\bar{q}} = -(-q)n_{\bar{q}}(\mu) \,.$$
Therefore, the chemical potential is related to V_q and $V_{\bar{q}}$ by
$$\mu = \frac{-\nabla^2 V_q - \nabla^2 V_{\bar{q}}}{2q\frac{\partial n_q}{\partial \mu}} \,.$$
Substituting this into Eq. (14), we get
$$\nabla^2 V_q + \nabla^2 V_{\bar{q}} + 12q^2 \left(\frac{\partial n_q}{\partial \mu}\right)^2 \left(\frac{\partial^2 \epsilon}{\partial \mu^2}\right)^{-1} V = 0 \,,$$
or
$$\nabla^2 (V_q + V_{\bar{q}}) + m_D^2 V = 0 \,, \tag{15}$$
where m_D is the Debye screening mass defined by
$$m_D^2 = 12q^2 \left(\frac{\partial n_q}{\partial \mu}\right)^2 \left(\frac{\partial^2 \epsilon}{\partial \mu^2}\right)^{-1}$$
$$= \frac{g_q q^2 T^2}{6} \tag{16}$$
$$= \frac{\pi^2}{9 \times 1.202} \frac{q^2 (n_q + n_{\bar{q}})}{T} \,.$$
Using Eq. (1), we can rewrite Eq. (15) as
$$\nabla^2 (V - V_0) + m_D^2 V = 0 \,. \tag{17}$$
The potential $V_0(\boldsymbol{r})$ satisfies the Poisson equation for a point source,
$$\nabla^2 V_0 = -q\delta(\boldsymbol{r}) \,. \tag{18}$$
Therefore, Eqs. (17) and (18) lead to the equation for V given by
$$\nabla^2 V + m_D^2 V = -q\delta(\boldsymbol{r}) \,. \tag{19}$$
This equation gives the well-known Yukawa potential
$$V(r) = \frac{q}{4\pi} \frac{e^{-m_D r}}{r}$$
$$= \frac{q}{4\pi} \frac{e^{-r/\lambda_D}}{r} \,,$$
where the Debye screening length λ_D is the inverse of the Debye screening mass,
$$\lambda_D{}^2 = 1/m_D^2$$
$$= \frac{6}{g_q} \frac{1}{q^2 T^2}$$
$$= \frac{9 \times 1.202 T}{\pi^2 q^2 (n_q + n_{\bar{q}})} \,.$$

§15.2 J/ψ Suppression in the Quark-Gluon Plasma

In the quark-gluon plasma, the string tension is zero. The only interaction between a charm quark and a charm antiquark is the Coulomb-type color interaction. From the last section, we learn that if we place a J/ψ particle in the quark-gluon plasma, the color charge of the charm quark c will be screened by the quarks, antiquarks and the gluons of the plasma. The effect of the Debye screening will be to modify the long-range Coulomb-type color interaction between c and $\bar{c}$ to turn it into a short-range Yukawa-type interaction, with the range given by the Debye screening length λ_D. Within this distance λ_D, the attractive interaction between the constituents c and $\bar{c}$ is effective but beyond this range the attractive interaction is ineffective, as the magnitude of the interaction diminishes exponentially with distance.

The Debye screening length λ_D is inversely proportional to the temperature. At high temperatures, the range of the attractive interaction becomes so small as to make it impossible for the $c\bar{c}$ pair to form a bound state. When this happens, the $c\bar{c}$ system dissociates into a separate c quark and a $\bar{c}$ antiquark in the plasma. The c quark and $\bar{c}$ antiquark subsequently hadronize by combining with light quarks or light antiquarks to emerge as 'open charm' mesons such as $D(c\bar{u}$, and $c\bar{d})$, $\bar{D}(\bar{c}u$, and $\bar{c}d)$, $D_s(c\bar{s})$, and $\bar{D}_s(\bar{c}s)$.

In nucleus-nucleus collisions, J/ψ particles are produced in the initial stage of the collision process, for example, by the hard-scattering processes. If a quark-gluon plasma is formed in the region of J/ψ production, then the effect of the plasma will be to make the J/ψ particle unbound, and the final yield of J/ψ particles will be suppressed as compared to the case when there is no quark-gluon plasma. Therefore, the suppression of J/ψ production may be used as a signature for the presence of the quark-gluon plasma [1].

We shall estimate in the nonrelativistic approximation the critical temperature T_c at which states formed by a $q\bar{q}$ pair in a quark-gluon plasma cannot become bound. The Hamiltonian for a $q\bar{q}$ system in the quark-gluon plasma is

$$H = \frac{p^2}{2\mu} - \frac{\alpha_{\text{eff}} e^{-r/\lambda_D}}{r}. \qquad (15.9)$$

The stability of the system can be investigated semiclassically. From the uncertainty relations, we have $<p^2> \sim 1/r^2$ and the energy of the $q\bar{q}$ system is

$$E(r) = \frac{1}{2\mu r^2} - \frac{\alpha_{\text{eff}} e^{-r/\lambda_D}}{r}.$$

A bound state is possible if $E(r)$ has a minimum with respect to r. The condition for an extremum of $E(r)$ is

$$-\frac{1}{\mu r^3} + \frac{\alpha_{\text{eff}}(1 + r/\lambda_D)e^{-r/\lambda_D}}{r^2} = 0, \qquad (15.10)$$

or

$$f(x) = x(1+x)e^{-x} = \frac{1}{\alpha_{\text{eff}}\mu\lambda_D}, \qquad (15.11)$$

where $x = r/\lambda_D$. The function $f(x)$ is zero at $x = 0$. It increases to a maximum value $f(x)|_{\text{max}} = 0.840$ at $x = 1.62$ and decreases monotonically thereafter, approaching zero as x goes to ∞. Therefore, Eq. (15.11) has solutions only when the right-hand-side of (15.11), $(1/\alpha_{\text{eff}}\mu\lambda_D)$, is less than 0.840. There will be no solutions of (15.11) if $(1/\alpha_{\text{eff}}\mu\lambda_D) > 0.840$. In other words,

$$q\bar{q} \text{ will not be bound if} \qquad \frac{1}{0.84\,\alpha_{\text{eff}}\mu} > \lambda_D. \qquad (15.12)$$

The above equation can be given an intuitive physical interpretation. If there is no screening, the radius of the $q\bar{q}$ system, which can be called the Bohr radius, can be obtained from Eq. (15.10) by setting the screening length λ_D to infinity. The Bohr radius r_{Bohr} of the $q\bar{q}$ system is therefore

$$r_{\text{Bohr}} = \frac{1}{\alpha_{\text{eff}}\mu}. \qquad (15.13)$$

Thus, Eq. (15.12) states that the $q\bar{q}$ system will not be bound if 1.19 times the Bohr radius r_{Bohr} of the $q\bar{q}$ system is greater than the Debye screening length λ_D,

$$q\bar{q} \text{ will not be bound if} \qquad 1.19\,r_{\text{Bohr}} > \lambda_D. \qquad (15.14)$$

By comparing the Bohr radius with the Debye screening length, we can infer whether or not a $q\bar{q}$ system can be bound in a quark-gluon plasma.

The Debye screening length depends on the temperature T. For the quark-gluon plasma with a flavor number $N_f = 3$, we use the result (15.8) from the lowest-order perturbative QCD theory to obtain

$$\lambda_D(\text{PQCD}) = \sqrt{\frac{2}{3g^2}}\frac{1}{T}. \qquad (15.15)$$

If we use an effective coupling constant $\alpha_{\text{eff}} = 0.52$ (as given for an isolated $c\bar{c}$ system not in the plasma [2]), then the Debye screening length at a temperature of 200 MeV is

$$\lambda_D(\text{PQCD}) = 0.36 \text{ fm} . \tag{15.16}$$

As we remarked earlier, the Debye length obtained from the lattice gauge theory is shorter than the Debye length obtained from the lattice gauge theory by a factor of 2. We estimate that λ_D (lattice gauge) ~ 0.18 fm.

For a $c\bar{c}$ system, the reduced mass is $\mu = 1840$ MeV/2 [3] and the Bohr radius for $\alpha_{\text{eff}} = 0.52$ is

$$r_{\text{Bohr}}(c\bar{c}) = 0.41 \text{ fm} ,$$

which indicates from Eq. (15.14) that a $c\bar{c}$ system cannot be a bound system in a quark-gluon plasma at $T = 200$ MeV.

From Eqs. (15.8) and (15.12) the critical dissociation temperature T_c, above which a $q\bar{q}$ system cannot be a bound system, is given for the lowest-order perturbative QCD theory by

$$T_c = \frac{\mu}{0.840} \sqrt{\frac{2\alpha_{\text{eff}}}{9\pi}} . \tag{15.17}$$

Using the quark mass of $m_c = 1.84$ GeV for the $c\bar{c}$ system [3], we have

$$T_c = 0.291\sqrt{\alpha_{\text{eff}}} \text{ GeV} .$$

If α_{eff} has the value of 0.52 [2], then $T_c = 209$ MeV. However, in the quark-gluon plasma, the QCD coupling constant decreases with temperature. At 1.5 times the critical temperature, α_{eff} is found to be about 0.2 [15] for which the dissociation temperature T_c is 130 MeV. Thus, the temperature above which the $c\bar{c}$ system cannot be bound is of the order of 100-200 MeV.

In summary, it is expected that a quark-gluon plasma with a temperature exceeding the dissociation temperature T_c will provide an environment where a $c\bar{c}$ system will not be bound. Composite $c\bar{c}$ bound states such as J/ψ particles produced initially by hard nucleon-nucleon collisions will become unbound in a high-temperature quark-gluon plasma. Consequently, the production of J/ψ particles will be suppressed as compared to the case when there is no quark-gluon plasma. As was first suggested by Matsui and Satz [1], the suppression of J/ψ production may be used as a signature for the presence of the quark-gluon plasma.

It has been suggested that near the critical transition temperature, there are other manifestations of the effects of the medium on
the properties and the abundance of hadrons such as J/ψ [11], ψ''
[22], scalar meson [23], kaons [24,25], ϕ [26], ρ, ω, and ϕ meson [27].
Interested readers may consult the relevant references for detailed
discussions.

⊕[**Supplement 15.2**
  The investigations on the dissociation of a $c\bar{c}$ system in a quark-gluon plasma
can be extended to quark systems with other flavors. It is instructive to study the
effect of the quark-gluon plasma on $s\bar{s}$ and light $q\bar{q}$ systems.
  From Eq. (15.16), the Debye screening length λ_D at $T = 200$ MeV is 0.36
fm according to the lowest-order perturbative QCD theory, and about half this
length according to the lattice gauge theory. For $q\bar{q}$ systems with u, d, $\bar{u}$ and $\bar{d}$ as
constituents, the values of the Bohr radii are substantially greater than λ_D. Systems
with these constituents cannot be bound in a quark-gluon plasma at $T = 200$ MeV.
  For an $s\bar{s}$ system, the reduced mass is $\mu = 199$ MeV/2 and the Bohr radius for
$\alpha_{\text{eff}} = 0.52$ is

$$r_{\text{Bohr}}(s\bar{s}) = 3.8 \text{ fm}.$$

We have $r_{\text{Bohr}}(s\bar{s}) \gg \lambda_D$, which indicates that an $s\bar{s}$ system cannot be a bound
system in a quark-gluon plasma at $T = 200$ MeV. As a consequence, systems of u, d
and s quarks cannot form bound states with antiquarks $\bar{u}, \bar{d}$ and $\bar{s}$ in a quark-gluon
plasma.   ]⊕

§15.3 Experimental Information on J/ψ Production and J/ψ Suppression

Experimental investigations to study J/ψ production with high-
energy heavy-ion beams have been carried out by the NA38 Collaboration at CERN [28]. The initial finding of a J/ψ suppression, relative
to the continuum background, provided much initial excitement. The
experimental results have been reviewed by many authors [29,30]. The
spectra of $\mu^+\mu^-$ pairs for the collision of ^{32}S on U at 200A GeV are
displayed in Fig. 15.2. The J/ψ resonance shows up as a peak of
$\mu^+\mu^-$ pair mass at $M_{\mu\mu} \sim 3.1$ GeV. The width of the resonance arises
mainly from instrumental resolution which is much greater than the
intrinsic width of the J/ψ particle. In this measurement, the muon
spectrometer is designed to accept $\mu^+\mu^-$ pairs in the rapidity range
$2.8 < y_{\text{lab}} < 4$. The transverse energy E_T^o deposited in a calorimeter
by the associated neutral pions in the central pseudorapidity range
$1.7 < \eta < 4.1$ is also measured. In Fig. 15.2, data points for events
with a small value of E_T^o ($E_T^o < 51$ GeV) are shown as the square points

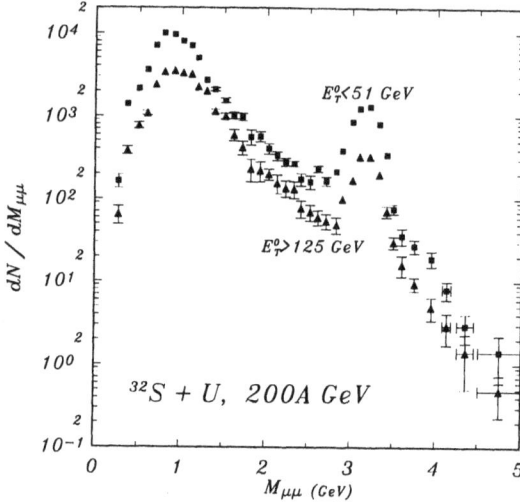

Fig. 15.2 The cross spectrum of $\mu^+\mu^-$ pairs produced in the collision of ^{32}S on U at an energy of 200 GeV per projectile nucleon. Data are from the NA38 Collaboration [30].

and the data points for events with a large value of E_T^o ($E_T^o > 125$ GeV) are shown as the triangular points. As the transverse energy E_T^o increases, the magnitude of the J/ψ peak relative to the magnitude of the background continuum under the J/ψ peak decreases. One can obtain the number of J/ψ particles in the resonance peak $N_{J/\psi}$ and the number of $\mu^+\mu^-$ pairs in the continuum N_{cont} under the resonance peak. One finds that the ratio $N_{J/\psi}/N_{\text{cont}}$ decreases with the magnitude of the transverse energy.

Experimental information on the relation between the ratio $N_{J/\psi}/N_{\text{cont}}$ and the average transverse energy of neutral particle $< E_T^o >$ is shown in Fig. 15.3. The quantity on the abscissa is

$$\frac{3 < E_T^0 >}{\Delta\eta \, \tau_0 \mathcal{A}} ,$$

where $\Delta\eta = 2.4$ is the range of pseudorapidity in the measurement, τ_0 is the initial time taken to be 1 fm/c, and $\mathcal{A}$ is the overlapping area of the colliding nuclei. This quantity can be labelled by ϵ and considered as a measure of the energy density of the created matter at the proper time τ_0, if one uses Bjorken's formula (13.5) to relate

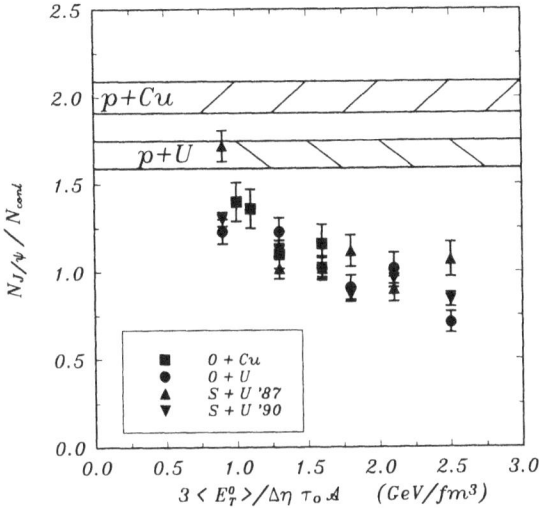

Fig. 15.3 The ratio of the number of J/ψ particles to the number of $\mu^+\mu^-$ pairs in the continuum, in the collision of ^{32}S on U at an energy of 200 GeV per projectile nucleon. Data are from the NA38 Collaboration [29].

the transverse energy to the initial energy density. Fig. 15.3 shows that the ratio $N_{J/\psi}/N_{cont}$ decreases with neutral transverse energy $<E_T^o>$. If a quark-gluon plasma is formed, because a higher initial energy density is associated with a greater initial temperature and a greater probability to form a quark-gluon plasma, the suppression of J/ψ production by the quark-gluon plasma is expected to lead to a ratio $N_{J/\psi}/N_{cont}$ which decreases with the initial energy density. It might appear that the results of Fig. 15.3 lend support to the presence of the quark-gluon plasma. However, to reach such a conclusion, it is necessary to rule out other sources of J/ψ suppression which are not the results of quark-gluon plasma formation. Suppression due to J/ψ-hadron interactions has been observed, and the experimental data can be explained in terms of the breakup of the produced J/ψ particles by collisions with hadrons, without invoking the presence of the quark-gluon plasma [32-38]. Unequivocal evidence for the presence of quark-gluon plasma must await further experimental tests and perhaps a different environment of production (greater energies and/or greater projectile-target masses).

§15.4 J/ψ Suppression in Hadron Environment

In a nucleus-nucleus collision, J/ψ particles are produced by hard-scattering processes in some of the many nucleon-nucleon collisions. The produced J/ψ particles may interact with hadrons and these J/ψ-hadron interactions may lead to the breakup of the J/ψ particles . For example, a J/ψ particle can interact with a hadron h via the reaction

$$J/\psi + h \rightarrow D + \bar{D} + X\,, \qquad (15.18)$$

which turns a J/ψ particle into a $D\bar{D}$ pair. Thus, J/ψ-hadron interactions will give rise to a suppression of J/ψ production.

We shall first discuss the cross section for producing J/ψ particles when there are no J/ψ-hadron interactions. In the collision of nucleus A and nucleus B, the theoretical considerations to obtain the cross section for J/ψ production are the same as those used to study the cross section for Drell-Yan processes in nucleus-nucleus collisions (see Section 14.2.2). The probability element for J/ψ production is

$$dP = \left(\rho_A(\boldsymbol{b}_A, z_A) db_A dz_A \right) \left(\rho_B(\boldsymbol{b}_B, z_B) db_B dz_B \right) \left(t(\boldsymbol{b} - \boldsymbol{b}_A - \boldsymbol{b}_B) \sigma_{J/\psi}^{NN} \right),$$

$$(15.19)$$

where t is the nucleon-nucleon thickness function and $\sigma_{J/\psi}^{NN}$ is the cross section for the production of a J/ψ particle in a nucleon-nucleon collision. The first two factors in Eq. (15.14) are the probability elements of picking a nucleon in A and in B respectively and the last factor $t\sigma_{J/\psi}^{NN}$ is the probability for the production of a J/ψ particle in a nucleon-nucleon collision. When one integrates Eq. (15.14) over the volume elements of nuclei A and B, one finds that the probability for the production of a J/ψ particle in a nucleus-nucleus collision is $T(\boldsymbol{b})\sigma_{J/\psi}^{NN}$, where the thickness function $T(\boldsymbol{b})$ is

$$T(\boldsymbol{b}) = \int \rho_A(\boldsymbol{b}_A, z_A) db_A dz_A\, \rho_B(\boldsymbol{b}_B, z_B) db_B dz_B\, t(\boldsymbol{b} - \boldsymbol{b}_A - \boldsymbol{b}_B)\,,$$

(see also Eq. (12.4) of Chapter 12). We shall consider spherically symmetric nuclear distributions for which $T(\boldsymbol{b})$ depends only on the magnitude $|\boldsymbol{b}| = b$. The total probability for producing a J/ψ particle in the collision of A and B at an impact parameter $\boldsymbol{b}$ is the sum

$$P_{J/\psi}^{AB}(b) = \sum_{n=1}^{AB} \binom{AB}{n} \left[T(b)\sigma_{J/\psi}^{NN} \right]^n \left[1 - T(b)\sigma_{J/\psi}^{NN} \right]^{AB-n}\,. \qquad (15.20)$$

The J/ψ production cross section in a nucleon-nucleon collision, $\sigma_{J/\psi}^{NN}$, is of the order of 10^{-30} cm^2 (or 10^{-4} fm^2) [31]. On the other hand, $T(b)$ as given by Eq. (12.26) is of the order of $1/(100 \text{ fm}^2)$ for the head-on collision of ^{32}S on U. Therefore, the probability for a nucleon-nucleon collision leading to J/ψ production, $T(b)\sigma_{J/\psi}^{NN}$, is a small quantity and the summation given by Eq. (15.20) is dominated by the first term with $n = 1$. Terms with $n > 1$ represent multiple J/ψ production processes and shadowing corrections, which are very small and can be neglected. To a good approximation, the probability for J/ψ production in a nucleus-nucleus collision is

$$P_{J/\psi}^{AB}(b) = AB[T(b)\sigma_{J/\psi}^{NN}]. \tag{15.21}$$

Each J/ψ-producing nucleon-nucleon collision yields one J/ψ particle. The differential number of J/ψ particles produced with a rapidity y is given by

$$\frac{dN_{J/\psi}^{AB}}{dy}(b) = AB\, T(b)\frac{d\sigma_{J/\psi}^{NN}}{dy}.$$

Therefore, when we integrate over the transverse area, we have

$$\frac{d\sigma_{J/\psi}^{AB}}{dy} = \int db\, \frac{dN_{J/\psi}}{dy}(b)$$

$$= AB\frac{d\sigma_{J/\psi}^{NN}}{dy}\int d\mathbf{b}\, T(b). \tag{15.22}$$

In many experimental investigations, one measures the integrated J/ψ production cross section over a given interval of rapidity,

$$\Delta\sigma_{J/\psi}^{AB} = \int_{y_0}^{y_1} \frac{d\sigma_{J/\psi}^{AB}}{dy}dy\,,$$

and

$$\Delta\sigma_{J/\psi}^{NN} = \int_{y_0}^{y_1} \frac{d\sigma_{J/\psi}^{NN}}{dy}dy\,.$$

(For example, in the NA38 Collaboration, the range of the rapidity of J/ψ particles lies in the central rapidity region in the interval $2.8 < y_{\text{lab}} < 4$. In the measurement of the total cross sections, the interval of y includes the whole range of rapidity and $\Delta\sigma_{J/\psi}^{AB}$ is just $\sigma_{J/\psi}^{AB}$). Eq. (15.22) leads to

$$\frac{\Delta\sigma_{J/\psi}^{AB}}{AB\Delta\sigma_{J/\psi}^{NN}} = \int d\mathbf{b}\, T(b)\,. \qquad (15.23)$$

The right hand side of the above equation is unity, which follows from the normalizations in Eqs. (12.5), (12.2a), and (12.2b). Therefore, in the absence of J/ψ-hadron interactions to break up the produced J/ψ particles, the ratio of the cross sections is

$$\frac{\Delta\sigma_{J/\psi}^{AB}}{\Delta\sigma_{J/\psi}^{NN}} = AB\,. \qquad (15.24)$$

What is the ratio of the cross sections $\Delta\sigma_{J/\psi}^{AB}/\Delta\sigma_{J/\psi}^{NN}$ when the produced J/ψ particles interact with projectile and target nucleons via reactions of the type (15.18), leading to the breakup of J/ψ particles and the suppression of J/ψ production?

To study the effects due to J/ψ-nucleon interactions, we need to know the momentum distribution of the produced J/ψ particles. The experimental differential invariant cross section for the reaction $h + N \to J/\psi + X$ can be parametrized in the form [31]

$$E\frac{d\sigma}{dp^3} = A(s)(1 - |x_F|)^m e^{-bp_T}\,, \qquad (15.25)$$

where

$$A = 1 + 0.15(\sqrt{s/\text{GeV}} - 21)\ \mu\text{b}\ \text{GeV}^{-2}\,,$$
$$b = 2\ \text{GeV}^{-1}$$

and

$$m = \begin{cases} 2 & \text{for} \quad h = \pi \\ 4 & \text{for} \quad h = p\,. \end{cases}$$

Thus, the produced J/ψ particles are found most likely in the central rapidity region for which $|x_F| \sim 0$. The relation (15.22) suggests that in nucleus-nucleus collisions, the produced J/ψ particles are also likely found in the central rapidity region, which is the region of acceptance for the produced J/ψ particles in many experimental measurements, as in the NA38 Collaboration.

For a J/ψ particle to interact with a nucleon by the reaction

$$J/\psi + N \to D + \bar{D} + X\,, \qquad (15.26)$$

the threshold energy of the J/ψ particle is $E_{J/\psi} = 6.34$ GeV in the nucleon rest frame. Thus, in high-energy nuclear collisions when the beam rapidity y_B is large enough to satisfy

$$m_{J/\psi} \cosh(y_B/2) \geq 6.34 \text{ GeV},$$

or when the beam energy satisfies the condition

$$\text{(beam energy per nucleon)} \geq 6.92 \text{ GeV}, \tag{15.27}$$

J/ψ particles produced in the central rapidity region have sufficient energies to exceed the threshold energy for the occurrence of reaction (15.26), and they can be broken up as they interact with target nucleons along their paths. We shall consider high energy collisions in which the above condition (15.27) is satisfied and we label the cross section for the breakup of a J/ψ in a J/ψ-nucleon collision by σ_{abs}. A produced J/ψ particle behaves as if it is absorbed on its way through the target matter with a J/ψ-nucleon absorption cross section σ_{abs}.

Let us study first the case of the collision of a proton p on a target nucleus A. To include the effect of absorption due to the passage of the produced J/ψ particle through the target nucleus, we write out Eq. (15.23) for the case without absorption, in terms of the density function of the target nucleus explicitly. Using Eq. (12.4) and treating the beam particle p as a point particle, we can write Eq. (15.23) as

$$\frac{\Delta\sigma_{J/\psi}^{pA}}{A\Delta\sigma_{J/\psi}^{NN}} = \int d\boldsymbol{b} \int d\boldsymbol{b}_A \, dz_A \rho_A(\boldsymbol{b}_A, z_A) \, t(\boldsymbol{b} - \boldsymbol{b}_A).$$

After integrating over the transverse coordinate $\boldsymbol{b}$ and using the relation $\int d\boldsymbol{b} \, t(\boldsymbol{b} - \boldsymbol{b}_A) = 1$, we obtain

$$\frac{\Delta\sigma_{J/\psi}^{pA}}{A\Delta\sigma_{J/\psi}^{NN}} = \int d\boldsymbol{b}_A \, dz_A \rho_A(\boldsymbol{b}_A, z_A). \tag{15.28}$$

We can interpret the above equation as identifying $\rho_A(\boldsymbol{b}_A, z_A) d\boldsymbol{b}_A \, dz_A$ to be the differential contribution to the ratio $\Delta\sigma_{J/\psi}^{pA}/A\Delta\sigma_{J/\psi}^{NN}$ arising from a J/ψ-producing nucleon-nucleon collision at the point $(\boldsymbol{b}_A, z_A)$. After this nucleon-nucleon collision, a composite object, which is presumably a $c\bar{c}$ pair and the predecessor of the J/ψ particle, is produced. In a subsequent collision of the produced $c\bar{c}$ pair (or the J/ψ particle) with a nucleon along its path, the cross section for the breakup of the produced particle is σ_{abs} and the survival probability for the produced particle is

$$(\text{survival probability}) = e^{-(\text{number of nucleons along path per unit area})\sigma_{abs}}$$
$$= e^{-A \int_{z_A}^{\infty} dz'_A \rho_A (b_A, z'_A)\sigma_{abs}}. \qquad (15.29)$$

Therefore, in the presence of J/ψ-nucleon interactions leading to the absorption of J/ψ particles (or their predecessors), the result (15.23) or (15.28) for no absorption must be amended by including the survival probability into the integrand:

$$\frac{\Delta\sigma_{J/\psi}^{pA}}{A\Delta\sigma_{J/\psi}^{NN}} = \int db_A \int_{-\infty}^{\infty} dz_A \rho_A (b_A, z_A) e^{-A \int_{z_A}^{\infty} dz'_A \rho_A (b_A, z'_A)\sigma_{abs}}. \qquad (15.30)$$

The evolution from the composite $c\bar{c}$ object into a J/ψ particle may require a finite proper time interval. In other words, there may be a formation time for the produced $c\bar{c}$ pair to acquire the property of a J/ψ particle. The cross section σ_{abs} for the absorption of the produced object may depend on the distance from the point of production, being initially small and increasing to a constant value after the time of formation. At present, the time evolution of the $c\bar{c}$ system into a J/ψ and the time dependence of the $c\bar{c}$-N (or J/ψ-N) cross section have not been characterized experimentally. The mechanism of how a J/ψ particle is produced in a nucleon-nucleon collision remains the subject of current research [39]. The behavior of the produced particles will certainly depend on the production mechanism. Model calculations to study the effects of the J/ψ formation time have been presented in Refs. [32], and [35-38].

⊕[Supplement 15.3

In the theoretical models [20,21,22] of J/ψ production in nucleon-nucleon collisions, the production cross section is generally considered to factorize into two factors.

The first factor corresponds to the production of a heavy quark c and an antiquark $\bar{c}$ from the hard collision of the incident particles. The participating partons in the hard-scattering process can be a gluon of one nucleon interacting with a gluon of the other nucleon, as depicted in Diagrams 14.7b, 14.7c, and 14.7d. They can also be a quark from one nucleon interacting an antiquark from another nucleon with the formation of an intermediate virtual gluon, as represented by Diagram 14.7a. After the gluon fusion or the quark-antiquark annihilation, an unbound $c\bar{c}$ pair is produced and forms the predecessor of the J/ψ bound state. Because the probability to find a gluon in a nucleon is much greater than the probability to find an antiquark among the seaquarks, gluon fusion is probably the more important contributor to the hard-scattering process, as compared to the quark-antiquark annihilation process. This first factor, representing the production of an unbound $c\bar{c}$ pair from the hard-scattering process, is concerned with the short-distance part of the production process which may be computable in perturbative QCD.

The second factor corresponds to the formation of the bound state from the unbound $c\bar{c}$ pair which, at the present time, can only be specified by phenomenological models of J/ψ formation. There is the color-singlet model [40] where one starts with the full $c\bar{c}$ production amplitude from which the color-singlet $c\bar{c}$ state is obtained via the radiation of a hard gluon. This color-singlet $c\bar{c}$ state is then projected out to yield the amplitudes for various color-singlet bound states with the proper spin, parity, and charge conjugation quantum numbers. There is another model, the color evaporation model [41], in which the color quantum numbers of the unbound $c\bar{c}$ pair from the hard-scattering process need not be a color-singlet state but are left unspecified. The observed color-singlet bound states are assumed to be formed from the unbound $c\bar{c}$ pair by multiple soft gluon radiation (which carry color). When one uses this model for J/ψ production, one first calculates the cross section for the production of an unbound $c\bar{c}$ pair with an invariant mass ranging from the bound state mass $m_{J/\psi}$ up to the mass threshold $(m_D + m_{\bar{D}})$ at which a $c\bar{c}$ system can dissociate into two open charm particles, D and $\bar{D}$. A fraction f of these unbound $c\bar{c}$ states within this mass range is then assumed to form the color-singlet J/ψ particle by multiple soft gluon radiation. The fraction f cannot be specified by the model but is obtainable by comparison with experiment. Both models are capable of explaining experimental data, and processes to differentiate the models of J/ψ formation have been proposed [39]. $\mathbb{I}\oplus$

To provide a simplified but illuminating picture of J/ψ suppression in nucleon-nucleus and nucleus-nucleus collisions due to hadron interactions, we shall follow Gerschel and Hüfner [33-34] and consider the simple case where the J/ψ-nucleon cross section leading to the breakup of the J/ψ particle is a constant value σ_{abs} and the formation time effects are neglected.

There is another amendment of the survival probability which is needed for small target nuclei. The survival probability (15.29) is based on the path concept for a large nucleus. For a small nucleus, the finite number of nucleons in the target nucleus must be taken into account. We define a thickness function $T_{A>}(b_A, z_A)$ appropriate for a J/ψ particle produced at z_A,

$$T_{A>}(b_A, z_A) = \int_{z_A}^{\infty} dz'_A \rho_A(b_A, z'_A) . \qquad (15.31)$$

With this thickness function, the probability for the breakup of a J/ψ particle produced initially at (b_A, z_A) is

$$T_{A>}(b_A, z_A)\sigma_{\text{abs}} .$$

The survival probability, which is the probability that the produced J/ψ particle survives after interacting with the target nucleons, is

$$\text{(survival probability)} = \left[1 - T_{A>}(b_A, z_A)\sigma_{\text{abs}}\right]^{A-1}, \qquad (15.32)$$

where the exponential index $A - 1$ arises because the J/ψ-producing target nucleon is excluded from interacting again. To take into account the finite number of nucleons in a nucleus, we shall use the power-law-type survival probability (15.32) in place of the exponential survival probability (15.29). The survival probability (15.32) has the advantage that it can be used for light target nuclei (including $A{=}1$) as well as for heavy target nuclei. Instead of Eq. (15.30), the ratio of the cross sections is then given by

$$\frac{\Delta\sigma_{J/\psi}^{pA}}{A\Delta\sigma_{J/\psi}^{NN}} = \int db_A \int_{-\infty}^{\infty} dz_A \rho_A(b_A, z_A) \left[1 - T_{A>}(b_A, z_A)\sigma_{\text{abs}}\right]^{A-1}.$$

$$(15.33)$$

We note that

$$d\left(\left[1 - T_{A>}(b_A, z_A)\sigma_{\text{abs}}\right]^A\right) = d\left(\left[1 - \int_{z_A}^{\infty} dz'_A \rho_A(b_A, z'_A)\sigma_{\text{abs}}\right]^A\right)$$

$$= A[1 - T_{A>}(b_A, z_A)\sigma_{\text{abs}}]^{A-1}\rho_A(b_A, z_A)\, dz_A \sigma_{\text{abs}}.$$

Substituting the above into Eq. (15.33), the ratio $\Delta\sigma_{J/\psi}^{pA}/\Delta\sigma_{J/\psi}^{NN}$ becomes

$$\frac{\Delta\sigma_{J/\psi}^{pA}}{\Delta\sigma_{J/\psi}^{NN}} = \frac{1}{\sigma_{\text{abs}}} \int db_A \int_{z_A=-\infty}^{z_A=\infty} d\left(\left[1 - \int_{z_A}^{\infty} dz'_A \rho_A(b_A, z'_A)\sigma_{\text{abs}}\right]^A\right)$$

$$= \frac{1}{\sigma_{\text{abs}}} \int db_A \left(1 - \left[1 - \int_{-\infty}^{\infty} dz'_A \rho_A(b_A, z'_A)\sigma_{\text{abs}}\right]^A\right)$$

$$= \frac{1}{\sigma_{\text{abs}}} \int db_A \left(1 - \left[1 - T_A(b_A)\sigma_{\text{abs}}\right]^A\right). \qquad (15.34)$$

Experimental data indicate that

$$\Delta\sigma_{J/\psi}^{pA}/\Delta\sigma_{J/\psi}^{NN} \approx A^{\alpha}, \qquad (15.35)$$

with α close to, but slightly less than, unity [42]. This implies that the degree of absorption is small. It is reasonable to expand Eq. (15.34) in powers of σ_{abs} which gives

$$\frac{\Delta\sigma_{J/\psi}^{pA}}{\Delta\sigma_{J/\psi}^{NN}} \approx \frac{1}{\sigma_{abs}}\int db_A\left(1-\left[1-AT_A(b_A)\sigma_{abs}+\frac{A(A-1)}{2}(T_A(b_A)\sigma_{abs})^2\right]\right)$$

$$= \frac{1}{\sigma_{abs}}\int db_A\left[AT_A(b_A)\sigma_{abs}-\frac{A(A-1)}{2}(T_A(b_A)\sigma_{abs})^2\right]$$

$$= A[1-\frac{A-1}{2}\sigma_{abs}\int db_A(T_A(b_A))^2]$$

$$= A\left[1-\left(\frac{A-1}{A}\right)\left(\frac{2\pi r_0^3 A}{3}\int db_A(T_A(b_A))^2\right)\left(\frac{3}{4\pi r_0^3}\right)\sigma_{abs}\right].$$

It is useful to write the above equation as

$$\frac{\Delta\sigma_{J/\psi}^{pA}}{\Delta\sigma_{J/\psi}^{NN}} = A\left(1-L\rho_0\sigma_{abs}\right), \qquad (15.36)$$

or

$$\frac{\Delta\sigma_{J/\psi}^{pA}}{\Delta\sigma_{J/\psi}^{NN}} \approx Ae^{-L\rho_0\sigma_{abs}}, \qquad (15.37)$$

where the effective path length L is

$$L = \frac{2\pi}{3}R_A^3\int db_A\ (T_A(b_A))^2\ \frac{A-1}{A}, \qquad (15.38)$$

$$R_A = r_0 A^{1/3},$$

and $\rho_0 = 3/4\pi r_0^3 \sim 0.14$ fm^{-3}, the equilibrium nuclear matter density.

From Eq. (15.37), the ratio of the cross sections can be obtained as an explicit function of the mass number A when the thickness function is properly parametrized. Depending on the size of the nucleus, there are two different parametrizations of the thickness function. For a heavy nucleus with a uniform density, the thickness function is given by Eq. (12.29),

$$T_A(b_A) = \frac{3}{2\pi R_A^3}\sqrt{R_A^2-b_A^2}\ \theta(R_A-b_A),$$

and we obtain

$$\int db_A(T_A(b_A))^2 = \frac{9}{8\pi R_A^2},$$

and

$$L = \frac{3}{4} R_A \frac{A-1}{A} \quad (\text{ heavy nucleus }). \tag{15.39}$$

For a light target nucleus with a Gaussian density distribution, the thickness function is given by Eq. (12.26),

$$T_A(b_A) = \exp(-b_A^2/2\beta_A^2)/2\pi\beta_A^2 \,,$$

where the root-mean-squared radius of A is given by

$$R_A(\text{rms}) = \sqrt{3}\beta_A = r_0' A^{1/3} \,,$$

and $r_o' \sim 1.05$ fm for light nuclei [43]. With the Gaussian density distribution, the effective path length L for a light nucleus is

$$L = \frac{1}{2} R_A \frac{r_0^2}{(r_0')^2} \frac{A-1}{A} \quad (\text{ light nucleus }). \tag{15.40}$$

Equations (15.37) (or (15.36)), (15.39), and (15.40) provide an explicit relation between $\Delta\sigma_{J/\psi}^{pA}/\Delta\sigma_{J/\psi}^{NN}$ and the mass number A.

We can compare Eq. (15.36) with the experimental parametrization in the form of A^α in Eq. (15.35) [42]. The function A^α can be expanded in powers of $(1-\alpha)$ as

$$\frac{\Delta\sigma_{J/\psi}^{pA}}{\Delta\sigma_{J/\psi}^{NN}} \approx A[1 - (1-\alpha)\ln A]. \tag{15.41}$$

A comparison of Eq. (15.41) and (15.36) shows that

$$\alpha = 1 - \frac{L\rho_0\sigma_{\text{abs}}}{\ln A}. \tag{15.42}$$

For heavy target nuclei for which a uniform density distribution and the thickness function Eq. (12.29) are good descriptions, this gives

$$\alpha = 1 - \frac{9}{16\pi} \frac{A^{1/3}}{\ln A} \frac{\sigma_{\text{abs}}}{r_0^2} \frac{A-1}{A}. \tag{15.43}$$

In the range of $50 < A < 200$, we have $A^{1/3}(A-1)/A\ln A \approx 1$. The experimental parametrization A^α and Eq. (15.36) (or Eq. (15.37)) are approximately equivalent.

Using the parametrization (15.37) and (15.38) to analyze many sets of experimental p-A data, Gerschel and Hüfner [34] found that the experimental J/ψ production data for p-A collisions can be fitted well with an effective J/ψ-nucleon cross section leading to the breakup of the J/ψ particle given by

$$\sigma_{\text{abs}} = 6.2 \pm 0.3 \text{ mb}, \tag{15.44}$$

corresponding to an α value of about 0.92 for heavy nuclei.

The above results for a nucleon-nucleus collision can be easily generalized to the case for the collision of a beam nucleus B on a target nucleus A. Gerschel and Hüfner found [33,34]

$$\frac{\Delta\sigma^{AB}_{J/\psi}}{\Delta\sigma^{NN}_{J/\psi}} \approx ABe^{-L\rho_0\sigma_{\text{abs}}}, \tag{15.45}$$

where

$$L = L_A + L_B, \tag{15.46}$$

and the effective path length L_A and L_B are given by formula (15.38) (see Supplement 15.4).

We use Eqs. (15.45)-(15.46) to study the integrated cross section $\sigma^{AB}_{J/\psi}$ for J/ψ production in nucleon-nucleus and nucleus-nucleus collisions. We assume nuclei with $A \leq 32$ to have a Gaussian density distribution and nuclei with $A > 32$ to have a uniform density distribution. The effective path lengths L_A and L_B for these collisions can then be obtained from Eqs. (15.39) and (15.40). We plot $B\sigma^{AB}_{J/\psi}/AB$ as a function of the effective path length $L_A + L_B$ in Fig. 15.4, where B is the branching ratio for the decay of the J/ψ into a pair of muons. The open points are data from nucleon-nucleus collisions and the solid points are from nucleus-nucleus collisions. The experimental pA and AB data can be represented well by the relation

$$\frac{B\sigma^{AB}_{J/\psi}}{AB} = 4.1 \; e^{-\rho_0\sigma_{\text{abs}}(L_A+L_B)} \text{ mb}, \tag{15.47}$$

which is shown as the solid curve in Fig. 15.4, obtained with the parameter

$$\sigma_{\text{abs}} = 5.2 \text{ mb}, \tag{15.48}$$

and $\rho_0 = 0.14$ fm^{-3}.

The experimental data follow approximately the relationship of Eq. (15.47). The good agreement of the experimental data with Eq.

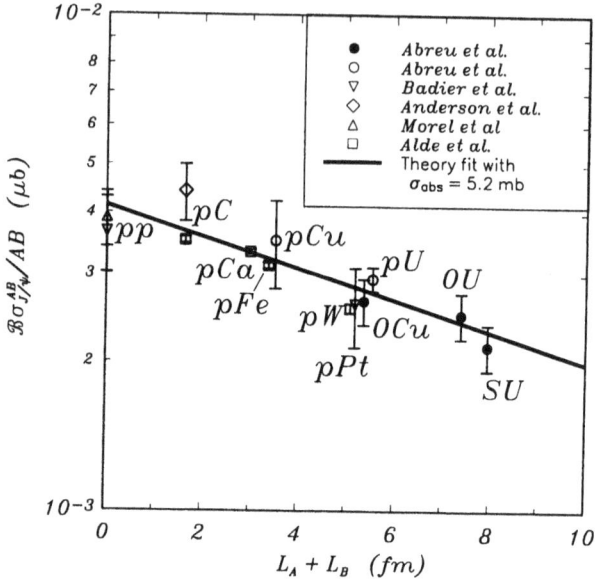

Fig. 15.4 The product of the branching ratio B and $\sigma_{J/\psi}^{AB}$, divided by AB, as a function of the effective path length $L_A + L_B$, which is calculated with Eqs. (15.46), (15.39), and (15.40). The data of Abreu *et al.* are from Ref. [29], Badier *et al.* from Ref. [44], Anderson *et al.* from Ref. [45], Morel *et al.* from Ref. [46], Alde *et al.* from Ref. [47].

(15.45) supports a phenomenological description of the absorption of produced J/ψ particles whose degree of absorption is dependent on their nuclear path lengths after they are produced, subject to ambiguities and uncertainties which we shall discuss below.

The phenomenological description of the experimental data indicates that J/ψ suppression arising from the interaction of the produced J/ψ particles with hadrons in nuclear collision environments is important. Therefore, these nuclear effects must be taken into account before one can extract the part of J/ψ suppression coming from the presence of the quark-gluon plasma.

⊕[Supplement 15.4

We show how to obtain the cross section (15.45) for J/ψ production with the inclusion of nuclear absorption effects, for the collision of a projectile nucleus B on a target nucleus A [33,34]. Again, we first write out Eq. (15.23) for the case without absorption, in terms of the density functions of the projectile and target nuclei explicitly. We have from Eqs. (15.23) and (12.4)

$$\frac{\Delta\sigma_{J/\psi}^{AB}}{AB\Delta\sigma_{J/\psi}^{NN}} = \int db \int db_A dz_A \rho_A(\boldsymbol{b}_A, z_A)\, db_B dz_B \rho_B(\boldsymbol{b}_B, z_B)\, t(\boldsymbol{b} - \boldsymbol{b}_A - \boldsymbol{b}_B). \quad (1)$$

Integrating out the transverse coordinate $\boldsymbol{b}$, we obtain

$$\frac{\Delta\sigma_{J/\psi}^{AB}}{AB\Delta\sigma_{J/\psi}^{NN}} = \int db_A dz_A \rho_A(\boldsymbol{b}_A, z_A)\, db_B dz_B \rho_B(\boldsymbol{b}_A, z_B). \quad (2)$$

We can interpret $\rho_A(\boldsymbol{b}_A, z_A)db_A\, dz_A\ \rho_B(\boldsymbol{b}_A, z_B)db_B\, dz_B$ in the above equation as specifying the differential contribution to the ratio $\Delta\sigma_{J/\psi}^{pA}/AB\Delta\sigma_{J/\psi}^{AB}$ arising from the production of a J/ψ particle in a J/ψ-producing nucleon-nucleon collision at the point $(\boldsymbol{b}_A, z_A)$ in nucleus A and the point $(\boldsymbol{b}_B, z_B)$ in nucleus B.

We shall consider J/ψ particles produced in the central rapidity region and neglect the effect of formation time. These J/ψ particles in the central rapidity region will travel forward through the target nucleus, starting from the longitudinal coordinate z_A. The survival probability associated with the path through the target nucleus is

$$\left[1 - T_{A>}(\boldsymbol{b}_A, z_A)\sigma_{\text{abs}}\right]^{A-1}.$$

The J/ψ particles produced in the central rapidity region will also be bombarded by projectile nucleons located at longitudinal coordinates from $-\infty$ to z_B in the projectile frame, having the rapidity of the projectile nucleus. For high energy collisions with a beam energy per nucleon greater than about 7 GeV [Eq. (15.27)], the projectile rapidity is much greater than the central rapidity value. The energy associated with the relative motion of the produced J/ψ particle with respect to the projectile nucleons is great enough to exceed the threshold energy for reactions of the type (15.18). The cross section for the breakup of the J/ψ particle in a collision of the projectile nucleon with the produced J/ψ particle is σ_{abs}. The survival probability of the J/ψ particles produced at z_B in the central rapidity region associated with the bombardment by the projectile nucleons is

$$\left[1 - T_{B<}(\boldsymbol{b}_B, z_B)\sigma_{\text{abs}}\right]^{B-1},$$

where

$$T_{B<}(\boldsymbol{b}_B, z_B) = \int_{-\infty}^{z_B} dz_B' \rho_B(\boldsymbol{b}_B, z_B').$$

Therefore, in the presence of J/ψ-nucleon interactions leading to the absorption of J/ψ particles (or their predecessors), the result of (2) for no absorption needs to be amended by including the survival probabilities into the integrand:

$$\frac{\Delta\sigma_{J/\psi}^{AB}}{AB\Delta\sigma_{J/\psi}^{NN}} = \int db_A \int_{-\infty}^{\infty} dz_A \rho_A(\boldsymbol{b}_A, z_A)[1 - T_{A>}(\boldsymbol{b}_A, z_A)\sigma_{\text{abs}}]^{A-1}$$

$$\times \int db_B \int_{-\infty}^{\infty} dz_B \rho_B(\boldsymbol{b}_B, z_B)[1 - T_{B<}(\boldsymbol{b}_B, z_B)\sigma_{\text{abs}}]^{B-1}.$$

Following the same steps as for the derivation of Eq. (15.37), we have

$$
\frac{\Delta\sigma_{J/\psi}^{AB}}{\Delta\sigma_{J/\psi}^{NN}} = \frac{1}{\sigma_{\text{abs}}^2} \int db_A \int_{z_A=-\infty}^{z_A=\infty} d\left[1 - \int_{z_A}^{\infty} dz'_A \rho_A(b_A, z'_A)\sigma_{\text{abs}}\right]^A
$$

$$
\times (-1) \int db_B \int_{z_B=-\infty}^{z_B=\infty} d\left[1 - \int_{-\infty}^{z_B} dz'_B \rho_B(b_B, z'_B)\sigma_{\text{abs}}\right]^B
$$

$$
= \frac{1}{\sigma_{\text{abs}}^2} \int db_A \left(1 - \left[1 - T_A(b_A)\sigma_{\text{abs}}\right]^A\right) \int db_B \left(1 - \left[1 - T_B(b_B)\sigma_{\text{abs}}\right]^B\right).
$$

Upon expanding the above in powers of σ_{abs}, we obtain

$$
\frac{\Delta\sigma_{J/\psi}^{AB}}{\Delta\sigma_{J/\psi}^{NN}} = \frac{1}{\sigma_{\text{abs}}^2} \int db_A \left[1 - \left(1 - AT_A(b_A)\sigma_{\text{abs}} + \frac{1}{2}A(A-1)(T_A(b_A)\sigma_{\text{abs}})^2\right)\right]
$$

$$
\times \int db_B \left[1 - \left(1 - BT_B(b_B)\sigma_{\text{abs}} + \frac{1}{2}B(B-1)(T_B(b_B)\sigma_{\text{abs}})^2\right)\right]
$$

$$
= AB\left[1 - \frac{1}{2}(A-1)\sigma_{\text{abs}}\int db_A(T_A(b_A))^2\right]\left[1 - \frac{1}{2}(B-1)\sigma_{\text{abs}}\int db_B(T_B(b_B))^2\right].
$$

It is useful to introduce effective path lengths L_A and L_B for the two colliding nuclei to write the above equation as

$$
\frac{\Delta\sigma_{J/\psi}^{AB}}{\Delta\sigma_{J/\psi}^{NN}} = AB(1 - L_A\rho_0\sigma_{\text{abs}})(1 - L_B\rho_0\sigma_{\text{abs}})
$$

$$
\approx AB[1 - (L_A + L_B)\rho_0\sigma_{\text{abs}}], \tag{3}
$$

where the effective path length L_A is

$$
L_A = \frac{2\pi}{3}R_A^3 \int db(T_A(b_A))^2\frac{A-1}{A}, \tag{4}
$$

and the path length L_B is similarly defined. Eq. (3) can be expressed approximately as

$$
\frac{\Delta\sigma_{J/\psi}^{AB}}{\Delta\sigma_{J/\psi}^{NN}} \approx ABe^{-L\rho_0\sigma_{\text{abs}}}, \tag{5}
$$

where

$$
L = L_A + L_B. \tag{6}
$$

Eqs. (4), (5), and (6) are the equations which had to be proved.

    The approximate agreement of the experimental data with Eq. (13.39) may suggest at first sight that the absorption of J/ψ particles arises solely from the collision of the produced J/ψ particles with the projectile or target nucleons.

There are however conceptual difficulties with this interpretation. One can deduce the total J/ψ-nucleon cross section by using virtual or real photons to produce J/ψ particles with the deep inelastic reaction

$$\mu + N \rightarrow \mu' + J/\psi + X \,,$$

or the reaction

$$\gamma + N \rightarrow J/\psi + X \,.$$

In the framework of the vector dominance model, one can relate the J/ψ photoproduction cross sections to the total cross section of J/ψ on a nucleon target. The results from the EMC Collaboration [48] give a total J/ψ-N cross section

$$\sigma_{\text{tot}}^{J/\psi-N} = 2.2 \pm 0.7 \ \text{mb} \,. \tag{15.49}$$

The J/ψ-N total cross section for the quasi-elastic process

$$J/\psi + N \rightarrow J/\psi + X$$

is [49]

$$\sigma_{\text{quasi-elastic}}^{J/\psi-N} = 0.079 \pm 0.012 \ \text{mb} \,. \tag{15.50}$$

The difference of the cross sections in Eq. (15.49) and Eq. (15.50) is 1.4 mb, which is the inelastic J/ψ-N cross section for the breakup of the J/ψ particle. This cross section is much smaller than the value of $\sigma_{\text{abs}} \sim 5.2$ mb (Eq. (15.48)) deduced from the phenomenological analysis of Fig. 15.4 assuming only nucleon absorption of J/ψ particles. Such a difference was noted by Peng *et al.* [50].

From the difference in the cross sections, it is reasonable to conclude that besides the absorption due to J/ψ-nucleon interactions, there are other additional sources of J/ψ absorption in nuclear collisions. We know that J/ψ particles can interact with pions by the reaction

$$J/\psi + \pi \rightarrow D + \bar{D} \,, \tag{15.51}$$

which has a threshold energy of $E_\pi = 0.633$ GeV in the J/ψ rest frame. Also, J/ψ particles can interact with ρ and ω mesons by the reactions

$$J/\psi + \left\{ \begin{matrix} \rho \\ \omega \end{matrix} \right\} \rightarrow D\bar{D} \,, \tag{15.52}$$

which yield an energy of 0.08 GeV in the center-of-mass of the J/ψ-meson system. These reactions lead to the breakup of J/ψ particles. Because π, ρ, and ω mesons are produced in nuclear collisions and may cross paths with the produced J/ψ particle, there can be J/ψ absorption arising from J/ψ-meson interactions of the type in Eqs. (15.51) and (15.52).

The suppression of J/ψ production due to the interactions of J/ψ particles with the produced mesons has been studied by many authors [35-38]. It has been pointed out in Ref. [37]-[38] that, in an approximate model, the suppression due to J/ψ-(produced meson) interactions leads to a ratio $\Delta\sigma_{J/\psi}^{pA}/AB\Delta\sigma_{J/\psi}^{NN}$ which depends on the effective path lengths L_A and L_B with the same form as Eq. (15.45) (see Supplement 15.5). Consequently, J/ψ-meson interactions contribute to J/ψ suppression in a way similar to J/ψ-nucleon interactions and may be one of the sources giving rise to the large value of σ_{abs} in the phenomenological analysis of the data of Fig. 15.4.

Our discussion on J/ψ suppression shows that because of final state J/ψ-hadron interactions, there is a J/ψ suppression arising from the interaction of the produced J/ψ particles with hadrons in the environments of nuclear collisions. These J/ψ-hadron interactions must be taken into account before one can extract that part of J/ψ suppression coming from the presence of a quark-gluon plasma.

⊕〚 Supplement 15.5

In this supplement, we follow the work of Ref. [37] and [38] to study the interactions of the produced J/ψ particles with the produced hadrons, which consist mostly of π, ρ, and ω mesons. We consider the collision of a projectile nucleus B with a target nucleus A at an impact parameter b, as is depicted in Fig. 12.2. We examine specifically the collision of a row of projectile nucleons at the projectile transverse coordinate b_B and a row of target nucleons at the target transverse coordinate $b_A = b - b_B$, with a transverse area σ_{in}. Among the many projectile and target nucleon-nucleon collisions in this row-on-row collision, there is a finite probability for J/ψ particles to be produced and these J/ψ particles are found most likely in the central rapidity region.

Mesons such as π, ρ, and ω are also produced in this row-on-row collision. We label the rapidity distribution from this row-on-row collision $dN/dy(b, b_B)$. There is a lapse of proper time from the time of the collision to the time of meson production, τ_{pro} (see Supplement 13.2 and Chapters 5, 6, and 7). After the proper time τ_{pro}, the produced mesons evolve as a hydrodynamical meson fluid until the freeze-out proper time τ_f. After the freeze-out time, the mesons are thermally disconnected and stream out of the collision region.

In the nucleon-nucleon center-of-mass system, the produced J/ψ particles are initially located near nucleon-nucleon collision points which are in the vicinity of each other due to Lorentz contraction. The produced J/ψ particles can interact with the produced mesons only after the mesons have formed at the proper time $\tau > \tau_{\mathrm{pro}}$. The rapidity variables of the produced mesons have a distribution. Those mesons with a rapidity much separated from the rapidity of the produced J/ψ particles will be spatially separated from the J/ψ particles when they emerge at τ_{pro}. Therefore mesons whose rapidity variables differ very much from the rapidity variables of the J/ψ particles cannot interact with the produced J/ψ particles. Only those produced mesons with rapidity variables close to the rapidity variables of the produced J/ψ particles can interact with the J/ψ particles when they emerge at τ_{pro}. We use the term *comover* to describe those produced mesons whose rapidity variables are close to the rapidity variables of the produced J/ψ particles [37,38]. Because the produced J/ψ particles are found mostly in the central rapidity region, we are interested in those comoving mesons produced in the central rapidity regions.

The produced comoving mesons can interact with the J/ψ particles by reactions of the type (15.51) or (15.52) (after the proper time τ_{pro}), leading to the breakup of the J/ψ particle. To be comoving, with the rapidity variables close to each other, the relative velocity between the J/ψ particles and the comovers is zero on the average. Their relative velocity comes from the thermal motion of mesons whose temperature is of the order of 100-200 MeV. Therefore, the average relative kinetic energy between a J/ψ particle and a comoving pion is much smaller than the threshold energy for the reaction (15.51) (E_π=0.633 GeV in the J/ψ rest frame). Consequently, the interaction of J/ψ particles with comoving pions is not likely to lead to the breakup of the J/ψ particles. On the other hand, the reactions listed in Eq. (15.52) for a J/ψ particle with a ρ or an ω meson lead to the breakup of the J/ψ particle, and these exothermic reactions do not have any energy threshold. Therefore, mainly only the ρ's and the ω's among the comoving mesons cause the breakup of the J/ψ particles. In this supplement, we shall speak of the comovers in the stricter sense as the produced comoving ρ's and ω's, which constitute only a fraction f_c of the produced hadrons. The comover density n_c is related to the hadron density n_h by

$$n_c = f_c\, n_h\,. \tag{1}$$

We envisage that a hadron fluid with a hadron density $n_h(\tau_{\mathrm{pro}})$ is formed at time τ_{pro} and evolves hydrodynamically. For simplicity, we shall assume Bjorken's hydrodynamics (see Section 13.3). The hadron density varies with the proper time according to (13.19),

$$n_h(\tau) = \frac{\tau_{\mathrm{pro}} n_h(\tau_{\mathrm{pro}})}{\tau}\,. \tag{2}$$

During the proper time interval from τ_{pro} to the freeze-out time τ_f, the J/ψ particles can interact with the comover fluid. As each J/ψ-comover collision leads to the breakup of the J/ψ particle, the survival probability arising from J/ψ-comover interactions is

$$(\text{survival probability}) = e^{-I} = e^{-\int_{\tau_{\mathrm{pro}}}^{\tau_f} d\tau\, \sigma_0\, v_{\mathrm{rel}}\, f_c\, n_h(\tau)}\,. \tag{3}$$

From Eqs. (1) and (2), the quantity I in the above equation is

$$\begin{aligned}
I &= \int_{\tau_{\mathrm{pro}}}^{\tau_f} d\tau\, \sigma_0\, v_{\mathrm{rel}}\, f_c\, n_h(\tau)\\
&= \sigma_0 v_{\mathrm{rel}} f_c \tau_{\mathrm{pro}} n_h(\tau_{\mathrm{pro}})\, \ln\!\left\{\frac{\tau_f}{\tau_{\mathrm{pro}}}\right\},
\end{aligned} \tag{4}$$

where σ_0 is the J/ψ-comover cross section leading to the breakup of the J/ψ particle.

We can relate the hadron density to the rapidity density of the hadrons produced at $\tau_{\rm pro}$ from the row-on-row collision of projectile and target nucleons. The transverse area of the row is σ_{in} and the longitudinal length element is $\tau_{\rm pro}dy$ at the proper time $\tau_{\rm pro}$. The hadron density for this row is obtained by dividing the number element dN by the volume element $\sigma_{in}\tau_{\rm pro}dy$:

$$n_h(\tau_{\rm pro}, b, b_B) = \frac{1}{\sigma_{in}\tau_{\rm pro}} \frac{dN}{dy}(b, b_B)\Bigg|_{y\sim 0}, \qquad (5)$$

where we have written out explicitly the dependence on b and b_B to indicate that these quantities refer to the row-on-row collision specified by the impact parameter b and the projectile transverse coordinate b_B. From the experimental information on transverse energy data from WA80 [51], we learn that the production of hadrons is proportional to the number of participants. For our row-on-row picture, the number of participants is

$$(\text{number of participants})(b, b_B) = \sigma_{in}[A \int_{-\infty}^{\infty} dz_A \rho(b - b_B, z_A) + B \int_{-\infty}^{\infty} dz_B \rho(b_B, z_A)]$$
$$= \sigma_{in}[AT_A(b - b_B) + BT_B(b_B)].$$

Thus, the rapidity density of hadrons is approximately

$$\frac{dN}{dy}(b, b_B) \sim \frac{1}{2}\frac{dN^{NN}}{dy}\sigma_{in}[AT_A(b - b_B) + BT_B(b_B)]. \qquad (6)$$

Hence, from Eqs. (3)-(6), the J/ψ survival probability is approximately

$$\sim e^{-c[AT_A(b-b_B)+BT_B(b_B)]}, \qquad (7)$$

where

$$c = \sigma_0 v_{\rm rel} \ln\{\frac{\tau_f}{\tau_{\rm pro}}\} f_c \frac{1}{2}\frac{dN^{NN}}{dy}\Bigg|_{y\sim 0}.$$

Eq. (7) shows that J/ψ-comover interactions lead to a survival probability which has a functional form similar to the survival probability of Eq. (15.29). We can follow the same steps below to calculate the ratio of cross sections as was done in Supplement 15.4.

In the presence of J/ψ-comover interactions, the ratio of cross sections in Eq. (15.23) becomes

$$\frac{\Delta\sigma_{J/\psi}^{AB}}{AB\Delta\sigma_{J/\psi}^{NN}} = \int db \int db_B dz_A \rho_A(b_A, z_A)\, db_B dz_B \rho_B(b_B, z_B)\, t(b - b_A - b_B)$$

$$\times e^{-c[AT_A(b_A)+BT_B(b_B)]}. \qquad (8)$$

Taking the function t to be a δ function, and integrating over b_A, we have

$$\frac{\Delta\sigma_{J/\psi}^{AB}}{AB\Delta\sigma_{J/\psi}^{NN}} = \int db \int db_B T_A(b - b_B)T_B(b_B)e^{-c[AT_A(b-b_B)+BT_B(b_B)]}.$$

Expanding the exponential function in powers of c, we have

$$\frac{\Delta\sigma_{J/\psi}^{AB}}{AB\Delta\sigma_{J/\psi}^{NN}} = \int d\boldsymbol{b} \int d\boldsymbol{b}_B T_A(\boldsymbol{b}-\boldsymbol{b}_B)T_B(\boldsymbol{b}_B)\left\{1 - c\left[AT_A(\boldsymbol{b}-\boldsymbol{b}_B)+BT_B(\boldsymbol{b}_B)\right]\right\}$$

$$= \left\{1 - c\left[A\int d\boldsymbol{b}_A[T_A(\boldsymbol{b}_A)]^2 + B\int d\boldsymbol{b}_B[T_B(\boldsymbol{b}_B)]^2\right]\right\}. \qquad (9)$$

Using the definition (15.46) for L_A and L_B, we have

$$A\int d\boldsymbol{b}_A[T_A(\boldsymbol{b}_A)]^2 = \left(\frac{3}{2\pi r_0^3}\right)\left(\frac{2\pi}{3}\right)R^3\int d\boldsymbol{b}_A[T_A(\boldsymbol{b}_A)]^2$$

$$\approx \frac{3}{2\pi r_0^3}L_A.$$

We therefore obtain

$$\frac{\Delta\sigma_{J/\psi}^{AB}}{\Delta\sigma_{J/\psi}^{NN}} \approx ABe^{-c'(L_A+L_B)}, \qquad (10)$$

where

$$c' = \frac{3c}{2\pi r_0^3}.$$

Eq. (10) has the same form as Eq. (15.45). The J/ψ-comover interactions gives rise to an absorption of J/ψ particles with the same functional form as the absorption of J/ψ particles due to J/ψ-nucleon interactions [37,38].　▮⊕

§15.5 Transverse Momentum Distribution of J/ψ Particles

There is much information about J/ψ production which one can obtain from its transverse momentum distribution. Results from the NA38 Collaboration indicate that $< P_T^2(J/\psi) >$ increases as the transverse energy of the reaction increases. This correlation is shown in Fig. 15.5a. Since the transverse energy of a reaction increases as the impact parameter decreases, the results of Fig. 15.5a can be interpreted as implying that the width of the J/ψ transverse momentum distribution increases with the degree of overlap of the colliding nuclei.

The NA38 Collaboration has also measured $< P_T^2(\text{cont}) >$ of $\mu^+\mu^-$ pairs with an invaraint mass in the neighborhood of the J/ψ peak in the continuum, as a function of the transverse energy. The results are shown in Fig. 15.5b. For the continuum dimuons, the width of the transverse momentum distribution is nearly flat, as a function of increasing transverse energy.

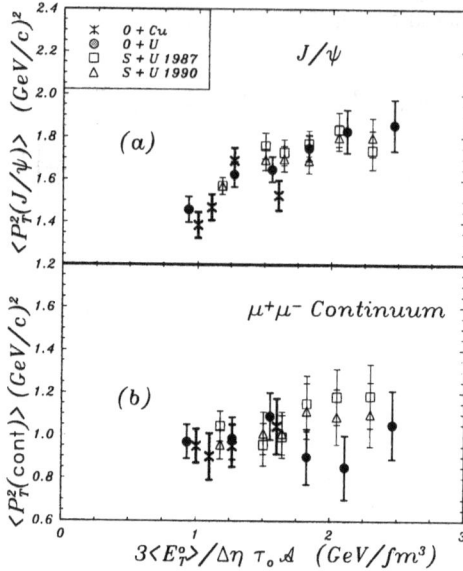

Fig. 15.5 (a) $< P_T^2(J/\psi) >$ of J/ψ particles produced in heavy-ion collisions at 200A GeV. (b) $< P_T^2(\text{cont}) >$ of $\mu^+\mu^-$ pairs in the continuum produced in heavy-ion collisions at 200A GeV. They are plotted as a function of $3 < E_T^o > /\Delta\eta\tau_o\mathcal{A}$.

From Bjorken's energy density formula (13.5), the quantity on the abscissa of Fig. 15.5, involving the transverse energy per unit pseudorapidity, can be considered as a measure of the initial energy density, if a quark-gluon plasma is formed. It may appear at first sight that if a quark-matter is created after the nucleus-nucleus collision, the correlation of $<P_T^2(J/\psi)>$ with $3<E_T^o>/\Delta\eta\tau_o\mathcal{A}$, as shown in Fig. 15.5a, implies the increase of transverse momentum with increasing initial energy density. In this scenario, a produced J/ψ with a larger P_T has a greater opportunity to escape the quark-gluon plasma and survives, while a J/ψ particle with a lower P_T is dissociated in the quark-gluon plasma [32], [35,36], [53-55].

The results of Fig. 15.5a and 15.5b can also be interpreted without invoking the scenario of quark-gluon plasma production. It is suggested that a parton which enters into the hard-scattering process to produce a J/ψ particle is subject to initial-state interactions [52]. The dominant process which produces the J/ψ particles comes from the fusion of gluons. The initial state interaction in J/ψ production involves the interaction of the gluon with nucleons. Multiple gluon-nucleon scattering will broaden the transverse momentum distribution of the gluon before its subsequent J/ψ-producing hard-scattering. The increase of the width of J/ψ transverse momentum distribution as a function of the transverse energy can be explained in terms of the increase of the width of the gluon when the overlap of the colliding nuclei increases. The gluon-nucleon cross section which is used to fit the transverse momentum data was found to be $\sigma_{gN} < P_T^2 >_{gN} = 0.39 \pm 0.04$ mb $(\text{GeV}/c)^2$ [52]. Following the same argument, initial-state interactions should also occur for the production of $\mu^+\mu^-$ pairs in the continuum, which come from the Drell-Yan process by quark-antiquark annihilation. There should also be an increase of the transverse momentum width as a function of the transverse energy due to the scattering of the quark with the nucleons before the occurrence of the Drell-Yan collision. However, the scattering cross section between a quark and a nucleon is smaller than the scattering cross section between a gluon and a nucleon. The quark-nucleon crosss section which is used to fit the transverse momentum data was found to be $\sigma_{qN} < P_T^2 >_{qN} = 0.13 \pm 0.07$ mb $(\text{GeV}/c)^2$ [29]. Hence, the quantity $< P_T^2(\text{cont}) >$ for the dimuon continuum increases only slowly with $< E_T^0 >$.

Although the above explanation appears plausible, it will be of interest to examine whether the extracted parton-nucleon cross sections are consistent with other considerations. Furthermore, from the concept of J/ψ-comover scattering [37,38], there are pions among the produced comovers. The kinetic energy of the pions in the comoving frame of reference is of the order of its temperature (about 200 MeV), but the threshold energy for a pion to break up a J/ψ particle into D and $\bar{D}$ is 0.633 GeV, in the J/ψ rest frame. Hence, the interaction between the produced pions and the produced J/ψ (or its predecessor) is not energetic enough to lead to the break-up of the J/ψ particle. The final-state interaction of J/ψ particle with these pion comovers will broaden the J/ψ transverse momentum distribution. It will be of interest to study quantitatively how the final-state interactions will broaden the transverse momentum distribution.

§**References for Chapter 15**

1. T. Matsui and H. Satz, Phys. Lett. B178, 416 (1986); T. Matsui, Zeit. Phys. C38, 245 (1988).
2. E. Eichten *et al.*, Phys. Rev. D21, 203 (1980); C. Quigg and J. L. Rosner, Phys. Rep. 56, 167 (1979).
3. Ta-Pei Cheng and Lin-Fong Li, *Gauge Theory of Elementary Particle Physics*,

Clarendon Press, Oxford, 1984.

4. L. D. McLerran and B. Svetitsky, Phys. Rev. D24, 450 (1981).

5. H. Satz, Nucl. Phys. A418, 447c (1984).

6. M. Fukugita, T. Kaneko, and Ukawa, Pjys. Lett. 154B, 185 (1985).

7. T. A. DeGrand and C. E. DeTar, Phys. Rev. D34, 2469 (1986).

8. C. Bernard, T. A. DeGrand, C. DeTar, S. Gottlieb, A. Krasnitz, M. C. Ogilvie, R. L. Sugar, and D. Tpoussaint, Nucl. Phys. A544, 519c (1992).

9. For a review of the status of finite temperature QCD lattice gauge theory, see A. Ukawa, Nucl. Phys. A498, 227c (1989), B. Petersson, Nucl. Phys. A525, 237c (1991), T. Hatsuda, Nucl. Phys. A544, 27c (1992), N. Christ, Nucl. Phys. A544, 81c (1992), and references cited in these references.

10. R. D. Pisarski and O. Alvarez, Phys. Rev. D26 3735 (1982).

11. T. Hashomoto, O. Miyamura, K. Hirose, and T. Kanki, Phys. Rev. Lett. 57, 2123 (1986); T. Hashomoto, K. Hirose, T. Kanki, and O. Miyamura, Zeit. Phys. C38, 251 (1988).

12. F. Takagi, Phys. Rev. D34, 1646 (1986).

13. D. Gross, R. D. Pisarski, and L. G. Yaffe, Rev. Mod. Phys. 53, 43 (1981).

14. K. Kanaya and H. Satz, Phys. Rev. D34, 3193 (1986).

15. T. A. DeGrand and C. E. DeTar, Phys. Rev. D34, 2469 (1986).

16. R. V. Gavai, M. Lev, B. Petersson, and H. Satz, Phys. Lett. 203B, 295 (1988).

17. F. Karsch and H. W. Wyld, Phys. Lett. 231B, 505 (1988).

18. F. Brown, N. H. Christ, Y. Deng, M. Gao, and T. J. Woch, Phys. Rev. Lett. 61, 2058 (1988).

19. A. Ukawa, Nucl. Phys. A498, 227c (1989).

20. L. D. Landau and E. M. Lifshitz, *Statistical Physics*, Pergamon Press, Oxford, Third Edition, 1980.

21. M. Abramowitz and I. A. Stegun, *Handbook of Mathematical Tables*, Dover Publications, New York, 1965.

22. R. Vogt and A. Jackson, Phys. Lett. 206, 333 (1988).

23. Y. Takahashi and S. Nagamiya, Nucl. Phys. A525, 623c (1991).

24. J. Rafelski, Phys. Rep. 88, 331 (1982).

25. C. M. Ko, Z. G. Wu, and L. H. Xia, and G. E. Brown, Phys. Rev. Lett. 66, 2577 (1991).

26. A. Shor, Phys. Rev. Lett. 54, 1122 (1985).

27. R. D. Pisarski, Phys. Lett. B110, 155 (1982).

28. C. Baglin *et al.*, NA38 Collaboration, Phys. Lett. B220, 471 (1989); C. Baglin *et al.*, NA38 Collaboration, Phys. Lett. B251, 465 (1990); C. Baglin *et al.*, NA38 Collaboration, Phys. Lett. B251, 472 (1990); C. Baglin *et al.*, NA38 Collaboration, Phys. Lett. B255, 459 (1991); C. Baglin *et al.*, NA38 Collaboration, Phys. Lett. B268, 453 (1991); C. Baglin *et al.*, NA38 Collaboration, Phys. Lett. B270, 105 (1991); C. Baglin *et al.*, NA38 Collaboration, Phys. Lett. B272, 449 (1991).

29. M. C. Abreu *et al.*, NA38 Collaboration, Nucl. Phys. A544, 209c (1992), and references cited therein.

30. J. Varela, Nucl. Phys. A525, 275c (1991).

31. N. S. Craigie, Phys. Rep. 47, 1 (1978), Table 1.

32. J.-P. Blaizot and J.-Y. Ollitrault, Phys. Lett. B199, 499 (1987); J.-P. Blaizot and J.-Y. Ollitrault, Phys. Lett. B217, 386 (1989).

33. C. Gerschel and J. Hüfner, Phys. Lett. B207, 253 (1988);

34. C. Gerschel and J. Hüfner, Nucl. Phys. A544, 513c (1992).

35. R. Vogt, M. Prakash, P. Koch and T. H. Hansson, Phys. Lett. B207, 263 (1988).

36. S. Gavin, M. Gyulassy and A. Jackson, Phys. Lett. B207, 257 (1988).
37. S. Gavin and R. Vogt, Nucl. Phys. B345, 104 (1990).
38. R. Vogt, S. J. Brodsky and P. Hoyer, Nucl. Phys. B360, 67 (1991); R. Vogt, Nucl. Phys. A544, 615c (1992).
39. K. Sridhar, Phys. Rev. Lett. 70, 1747 (1993).
40. M. B. Einhorn and . D. Ellis, Phys. Rev. D12, 2007 (1975); C. E. Carlson and Suaya, Phys. Rev. D14, 3115 (1976); E. L. Berger and D. Jones, Phys. Rev. D23, 1521 (1981); R. Baier and Rückl, Zeit. Phys. C19, 251 (1983). R. Baier and Rückl, Nucl. Phys. B218, 289 (1983).
41. H. Fritzsch, Phys. Lett. B67, 217 (1977); M. Glück and E. Reya, Phys. Lett. B79, 453 (1978); M. Glück, J. F. Owens and E. Reya, Phys. Rev. D17, 2324 (1978).
42. J. Badier et al., Z. Phys. C20, 101 (1983); M. D. Sokoloff et al., Phys. Rev. Lett. 57, 3003 (1986); S. Katsanevas et al., Phys. Rev. Lett. 60, 2121 (1988); S. Kartik et al., Phys. Rev. D41, 1 (1990); D. M. Alde et al., Phys. Rev. Lett. 66, 133 (1991).
43. H. R. Collard, L. R. B. Elton, and R. Hofstadter, *Nuclear Radii*, Springer Verlag, Berlin 1967. In Fig. 3 of this reference, nuclear root-mean-squared radii are compared with $r'_0 = 1.00$ fm. The value of $r'_0 = 1.05$ fm gives a better representation for the average root-mean-squared radii of light nuclei.
44. J. Badier et al., Z. Phys. C20, 101 (1983).
45. K. J. Anderson et al., Phys. Rev. Lett. 42, 944 (1979).
46. C. Morel et al., Phys. Lett. B252, 505 (1990).
47. D. M. Alde et al., Phys. Rev. Lett. 66, 133 (1991).
48. J. J. Aubert et al., Nucl. Phys. B213, 1 (1983).
49. R. L. Anderson, SLAC-Pub 1741 (1976).
50. J.-C. Peng et al., in Proceedings of Workshop on Nuclear Physics on the Light Cone, July 1988, World Scientific Publisher, p. 65.
51. S. Sorensen et al, WA80 Collaboration, Zeit. Phys. C38, 3 (1988).
52. J. Hüfner, Y. Kurihara, and H. Pirner, Phys. Lett. B215, 218 (1988).
53. P. V. Ruuskanen and H. satz, Zeit. Phys. C37, 623 (1988).
54. F. Karsch and P. Petronzio, Zeit. Phys. C37, 627 (1988).
55. M. C. Chu and T. Matsui, Phys. Rev. D37, 1851 (1988).

16. Signatures for the Quark-Gluon Plasma (III)

§16.1 Photon Production in the Quark-Gluon Plasma

In the quark-gluon plasma, a quark can interact with an antiquark to produce a photon and a gluon,

$$q + \bar{q} \rightarrow \gamma + g. \tag{16.1}$$

This process is called the *annihilation* process represented by the Feynman diagrams 16.1(a) and 16.1(b). The analogous electromagnetic process $q\bar{q} \rightarrow \gamma\gamma$ is allowed but the probability for its occurrence is smaller than the probability for $q\bar{q} \rightarrow \gamma g$ by a factor of the order of (α_e/α_s), or about 0.02. Here, $\alpha_e(= 1/137.0359895)$ is the electromagnetic fine-structure constant

$$\alpha_e = \frac{e^2}{4\pi}, \tag{16.2}$$

and α_s is related to the strong interaction coupling constant g by

$$\alpha_s = \frac{g^2}{4\pi}. \tag{16.3}$$

Because $\alpha_e \ll \alpha_s$, we shall be interested only in the $q\bar{q} \rightarrow \gamma g$ process, and not the $q\bar{q} \rightarrow \gamma\gamma$ process when we study photon production by $q\bar{q}$ annihilation in the quark-gluon plasma.

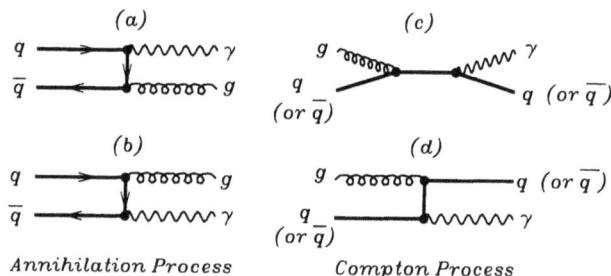

Fig. 16.1   The Feynman diagrams (a) and (b) represent the annihilation process, $q + \bar{q} \rightarrow \gamma + g$, and the Feynman diagrams (c) and (d) represent the Compton process, $g + q \rightarrow \gamma + q$ (or $g + \bar{q} \rightarrow \gamma + \bar{q}$).

A gluon can interact with a quark to produce a photon by the reaction

$$g + q \to \gamma + q , \qquad (16.4a)$$

or with an antiquark to produce a photon by the reaction

$$g + \bar{q} \to \gamma + \bar{q} . \qquad (16.4b)$$

These reactions are represented by the Feynman diagrams 16.1(c) and 16.1(d). They are analogous to the scattering of a photon off a charged particle, which is called the Compton scattering. The incident gluon in (16.4a) or (16.4b) plays the role analogous to the incident photon in the Compton scattering of the photon. Hence, the processes (16.4a) and (16.4b) are called *Compton* processes.

After a photon is produced, it must exit the collision region, in order to be detected. Since the photon interacts with the particles in the collision region only through the electromagnetic interaction, the interaction is not strong. Consequently, the mean-free path of the photon is expected to be quite large and the photon may not suffer a collision after it is produced. On the other hand, the photon production rate and the photon momentum distribution depend on the momentum distributions of the quarks, antiquarks, and gluons in the plasma, which are governed by the thermodynamical condition of the plasma. Therefore, photons produced in the quark-gluon plasma carry information on the thermodynamical state of the medium at the moment of their production [1-5].

⊕[Supplement 16.1

The annihilation of a q and a $\bar{q}$ to lead to a single real photon is forbidden by the law of energy-momentum conservation. If $q + \bar{q} \to \gamma$ were allowed, the photon would have the momentum $p_\gamma = q + \bar{q}$ and the square of the invariant mass of the photon would be

$$p_\gamma^2 = (q + \bar{q})^2$$
$$= 2m^2 + 2(q^0 \bar{q}^0 - |\mathbf{q}| \, |\bar{\mathbf{q}}| \cos\theta),$$

where m is the rest mass of the quark and θ is the angle between $\mathbf{q}$ and $\bar{\mathbf{q}}$. Because $\cos\theta \le 1$, $|\mathbf{q}| < q^0$, and $|\bar{\mathbf{q}}| < \bar{q}^0$, we always have

$$q^0 \bar{q}^0 - |\mathbf{q}| \, |\bar{\mathbf{q}}| \cos\theta > 0$$

and thus

$$p_\gamma^2 > 0. \qquad (1)$$

On the other hand, a real photon is characterised by a zero rest mass and

$$p_\gamma^2 = 0. \qquad (2)$$

Because Eq. (1) and Eq. (2) contradict each other, the annihilation of a quark and an antiquark into a single real photon is impossible. However, a quark and an antiquark can annihilate to form a virtual photon, which is the dilepton production process discussed in Fig. 14.1.   ]⊕

§16.2 Photon Production by Quark-Antiquark Annihilation

To obtain the production rate and the momentum distribution of photons by quark-antiquark annihilation we need the cross section for the process $q + \bar{q} \to \gamma + g$. The differential cross section for this process can be worked out using Feynman diagram techniques as in Chapter 14. We shall not repeat similar steps here. It suffices to mention that within the leading-order Feynman diagrams, the photon differential cross section for the $q\bar{q} \to \gamma g$ process is related to the photon differential cross section for the $q\bar{q} \to \gamma\gamma$ process. The two processes have the same properties with regard to particle momenta, particle spins and particle polarizations. They differ only by replacing each of the two electromagnetic photon-quark coupling vertices in $q\bar{q} \to \gamma\gamma$ with the quark-gluon vertex in $q\bar{q} \to \gamma g$ (see Fig. 16.2).

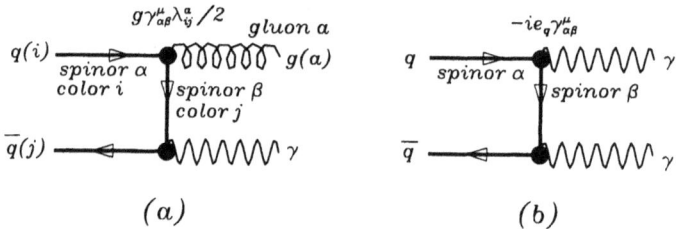

Fig. 16.2   (a) The Feynman diagram for $q(i) + \bar{q}(j) \to g(a) + \gamma$, and (b) the Feynman diagram for $q + \bar{q} \to \gamma + \gamma$.

From the rules of Feynman diagrams [6-9], we note that in the $q(i)\bar{q}(j) \to \gamma g(a)$ process the gluon-quark vertex in Fig. 16.2a is associated with the quantity

$$g\gamma^\mu_{\alpha\beta}\frac{\lambda^a_{ij}}{2} \qquad (16.5)$$

where i and j are the color indices of the two quark lines joining the vertex, a is the index of the gluon, and λ^a is a Gell-Mann matrix as defined in Eq. (10.34). In Eq. (16.5), α and β are the spinor indices of the quark lines joining the vertex, and γ^μ is a Dirac gamma matrix. The corresponding photon-quark vertex in Fig. 16.2b for the analogous $q\bar{q} \to \gamma\gamma$ process is associated with the quantity [6,7]

$$-ie_q\gamma^\mu_{\alpha\beta}, \qquad (16.6)$$

where e_q is the electric charge of the quark. Consequently, $q(i)\bar{q}(j) \to \gamma g(a)$ and $q\bar{q} \to \gamma\gamma$ differ in the coupling constants (g versus e_q), in the additional color factor $\lambda_{ij}^a/2$ for the $q\bar{q} \to \gamma g$ reaction, and in the unimportant phase $-i$. The two diagrams 16.2a and 16.2b are otherwise identical with respect to particle momenta, particle spins, and particle polarizations.

It is easy to separate out the color-dependence of the differential cross section by comparing expression (16.5) with expression (16.6). The differential cross section is proportional to the absolute square of the Feynman amplitude which includes the factor (16.5) or (16.6). Therefore, the photon differential cross section for the process $q(i)\bar{q}(j) \to \gamma g(a)$ is

$$E_\gamma \frac{d\sigma}{d\boldsymbol{p}_\gamma}[q(i)\bar{q}(j) \to \gamma g(a)] = \left|\frac{\lambda_{ij}^a}{2}\right|^2 E_\gamma \frac{d\bar{\sigma}}{d\boldsymbol{p}_\gamma}(q\bar{q} \to \gamma g), \quad (16.7)$$

where the bar symbol above the symbol σ is to indicate that $E_\gamma d\bar{\sigma}/d\boldsymbol{p}_\gamma(q\bar{q} \to \gamma g)$ is the color-independent part of the differential cross section. This color-independent part of the differential cross section $E_\gamma d\bar{\sigma}/d\boldsymbol{p}_\gamma(q\bar{q} \to \gamma g)$ is related to the differential cross section $E_\gamma d\sigma/d\boldsymbol{p}_\gamma(q\bar{q} \to \gamma\gamma)$ by the ratio of the coupling constants,

$$E_\gamma \frac{d\bar{\sigma}}{d\boldsymbol{p}_\gamma}(q\bar{q} \to \gamma g) = \left(\frac{g}{e_q}\right)^2 E_\gamma \frac{d\sigma}{d\boldsymbol{p}_\gamma}(q\bar{q} \to \gamma\gamma)$$

$$= \frac{\alpha_s}{\alpha_e}\left(\frac{e}{e_q}\right)^2 E_\gamma \frac{d\sigma}{d\boldsymbol{p}_\gamma}(q\bar{q} \to \gamma\gamma). \quad (16.8)$$

We shall show in Supplement 16.2 that $E_\gamma d\sigma/d\boldsymbol{p}_\gamma$ and $d\sigma/dt$ are related by

$$E_\gamma \frac{d\sigma}{d\boldsymbol{p}_\gamma}(q\bar{q} \to \gamma\gamma) = \frac{\sqrt{s-4m^2}}{2\pi}\frac{d\sigma}{dt}(q\bar{q} \to \gamma\gamma)\,\delta\left(\frac{\boldsymbol{p}_\gamma \cdot (\boldsymbol{p}_q + \boldsymbol{p}_{\bar{q}})}{\sqrt{s}} - \frac{\sqrt{s}}{2}\right).$$
$$(16.9a)$$

and similarly

$$E_\gamma \frac{d\bar{\sigma}}{d\boldsymbol{p}_\gamma}(q\bar{q} \to \gamma g) = \frac{\sqrt{s-4m^2}}{2\pi}\frac{d\bar{\sigma}}{dt}(q\bar{q} \to \gamma g)\,\delta\left(\frac{\boldsymbol{p}_\gamma \cdot (\boldsymbol{p}_q + \boldsymbol{p}_{\bar{q}})}{\sqrt{s}} - \frac{\sqrt{s}}{2}\right).$$
$$(16.9b)$$

Hence, from Eq. (16.8) and Eq. (16.9), we have

$$\frac{d\bar{\sigma}}{dt}(q\bar{q} \to \gamma g) = \frac{\alpha_s}{\alpha_e}\left(\frac{e}{e_q}\right)^2 \frac{d\sigma}{dt}(q\bar{q} \to \gamma\gamma). \quad (16.10)$$

The photon differential cross section $d\sigma/dt(q\bar{q} \to \gamma\gamma)$ on the righthand side of the above equation is related to $d\sigma/dt(e^+e^- \to \gamma\gamma)$ by

$$\frac{d\sigma}{dt}(q\bar{q} \to \gamma\gamma) = \left(\frac{e_q}{e}\right)^4 \frac{d\sigma}{dt}(e^+e^- \to \gamma\gamma). \tag{16.11}$$

Since the differential cross section $d\sigma/dt(e^+e^- \to \gamma\gamma)$ has been worked out in detail in Ref. 8, Eqs. (16.10) and (16.11) can be used to write out explicitly the analogous differential cross section $d\bar{\sigma}/dt(q\bar{q} \to \gamma g)$. We note that the reactions on the two sides of Eq. (16.10) are two-particle reactions of the form (particle 1)+(particle 2) → (particle 3)+(particle 4), with the particle momenta described by $p_1 + p_2 \to p_3 + p_4$. The equality of Eq. (16.10) states that the differential cross sections for the $q\bar{q} \to \gamma g$ reaction on the lefthand side of Eq. (16.10) is the same as the differential cross sections for the $q\bar{q} \to \gamma\gamma$ reaction on the righthand side of Eq. (16.10), when the momenta, the spins and the polarizations of corresponding ith particle in the two different reactions are the same. We can label q, $\bar{q}$ and the photon in these reactions as particles 1, 2, and 3 respectively and identify p_1 as the quark momentum p_q, p_2 as the antiquark momentum $p_{\bar{q}}$, and p_3 as the photon momentum p_γ. Particle 4 differs for the reactions on the two sides of Eq. (16.10). For the $q\bar{q} \to \gamma g$ reaction on the lefthand side, particle 4 is the gluon, and for the $q\bar{q} \to \gamma\gamma$ reaction on the righthand side, particle 4 is the second photon γ'. Consequently, we identify p_4 as the gluon momentum p_g for the $q\bar{q} \to \gamma g$ reaction, and p_4 as the momentum of the second photon $p_{\gamma'}$ for the $q\bar{q} \to \gamma\gamma'$ reaction.

We use the Mandelstam variables

$$s = (p_1 + p_2)^2 = (p_q + p_{\bar{q}})^2, \tag{16.12a}$$

$$t = (p_1 - p_3)^2 = (p_q - p_\gamma)^2, \tag{16.12b}$$

and

$$u = (p_1 - p_4)^2 = (p_2 - p_3)^2 = (p_{\bar{q}} - p_\gamma)^2. \tag{16.12c}$$

The photon differential cross section $d\sigma/dt(q\bar{q} \to \gamma\gamma)$, averaged over initial quark and antiquark spin states and summed over final photon polarizations, is given according to Eq. (88.4) of Ref. [8] by

$$\frac{d\sigma}{dt}(q\bar{q} \to \gamma\gamma) = \left(\frac{e_q}{e}\right)^4 \frac{8\pi\alpha_e^2}{s(s-4m^2)}\left\{\left(\frac{m^2}{t-m^2} + \frac{m^2}{u-m^2}\right)^2 \right.$$
$$\left. + \left(\frac{m^2}{t-m^2} + \frac{m^2}{u-m^2}\right) - \frac{1}{4}\left(\frac{t-m^2}{u-m^2} + \frac{u-m^2}{t-m^2}\right)\right\} \tag{16.13}$$

where the factor $(e_q/e)^4$ takes into account the electric charge e_q of the quark, and we have made a change of notation from Ref. 8. From Eqs. (16.11) and (16.13), the color-independent part of the differential cross section $d\bar{\sigma}/dt(q\bar{q} \to \gamma g)$, averaged over initial quark and antiquark spins and summed over photon and gluon polarizations, is given by

$$\frac{d\bar{\sigma}}{dt}(q\bar{q} \to \gamma g) = \left(\frac{e_q}{e}\right)^2 \frac{8\pi\alpha_s\alpha_e}{s(s-4m^2)} \left\{ \left(\frac{m^2}{t-m^2} + \frac{m^2}{u-m^2}\right)^2 \right.$$
$$\left. + \left(\frac{m^2}{t-m^2} + \frac{m^2}{u-m^2}\right) - \frac{1}{4}\left(\frac{t-m^2}{u-m^2} + \frac{u-m^2}{t-m^2}\right) \right\}. \quad (16.14)$$

⊕〚 Supplement 16.2

It is useful to write out the invariant photon differential cross section $E_\gamma d\sigma/d\boldsymbol{p}_\gamma$ explicitly in terms of $d\sigma/dt$. Using the superscript '∗' to represent quantities in the center-of-mass system, we have from Eq. (16.12b)

$$t = m^2 - 2E_\gamma^*\{E_q^* - |\boldsymbol{p}_q^*|\mu(\boldsymbol{p}_\gamma^*, \boldsymbol{p}_q^*)\}, \quad (1)$$

where $\mu(\boldsymbol{p}_\gamma^*, \boldsymbol{p}_q^*) = \cos\theta(\boldsymbol{p}_\gamma^*, \boldsymbol{p}_q^*)$. Therefore, we have

$$dt = 2E_\gamma^*|\boldsymbol{p}_q^*|d\mu,$$

and the invariant cross section is

$$E_\gamma \frac{d\sigma}{d\boldsymbol{p}_\gamma} = E_\gamma^* \frac{d\sigma}{(E_\gamma^*)^2 \, dE_\gamma^* \, d\mu \, d\phi}$$
$$= \frac{d\sigma \, 2E_\gamma^*|\boldsymbol{p}_q^*|}{E_\gamma^* \, dE_\gamma^* \, dt \, d\phi}$$
$$= \sqrt{s-4m^2} \frac{d\sigma}{dE_\gamma^* \, dt \, d\phi}.$$

In the center-of-mass system, each photon carries one-half of the center-of-mass energy. Therefore, the invariant cross section is

$$E_\gamma \frac{d\sigma}{d\boldsymbol{p}_\gamma}(q\bar{q} \to \gamma\gamma) = \sqrt{s-4m^2} \frac{d\sigma}{dt \, d\phi}(q\bar{q} \to \gamma\gamma)\,\delta(E_\gamma^* - \frac{\sqrt{s}}{2}). \quad (2)$$

To express the center-of-mass energy E_γ^* in terms of p_γ, p_q and $p_{\bar{q}}$ in any frame, we take the scalar product of p_γ with $p_q + p_{\bar{q}}$ which is independent of the Lorentz frames. We evaluate this scalar product in the center-of-mass frame and we have

$$p_\gamma \cdot (p_q + p_{\bar{q}}) = E_\gamma^*(E_q^* + E_{\bar{q}}^*) - \boldsymbol{p}_\gamma^* \cdot (\boldsymbol{p}_q^* + \boldsymbol{p}_{\bar{q}}^*). \quad (3)$$

But, in the center-of-mass system, the total momentum is zero

$$p_q^* + p_{\bar{q}}^* = 0, \tag{4}$$

and the sum $E_q^* + E_{\bar{q}}^*$ is just the center-of-mass energy $\sqrt{s}$ which is equal to $\sqrt{(p_q + p_{\bar{q}})^2}$. Therefore, from Eq. (3) and Eq. (4), the photon energy in the center-of-mass system is

$$E_\gamma^* = \frac{p_\gamma \cdot (p_q + p_{\bar{q}})}{\sqrt{s}}. \tag{5}$$

From Eqs. (2) and (5), the photon differential cross section $E_\gamma d\sigma/d\mathbf{p}_\gamma$ in terms of $d\sigma/dt$ is

$$E_\gamma \frac{d\sigma}{d\mathbf{p}_\gamma}(q\bar{q} \to \gamma\gamma) = \sqrt{s - 4m^2} \frac{d\sigma}{dt\, d\phi}(q\bar{q} \to \gamma\gamma)\delta\left(\frac{p_\gamma \cdot (p_q + p_{\bar{q}})}{\sqrt{s}} - \frac{\sqrt{s}}{2}\right). \tag{6}$$

For unpolarized quark states with no preferred direction, there is complete azimuthal symmetry and the azimuthal differential distribution is just $1/2\pi$. Therefore, with unpolarized quarks and antiquarks, the photon differential cross section $E_\gamma d\sigma/d\mathbf{p}_\gamma$ can be written in terms of $d\sigma/dt$ by

$$E_\gamma \frac{d\sigma}{d\mathbf{p}_\gamma}(q\bar{q} \to \gamma\gamma) = \frac{\sqrt{s - 4m^2}}{2\pi} \frac{d\sigma}{dt}(q\bar{q} \to \gamma\gamma)\delta\left(\frac{p_\gamma \cdot (p_q + p_{\bar{q}})}{\sqrt{s}} - \frac{\sqrt{s}}{2}\right). \tag{7}$$

Similarly, for the reaction $q\bar{q} \to \gamma g$, the color-independent part of the differential cross section is related to the corresponding $d\bar{\sigma}/dt$ by

$$E_\gamma \frac{d\bar{\sigma}}{d\mathbf{p}_\gamma}(q\bar{q} \to \gamma g) = \frac{\sqrt{s - 4m^2}}{2\pi} \frac{d\bar{\sigma}}{dt}(q\bar{q} \to \gamma g)\delta\left(\frac{p_\gamma \cdot (p_q + p_{\bar{q}})}{\sqrt{s}} - \frac{\sqrt{s}}{2}\right). \tag{8}$$

It is useful to study some characteristics of the photon differential cross section for $q\bar{q} \to \gamma g$. We observe from Eq. (16.14) that for a given center-of-mass collision energy $\sqrt{s}$, the photon differential cross section contains terms which vary as $(u - m^2)/(t - m^2)$ and $(t - m^2)/(u - m^2)$. The factor $1/(t - m^2)$ arises from the quark propagator between the two vertices in the Feynman diagram 16.1a and the factor $1/(u - m^2)$ arises from the quark propagator in the Feynman diagram 16.1b. These terms are dominant when the magnitudes of their denominators become small. As a consequence, the photon differential cross section is a maximum at the photon momenta for which $|t - m^2|$ or $|u - m^2|$ is a minimum. From Eq. (16.12b), the quantity $t - m^2$ is

$$t - m^2 = -2p_\gamma \cdot p_q$$
$$= -2E_\gamma(E_q - |\mathbf{p}_q| \cos\theta_{\gamma q}).$$

The quantity $|t - m^2|$ is therefore a minimum when

$$\theta_{\gamma q} = 0 \, . \tag{16.15}$$

This occurs when the photon momentum $\boldsymbol{p}_\gamma$ lines up with the momentum of the quark $\boldsymbol{p}_q$. Similarly, we have

$$u - m^2 = -2E_\gamma(E_{\bar{q}} - |\boldsymbol{p}_{\bar{q}}|\cos\theta_{\gamma\bar{q}}) \, .$$

The quantity $|u - m^2|$ is a minimum when

$$\theta_{\gamma\bar{q}} = 0 \, , \tag{16.16}$$

which occurs when the photon momentum $\boldsymbol{p}_\gamma$ lines up with the momentum of the antiquark $\boldsymbol{p}_{\bar{q}}$. Thus, the photon differential cross section for $q\bar{q} \to \gamma g$ have two peaks, one in the direction of the quark and one in the direction of the antiquark.

What are the angular widths $\Delta\theta$ of the photon differential cross sections at these two peaks? We can expand $(t - m^2)^{-1}$ about $\theta_{\gamma q}$ equal to zero, where the photon differential cross section is maximum, and we have

$$\frac{1}{t - m^2} \approx \frac{1}{-2E_\gamma[E_q - |\boldsymbol{p}_q| + |\boldsymbol{p}_q|\frac{1}{2}(\theta_{\gamma q})^2]}$$

$$\approx -\frac{1}{E_\gamma E_q\left[\left(\frac{m}{E_q}\right)^2 + (\theta_{\gamma q})^2\right]} \, .$$

Therefore, around the direction of $\boldsymbol{p}_q$, the photon production cross section drops rapidly beyond the angular cone

$$\Delta\theta_{\gamma q} = \frac{m}{E_q} \, , \tag{16.17a}$$

and similarly around the direction of $\boldsymbol{p}_{\bar{q}}$, the photon production cross section drops rapidly beyond the angular cone

$$\Delta\theta_{\gamma\bar{q}} = \frac{m}{E_{\bar{q}}} \, . \tag{16.17b}$$

When both annihilating particles are relativistic with $E_q, E_{\bar{q}} \gg m$, the produced photon in the $q\bar{q} \to \gamma g$ process is most likely found in narrow cones along the directions of the two annihilating particles.

We have thus established the directions along which the photons are most likely to be produced. We next determine the most likely photon energy. The photon energy in the center-of-mass system is one-half of the total center-of-mass energy $\sqrt{s}$. It can be expressed in terms of p_γ, p_q, and $p_{\bar{q}}$ in any Lorentz frame, as given by Eq. (5) of Supplement 16.2. Therefore, the condition which determines the photon energy is

$$p_\gamma \cdot p_q + p_\gamma \cdot p_{\bar{q}} - s/2 = 0. \tag{16.18}$$

The scalar product $p_\gamma \cdot p_q$ is

$$p_\gamma \cdot p_q = E_\gamma (E_q - |\mathbf{p}_q| \cos \theta_{\gamma q}).$$

We first examine the case when the photon momentum $\mathbf{p}_\gamma$ is colinear with $\mathbf{p}_q$. We have $\theta_{\gamma q} = 0$,

$$p_\gamma \cdot p_q \approx \frac{E_\gamma}{2E_q} m^2 ,$$

and consequently the left-hand-side of Eq. (16.18) becomes

$$\begin{aligned}
p_\gamma \cdot (p_q + p_{\bar{q}}) - s/2 &= p_\gamma \cdot p_q + p_\gamma \cdot p_{\bar{q}} - (2m^2 + 2p_q \cdot p_{\bar{q}})/2 \\
&= p_\gamma \cdot p_q + p_\gamma \cdot p_{\bar{q}} - m^2 - p_q \cdot p_{\bar{q}} \\
&\approx (p_\gamma - p_q) \cdot p_{\bar{q}} + \frac{E_\gamma}{2E_q} m^2 - m^2 .
\end{aligned} \tag{16.19}$$

For the relativistic case when the rest mass m is small and negligible compared to other energy scales, Eq. (16.19) shows that because $p_{\bar{q}}$ can be any momentum 4-vector, the condition Eq. (16.18) is approximately satisfied if the photon momentum 4-vector is approximately the same as the momentum 4-vector of the quark,

$$p_\gamma \approx p_q , \tag{16.20a}$$

when the photon is colinear with $\mathbf{p}_q$. Similarly, for the relativistic case with the photon momentum $\mathbf{p}_\gamma$ colinear with $\mathbf{p}_{\bar{q}}$, the condition Eq. (16.18) is approximately satisfied when the photon momentum 4-vector is approximately equal to the momentum 4-vector of the antiquark,

$$p_\gamma \approx p_{\bar{q}} . \tag{16.20b}$$

The results of Eqs. (16.15), (16.16), (16.17) and (16.20) provide the following pictorial description for the process of photon production by quark-antiquark annihilation at relativistic energies. After the annihilation, one of the two annihilating particles turns into a photon which maintains approximately the same momentum and energy as this particle, while the other particle turns into a gluon with approximately the same momentum and energy of the other particle. In the context of the Feynman diagrams 16.1(a) and 16.1(b), the peaks of the cross sections occur when the boson lines of the γ and the gluon g join onto the fermion lines of the q and the $\bar{q}$ without any significant change of the particle momenta, under the 'exchange' of a fermion with very low momentum.

From this pictorial description of the process and the narrowness of the photon differential cross section when the energies are much greater than the quark masses, we can approximate the invariant photon differential cross section for the relativistic case as

$$E_\gamma \frac{d\bar{\sigma}}{d\boldsymbol{p}_\gamma}(q\bar{q} \to \gamma g) \approx \bar{\sigma}_{q\bar{q}\to\gamma g}(s)\frac{1}{2}E_\gamma[\delta(\boldsymbol{p}_\gamma - \boldsymbol{p}_q) + \delta(\boldsymbol{p}_\gamma - \boldsymbol{p}_{\bar{q}})]. \quad (16.21)$$

Clearly, the integration of the above equation over $d\boldsymbol{p}_\gamma/E_\gamma$ gives back the total cross section. This result will be used later to study photon production in the quark-gluon plasma.

The total cross section $\bar{\sigma}_{q\bar{q}\to\gamma g}(s)$ appears in Eq. (16.21), but what is this total cross section? It can be obtained by integrating $d\bar{\sigma}/dt$ over the allowed range of t. The range of t, as given by Eq. (27) of Exercise 14.1, is

$$m^2 - \frac{1}{2}s - \frac{1}{2}\sqrt{s}\sqrt{s - 4m^2} \le t \le m^2 - \frac{1}{2}s + \frac{1}{2}\sqrt{s}\sqrt{s - 4m^2}.$$

The total cross section for the $e^+e^- \to \gamma\gamma$ process is given explicitly by Eq. (88.6) of Ref. [8]. Using Eq. (16.14), the corresponding color-independent part of the total cross section for the $q\bar{q} \to \gamma g$ process is

$$\bar{\sigma}_{q\bar{q}\to\gamma g}(s) = \left(\frac{e_q}{e}\right)^2 \bar{\sigma}_{\text{ann}}(s), \quad (16.22a)$$

where

$$\bar{\sigma}_{\text{ann}}(s) = \frac{4\pi\alpha_e\alpha_s}{s - 4m^2}\left\{\left(1 + \frac{4m^2}{s} - \frac{16m^4}{s^2}\right)\ln\left(\frac{\sqrt{s} + \sqrt{s - 4m^2}}{\sqrt{s} - \sqrt{s - 4m^2}}\right)\right.$$
$$\left. - \left(1 + \frac{4m^2}{s}\right)\sqrt{1 - \frac{4m^2}{s}}\right\}. \quad (16.22b)$$

In the non-relativistic limit of $s \sim 4m^2$, the total cross section $\overline{\sigma}_{\text{ann}}(s)$ is

$$\overline{\sigma}_{\text{ann}}(s) = \frac{2\pi\alpha_e\alpha_s}{2m^2\sqrt{(s/4m^2) - 1}}, \tag{16.23}$$

and in the relativistic case with $s \gg 4m^2$, it is

$$\overline{\sigma}_{\text{ann}}(s) = \frac{4\pi\alpha_e\alpha_s}{s}\left\{\ln\left(\frac{s}{m^2}\right) - 1\right\}. \tag{16.24}$$

§16.3 Photon Production by $q\overline{q}$ Annihilation in the Plasma

The evaluation of the production rate and the momentum distribution of photons can be carried out in the same way as in the evaluation of the rate for dilepton production (see Section §14.1). We consider for simplicity again a plasma where the net baryon density is zero so that the quark distribution $f_q(\boldsymbol{p}_q)$ can be taken to be the same as the antiquark distribution $f_{\overline{q}}(\boldsymbol{p}_{\overline{q}})$. The results obtained here can be easily generalized to the cases when $f_{\overline{q}}(\boldsymbol{p}_{\overline{q}})$ and the distributions $f_q(\boldsymbol{p}_q)$ are different.

Using the approximate photon differential cross section (16.21), we shall show in Supplement 16.3 that the rate and the momentum distribution of produced photons due to quark-antiquark annihilation in a quark-gluon plasma can be written as

$$E_\gamma \frac{dN_\gamma^{\text{ann}}}{d\boldsymbol{p}_\gamma d^4x} = \frac{4N_s^2}{(2\pi)^6}f_q(\boldsymbol{p}_\gamma)$$

$$\times \sum_{f=1}^{N_f}\int_{\boldsymbol{p}_\gamma \approx \boldsymbol{p}_q} d^3p_{\overline{q}}f_{\overline{q}}(\boldsymbol{p}_{\overline{q}})[1 + f_g(\boldsymbol{p}_{\overline{q}})]\overline{\sigma}_{q_f\overline{q}_f \to \gamma g}(s)\frac{\sqrt{s(s - 4m^2)}}{2E_{\overline{q}}}. \tag{16.25}$$

The quantity $E_\gamma dN_\gamma^{\text{ann}}/d\boldsymbol{p}_\gamma d^4x$ is a function of the photon momentum $\boldsymbol{p}_\gamma$, and it is a joint distribution which provides information on the rate of photon production and the momentum distribution of the produced photons. The integral of this quantity over the space-time volume element d^4x gives the momentum distribution of the photons and the integral of this quantity over the invariant momentum volume element $d\boldsymbol{p}_\gamma/E_\gamma$ gives the rate of photon production per unit space-time 4-volume. We shall call this quantity $E_\gamma dN_\gamma^{\text{ann}}/d\boldsymbol{p}_\gamma d^4x$

the *annihilation photon distribution* and the superscript label 'ann' is to indicate that this is the photon distribution arising from the annihilation process $q\bar{q} \to \gamma g$.

Eq. (16.25) gives the photon distribution as an explicit function of the quark distribution and the gluon distribution of the plasma. In particular, in this case when the quark distribution is taken to be the same as that of the antiquark, the photon distribution is directly proportional to the quark momentum distribution $f_q(\boldsymbol{p}_\gamma)$ evaluated at the photon momentum $\boldsymbol{p}_\gamma$. The physical reason for such a dependence is easy to understand, as we discussed in the last section. The annihilation process at relativistic energies occurs as if either the annihilating quark or antiquark is converted into a photon of approximately the same energy and momentum as that particle, (the rest mass of the quark being negligible). Hence, the photon distribution is directly proportional to the quark distribution. We shall see later that the integral over d^3p_q in Eq. (16.25) gives a function proportional to $(\ln E_\gamma + \text{constant})$, which is a slowly-varying function of the photon energy E_γ. Hence, the dominant variation of the photon distribution comes from the quark distribution $f_q(\boldsymbol{p}_\gamma)$.

Depending on the state of the system, the quark distribution $f_q(\boldsymbol{p}_\gamma)$ can be a general function of the momentum. If there is no thermal equilibrium and the momentum distribution of the quarks has peculiar features which carry this information, these features will show up in the photon spectra. Even for a system in thermal equilibrium, the quark momentum distribution may have a greatly enhanced high-momentum component because of the boundary of the plasma [10], and the enhanced high-momentum component will show up in the photon spectra.

To carry out the integral in Eq. (16.25), we assume that $f_{\bar{q}}(\boldsymbol{p}_{\bar{q}})$ and $f_g(\boldsymbol{p}_g)$ depend only on energy. We shall show in Supplement 16.3 that the photon distribution (16.25) becomes

$$E_\gamma \frac{dN_\gamma^{\text{ann}}}{d\boldsymbol{p}_\gamma d^4x} = \frac{4N_s^2}{(2\pi)^5} \frac{f_q(\boldsymbol{p}_\gamma)}{4E_\gamma} \sum_{f=1}^{N_f} \left(\frac{e_f}{e}\right)^2$$

$$\times \int ds\, dE_{\bar{q}} f_{\bar{q}}(E_{\bar{q}})[1+f_g(E_{\bar{q}})]\sqrt{s(s-4m^2)}\bar{\sigma}_{\text{ann}}(s). \qquad (16.26)$$

In the quark-gluon plasma, $f_{\bar{q}}(E_{\bar{q}})$ is a Fermi-Dirac distribution and $f_g(E_g)$ is a Bose-Einstein distribution, the integral in $E_{\bar{q}}$ can be carried out analytically, and we obtain the annihilation photon distribution as a one-dimensional integral

$$
E_\gamma \frac{dN_\gamma^{\mathrm{ann}}}{d\boldsymbol{p}_\gamma d^4 x} = \frac{4N_s^2}{(2\pi)^5} f_q(\boldsymbol{p}_\gamma) \frac{T}{4E_\gamma} \sum_{f=1}^{N_f} \left(\frac{e_f}{e}\right)^2
$$

$$
\times \sum_{n=0}^{\infty} \frac{1}{2n+1} \int ds \, e^{-(2n+1)s/4E_\gamma T} \sqrt{s(s-4m^2)} \overline{\sigma}_{\mathrm{ann}}(s). \quad (16.27)
$$

⊕[Supplement 16.3

To establish the results of Eqs. (16.25) − (16.27), we follow closely the steps of Exercise 14.2. The number of quarks with flavor f, color i, momentum $\boldsymbol{p}_q$, and energy $E_q = (\boldsymbol{p}_q^2 + m^2)^{1/2}$, in the volume element d^3x and the momentum element d^3p_q, is

$$
\frac{d^3x \, d^3p_q}{(2\pi)^3} \, N_s \, f_q(\boldsymbol{p}_q). \quad (1)
$$

The number of antiquarks having color j, momentum $\boldsymbol{p}_{\bar{q}}$ and energy $E_{\bar{q}} = (\boldsymbol{p}_{\bar{q}}^2 + m^2)^{1/2}$ in the momentum element $d^3p_{\bar{q}}$, with which a quark with flavor f and color i can interact in a unit time, to produce a photon and a gluon $g(a)$ by the reaction $q(i) + \bar{q}(j) \to \gamma + g(a)$, is

$$
\sigma_{q_f(i)\bar{q}_f(j)\to\gamma g(a)}(s) v_{q\bar{q}} \frac{d^3p_{\bar{q}}}{(2\pi)^3} N_s \, f_{\bar{q}}(\boldsymbol{p}_{\bar{q}}). \quad (2)
$$

We separate out the color dependence of the cross section $\sigma_{q_f(i)\bar{q}_f(j)\to\gamma g(a)}(s)$ by using Eq. (16.7),

$$
\sigma_{q_f(i)\bar{q}_f(j)\to\gamma g(a)}(s) = \left|\frac{\lambda_{ij}^a}{2}\right|^2 \overline{\sigma}_{q_f\bar{q}_f\to\gamma g}(s),
$$

where $\overline{\sigma}_{q_f\bar{q}_f\to\gamma g}(s)$ is the color-independent part of the total $q_f\bar{q}_f \to \gamma g$ cross section.

The number of photons produced per unit time is the integral of the product of Eqs. (1) and (2), summed over the flavor f, the color indices of the quark i, all color indices of the antiquark j, and all gluon indices a,

$$
\frac{dN_\gamma}{dt} = N_s^2 \sum_{f=1}^{N_f} \sum_{i=1}^{3} \sum_{j=1}^{3} \sum_{a=1}^{8} \left|\frac{\lambda_{ij}^a}{2}\right|^2
$$

$$
\times \int \frac{d^3x \, d^3p_q \, d^3p_{\bar{q}}}{(2\pi)^6} f_q(\boldsymbol{p}_q) f_{\bar{q}}(\boldsymbol{p}_{\bar{q}}) [1 + f_g(\boldsymbol{p}_q + \boldsymbol{p}_{\bar{q}} - \boldsymbol{p}_\gamma)] \overline{\sigma}_{q_f\bar{q}_f\to\gamma g}(s) v_{q\bar{q}}, \quad (3)
$$

where $[1 + f_g(\boldsymbol{p}_g)]$ is included because the gluon produced simultaneously with the photon in the $q\bar{q} \to \gamma g$ reaction obeys Bose-Einstein statistics. The probability for the production of the gluon with a momentum $\boldsymbol{p}_g = \boldsymbol{p}_q + \boldsymbol{p}_{\bar{q}} - \boldsymbol{p}_\gamma$ is enhanced by the factor $[1 + f_g(\boldsymbol{p}_g)]$, where $f_g(\boldsymbol{p}_g)$ is the probability that the gluon state at $\boldsymbol{p}_g$ is already occupied.

Using the explicit form of the Gell-Mann matrices in Eq. (10.34), the summation over i, j, and a can be easily carried out. We obtain

$$\sum_{i=1}^{3}\sum_{j=1}^{3}\sum_{a=1}^{8}\left|\frac{\lambda_{ij}^a}{2}\right|^2 = \frac{8 \times 2}{4} = 4\,. \tag{4}$$

From Eq. (3) and Eq. (4), the number of photons produced per unit time is

$$\frac{dN_\gamma}{dt} = 4N_s^2\sum_{f=1}^{N_f}\int\frac{d^3x\,d^3p_q\,d^3p_{\bar{q}}}{(2\pi)^6}f_q(\boldsymbol{p}_q)f_{\bar{q}}(\boldsymbol{p}_{\bar{q}})[1 + f_g(\boldsymbol{p}_q + \boldsymbol{p}_{\bar{q}} - \boldsymbol{p}_\gamma)]\bar{\sigma}_{q_f\bar{q}_f\to\gamma g}(s)v_{q\bar{q}}, \tag{5}$$

To obtain the distribution of the annihilation photons, we decompose the total cross section in terms of its contributions from various photon momenta,

$$\bar{\sigma}_{q_f\bar{q}_f\to\gamma g}(s) = \int E_\gamma\frac{d\bar{\sigma}}{d\boldsymbol{p}_\gamma}(q_f\bar{q}_f \to \gamma g)\frac{d\boldsymbol{p}_\gamma}{E_\gamma}\,. \tag{6}$$

Eqs. (5) and (6) lead to

$$\frac{dN_\gamma}{d^4x} = 4N_s^2\sum_{f=1}^{N_f}\int\frac{d^3p_q\,d^3p_{\bar{q}}}{(2\pi)^6}f_q(\boldsymbol{p}_q)f_{\bar{q}}(\boldsymbol{p}_{\bar{q}})[1 + f_g(\boldsymbol{p}_g)]E_\gamma\frac{d\bar{\sigma}}{d\boldsymbol{p}_\gamma}(q_f\bar{q}_f \to \gamma g)\frac{d\boldsymbol{p}_\gamma}{E_\gamma}v_{q\bar{q}}, \tag{7}$$

From the above equation, we get the annihilation photon distribution

$$E_\gamma\frac{dN_\gamma^{\text{ann}}}{d\boldsymbol{p}_\gamma d^4x} = 4N_s^2\sum_{f=1}^{N_f}\int\frac{d^3p_q\,d^3p_{\bar{q}}}{(2\pi)^6}f_q(\boldsymbol{p}_q)f_{\bar{q}}(\boldsymbol{p}_{\bar{q}})[1 + f_g(\boldsymbol{p}_g)]E_\gamma\frac{d\bar{\sigma}}{d\boldsymbol{p}_\gamma}(q_f\bar{q}_f\to\gamma g)v_{q\bar{q}}, \tag{8}$$

The integral in Eq. (8) is a six-fold integration. One of the two azimuthal angles can be trivially integrated out. Because the photon differential cross section contains a delta function in energy, as in Eq. (2) or Eq. (8) of Supplement 16.2, we can further integrate out the delta function in the photon energy to yield a four-dimensional integral, with non-trivial integration boundaries (see Ref. [11] for a detailed discussion of the integration boundaries).

To gain a general idea of the gross features of photon production in the quark-gluon plasma, we seek an approximate analytical expression for the integral in Eq. (8). We note that for a relativistic system of quarks, antiquarks and gluons, the photon is most likely emitted along the direction of the annihilating quark or antiquark, with the momentum and the energy of the annihilated particle. The photon differential cross section can be approximated by Eq. (16.21) as

$$E_\gamma\frac{d\bar{\sigma}}{d\boldsymbol{p}_\gamma}(q_f\bar{q}_f \to \gamma g) \approx \bar{\sigma}_{q_f\bar{q}_f\to\gamma g}(s)\frac{1}{2}E_\gamma[\delta(\boldsymbol{p}_\gamma - \boldsymbol{p}_q) + \delta(\boldsymbol{p}_\gamma - \boldsymbol{p}_{\bar{q}})]\,. \tag{9}$$

Equation (12) of Supplement 4.1 gives $v_{q\bar{q}} = |v_q - v_{\bar{q}}|$ in terms of other kinematic variables,

$$
\begin{aligned}
v_{q\bar{q}} &= \frac{\sqrt{(s - p_q^2 - p_{\bar{q}}^2)^2/4 - p_q^2 p_{\bar{q}}^2}}{E_q E_{\bar{q}}} \\
&= \frac{\sqrt{s(s - 4m^2)}}{2 E_q E_{\bar{q}}} .
\end{aligned} \tag{10}
$$

Substituting Eq. (9) and Eq. (10) into Eq. (8), we have

$$
E_\gamma \frac{dN_\gamma^{\mathrm{ann}}}{d\boldsymbol{p}_\gamma d^4 x} = 4 N_s^2 \sum_{f=1}^{N_f} \int \frac{d^3 p_q \, d^3 p_{\bar{q}}}{(2\pi)^6} f_q(\boldsymbol{p}_q) f_{\bar{q}}(\boldsymbol{p}_{\bar{q}})[1 + f_g(\boldsymbol{p}_1 + \boldsymbol{p}_{\bar{q}} - \boldsymbol{p}_\gamma)]
$$

$$
\times \, \bar{\sigma}_{q_f \bar{q}_f \to \gamma g}(s) \frac{E_\gamma}{2} [\delta(\boldsymbol{p}_\gamma - \boldsymbol{p}_q) + \delta(\boldsymbol{p}_\gamma - \boldsymbol{p}_{\bar{q}})] \frac{\sqrt{s(s - 4m^2)}}{2 E_q E_{\bar{q}}} . \tag{11}
$$

The integration over the delta functions yields

$$
\begin{aligned}
E_\gamma \frac{dN_\gamma^{\mathrm{ann}}}{d\boldsymbol{p}_\gamma d^4 x} = \frac{4 N_s^2 E_\gamma}{2(2\pi)^6} \sum_{f=1}^{N_f} \Bigg\{ & \int_{\boldsymbol{p}_\gamma \approx \boldsymbol{p}_q} d^3 p_{\bar{q}} f_q(\boldsymbol{p}_\gamma) f_{\bar{q}}(\boldsymbol{p}_{\bar{q}})[1 + f_g(\boldsymbol{p}_{\bar{q}})] \bar{\sigma}_{q_f \bar{q}_f \to \gamma g}(s) \frac{\sqrt{s(s - 4m^2)}}{2 E_\gamma E_{\bar{q}}} \\
& + \int_{\boldsymbol{p}_\gamma \approx \boldsymbol{p}_{\bar{q}}} d^3 p_q f_q(\boldsymbol{p}_q) f_{\bar{q}}(\boldsymbol{p}_\gamma)[1 + f_g(\boldsymbol{p}_q)] \bar{\sigma}_{q_f \bar{q}_f \to \gamma g}(s) \frac{\sqrt{s(s - 4m^2)}}{2 E_q E_\gamma} \Bigg\} .
\end{aligned}
$$

Because the distributions for quarks and antiquarks are taken to be the same, the two terms in the curly bracket of the above equation give the same contribution. The total contribution is just two times one of the terms. The above equation becomes

$$
E_\gamma \frac{dN_\gamma^{\mathrm{ann}}}{d\boldsymbol{p}_\gamma d^4 x} = \frac{4 N_s^2}{(2\pi)^6} f_q(\boldsymbol{p}_\gamma) \sum_{f=1}^{N_f} \int_{\boldsymbol{p}_\gamma \approx \boldsymbol{p}_{\bar{q}}} d^3 p_{\bar{q}} f_{\bar{q}}(\boldsymbol{p}_{\bar{q}})[1 + f_g(\boldsymbol{p}_{\bar{q}})] \bar{\sigma}_{q_f \bar{q}_f \to \gamma g}(s) \frac{\sqrt{s(s - 4m^2)}}{2 E_{\bar{q}}}, \tag{12}
$$

which is Eq. (16.25). To carry out the integration over $d^3 p_{\bar{q}}$, we shall assume that $f_{\bar{q}}(\boldsymbol{p}_{\bar{q}})$ and $f_g(\boldsymbol{p}_{\bar{q}})$ depend only on the energy $E_{\bar{q}}$. The differential element $d^3 p_{\bar{q}}$ can be written as

$$
\begin{aligned}
d^3 p_{\bar{q}} &= |\boldsymbol{p}_{\bar{q}}|^2 \, d|\boldsymbol{p}_{\bar{q}}| d\mu(\boldsymbol{p}_\gamma, \boldsymbol{p}_{\bar{q}}) \, d\phi \\
&\approx E_{\bar{q}}^2 dE_{\bar{q}} d\mu(\boldsymbol{p}_\gamma, \boldsymbol{p}_{\bar{q}}) \, d\phi .
\end{aligned}
$$

However, because $\boldsymbol{p}_\gamma \approx -\boldsymbol{p}_{\bar{q}}$, we have

$$
\begin{aligned}
s &= (p_q + p_{\bar{q}})^2 \\
&\approx (p_\gamma + p_{\bar{q}})^2 \\
&\approx 2 E_\gamma \, E_{\bar{q}}[1 + \mu(\boldsymbol{p}_\gamma, \boldsymbol{p}_{\bar{q}})] . \tag{13}
\end{aligned}
$$

Therefore, we have

$$ds \approx 2E_{\bar{q}}E_{\gamma}d\mu(\boldsymbol{p}_{\gamma}, \boldsymbol{p}_{\bar{q}}),$$

and

$$d^3 p_{\bar{q}} \approx \frac{E_{\bar{q}}}{2E_{\gamma}} dE_{\bar{q}} ds\, d\phi. \tag{14}$$

Substituting Eq. (14) into (12), we get

$$
E_{\gamma} \frac{dN_{\gamma}^{\mathrm{ann}}}{d\boldsymbol{p}_{\gamma} d^4 x}
$$

$$
= \frac{4N_s^2}{(2\pi)^6} f_q(\boldsymbol{p}_{\gamma}) \sum_{f=1}^{N_f} \int \frac{E_{\bar{q}}}{2E_{\gamma}} dE_{\bar{q}} ds\, d\phi f_{\bar{q}}(E_{\bar{q}})[1 + f_g(E_{\bar{q}})]\overline{\sigma}_{q_f \bar{q}_f \to \gamma g}(s) \frac{\sqrt{s(s - 4m^2)}}{2E_{\bar{q}}}
$$

$$
= \frac{4N_s^2}{(2\pi)^6} \frac{f_q(\boldsymbol{p}_{\gamma})}{4E_{\gamma}} \sum_{f=1}^{N_f} \int ds\, d\phi dE_{\bar{q}} f_{\bar{q}}(E_{\bar{q}})[1 + f_g(E_{\bar{q}})]\sqrt{s(s - 4m^2)}\,\overline{\sigma}_{q_f \bar{q}_f \to \gamma g}(s)
$$

$$
= \frac{4N_s^2}{(2\pi)^5} \frac{f_q(\boldsymbol{p}_{\gamma})}{4E_{\gamma}} \sum_{f=1}^{N_f} \int ds\, dE_{\bar{q}} f_{\bar{q}}(E_{\bar{q}})[1 + f_g(E_{\bar{q}})]\sqrt{s(s - 4m^2)}\,\overline{\sigma}_{q_f \bar{q}_f \to \gamma g}(s). \tag{15}
$$

This is Eq. (16.26) after we rewrite $\overline{\sigma}_{q\bar{q} \to \gamma g}(s)$ as $(e_q/e)^2 \overline{\sigma}_{\mathrm{ann}}(s)$, as defined by Eq. (16.22).

We can carry out the integration over $E_{\bar{q}}$. The lower limit of $E_{\bar{q}}$ is determined by the condition that because s is given by Eq. (13) and $|\mu(\boldsymbol{p}_{\gamma}, \boldsymbol{p}_{\bar{q}})| \leq 1$, we have

$$s \leq 4E_{\bar{q}}E_{\gamma},$$

or

$$E_{\bar{q}} \geq \frac{s}{4E_{\gamma}}.$$

For this case with $p_{\gamma} \approx p_q$ and a fixed value of s, the lower limit of $E_{\bar{q}}$ is $s/4E_{\gamma}$. To perform the integration over $E_{\bar{q}}$ in Eq. (15), we assume a Fermi-Dirac distribution for $f_{\bar{q}}(E_{\bar{q}})$

$$f_{\bar{q}}(E) = \frac{1}{e^{E/T} + 1} \tag{16}$$

and a Bose-Einstein distribution for the gluons

$$f_g(E) = \frac{1}{e^{E/T} - 1}. \tag{17}$$

The integrand in Eq. (15) contains the term

$$
f_{\bar{q}}(E_{\bar{q}})[1 + f_g(E_{\bar{q}})] = \frac{1}{e^{E_{\bar{q}}/T} + 1}\left[1 + \frac{1}{e^{E_{\bar{q}}/T} - 1}\right]
$$

$$
= \frac{1}{e^{E_{\bar{q}}/T} + 1} \frac{e^{E_{\bar{q}}/T}}{e^{E_{\bar{q}}/T} - 1}
$$

$$
= \frac{1}{2}\left[\frac{1}{e^{E_{\bar{q}}/T} + 1} + \frac{1}{e^{E_{\bar{q}}/T} - 1}\right]. \tag{18}
$$

Hence, the integral over $E_{\bar{q}}$ in Eq. (15) becomes

$$
\int_{s/4E_\gamma}^\infty dE_{\bar{q}} f_{\bar{q}}(E_{\bar{q}})[1 + f_g(E_{\bar{q}})] = \frac{T}{2}\left[\ln(1 + e^{-s/4E_\gamma T}) - \ln(1 - e^{-s/4E_\gamma T})\right]
$$

$$
= \frac{T}{2}\left[\sum_{n=1}^\infty \frac{(-1)^{n+1}(e^{-s/4E_\gamma T})^n}{n} - \sum_{n=1}^\infty \frac{(-1)^{n+1}(-e^{-s/4E_\gamma T})^n}{n}\right]
$$

$$
= \frac{T}{2}\left[\sum_{n=1}^\infty \frac{(-1)^{n+1}}{n}\left\{e^{-ns/4E_\gamma T} - (-1)^n e^{-ns/4E_\gamma T}\right\}\right]
$$

$$
= T \sum_{n=0,1,2,\ldots}^\infty \frac{e^{-(2n+1)s/4E_\gamma T}}{2n+1}. \tag{19}
$$

Eq. (15) and Eq. (19) lead to the annihilation photon distribution

$$
E_\gamma \frac{dN_\gamma^{\mathrm{ann}}}{d\boldsymbol{p}_\gamma d^4 x}
$$

$$
= \frac{4N_s^2}{(2\pi)^5} f_q(\boldsymbol{p}_\gamma) \frac{T}{4E_\gamma} \sum_{f=1}^{N_f} \int ds \sqrt{s(s-4m^2)}\, \bar{\sigma}_{q_f \bar{q}_f \to \gamma g}(s) \sum_{n=0}^\infty \frac{e^{-(2n+1)s/4E_\gamma T}}{2n+1}
$$

$$
= \frac{4N_s^2}{(2\pi)^5} f_q(\boldsymbol{p}_\gamma) \frac{T}{4E_\gamma} \sum_{f=1}^{N_f} \sum_{n=0}^\infty \frac{1}{2n+1} \int ds\, e^{-(2n+1)s/4E_\gamma T} \sqrt{s(s-4m^2)}\, \bar{\sigma}_{q_f \bar{q}_f \to \gamma g}(s).
$$

The total color-independent part of the cross section $\bar{\sigma}_{q_f \bar{q}_f \to \gamma g}(s)$ depends on the quark electric charge e_f according to Eq. (16.22a). Therefore, we can rewrite the above equation as

$$
E_\gamma \frac{dN_\gamma^{\mathrm{ann}}}{d\boldsymbol{p}_\gamma d^4 x} = \frac{4N_s^2}{(2\pi)^5} f_q(\boldsymbol{p}_\gamma) \frac{T}{4E_\gamma} \sum_{f=1}^{N_f}\left(\frac{e_f}{e}\right)^2
$$

$$
\times \sum_{n=0}^\infty \frac{1}{2n+1} \int ds\, e^{-(2n+1)s/4E_\gamma T} \sqrt{s(s-4m^2)}\, \bar{\sigma}_{\mathrm{ann}}(s),
$$

which is Eq. (16.27).   I⊕

Eq. (16.27) and the general formula Eq. (16.22b) for the annihilation cross section $\bar{\sigma}_{\mathrm{ann}}(s)$ can be used to evaluate the annihilation photon distribution in the quark-gluon plasma. Before we carry out this evaluation, we can study the magnitude of the quantity $\sqrt{s(s-4m^2)}\bar{\sigma}_{\mathrm{ann}}(s)$ which appears in the integrand of Eq. (16.27).

From Eq. (16.23) and Eq. (16.24), this quantity has the nonrelativistic limit

$$\sqrt{s(s-4m^2)}\,\bar{\sigma}_{\rm ann}(s) = 2\pi\alpha_e\alpha_s \frac{\sqrt{s}}{m}. \tag{16.28}$$

and the relativistic limit

$$\sqrt{s(s-4m^2)}\,\bar{\sigma}_{\rm ann}(s) = 4\pi\alpha_e\alpha_s\left\{\ln\left(\frac{s}{m^2}\right)-1\right\}. \tag{16.29}$$

These results indicate that $\sqrt{s(s-4m^2)}\,\bar{\sigma}_{\rm ann}(s)$ for $s \gg 4m^2$ is much greater than its corresponding value at $s \sim 4m^2$. The quantity $\sqrt{s(s-4m^2)}\,\bar{\sigma}_{\rm ann}(s)$ is an increasing function of s. The integrand in Eq. (16.27) contains also the distribution $\exp\{-(2n+1)s/4E_\gamma T\}$. The average value of s under this distribution is of the order of $4E_\gamma T/(2n+1)$. For high temperatures and high photon energies such that $E_\gamma T$ is much greater than m^2, the average value of $\sqrt{s}$ is much greater than the quark rest mass and we are in the relativistic regime. We are justified to use the relativistic approximation (16.24) or the corresponding Eq. (16.29) to evaluate the integral in (16.27). Using this relativistic approximation, we shall show in Supplement 16.4 that the annihilation photon distribution in the plasma is given by

$$E_\gamma \frac{dN_\gamma^{\rm ann}}{d\mathbf{p}_\gamma d^4x} = \frac{N_s^2\alpha_e\alpha_s}{16\pi^2}f_q(\mathbf{p}_\gamma)T^2\sum_{f=1}^{N_f}\left(\frac{e_f}{e}\right)^2\left\{\ln\left(\frac{4E_\gamma T}{m^2}\right)+C_{\rm ann}\right\}, \tag{16.30a}$$

where

$$C_{\rm ann} = -C_{\rm Euler} - 1 - \frac{8}{\pi^2}\sum_{n=0}^{\infty}\frac{\ln(2n+1)}{(2n+1)^2}, \tag{16.30b}$$

and $C_{\rm Euler}$ is the Euler number 0.577215. Numerical evaluation of the series gives

$$\sum_{n=0}^{\infty}\frac{\ln(2n+1)}{(2n+1)^2} = 0.41811584,$$

and the constant $C_{\rm ann}$ is

$$C_{\rm ann} = -1.91613. \tag{16.30c}$$

The annihilation photon distribution depends on T and the mass of the quark m.

For a quark-gluon plasma with u and d quarks, $N_f = 2$ and we have

$$\sum_{f=1}^{N_f} \left(\frac{e_f}{e}\right)^2 = \frac{4}{9} + \frac{1}{9} = \frac{5}{9}.$$

Eq. (16.30) becomes

$$E_\gamma \frac{dN_\gamma^{\text{ann}}}{d\boldsymbol{p}_\gamma d^4 x} = \frac{5}{9} \frac{\alpha_e \alpha_s}{4\pi^2} f_q(\boldsymbol{p}_\gamma) T^2 \left\{ \ln\left(\frac{4E_\gamma T}{m^2}\right) + C_{\text{ann}} \right\}. \qquad (16.31)$$

The annihilation photon distribution, Eq. (16.30), is obtained with the Fermi-Dirac distribution for $f_{\bar{q}}(E_{\bar{q}})$ and the Bose-Einstein distribution for $f_g(E_g)$. It is of interest to compare the photon distribution if we use the Boltzmann distribution for $f_{\bar{q}}(E_{\bar{q}})$ and the Bose-Einstein distribution for $f_g(E_g)$, as in Ref. 5. We shall show in Supplement 16.4 that the annihilation photon distribution for that case is

$$E_\gamma \frac{dN_\gamma^{\text{ann}}}{d\boldsymbol{p}_\gamma d^4 x} = \frac{5}{9} \frac{\alpha_e \alpha_s}{3\pi^2} f_q(\boldsymbol{p}_\gamma) T^2 \left\{ \ln\left(\frac{4E_\gamma T}{m^2}\right) + C'_{\text{ann}} \right\}, \qquad (16.32a)$$

where the constant C'_{ann} is

$$C'_{\text{ann}} = -C_{\text{Euler}} - 1 - \frac{6}{\pi^2} \sum_{n=1}^{\infty} \frac{\ln n}{n^2}. \qquad (16.32b)$$

The above results of Eq. (16.32a) and (16.32b) agree with Eq. (14) of Kapusta *et al.* [5], except that in Ref. [5] quarks are assumed to be massless and an infrared cut-off k_c is used, with the quark-mass m in Eq. (16.32a) replaced by k_c.

By comparing Eq. (16.31) and Eq. (16.32), we note that when we use different antiquark distributions, the leading-log term of the photon distributions differ by a factor of $(1/4) : (1/3)$.

$\oplus$[Supplement 16.4

We show how the result of Eq. (16.30) is obtained. Upon using the relativistic approximation (16.29) for $\sqrt{s(s-4m^2)}\bar{\sigma}_{\text{ann}}(s)$, the integral in Eq. (16.27) becomes

$$\int ds \, e^{-(2n+1)s/4E_\gamma T} \sqrt{s(s-4m^2)}\bar{\sigma}_{\text{ann}}(s)$$

$$= 4\pi\alpha_e\alpha_s \int ds \, e^{-(2n+1)s/4E_\gamma T} \left\{ \ln\left(\frac{s}{m^2}\right) - 1 \right\}. \qquad (1)$$

The lower limit of s, as determined from $(p_q + p_{\bar{q}})^2$, is $4m^2$. In the relativistic approximation (16.24), the cross section at this lower limit of energy is a poor expression for very low energies. However, the dominant contribution to the integral comes from the region around $s \sim 4E_\gamma T/(2n+1)$, which is far from the low energy region. The low energy region does not need to be too accurate. We make the change of variable $s = 4m^2 z$. Integral (1) becomes

$$\int ds\, e^{-(2n+1)s/4E_\gamma T} \sqrt{s(s-4m^2)}\,\overline{\sigma}_{\text{ann}}(s)$$

$$\approx 4\pi\alpha_e\alpha_s \int_{4m^2}^{\infty} ds\, e^{-(2n+1)s/4E_\gamma T} \left\{ \ln\left(\frac{s}{m^2}\right) - 1 \right\}$$

$$= 4\pi\alpha_e\alpha_s 4m^2 \int_{1}^{\infty} dz\, e^{-(2n+1)4m^2 z/4E_\gamma T} \left\{ \ln\left(\frac{4m^2 z}{m^2}\right) - 1 \right\}$$

$$\cong 4\pi\alpha_e\alpha_s 4m^2 \int_{1}^{\infty} dz\, e^{-(2n+1)4m^2 z/4E_\gamma T} \left\{ \ln(z) + \ln 4 - 1 \right\}. \tag{2}$$

From Eq. (8.212.16) and Eq. (8.214.1) of Ref. [12], we have

$$\int_{1}^{\infty} dz\, e^{-xz}\, \ln(z) = \frac{Ei(-x)}{-x} \qquad \text{for } 0 < x < 1, \tag{3}$$

where the exponential-integral function Ei is

$$Ei(-x) = C_{\text{Euler}} + \ln(x) + \sum_{k=1}^{\infty} \frac{(-x)^k}{k \cdot k!} \qquad \text{for } 0 < x. \tag{4}$$

Using these results for Eq. (2), we identify x as

$$x = \frac{(2n+1)4m^2}{4E_\gamma T},$$

and we obtain

$$\int ds\, e^{-(2n+1)s/4E_\gamma T} \sqrt{s(s-4m^2)}\,\overline{\sigma}_{\text{ann}}(s)$$

$$= -4\pi\alpha_e\alpha_s\, 4m^2 \frac{4E_\gamma T}{(2n+1)4m^2} \left[C_{\text{Euler}} + \ln\left\{ \frac{(2n+1)4m^2}{4E_\gamma T} \right\} + \sum_{k=1}^{\infty} \frac{(-x)^k}{k \cdot k!} \right.$$

$$\left. - (\ln 4 - 1)\left(1 + \sum_{k=1}^{\infty} \frac{(-x)^k}{\cdot k!} \right) \right].$$

For $m^2 \ll E_\gamma T$, the last summation over $(-x)^k = (-m^2/4E_\gamma T)^k$ can be neglected and the above equation becomes

$$\int ds\, e^{-(2n+1)s/4E_\gamma T} \sqrt{s(s-4m^2)}\,\overline{\sigma}_{\text{ann}}(s)$$

$$= -4\pi\alpha_e\alpha_s \frac{4E_\gamma T}{(2n+1)} \left[\ln\left\{ \frac{m^2}{4E_\gamma T} \right\} + C_{\text{Euler}} + 1 + \ln(2n+1) \right]$$

$$= 4\pi\alpha_e\alpha_s \frac{4E_\gamma T}{(2n+1)} \left[\ln\left\{ \frac{4E_\gamma T}{m^2} \right\} - C_{\text{Euler}} - 1 - \ln(2n+1) \right]. \tag{5}$$

Substituting Eq. (5) into Eq. (16.27), we get

$$
E_\gamma \frac{dN_\gamma^{\rm ann}}{d\boldsymbol{p}_\gamma d^4 x} = \frac{4N_s^2}{(2\pi)^5} f_q(\boldsymbol{p}_\gamma) \frac{T}{4E_\gamma} \sum_{f=1}^{N_f} \left(\frac{e_f}{e}\right)^2
$$
$$
\times \sum_{n=0}^{\infty} \frac{1}{2n+1} 4\pi \alpha_e \alpha_s \frac{4E_\gamma T}{(2n+1)} \left[\ln\left\{ \frac{4E_\gamma T}{m^2} \right\} - C_{\rm Euler} - 1 - \ln(2n+1) \right]. \quad (6)
$$

From Eq. (23.2.20) of Ref. [13], we have

$$
\sum_{n=0}^{\infty} \frac{1}{(2n+1)^2} = (1 - 2^{-2})\zeta(2)
$$
$$
= \frac{3}{4} \cdot \frac{\pi^2}{6}. \quad (7)
$$

Hence, Eq. (6) becomes

$$
E_\gamma \frac{dN_\gamma^{\rm ann}}{d\boldsymbol{p}_\gamma d^4 x} = \frac{N_s^2 \alpha_e \alpha_s}{16\pi^2} f_q(\boldsymbol{p}_\gamma) T^2 \sum_{f=1}^{N_f} \left(\frac{e_f}{e}\right)^2 \left[\ln\left\{ \frac{4E_\gamma T}{m^2} \right\} - C_{\rm Euler} - 1 - \frac{8}{\pi^2} \sum_{n=1}^{\infty} \frac{\ln(2n+1)}{(2n+1)^2} \right],
$$

which is Eq. (16.30).

Equation (16.30) is obtained with the Fermi-Dirac distribution for $f_{\bar{q}}(E_{\bar{q}})$ and the Bose-Einstein distribution for $f_g(E_g)$. What is the photon distribution if we use the Boltzmann distribution for $f_{\bar{q}}(E_{\bar{q}})$ and the Bose-Einstein distribution for $f_g(E_g)$, as in Ref. 5?

If we follow Ref. 5, then in Eq. (16.25), we have

$$
f_{\bar{q}}(E_{\bar{q}})[1 + f_g(E_{\bar{q}})] = \frac{e^{-E_{\bar{q}}/T}}{1 - e^{-E_{\bar{q}}/T}}.
$$

The integral in $E_{\bar{q}}$ of Eq. (16.25) becomes

$$
\int_{s/4E_\gamma}^{\infty} dE_{\bar{q}} f_{\bar{q}}(E_{\bar{q}})[1 + f_g(E_{\bar{q}})] = -T \ln\left(1 - e^{-s/4E_\gamma T} \right)
$$
$$
= T \sum_{n=1}^{\infty} \frac{e^{-ns/4E_\gamma T}}{n}.
$$

The annihilation photon distribution (16.26) becomes

$$
E_\gamma \frac{dN_\gamma^{\rm ann}}{d\boldsymbol{p}_\gamma d^4 x} = \frac{4N_s^2}{(2\pi)^5} f_q(\boldsymbol{p}_\gamma) \frac{T}{4E_\gamma} \sum_{f=1}^{N_f} \left(\frac{e_f}{e}\right)^2 \sum_{n=1}^{\infty} \frac{1}{n} \int ds \, e^{-ns/4E_\gamma T} \sqrt{s(s-4m^2)} \, \bar{\sigma}_{\rm ann}(s). \quad (8)
$$

Upon using the result of Eq. (5), we have

$$\int ds\, e^{-ns/4E_\gamma T}\sqrt{s(s-4m^2)}\,\overline{\sigma}_{\text{ann}}(s)=4\pi\alpha_e\alpha_s\frac{4E_\gamma T}{n}\left[\ln\left\{\frac{4E_\gamma T}{m^2}\right\}-C_{\text{Euler}}-1-\ln n\right].$$

Using Eq. (23.2.24) of Ref. [13],

$$\sum_{n=1}^{\infty}\frac{1}{n^2}=\zeta(2)=\frac{\pi^2}{6},$$

we can rewrite the annihilation photon distribution, Eq. (8), as

$$E_\gamma\frac{dN_\gamma^{\text{ann}}}{d\boldsymbol{p}_\gamma d^4 x}=\frac{4N_s^2\alpha_e\alpha_s}{48\pi^2}f_q(\boldsymbol{p}_\gamma)T^2\sum_{f=1}^{N_f}\left(\frac{e_f}{e}\right)^2\left[\ln\left\{\frac{4E_\gamma T}{m^2}\right\}-C_{\text{Euler}}-1-\frac{6}{\pi^2}\sum_{n=1}^{\infty}\frac{\ln n}{n^2}\right].$$

For a system of u and d quarks, $N_f = 2$ and $\sum_f(e_f/e)^2 = 5/9$. When we use a Boltzmann distribution for $f_{\overline{q}}(E_{\overline{q}})$ and a Bose-Einstein distribution for $f_g(E_{\overline{q}})$, the annihilation photon distribution is

$$E_\gamma\frac{dN_\gamma^{\text{ann}}}{d\boldsymbol{p}_\gamma d^4 x}=\frac{5}{9}\frac{\alpha_e\alpha_s}{3\pi^2}f_q(\boldsymbol{p}_\gamma)T^2\left[\ln\left\{\frac{4E_\gamma T}{m^2}\right\}-C_{\text{Euler}}-1-\frac{6}{\pi^2}\sum_{n=1}^{\infty}\frac{\ln n}{n^2}\right], \quad (9)$$

which is Eq. (16.32), as had to be proved. It agrees with Eq. (14) of Kapusta *et al.* [5], when the quark mass m in Eq. (9) is replaced by the infrared momentum cut-off k_c for massless quarks. ⊕

§16.4 Photon Production by the Compton Process

We turn now to the second process of photon production in which a gluon interacts with a quark to produce a photon by the reaction

$$g + q \rightarrow \gamma + q, \quad (16.33a)$$

or with an antiquark $\overline{q}$ to produce a photon by the reaction

$$g + \overline{q} \rightarrow \gamma + \overline{q}. \quad (16.33b)$$

These reactions are represented by the Feynman diagrams 16.1(c) and 16.1(d). The scattering of a gluon off a quark or an antiquark to produce a photon is analogous to the Compton reaction which involves the scattering of a photon off a charged particle. The processes (16.33a) and (16.33b) are therefore called *Compton processes*.

The photon differential cross section for the $gq \rightarrow \gamma q$ reaction (16.33a) and photon differential cross sections for the $g\overline{q} \rightarrow \gamma\overline{q}$ reaction

(16.33b) are the same. It suffices to discuss only the $gq \to \gamma q$ process, while the results for the other $g\bar{q} \to \gamma\bar{q}$ process can be considered to behave in the same way.

To study the $gq \to \gamma q$ process, it is useful to utilize well-known results for the analogous Compton process $\gamma q \to \gamma q$ which appear in standard textbooks [6,8,9]. Consider the scattering between a gluon of index a and a quark $q(i)$ with a color index i to result in a photon γ and a quark $q(j)$ with a color index j, in the $g(a)\,q(i) \to \gamma\,q(j)$ reaction. From the rules of the Feynman diagrams [6,7], the gluon-quark vertex in this reaction is associated with the quantity

$$g\gamma_{\alpha\beta}^{\mu}\frac{\lambda_{ij}^{a}}{2}. \qquad (16.34)$$

In the analogous $\gamma q \to \gamma q$ process, the corresponding photon-quark vertex is associated with the quantity [6-9]

$$-ie_{q}\gamma_{\alpha\beta}^{\mu}. \qquad (16.35)$$

Except for these differences, the two processes are identical with respect to particle momenta, particle spins and particle polarizations. Consequently, just as in the case of photon production by $q\bar{q}$ annihilation discussed earlier in Section §16.2, we can separate out the color-dependence of the differential cross section by comparing Eq. (16.34) with Eq. (16.35). The photon differential cross section for $g(a)\,q(i) \to \gamma\,q(j)$ is

$$E_{\gamma}\frac{d\sigma}{d\boldsymbol{p}_{\gamma}}[g(a)\,q(i) \to \gamma\,q(j)] = \left|\frac{\lambda_{ij}^{a}}{2}\right|^{2} E_{\gamma}\frac{d\bar{\sigma}}{d\boldsymbol{p}_{\gamma}}(gq \to \gamma q), \qquad (16.36)$$

where $E_{\gamma}d\bar{\sigma}/d\boldsymbol{p}_{\gamma}(gq \to \gamma q)$ is the color-independent part of the photon differential cross section. This photon differential cross section $E_{\gamma}d\bar{\sigma}/d\boldsymbol{p}_{\gamma}(gq \to \gamma q)$ is related to the photon differential section for the $\gamma q \to \gamma q$ process, pertaining to the same particle momenta, spins, and polarizations, by

$$E_{\gamma}\frac{d\bar{\sigma}}{d\boldsymbol{p}_{\gamma}}(gq \to \gamma q) = \left(\frac{g}{e_{q}}\right)^{2} E_{\gamma}\frac{d\sigma}{d\boldsymbol{p}_{\gamma}}(\gamma q \to \gamma q)$$

$$= \frac{\alpha_{s}}{\alpha_{e}}\left(\frac{e}{e_{q}}\right)^{2} E_{\gamma}\frac{d\sigma}{d\boldsymbol{p}_{\gamma}}(\gamma q \to \gamma q). \qquad (16.37)$$

To write out the photon differential cross section explicitly using the above equation, we label the reactions on the two sides of the equation as $p_1 + p_2 \rightarrow p_3 + p_4$. The identities of p_1 for the two reactions on the two sides of Eq. (16.37) are different. We identify the gluon momentum p_g as p_1 for the $gq \rightarrow \gamma q$ reaction on the lefthand side of Eq. (16.37), and identify the incident photon momentum p_γ as p_1 for the $\gamma q \rightarrow \gamma q$ reaction on the righthand side. We identify the initial quark momentum p_q as p_2, the final photon momentum p_γ as p_3, the final quark momentum p'_q as p_4.

We use the Mandelstam variables

$$s = (p_1 + p_2)^2 = (p_g + p_q)^2, \tag{16.38a}$$

$$t = (p_1 - p_3)^2 = (p_g - p_\gamma)^2, \tag{16.38b}$$

and

$$u = (p_1 - p_4)^2 = (p_2 - p_3)^2 = (p_q - p_\gamma)^2. \tag{16.38c}$$

The photon differential cross section for the $\gamma q \rightarrow \gamma q$ reaction, averaged over initial photon or gluon polarizations and quark spins and summed over final photon polarizations and quark spins, is given by Eq. (86.6) of Ref. [8] as

$$\frac{d\sigma}{dt}(\gamma q \rightarrow \gamma q) = \left(\frac{e_q}{e}\right)^4 \frac{8\pi\alpha_e^2}{(s-m^2)^2}\Bigg\{\left(\frac{m^2}{s-m^2} + \frac{m^2}{u-m^2}\right)^2$$
$$+ \left(\frac{m^2}{s-m^2} + \frac{m^2}{u-m^2}\right) - \frac{1}{4}\left(\frac{s-m^2}{u-m^2} + \frac{u-m^2}{s-m^2}\right)\Bigg\}, \tag{16.39}$$

where the additional factor $(e_q/e)^4$ is included here to take into account the electric charge of the quark. From Eqs. (16.37) and (16.39), we have

$$\frac{d\bar{\sigma}}{dt}(gq \rightarrow \gamma q) = \left(\frac{e_q}{e}\right)^2 \frac{8\pi\alpha_s\alpha_e}{(s-m^2)^2}\Bigg\{\left(\frac{m^2}{s-m^2} + \frac{m^2}{u-m^2}\right)^2$$
$$+ \left(\frac{m^2}{s-m^2} + \frac{m^2}{u-m^2}\right) - \frac{1}{4}\left(\frac{s-m^2}{u-m^2} + \frac{u-m^2}{s-m^2}\right)\Bigg\}. \tag{16.40}$$

We observe from the above equations that the photon differential cross section contains the term $(s-m^2)/(u-m^2)$ where the factor $1/(u-m^2)$ arises from the quark propagator between the two vertices in the Feynman diagram 16.1(d). The term $(s-m^2)/(u-m^2)$ becomes dominant when $|u-m^2|$ becomes much smaller than $s-m^2$. The situation is similar to the previous case of photon production by $q\bar{q}$

annihilation discussed in Section §16.2. Therefore, we can repeat similar arguments and approximations here. The photon differential cross section at high energies is a maximum at the photon momenta for which $|u - m^2|$ is a minimum. The quantity $|u - m^2|$ is

$$|u - m^2| = 2E_\gamma(E_q - |\boldsymbol{p}_q| \cos \theta_{\gamma q}),$$

which has a minimum when

$$\theta_{\gamma q} = 0. \tag{16.41}$$

This occurs when the photon momentum $\boldsymbol{p}_\gamma$ lines up with the initial quark momentum $\boldsymbol{p}_q$. Thus, the photon differential cross section for $gq \to \gamma q$ has a peak in the direction of the initial quark q.

What is the angular width of the photon differential cross section at this peak? We can again expand $(u - m^2)^{-1}$ about $\theta_{\gamma q}$ where the differential cross section is maximum. We have

$$\frac{1}{u - m^2} \approx \frac{1}{-2E_\gamma[E_q - |\boldsymbol{p}_q| + |\boldsymbol{p}_q|\frac{1}{2}(\theta_{\gamma q})^2]}$$

$$\approx -\frac{1}{E_\gamma E_q\left[\left(\frac{m}{E_q}\right)^2 + (\theta_{\gamma q})^2\right]}.$$

Therefore, around the direction of the quark momentum $\boldsymbol{p}_q$, the photon production cross section drops down rapidly beyond the angular cone

$$\Delta\theta_{\gamma q} = \frac{m}{E_q}. \tag{16.42}$$

When the interacting quark is relativistic with $E_q \gg m$, the produced photon in the $gq \to \gamma q$ process is most likely found in a narrow cone $\Delta\theta$ along the direction of the initial quark q.

⊕⟦ Supplement 16.5

It is useful to obtain the energies of the particles in the center-of-mass frame in terms of $\sqrt{s}$ for the reaction $gq \to \gamma q$. The center-of-mass energy is

$$\sqrt{s} = E_g^* + E_q^*.$$

and the magnitude of the momentum of g is the same as the magnitude of the momentum of q,

$$|\boldsymbol{p}_g^*| = E_g^* = |\boldsymbol{p}_q^*| = \sqrt{(E_q^*)^2 - m^2}.$$

Consequently, the equation for E_g^* is

$$\sqrt{s} = E_g^* + \sqrt{(E_g^*)^2 + m^2} \,,$$

which can be solved for E_g^* as a function of s. We find

$$E_g^* = \frac{s - m^2}{2\sqrt{s}} \,, \tag{1}$$

and

$$E_q^* = \sqrt{s} - E_g^* = \frac{s + m^2}{2\sqrt{s}} \,. \tag{2}$$

It can be similarly shown that for the final particles γ and q, we have

$$E_\gamma^* = \frac{s - m^2}{2\sqrt{s}} \,, \tag{3}$$

and the final energy of the quark in the center-of-mass frame is

$$E_q^{*\prime} = \sqrt{s} - E_g^* = \frac{s + m^2}{2\sqrt{s}} \,. \tag{4}$$

We would like to obtain the upper and lower limits of t and u for a given value of s. The variable t is given from Eq. (16.38b) by

$$\begin{aligned}
t &= (p_g - p_\gamma)^2 \\
&= -2E_g^* E_\gamma^*[1 - \cos\theta_{\gamma^*g^*}] \,.
\end{aligned} \tag{5}$$

The upper limit of t, which we label as t_+, occurs at $\theta_{\gamma^*g^*} = 0$ and the lower limit of t, which we denote as t_-, occurs at $\theta_{\gamma^*g^*} = \pi$. From Eqs. (1), (3), and (5), we find

$$t_+ = 0$$

and

$$\begin{aligned}
t_- &= -4E_g^* E_\gamma^* \\
&= -\frac{(s - m^2)^2}{s} \,.
\end{aligned}$$

Because $s + t + u = 2m^2$, the corresponding upper limit of u is given by

$$u_+ = \frac{m^4}{s} \,,$$

and the lower limit of u is

$$u_- = 2m^2 - s \,. \qquad\qquad]\oplus$$

We can use arguments similar to those in Section §16.2 to determine the most likely energy of the photon. From Supplement 16.5, the photon energy in the center-of-mass system is

$$E_\gamma^* = \frac{s - m^2}{2\sqrt{s}}.$$

(16.43)

On the other hand, the photon energy in the center-of-mass system is related to the photon momentum p_γ in any frame by Eq. (5) of Supplement 16.2,

$$E_\gamma^* = \frac{p_\gamma \cdot (p_1 + p_2)}{\sqrt{s}}.$$

We have identified p_1 with the gluon momentum p_g, and p_2 with the quark momentum p_q. Therefore, the condition which determines the photon energy is

$$p_\gamma \cdot p_g + p_\gamma \cdot p_q - \frac{\sqrt{s}(s - m^2)}{2\sqrt{s}} = 0.$$

(16.44)

When the photon momentum $\boldsymbol{p}_\gamma$ is colinear with $\boldsymbol{p}_q$, we have $\theta_{\gamma q} = 0$ and the scalar product $p_\gamma \cdot p_q$ in the above equation is

$$p_\gamma \cdot p_q \approx \frac{E_\gamma}{2E_q} m^2.$$

Consequently the left-hand-side of Eq. (16.44) becomes

$$
\begin{aligned}
p_\gamma \cdot (p_g + p_q) - \frac{s - m^2}{2} &= p_\gamma \cdot p_g + p_\gamma \cdot p_q - \frac{m^2 + 2p_g \cdot p_q - m^2}{2} \\
&= p_\gamma \cdot p_g + p_\gamma \cdot p_q - p_g \cdot p_q \\
&\approx (p_\gamma - p_q) \cdot p_g + \frac{E_\gamma}{2E_q} m^2.
\end{aligned}
$$

(16.45)

For the relativistic case when the rest mass m is small and negligible compared to other energy scales, Eq. (16.45) shows that because p_g can be any momentum 4-vector, the condition Eq. (16.44) is approximately satisfied if the photon momentum 4-vector is approximately the same as the momentum 4-vector of the quark,

$$p_\gamma \approx p_q.$$

(16.46)

From the results of Eqs. (16.41), (16.42), and (16.46), the process of Compton scattering $gq \to \gamma q$ at relativistic energies can be viewed as if the initial quark q in this reaction turns itself into a photon γ with approximately the same energy and momentum, while the initial gluon g turns into the final quark q with approximately the same energy and momentum.

From this pictorial description of the process, we can approximate the color-independent part of the photon differential cross section for the relativistic case as

$$E_\gamma \frac{d\bar\sigma}{d\boldsymbol{p}_\gamma}(gq \to \gamma q) \approx \bar\sigma_{gq \to \gamma q}(s) E_\gamma \delta(\boldsymbol{p}_\gamma - \boldsymbol{p}_q). \tag{16.47}$$

The total cross section can be obtained by integrating $d\bar\sigma/dt$ over the allowed range $t_- < t < t_+$. The total cross section for the analogous Compton $\gamma e \to \gamma e$ process is given explicitly by Eq. (86.16) of Ref. [8]. The corresponding total cross section for the $gq \to \gamma q$ process is

$$\bar\sigma_{gq \to \gamma q}(s) = \left(\frac{e_q}{e}\right)^2 \bar\sigma_{\mathrm{Comp}}(s), \tag{16.48}$$

where

$$\bar\sigma_{\mathrm{Comp}}(s) = \frac{2\pi\alpha_e\alpha_s}{s - m^2} \left\{ \left(1 - \frac{4m^2}{s - m^2} - \frac{8m^4}{(s - m^2)^2}\right) \ln\left\{\frac{s}{m^2}\right\} \right.$$
$$\left. + \frac{1}{2} + \frac{8m^2}{s - m^2} - \frac{m^4}{2s^2} \right\}. \tag{16.49}$$

In the non-relativistic limit of $s \sim m^2$, the total cross section $\bar\sigma_{\mathrm{Comp}}(s)$ is

$$\bar\sigma_{\mathrm{Comp}}(s) = \frac{8\pi\alpha_e\alpha_s}{3m^2} \left\{ 1 - \frac{s - m^2}{m^2} \right\}, \tag{16.50}$$

and in the relativistic case with $s \gg m^2$, it is

$$\bar\sigma_{\mathrm{Comp}}(s) = \frac{2\pi\alpha_e\alpha_s}{s} \left\{ \ln\left(\frac{s}{m^2}\right) + \frac{1}{2} \right\}. \tag{16.51}$$

§16.5 Photon Production in the Plasma due to the Compton Process

The evaluation of the photon distribution due to the Compton process can be carried out in the same way as the previous evaluation of the annihilation photon distribution (see Section §16.3). Using the approximate differential cross section (16.47), we shall show in Supplement 16.6 that the Compton photon distribution from gq collisions in a quark-gluon plasma can be written as

$$E_\gamma \frac{dN_\gamma(gq \to \gamma q)}{dp_\gamma d^4 x} = \frac{4N_s N_\epsilon}{(2\pi)^6} f_q(\boldsymbol{p}_\gamma)$$

$$\times \sum_{f=1}^{N_f} \int_{p_\gamma \approx p_{\bar{q}}} d^3 p_g f_g(\boldsymbol{p}_g)[1 - f_q(\boldsymbol{p}_g)]\bar{\sigma}_{gq_f \to \gamma q_f}(s)\frac{(s - m^2)}{2E_g}. \quad (16.52)$$

where $N_s(= 2)$ is the spin degeneracy of the quarks and $N_\epsilon(= 2)$ is the number of polarizations of the gluons.

Eq. (16.52) gives the result that the photon distribution is directly proportional to the quark momentum distribution $f_q(\boldsymbol{p}_q)$ evaluated at the photon momentum $\boldsymbol{p}_\gamma = \boldsymbol{p}_q$. As in the annihilation case, we can understand such a dependence as discussed already in the last section. The Compton process $gq \to \gamma q$ at relativistic energies occurs as if the quark is converted into a photon of approximately the same energy and momentum.

We shall see later that the integral over $d^3 p_g$ in Eq. (16.52) gives a function proportional to $(\ln E_\gamma + \text{constant})$, which is a slowly-varying function of the photon energy E_γ. Hence, the photon distribution is rather insensitive to the gluon distribution. The dominant variation of the photon distribution in the $gq \to \gamma q$ reaction comes from the quark distribution $f_q(\boldsymbol{p}_q)$. Similarly the dominant variation of the photon distribution in the $g\bar{q} \to \gamma\bar{q}$ reaction comes from the antiquark distribution $f_{\bar{q}}(\boldsymbol{p}_{\bar{q}})$. The photon distribution carries information on the momentum distribution of the quarks and antiquarks in the plasma.

It is worth noting that Eq. (16.52) reveals that the momentum distribution of energetic photons depend more sensitively on the quark and the antiquark distributions, and much less sensitively on the gluon distribution. The reason for this difference arises from the differential cross section (16.47) where the photon momentum take on the quark momentum or the antiquark momentum, in a gluon Compton scattering process at high energies.

To carry out the integral in Eq. (16.52), we assume that the gluon distribution $f_g(\boldsymbol{p}_g)$ and the quark distribution $f_q(\boldsymbol{p}_q)$ depend only on energy. We shall show in Supplement 16.6 that the photon distribution from the $gq \to \gamma q$ reaction becomes

$$E_\gamma \frac{dN_\gamma(gq \rightarrow \gamma q)}{d\boldsymbol{p}_\gamma d^4 x} = \frac{4N_s N_\epsilon}{(2\pi)^5} \frac{f_q(\boldsymbol{p}_\gamma)}{4E_\gamma} \sum_{f=1}^{N_f} \left(\frac{e_f}{e}\right)^2$$

$$\times \int ds \, dE_g f_g(E_g)[1 - f_q(E_g)](s - m^2)\overline{\sigma}_{\text{Comp}}(s). \qquad (16.53)$$

For photon production by the Compton process, $f_g(E_g)$ is a Bose-Einstein distribution and $f_q(E_q)$ is a Fermi-Dirac distribution, the integral in E_g can be carried out analytically and we obtain the Compton photon distribution from the $gq \rightarrow \gamma q$ reaction as a one-dimensional integral

$$E_\gamma \frac{dN_\gamma(gq \rightarrow \gamma q)}{d\boldsymbol{p}_\gamma d^4 x} = \frac{4N_s N_\epsilon}{(2\pi)^5} f_q(\boldsymbol{p}_\gamma) \frac{T}{4E_\gamma} \sum_{f=1}^{N_f} \left(\frac{e_f}{e}\right)^2$$

$$\times \sum_{n=0}^{\infty} \frac{1}{2n+1} \int ds \, e^{-(2n+1)s/4E_\gamma T}(s - m^2)\overline{\sigma}_{\text{Comp}}(s). \qquad (16.54)$$

$\oplus$[Supplement 16.6

To establish the results of Eqs. (16.52) − (16.54), we follow closely the steps of Exercise 16.3 for photon production by the annihilation process $q\bar{q} \rightarrow \gamma g$ process, which can be consulted for comparison.

The number of gluons of type a with momentum $\boldsymbol{p}_g$ and energy $E_g = |\boldsymbol{p}_g|$, in the volume element $d^3 x$ and the momentum element $d^3 p_g$, is

$$\frac{d^3 x \, d^3 p_g}{(2\pi)^3} N_\epsilon f_g(\boldsymbol{p}_g), \qquad (1)$$

where $N_\epsilon (= 2)$ is the number of polarizations of the gluon. The number of quarks of flavor f, color index i, momentum $\boldsymbol{p}_q$, and energy $E_q = (\boldsymbol{p}_q^2 + m^2)^{1/2}$ in the momentum element $d^3 p_q$, with which a gluon of type a can interact in a unit time to produce a photon and a quark of flavor f and color index j by the reaction $g(a) q_f(i) \rightarrow \gamma q_f(j)$, is

$$\sigma_{g(a)q_f(i) \rightarrow \gamma q_f(j)}(s) v_{gq} \frac{d^3 p_q}{(2\pi)^3} N_s f_q(\boldsymbol{p}_q). \qquad (2)$$

We separate out the color dependence of the cross section $\sigma_{g(a)q_f(i) \rightarrow \gamma q_f(j)}(s)$ by using Eq. (16.36),

$$\sigma_{g(a)q_f(i) \rightarrow \gamma \bar{q}_f(j)}(s) = \left|\frac{\lambda_{ij}^a}{2}\right|^2 \overline{\sigma}_{gq_f \rightarrow \gamma q_f}(s),$$

where $\overline{\sigma}_{gq_f \to \gamma q_f}(s)$ is the color-independent part of the total cross section for the $g(a)q_f(i) \to \gamma q_f(j)$ reaction.

The number of photons produced per unit time is the integral of the product of Eqs. (1) and (2), summed over all gluon indices a, color indices i and j, and flavor f:

$$\frac{dN_\gamma}{dt} = N_s N_\epsilon \sum_{a=1}^{8} \sum_{i=1}^{3} \sum_{j=1}^{3} \sum_{f=1}^{N_f} \left| \frac{\lambda_{ij}^a}{2} \right|^2$$

$$\times \int \frac{d^3x d^3p_g d^3p_q}{(2\pi)^6} f_g(\boldsymbol{p}_g) f_q(\boldsymbol{p}_q) [1 - f_q(\boldsymbol{p}_q')] \overline{\sigma}_{gq_f \to \gamma q_f}(s) v_{gq}, \quad (3)$$

where the factor containing $f_q(\boldsymbol{p}_q')$ is included to take into account the fact that the quark q_f produced simultaneously with the photon in the $gq_f \to q_f\gamma$ reaction obeys Fermi-Dirac statistics. The probability for the production of the quark q_f with a momentum $\boldsymbol{p}_q' = \boldsymbol{p}_g + \boldsymbol{p}_q - \boldsymbol{p}_\gamma$ is reduced by the factor $[1 - f_q(\boldsymbol{p}_q')]$, where $f_q(\boldsymbol{p}_q')$ is the probability that the quark state at $\boldsymbol{p}_q'$ is already occupied.

Using the explicit expression for the Gell-Mann matrices in Eq. (10.34), we obtain

$$\sum_{a=1}^{8} \sum_{i=1}^{3} \sum_{j=1}^{3} \left| \frac{\lambda_{ij}^a}{2} \right|^2 = \frac{8 \times 2}{4} = 4. \quad (4)$$

From Eq. (3) and Eq. (4), the number of photons produced per unit time is

$$\frac{dN_\gamma}{dt} = 4 N_s N_\epsilon \sum_{f=1}^{N_f} \int \frac{d^3x d^3p_g d^3p_q}{(2\pi)^6} f_g(\boldsymbol{p}_g) f_q(\boldsymbol{p}_q) [1 - f_q(\boldsymbol{p}_q')] \overline{\sigma}_{gq_f \to \gamma q_f}(s) v_{gq}. \quad (5)$$

Decompose the total cross section in terms of its contributions from various photon momenta,

$$\overline{\sigma}_{gq_f \to \gamma q_f}(s) = \int E_\gamma \frac{d\overline{\sigma}}{d\boldsymbol{p}_\gamma}(gq_f \to \gamma q_f) \frac{d\boldsymbol{p}_\gamma}{E_\gamma}. \quad (6)$$

Eq. (5) and Eq. (6) lead to

$$\frac{dN_\gamma}{d^4x} = 4 N_s N_\epsilon \sum_{f=1}^{N_f} \int \frac{d^3p_g d^3p_q}{(2\pi)^6} f_g(\boldsymbol{p}_g) f_q(\boldsymbol{p}_q) [1 - f_q(\boldsymbol{p}_q')] E_\gamma \frac{d\overline{\sigma}}{d\boldsymbol{p}_\gamma}(gq_f \to \gamma q_f) \frac{d\boldsymbol{p}_\gamma}{E_\gamma} v_{gq}.$$

$$(7)$$

From the above equation, we obtain the Compton photon distribution arising from the scattering of a gluon off a quark as

$$E_\gamma \frac{dN_\gamma(gq \to \gamma q)}{d\boldsymbol{p}_\gamma d^4x} = 4 N_s N_\epsilon \sum_{f=1}^{N_f} \int \frac{d^3p_g d^3p_q}{(2\pi)^6} f_g(\boldsymbol{p}_g) f_q(\boldsymbol{p}_q) [1 - f_q(\boldsymbol{p}_q')]$$

$$\times E_\gamma \frac{d\overline{\sigma}}{d\boldsymbol{p}_\gamma}(gq_f \to \gamma q_f) v_{gq}. \quad (8)$$

For a relativistic systems of quarks, antiquarks, and qluons, the photon is emitted most likely along the direction of the colliding quark or antiquark with the same momentum and the energy. The differential cross section can be approximated by Eq. (16.47) as

$$E_\gamma \frac{d\bar\sigma}{d\mathbf{p}_\gamma}(gq_f \to \gamma q_f) \approx \bar\sigma_{gq_f \to \gamma q_f}(s) E_\gamma \delta(\mathbf{p}_\gamma - \mathbf{p}_{q_f}). \tag{9}$$

From Eq. (12) of Supplement 4.1, we can write $v_{gq} = |v_g - v_q|$ in terms of other kinematic variables,

$$v_{gq} = \frac{\sqrt{(s - p_g^2 - p_q^2)^2/4 - p_g^2 p_q^2}}{E_g E_q}$$

$$= \frac{(s - m^2)}{2 E_g E_q}. \tag{10}$$

Substituting Eq. (9) and Eq. (10) into Eq. (8) and integrating over the delta function, we get

$$E_\gamma \frac{dN_\gamma(gq \to \gamma q)}{d\mathbf{p}_\gamma d^4 x} = \frac{4 N_s N_\epsilon E_\gamma}{(2\pi)^6} \sum_{f=1}^{N_f} \int_{\mathbf{p}_\gamma \approx \mathbf{p}_q} d^3 p_g f_g(\mathbf{p}_g) f_q(\mathbf{p}_\gamma)[1 - f_q(\mathbf{p}_g)]$$

$$\times \bar\sigma_{gq_f \to q_f \gamma}(s) \frac{(s - m^2)}{2 E_g E_\gamma}. \tag{11}$$

The differential element $d^3 p_g$ is (see Eq. (14) of Supplement 16.3)

$$d^3 p_g \approx \frac{E_g}{2 E_\gamma} dE_g \, ds \, d\phi. \tag{12}$$

Substituting Eq. (12) into (11), we obtain the photon distribution arising from gq collisions as

$$E_\gamma \frac{dN_\gamma(gq \to \gamma q)}{d\mathbf{p}_\gamma d^4 x}$$

$$= \frac{4 N_s N_\epsilon}{(2\pi)^6} f_q(\mathbf{p}_\gamma) \sum_{f=1}^{N_f} \int \frac{E_g}{2} dE_g \, ds \, d\phi f_g(E_g)[1 - f_q(E_g)] \bar\sigma_{gq_f \to \gamma q_f}(s) \frac{(s - m^2)}{2 E_g E_\gamma}$$

$$= \frac{4 N_s N_\epsilon}{(2\pi)^6} \frac{f_q(\mathbf{p}_\gamma)}{4 E_\gamma} \sum_{f=1}^{N_f} \int ds \, d\phi dE_g f_g(E_g)[1 - f_q(E_g)](s - m^2) \bar\sigma_{gq_f \to \gamma q_f}(s). \tag{13}$$

After integrating over the azimuthal angle ϕ, the Compton photon distribution arising from the interaction of gluons with quarks is

$$E_\gamma \frac{dN_\gamma(gq \to \gamma q)}{d\mathbf{p}_\gamma d^4 x} = \frac{4 N_s N_\epsilon}{(2\pi)^5} f_q(\mathbf{p}_\gamma) \sum_{f=1}^{N_f} \int ds \frac{dE_g}{4 E_\gamma} f_g(E_g)[1 - f_q(E_g)](s - m^2) \bar\sigma_{gq_f \to \gamma q_f}(s),$$

$$\tag{14}$$

which is Eq. (16.53) after we express $\overline{\sigma}_{gq_f \to \gamma q_f}(s)$ in terms of $\overline{\sigma}_{\text{Comp}}$ by Eq. (16.48).

We can carry out the integration over E_g. The lower limit of E_g is $s/4E_\gamma$. To perform the integration in E_g in Eq. (14), we assume a Bose-Einstein distribution for the gluon $f_g(E_g)$

$$f_g(E) = \frac{1}{e^{E/T} - 1} \tag{15}$$

and a Fermi-Dirac distribution for the quark

$$f_q(E) = \frac{1}{e^{E/T} + 1}. \tag{16}$$

The integrand in Eq. (14) contains the term

$$
\begin{aligned}
f_g(E_g)[1 - f_q(E_g)] &= \frac{1}{e^{E_g/T} - 1}\left[1 - \frac{1}{e^{E_g/T} + 1}\right] \\
&= \frac{1}{e^{E_g/T} - 1} \frac{e^{E_g/T}}{e^{E_g/T} + 1} \\
&= \frac{1}{2}\left[\frac{1}{e^{E_g/T} + 1} + \frac{1}{e^{E_g/T} - 1}\right].
\end{aligned} \tag{17}
$$

Following the same expansion as in Eq. (19) of Supplement 16.3, the integral of E_g in Eq. (14) becomes

$$\int_{s/4E_\gamma}^{\infty} dE_g\, f_g(E_g)[1 - f_q(E_g)] = T \sum_{n=0,1,2,\dots}^{\infty} \frac{e^{-(2n+1)s/4E_\gamma T}}{2n + 1}. \tag{18}$$

Eq. (14) and Eq. (18) lead to the Compton photon distribution from the process $gq \to \gamma q$

$$
\begin{aligned}
E_\gamma \frac{dN_\gamma(gq \to \gamma q)}{d\mathbf{p}_\gamma d^4x} &= \frac{4N_s N_\epsilon}{(2\pi)^5} f_q(\mathbf{p}_\gamma) \frac{T}{4E_\gamma} \sum_{f=1}^{N_f} \int ds \sum_{n=0}^{\infty} \frac{e^{-(2n+1)s/4E_\gamma T}}{2n+1}(s - m^2)\overline{\sigma}_{gq_f \to \gamma q_f}(s) \\
&= \frac{4N_s N_\epsilon}{(2\pi)^5} f_q(\mathbf{p}_\gamma) \frac{T}{4E_\gamma} \sum_{f=1}^{N_f} \sum_{n=0}^{\infty} \frac{1}{2n+1} \int ds\, e^{-(2n+1)s/4E_\gamma T}(s - m^2)\overline{\sigma}_{gq_f \to \gamma q_f}(s).
\end{aligned}
$$

The total cross section $\overline{\sigma}_{gq \to \gamma q}(s)$ depends on the quark electric charge e_f according to Eq. (16.48). Therefore, we can rewrite the above equation as

$$
\begin{aligned}
E_\gamma \frac{dN_\gamma(gq \to \gamma q)}{d\mathbf{p}_\gamma d^4x} &= \frac{4N_s N_\epsilon}{(2\pi)^5} f_q(\mathbf{p}_\gamma) \frac{T}{4E_\gamma} \sum_{f=1}^{N_f}\left(\frac{e_f}{e}\right)^2 \\
&\quad \times \sum_{n=0}^{\infty} \frac{1}{2n+1} \int ds\, e^{-(2n+1)s/4E_\gamma T}(s - m^2)\overline{\sigma}_{\text{Comp}}(s),
\end{aligned}
$$

which is Eq. (16.54). $\mathbb{I}\oplus$

Eq. (16.54) and the general formula Eq. (16.49) for the Compton cross section $\overline{\sigma}_{\text{Comp}}(s)$ can be used to evaluate the Compton photon distribution in the quark-gluon plasma. As in the annihilation photon case in Section §16.3, the integrand in Eq. (16.54) contains the distribution $\exp\{-(2n+1)s/4E_\gamma T\}$. The average value of s under this distribution is of the order of $4E_\gamma T/(2n+1)$. For high temperatures and high photon energies such that T and E_γ are much greater than m, the average value of s is much greater than square of the rest mass m^2 and we are in the relativistic regime. We are justified to use the relativistic approximation (16.51) to evaluate the integral (16.54). Using this relativistic approximation following the procedures in Section §16.3, we shall show in Supplement 16.7 that the Compton photon distribution in the plasma for the $gq \to \gamma q$ process is given by

$$E_\gamma \frac{dN_\gamma(gq \to \gamma q)}{d\boldsymbol{p}_\gamma d^4 x} = \frac{N_s N_\epsilon \alpha_e \alpha_s}{32\pi^2} f_q(\boldsymbol{p}_\gamma) T^2 \sum_{f=1}^{N_f} \left(\frac{e_f}{e}\right)^2 \left\{ \ln\left(\frac{4E_\gamma T}{m^2}\right) + C_{\text{Comp}} \right\},$$

$$(16.55a)$$

where the constant C_{Comp} is given as in Eqs. (16.30b) and (16.30c) by

$$C_{\text{Comp}} = -C_{\text{Euler}} + \frac{1}{2} - \frac{8}{\pi^2} \sum_{n=0}^{\infty} \frac{\ln(2n+1)}{(2n+1)^2} = -0.41613. \qquad (16.55b)$$

For the $g\overline{q} \to \gamma\overline{q}$ process, one can use similar arguments to show that the photon distribution from the $g\overline{q} \to \gamma\overline{q}$ is given by

$$E_\gamma \frac{dN_\gamma(g\overline{q} \to \gamma\overline{q})}{d\boldsymbol{p}_\gamma d^4 x} = \frac{N_s N_\epsilon \alpha_e \alpha_s}{32\pi^2} f_{\overline{q}}(\boldsymbol{p}_\gamma) T^2 \sum_{f=1}^{N_f} \left(\frac{e_f}{e}\right)^2 \left\{ \ln\left(\frac{4E_\gamma T}{m^2}\right) + C_{\text{Comp}} \right\}.$$

$$(16.56)$$

The Compton photon distribution is the sum of the photon distribution from $gq \to \gamma q$ in Eq. (16.55) and from $g\overline{q} \to \gamma\overline{q}$ in Eq. (16.56). Hence, the total Compton photon distribution is

$$E_\gamma \frac{dN_\gamma^{\text{Comp}}}{d\boldsymbol{p}_\gamma d^4 x} = \frac{N_s N_\epsilon \alpha_e \alpha_s}{32\pi^2} [f_q(\boldsymbol{p}_\gamma) + f_{\overline{q}}(\boldsymbol{p}_\gamma)] T^2 \sum_{f=1}^{N_f} \left(\frac{e_f}{e}\right)^2 \left\{ \ln\left(\frac{4E_\gamma T}{m^2}\right) + C_{\text{Comp}} \right\}.$$

$$(16.57)$$

For a plasma with a zero net baryon density, $f_q(\boldsymbol{p}_q)$ is the same as $f_{\bar{q}}(\boldsymbol{p}_{\bar{q}})$ and we have

$$
E_\gamma \frac{dN_\gamma^{\text{Comp}}}{d\boldsymbol{p}_\gamma d^4x} = \frac{N_s N_\epsilon \alpha_e \alpha_s}{16\pi^2} f_q(\boldsymbol{p}_\gamma) T^2 \sum_{f=1}^{N_f} \left(\frac{e_f}{e}\right)^2 \left\{ \ln\left(\frac{4E_\gamma T}{m^2}\right) + C_{\text{Comp}} \right\}.
$$

$$(16.58)$$

For a quark-gluon plasma with u quarks and d quarks, $N_f = 2$ and we have

$$
\sum_{f=1}^{N_f} \left(\frac{e_f}{e}\right)^2 = \frac{5}{9}.
$$

The Compton photon distribution becomes been

$$
E_\gamma \frac{dN_\gamma^{\text{Comp}}}{d\boldsymbol{p}_\gamma d^4x} = \frac{5}{9} \frac{\alpha_e \alpha_s}{4\pi^2} f_q(\boldsymbol{p}_\gamma) T^2 \left\{ \ln\left(\frac{4E_\gamma T}{m^2}\right) + C_{\text{Comp}} \right\}. \qquad (16.59)
$$

The above Compton photon distribution has been obtained with the Bose-Einstein distribution $f_g(E_g)$ for the gluons and the Fermi-Dirac distribution $f_q(E_q)$ for the quark (or antiquark). Ref. [5] uses the Boltzmann distribution for $f_g(E_g)$ and the Fermi-Dirac distribution for $f_q(E_q)$. It can be shown in Supplement 16.7 that the photon distribution then becomes

$$
E_\gamma \frac{dN_\gamma^{\text{Comp}}}{d\boldsymbol{p}_\gamma d^4x} = \frac{5}{9} \frac{\alpha_e \alpha_s}{6\pi^2} f_q(\boldsymbol{p}_\gamma) T^2 \left\{ \ln\left(\frac{4E_\gamma T}{m^2}\right) + C'_{\text{Comp}} \right\}, \qquad (16.60a)
$$

where

$$
C'_{\text{Comp}} = -C_{\text{Euler}} + \frac{1}{2} + \frac{12}{\pi^2} \sum_{n=1}^{\infty} \frac{(-1)^n \ln n}{n^2}. \qquad (16.60b)
$$

The above Eq. (16.60) agrees with Eq. (13) of Kapusta et $al.$ [5], except that in Ref. [5] the quarks are assumed to be massless and an infrared cut-off k_c is used, with the quark-mass m in Eq. (16.60a) replaced by k_c.

A comparison of Eq. (16.59) and (16.60) indicates that they are in the ratio of $(1/4) : (1/6)$, when we use two different gluon distributions.

We are now in a position to calculate the total photon distribution in the quark gluon plasma as the sum of contributions from the annihilation process and from the Compton process. Equation (16.30) and equation (16.58) lead to the total photon distribution

$$E_\gamma \frac{dN_\gamma}{d\mathbf{p}_\gamma d^4 x} = E_\gamma \frac{dN_\gamma^{\text{ann}}}{d\mathbf{p}_\gamma d^4 x} + E_\gamma \frac{dN_\gamma^{\text{Comp}}}{d\mathbf{p}_\gamma d^4 x}$$

$$= \frac{4\alpha_e \alpha_s}{8\pi^2} f_q(\mathbf{p}_\gamma) T^2 \sum_{f=1}^{N_f} \left(\frac{e_f}{e}\right)^2 \left\{ \ln\left(\frac{4E_\gamma T}{m^2}\right) + \frac{C_{\text{ann}} + C_{\text{Comp}}}{2} \right\}, \quad (16.61)$$

where we have used $N_s = 2$ and $N_\epsilon = 2$. For a quark-gluon system with u and d quarks, the total photon distribution is therefore,

$$E_\gamma \frac{dN_\gamma}{d\mathbf{p}_\gamma d^4 x} = \frac{5}{9} \frac{\alpha_e \alpha_s}{2\pi^2} f_q(\mathbf{p}_\gamma) T^2 \left\{ \ln\left(\frac{4E_\gamma T}{m^2}\right) + \frac{C_{\text{ann}} + C_{\text{Comp}}}{2} \right\}. \quad (16.62)$$

We can compare the total photon distribution obtained in Ref. [5] using the approximate Boltzmann distribution for $f_g(E_g)$ and $f_q(E_q)$ and the Fermi-Dirac distribution for the distribution of the final quark $f_q(E_{q'})$. The total photon distribution obtained in Ref. [5], for a quark-gluon plasma with u and d quarks, is

$$E_\gamma \frac{dN_\gamma}{d\mathbf{p}_\gamma d^4 x} = \frac{5}{9} \frac{\alpha_e \alpha_s}{2\pi^2} f_q(\mathbf{p}_\gamma) T^2 \left\{ \ln\left(\frac{4E_\gamma T}{k_c^2}\right) + \frac{2C'_{\text{ann}}}{3} + \frac{C'_{\text{Comp}}}{3} \right\}. \quad (16.63)$$

Upon comparing Eq. (16.62) with Eq. (16.63), we note that except for the difference of the infrared cut-off k_c for massless quarks in contrast to the case of quarks with a finite mass m, the two photon distributions have the same leading-log terms. It is remarkable that as far as the the leading-log term is concerned, the total sums (16.62) and (16.63) obtained by adding the annihilation and the Compton contributions the two contributions are the same, even though the corresponding individual annihilation and Compton photon contributions are different.

⊕[Supplement 16.7
    We show below how the result of Eq. (16.55) is obtained. Upon using the relativistic approximation (16.51) for $\bar{\sigma}_{\text{Comp}}(s)$, the integral in Eq. (16.54) becomes

$$\int ds\, e^{-(2n+1)s/4E_\gamma T}(s-m^2)\bar{\sigma}_{\text{Comp}}(s) = 2\pi\alpha_e\alpha_s \int ds\, e^{-(2n+1)s/4E_\gamma T}\left\{\ln\left(\frac{s}{m^2}\right) + \frac{1}{2}\right\}. \quad (1)$$

The lower limit of s, as determined from $(p_g + p_q)^2$, is m^2. We make the change of variable $s = m^2 z$. Integral (1) becomes

$$\int ds\, e^{-(2n+1)s/4E_\gamma T}(s - m^2)\overline{\sigma}_{\text{Comp}}(s)$$

$$= 2\pi\alpha_e\alpha_s \int_{m^2}^{\infty} ds\, e^{-(2n+1)s/4E_\gamma T}\left\{\ln\left(\frac{s}{m^2}\right) + \frac{1}{2}\right\}$$

$$= 2\pi\alpha_e\alpha_s m^2 \int_{1}^{\infty} dz\, e^{-(2n+1)m^2 z/4E_\gamma T}\left\{\ln\left(\frac{m^2 z}{m^2}\right) + \frac{1}{2}\right\}$$

$$= 2\pi\alpha_e\alpha_s m^2 \int_{1}^{\infty} dz\, e^{-(2n+1)m^2 z/4E_\gamma T}\left\{\ln(z) + \frac{1}{2}\right\}. \tag{2}$$

Using Eqs. (3) and (4) of Supplement 16.4 and writing x as

$$x = \frac{(2n+1)m^2}{4E_\gamma T},$$

we obtain

$$\int ds\, e^{-(2n+1)s/4E_\gamma T}(s - m^2)\overline{\sigma}_{\text{Comp}}(s)$$

$$= -2\pi\alpha_e\alpha_s m^2 \frac{4E_\gamma T}{(2n+1)m^2}\left[C_{\text{Euler}} + \ln\left\{\frac{(2n+1)m^2}{4E_\gamma T}\right\} + \sum_{k=1}^{\infty}\frac{(-x)^k}{k\cdot k!} - \frac{1}{2}e^{-x}\right].$$

For $m^2 \ll E_\gamma T$, the last summation over $(-x)^k = (-m^2/4E_\gamma T)^k$ can be neglected and $e^{-x} \to 1$. The above equation becomes

$$\int ds\, e^{-(2n+1)s/4E_\gamma T}\sqrt{s(s - 4m^2)}\overline{\sigma}_{\text{Comp}}(s)$$

$$= -2\pi\alpha_e\alpha_s m^2 \frac{4E_\gamma T}{(2n+1)m^2}\left[\ln\left\{\frac{m^2}{4E_\gamma T}\right\} + C_{\text{Euler}} - \frac{1}{2} + \ln(2n+1)\right]$$

$$= 2\pi\alpha_e\alpha_s \frac{4E_\gamma T}{(2n+1)}\left[\ln\left\{\frac{4E_\gamma T}{m^2}\right\} - C_{\text{Euler}} + \frac{1}{2} - \ln(2n+1)\right]. \tag{3}$$

Substituting Eq. (3) into Eq. (16.54), we get

$$E_\gamma \frac{dN_\gamma(gq \to \gamma q)}{dp_\gamma d^4 x} = \frac{4N_s N_\epsilon}{(2\pi)^5}f_q(p_\gamma)\frac{T}{4E_\gamma}\sum_{f=1}^{N_f}\left(\frac{e_f}{e}\right)^2$$

$$\times \sum_{n=0}^{\infty}\frac{1}{2n+1}2\pi\alpha_e\alpha_s\frac{4E_\gamma T}{(2n+1)}\left[\ln\left\{\frac{4E_\gamma T}{m^2}\right\} - C_{\text{Euler}} + \frac{1}{2} - \ln(2n+1)\right]. \tag{4}$$

From Eq. (23.2.20) of Ref. [13], we have

$$\sum_{n=0}^{\infty}\frac{1}{(2n+1)^2} = \frac{3}{4}\cdot\frac{\pi^2}{6}.$$

Hence, Eq. (4) becomes

$$E_\gamma \frac{dN_\gamma(gq \to \gamma q)}{d\mathbf{p}_\gamma d^4 x} = \frac{4 N_s N_\epsilon \alpha_e \alpha_s}{128 \pi^2} f_q(\mathbf{p}_\gamma) T^2 \sum_{f=1}^{N_f} \left(\frac{e_f}{e}\right)^2$$

$$\times \left[\ln\left\{ \frac{4 E_\gamma T}{m^2} \right\} - C_{\text{Euler}} + \frac{1}{2} - \frac{8}{\pi^2} \sum_{n=0}^{\infty} \frac{\ln(2n+1)}{(2n+1)^2} \right],$$

which is Eq. (16.55).

The Compton distribution in Eq. (16.55) has been obtained with the Bose-Einstein distribution for $f_g(E_g)$ and the Fermi-Dirac distribution for $f_q(E_q)$. What is the photon distribution if we use the Boltzmann distribution for $f_g(E_g)$ and the Fermi-Dirac distribution for $f_q(E_q)$, as in Ref. 5?

If we follow Ref. 5, then in Eq. (16.53), we have

$$f_g(E_g)[1 - f_q(E_g)] = \frac{e^{-E_g/T}}{1 + e^{-E_g/T}}.$$

The integral of E_g in Eq. (16.53) becomes

$$\int_{s/4E_\gamma}^{\infty} dE_g f_g(E_g)[1 - f_q(E_g)] = T \ln\left(1 + e^{-s/4E_\gamma T} \right)$$

$$= T \sum_{n=1}^{\infty} \frac{(-1)^{n+1} e^{-ns/4E_\gamma T}}{n}.$$

The photon distribution for the $gq \to \gamma q$ process becomes

$$E_\gamma \frac{dN_\gamma(gq \to \gamma q)}{d\mathbf{p}_\gamma d^4 x} = \frac{4 N_s N_\epsilon}{(2\pi)^5} \sum_{f=1}^{N_f} \left(\frac{e_f}{e}\right)^2 f_q(\mathbf{p}_\gamma) \frac{T}{4 E_\gamma}$$

$$\times \sum_{n=1}^{\infty} \frac{(-1)^{n+1}}{n} \int ds \, e^{-ns/4E_\gamma T} \sqrt{s(s - 4m^2)} \bar{\sigma}_{\text{Comp}}(s).$$

Using the result of Eq. (3), and Eq. (23.2.24) of Ref. [13],

$$\sum_{n=1}^{\infty} \frac{(-1)^2}{n^2} = (1 - 2^{1-2})\zeta(2) = \frac{\pi^2}{12},$$

the photon distribution for the $gq \to \gamma q$ process becomes

$$E_\gamma \frac{dN_\gamma(gq \to \gamma q)}{d\mathbf{p}_\gamma d^4 x} = \frac{4 N_s N_\epsilon \alpha_e \alpha_s}{192 \pi^2} f_q(\mathbf{p}_\gamma) T^2 \sum_{f=1}^{N_f} \left(\frac{e_f}{e}\right)^2$$

$$\times \left[\ln\left\{ \frac{4 E_\gamma T}{m^2} \right\} - C_{\text{Euler}} + \frac{1}{2} - \frac{12}{\pi^2} \sum_{n=1}^{\infty} \frac{(-1)^{n+1} \ln n}{n^2} \right].$$

For a system of quark-gluon plasma with u and d quarks, and $N_f = 2$, the photon distribution for the $gq \to \gamma q$ process becomes

$$E_\gamma \frac{dN_\gamma(gq \to \gamma q)}{d\mathbf{p}_\gamma d^4 x} = \frac{5}{9} \frac{\alpha_e \alpha_s}{12\pi^2} f_q(\mathbf{p}_\gamma) T^2 \left[\ln \left\{ \frac{4E_\gamma T}{m^2} \right\} - C_{\text{Euler}} + \frac{1}{2} + \frac{12}{\pi^2} \sum_{n=1}^{\infty} \frac{(-1)^n \ln n}{n^2} \right].$$
$$(5)$$

For a plasma with a zero net baryon density the Compton photon distribution from the $g\bar{q} \to \gamma\bar{q}$ process is the same as the photon distribution from the $gq \to \gamma q$ process. The total Compton photon distribution comes from both $gq \to \gamma q$ and $g\bar{q} \to \gamma\bar{q}$. Hence, when we use a Boltzmann distribution for $f_g(E_g)$ and a Fermi-Dirac distribution for $f_q(E_g)$, the total Compton photon distribution is

$$E_\gamma \frac{dN_\gamma^{Comp}}{d\mathbf{p}_\gamma d^4 x} = \frac{5}{9} \frac{\alpha_e \alpha_s}{6\pi^2} f_q(\mathbf{p}_\gamma) T^2 \left[\ln \left\{ \frac{4E_\gamma T}{m^2} \right\} - C_{\text{Euler}} + \frac{1}{2} + \frac{12}{\pi^2} \sum_{n=1}^{\infty} \frac{(-1)^n \ln n}{n^2} \right], \quad (6)$$

which is just Eq. (16.60), as had to be proved.   I⊕

The photon distribution (16.61) arising from the annihilation and the Compton processes in the quark-gluon plasma depends on the quark rest mass m, which appears in the quark propagator in Feynman diagrams representing these processes (Fig. 16.1). If the quark masses were vanishingly small, the cross sections would diverge. A non-vanishing quark mass serves to regulate this divergence. Although the mass of the quark without interaction is quite small (see Table 9.1), the effective mass of a quark propagating in an interacting thermal quark-gluon plasma is significantly modified when one includes the many-body effects of the medium at finite temperatures. For example, there can be the phenomenon of Debye screening as discussed in Chapter 14. There can also be the phenomenon of 'Landau damping' which involves the absorption and emission of field quanta from the thermal medium. By resumming an infinite series of 'hard thermal loops' at high temperature, Braaten and Pisarski found that a quark acquires an effective thermal mass m_{th} in a thermal medium of temperature T given by [14]

$$m_{\text{th}} = \frac{gT}{\sqrt{6}}. \quad (16.64)$$

A careful analysis of the consequences of the thermal medium by Kapusta et al. [5] showed that the net effect of the medium on photon production is to replace the term $(\ln 4E_\gamma T/m^2)$ in the photon distribution by $(\ln 4E_\gamma T/2m_{\text{th}}^2)$, where the effective mass $\sqrt{2}m_{\text{th}}$ arises from the behavior of the energy of the quark in the medium as a function of its momentum. If we follow this procedure of Kapusta et al., the total photon distribution from the interactions of the constituents of the quark-gluon plasma will be

$$E_\gamma \frac{dN_\gamma}{dp_\gamma d^4x} = \frac{4\alpha_e\alpha_s}{8\pi^2} f_q(p_\gamma) T^2 \sum_{f=1}^{N_f} \left(\frac{e_f}{e}\right)^2 \left\{\ln\left(\frac{4E_\gamma T}{2m_{th}^2}\right) + \frac{C_{ann} + C_{Comp}}{2}\right\}$$

$$= \frac{4\alpha_e\alpha_s}{8\pi^2} f_q(p_\gamma) T^2 \sum_{f=1}^{N_f} \left(\frac{e_f}{e}\right)^2 \ln\left(\frac{3.7388 E_\gamma}{g^2 T}\right). \tag{16.65}$$

For a quark-gluon plasma with u and d quarks, the above equation gives

$$E_\gamma \frac{dN_\gamma}{dp_\gamma d^4x} = \frac{5}{9} \frac{\alpha_e\alpha_s}{2\pi^2} f_q(p_\gamma) T^2 \ln\left\{\frac{3.7388 E_\gamma}{g^2 T}\right\}. \tag{16.66}$$

⊕⟦ Supplement 16.8
   It is instructive to follow Kapusta et al. [5] to show that the rate of photon production obtained in Eqs. (16.65) and (16.66) allows an estimate of the time for a system of strongly interacting quarks and gluons to equilibrate with the produced photons.
   We consider a quark-gluon plasma at a constant temperature and envisage that the system of quarks, antiquarks, and gluons initially has no photon. The equilibrium photon (phase-space) density is

$$\mathcal{F}_{eq}(p_\gamma, x) \equiv \frac{dN_\gamma}{dp_\gamma d^3x}(\text{equilibrium}) = \frac{2}{(2\pi)^3} \frac{1}{e^{E/T} - 1}. \tag{1}$$

By detailed balance, the rate of net photon production is proportional to the difference of the equilibrium photon density and the instantaneous photon density so that when the density reaches the equilibrium density, there is no net photon production. The equation for the time-dependence of the photon density is

$$\frac{d}{dt} \mathcal{F}(p_\gamma, xt) = \frac{1}{\tau}[\mathcal{F}_{eq}(p_\gamma, x) - \mathcal{F}(p_\gamma, xt)] \tag{2}$$

where the constant τ is the equilibrium density divided by the rate of change of photon density $dN_\gamma/dp_\gamma d^4x$,

$$\tau = \frac{\mathcal{F}_{eq}(p_\gamma, x)}{dN_\gamma/dp_\gamma d^4x}. \tag{3}$$

The solution of Eq. (2) gives the time dependence of the phase-space density of the photon

$$\mathcal{F}(p_\gamma, xt) = \mathcal{F}_{eq}(p_\gamma, x)(1 - e^{-t/\tau}). \tag{4}$$

Hence, τ is the length of time for the photon density to reach a substantial fraction of the equilibrium density. The length of time τ can be called the photon equilibration time in a system of quarks, antiquarks, gluons and photons.

With the rate of photon production as given by Eq. (16.66) for our system of quarks, antiquarks, gluons and photons, the equilibration time from Eqs. (1), (3), and (16.66) is

$$
\tau = \frac{2}{(2\pi)^3} \frac{1}{e^{E/T} - 1} \div \left[\frac{5}{9} \frac{\alpha_e \alpha_s}{2\pi^2} f_q(\boldsymbol{p}_\gamma) T^2 \ln\left\{ \frac{3.7388 E_\gamma}{g^2 T} \right\} \right]
$$

$$
= \frac{9}{10\pi \alpha_e \alpha_s} \cdot \frac{E_\gamma}{T^2} \frac{e^{E_\gamma/T} + 1}{e^{E_\gamma/T} - 1} \frac{1}{\ln(3.7388 E_\gamma / 4\pi \alpha_s T)} . \tag{5}
$$

When we take $\alpha_s = 0.4$, we find that for a temperature of 200 MeV, the equilibration times for $E_\gamma = 1, 2$, and 3 GeV photons are $\tau = 373, 481$, and 601 fm/c. The length of time is much larger than the length of time of about 10 fm/c we usually associate with high-energy heavy-ion collisions. Hence, the equilibrium of photons with the quarks, antiquarks and gluons is unlikely. In other words, a photon has a very long mean-free path and is unlikely to interact with quark matter on its way to the detector. Therefore, it carries the information of the state of the matter when it is created.   ]⊕

§16.6 Photon Production by Hadrons

Besides the emission of photons from the quark-gluon plasma, photons can also be emitted from from the hot hadron gas. It is necessary to determine the photon contribution from the different sources so as to extract information on photon emission from the quark-gluon plasma.

A π^+ meson can annihilate a π^- meson to lead to the production of a photon and a ρ^0 meson

$$
\pi^+ + \pi^- \rightarrow \gamma + \rho^0 . \tag{16.67a}
$$

This reaction is represented by the Feynman diagrams in Fig. 16.3.

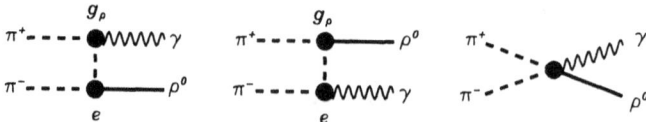

Fig. 16.3   The Feynman diagrams for the $\pi^+ + \pi^- \rightarrow \gamma + \rho^0$ reaction.

A π^+ or a π^- meson can also interact with a π^o meson to produce a photon and a ρ meson,

$$\pi^{\pm} + \pi^0 \rightarrow \gamma + \rho^{\pm}, \tag{16.67b}$$

whose Feynman diagrams are similar to those in Fig. 16.3. The $\pi\pi \rightarrow \gamma\rho$ reactions specified by Eqs. (16.67a) and (16.67b) are analogous to the $q\bar{q} \rightarrow \gamma + g$ annihilation reaction discussed in Section §16.2 and Section §16.3.

A pion can also interact with a ρ^0 meson to lead to a photon and a scattered pion as in the reaction

$$\pi^{\pm} + \rho^0 \rightarrow \gamma + \pi^{\pm}, \tag{16.68a}$$

$$\pi^{\pm} + \rho^{\mp} \rightarrow \gamma + \pi^0, \tag{16.68b}$$

and

$$\pi^0 + \rho^{\pm} \rightarrow \gamma + \pi^{\pm}. \tag{16.68c}$$

These reactions are analogous to the Compton processes in Section §16.4 and Section §16.5.

The differential cross sections for the processes (16.67)-(16.68) have been given explicitly by Kapusta *et al.* The results for the $\pi\pi \rightarrow \gamma\rho$ reactions are [5]

$$\frac{d\sigma}{dt}(\pi^+\pi^- \rightarrow \gamma\rho^0) = \frac{8\pi\alpha_e\alpha_\rho}{s(s-4m_\pi^2)} \left\{ 2 - (m_\rho^2 - 4m_\pi^2) \left[\frac{s-2m_\pi^2}{s-m_\rho^2} \left(\frac{1}{t-m_\pi^2} \right. \right. \right.$$
$$\left. \left. \left. + \frac{1}{u-m_\pi^2} \right) + \frac{m_\pi^2}{(t-m_\pi^2)^2} + \frac{m_\pi^2}{(u-m_\pi^2)^2} \right] \right\}. \tag{16.69}$$

and

$$\frac{d\sigma}{dt}(\pi^{\pm}\pi^0 \rightarrow \gamma\rho^{\pm}) = \frac{-\pi\alpha_e\alpha_\rho}{s(s-4m_\pi^2)} \left[\frac{(s-2m_\rho^2)(t-m_\pi^2)^2}{m_\rho^2(s-m_\rho^2)^2} + \frac{(s-6m_\rho^2)(t-m_\pi^2)}{m_\rho^2(s-m_\rho^2)} \right.$$
$$\left. + \frac{m_\pi^2}{m_\rho^2} - \frac{9}{2} + \frac{4(s-4m_\pi^2)s}{(s-m_\rho^2)^2} + \frac{4(s-4m_\pi^2)s}{t-m_\pi^2} \left(\frac{s}{s-m_\rho^2} + \frac{m_\pi^2}{t-m_\pi^2} \right) \right]. \tag{16.70}$$

where α_ρ is related to the π-ρ coupling constant g_ρ by

$$\alpha_\rho = \frac{g_\rho^2}{4\pi}. \tag{16.71}$$

The explicit forms of the differential cross sections for the $\pi\rho \rightarrow \gamma\pi$ reactions can be found in Ref. [5].

If we are interested in photons with energies much greater than the rest masses m_π and m_ρ of the hadrons in the gas, relativistic approximations can be used to study photon production in hadron matter. One can use procedures similar to those in the last few sections, with appropriate modifications to take into account the different degeneracies and Bose-Einstein distributions for particles in the hadron gas.

The differential cross sections (16.70) and (16.71) for $\pi\pi \to \gamma + \rho$ contain terms of the type $1/(t - m_\pi^2)$ and/or $1/(u - m_\pi^2)$ which arise from the pion propagator in various Feynman diagrams. Terms containing these factors become dominant when $|t - m^2|$ and/or $|u - m^2|$ attain minimum values. The situation is similar to the previous case of photon production by $q\bar{q}$ annihilation discussed in Section §16.2. Therefore, we can repeat similar arguments here. For high-energy hadrons, the relativistic approximation for $\pi\pi \to \gamma\rho$ consists of approximating the differential cross section by delta functions as

$$E_\gamma \frac{d\sigma}{d\mathbf{p}_\gamma}(\pi^+\pi^- \to \gamma\rho^0) \approx \frac{1}{2}\sigma_{\pi^+\pi^- \to \gamma\rho^0}(s)E_\gamma[\delta(\mathbf{p}_\gamma - \mathbf{p}_{\pi^+}) + \delta(\mathbf{p}_\gamma - \mathbf{p}_{\pi^-})].$$

$$\text{(16.72)}$$

and

$$E_\gamma \frac{d\sigma}{d\mathbf{p}_\gamma}(\pi^\pm\pi^0 \to \gamma\rho^\pm) \approx \sigma_{\pi^\pm\pi^0 \to \gamma\rho^\pm}(s)E_\gamma\delta(\mathbf{p}_\gamma - \mathbf{p}_{\pi^\pm}), \qquad \text{(16.73)}$$

where $\sigma_{\pi^i\pi^j \to \gamma\rho^k}(s)$ is the total cross section for $\pi^i\pi^j \to \gamma\rho^k$.

When we use these relativistic approximations to calculate the photon distributions in the hadron gas, we follow the same procedures as in Section §16.3, and find the photon distribution from $\pi\pi$ interactions in the hadron gas as given by

$$E_\gamma \frac{dN_\gamma(\pi^i\pi^j \to \gamma\rho^k)}{d\mathbf{p}_\gamma d^4x} = \frac{1}{(2\pi)^5}f_{\pi^i}(\mathbf{p}_\gamma)\int ds \frac{dE_{\pi^j}}{4E_\gamma}f_{\pi^j}(E_{\pi^j})[1 + f_\rho(E_{\pi^j})]$$
$$\times \sqrt{s(s - 4m^2)}\sigma_{\pi^i\pi^j \to \gamma\rho^k}(s). \qquad \text{(16.74)}$$

Similar approximations can be made for the reactions (16.68a)-(16.68c) when we consider photons with energies much greater than the rest mass of the ρ mesons.

From the previous results in Sections §16.3 and §16.5, we expect that the integral over E_1 and s in Eq. (16.73) will result in a function of the form $(\ln E_\gamma + \text{constant})$ which is a slowly-varying function of E_γ. Hence, the photon distribution from the interaction of energetic hadrons in a hadron gas follows approximately the shape of the hadron distribution, $f_\pi(\boldsymbol{p}_\pi)$, which for high photon energies has the form of $e^{-E/T_{\text{hadron}}}$ characterized by the hadron temperature T_{hadron}.

On the other hand, from the results of Eq. (16.59), we can infer that the photon spectrum from the quark-gluon plasma follows approximately the shape of the quark distribution $f_q(\boldsymbol{p}_q)$, which at high energies has the form of $\exp\{-E/T_{\text{quark}}\}$ characterised by the quark-gluon plasma temperature T_{quark}. Because the quark-gluon plasma is formed at a temperature greater than the critical temperature for the phase transition and hadrons will have a temperature lower than the critical temperature, we expect that photons from the quark-gluon plasma will have an energy distribution characterized by a higher temperature, as distinct from photons emitted from the hadron gas characterised by a lower temperature. A measurement of the energy spectrum of photons will reveal much on the source from which they originate.

It is an interesting question to ask that if the temperatures are the same, are there differences in the photon spectra from the quark-gluon plasma and from the hadron gas? We note from our results that both photon distributions from the quark-gluon plasma and from the hadron gas are proportional to the constituent distributions, and the distinctions between the constituent distributions (the Fermi-Dirac distribution for the quarks and the Bose-Einstein distributions for the hadrons) are small at high photon or constituent energies. Therefore, we expect that the photon spectra from the quark-gluon plasma and from the hadron gas at the same temperature will have shapes similar to each other at high photon energies. For hadron matter, the coupling constant α_ρ can be taken to be 2.9 which is large compared to the value of $\alpha_s = 0.4$ for the quark gluon plasma [5]. There are a large number of constituents which can interact and can produce photons in the hadron gas. In addition to the annihilation and the Compton processes, there are also the decays of vector mesons, such as $\rho \to \gamma\pi\pi$, in the hadron gas which however contribute only a small fraction of the photons at high energies. By direct numerical calculations, Kapusta *et al.* found that at the same temperature of 200 MeV, the shapes of the spectrum of photons from the quark-gluon plasma and from the hadron gas are nearly the same at high photon energies, as expected. The magnitude of the distribution of photons from a hadron gas is slightly greater than that from the quark-gluon plasma at high photon energies. A hadron gas "shines as brightly as (or even slightly brighter than) a quark-gluon plasma" [5]. While it is difficult to distinguish photons from two different sources if they are placed at the same temperature, the occurrence of the quark-gluon plasma will be accompanied by photons with a greater temperature. The photon energy distribution exhibiting a much higher temperature may allow one to distinguish the quark-gluon plasma source from the colder hadron source.

§16.7 Photon Production by Parton Collisions

Besides the emission of photons from the quark-gluon plasma and from the interactions of hadrons, photons can also be emitted by the initial-state collision of a constituent of one nucleon with the constituent of the other nucleon [15]-[18]. For example, a quark of one nucleon can interact with an antiquark of the other nucleon, to produce a photon and a gluon by the annihilation process represented by Eq. (16.1) and Figs. 16.1a and 16.1b. A gluon of one nucleon can interact with a quark or an antiquark in the other nucleon, to produce a photon by the Compton process, as in Eqs. (16.4a) and (16.4b) and represented by Feynman diagrams in Figs. 16.1c and 16.1d.

The cross section for photon production due to parton collisions can be calculated with the hard-scattering model (Eq. (4.16)), which is mathematically similar to the integrals involved in determining the photon distribution in the quark-gluon plasma (Eq. (8) of Supplement 16.3 and Eq. (8) of Supplement 16.6). The major difference between the two calculations consists in the different constituent distributions. For the photon distribution from the quark-gluon plasma, the constituent distributions are the thermal distributions of q, $\bar{q}$ and gluons. For the photon distribution by parton collisions, the initial constituent distributions are the structure functions of q, $\bar{q}$ and gluons in a nucleon. Because of this similarity, some of the formulas such as Eqs. (16.26) and (16.53), obtained for the quark-gluon plasma case, can be used to infer the functional dependence of the photon spectrum.

We are interested in photons with an energy much greater than the quark masses. Approximate differential cross sections Eq. (16.21) and (16.47) can be used to obtain the results of Eq. (16.26) and (16.53). From Eq. (16.26), we expect that the annihilation process leads to a photon distribution which is approximately proportional to the quark distribution, multiplied by a logarithmic function involving the antiquark distribution. Similarly, from Eq. (16.53), the Compton process leads to a photon distribution approximately proportional to the quark distribution multiplied by a logarithmic function involving the gluon distribution. There is also a contribution proportional to the antiquark distribution, which for large photon energies, gives only a small contribution in nucleon-nucleon collisions. The result is a photon distribution approximately proportional to the quark distribution in a nucleon, multiplied by a slowly varying function of the photon energy. Therefore, the photon distribution from parton collisions in a nucleon-nucleon collision is approximately

$$E_\gamma \frac{dN_\gamma}{d\boldsymbol{p}_\gamma} \propto f_q(\boldsymbol{p}_\gamma) \,. \tag{16.75}$$

In order to get an intuitive insight into the photon spectrum from parton collisions, we can consider a valence quark distribution function as given by Eq. (14.20) to write Eq. (16.75) as

$$E_\gamma \frac{dN_\gamma}{d\boldsymbol{p}_\gamma} \propto x_\gamma^{A_1^q}(1 - x_\gamma)^{A_2^q} \,,$$

where the variation of x_γ in the smooth function $P^q(x)$ is neglected. The photon light-cone variable x_γ is equal to $E_\gamma^*/(\sqrt{s}/2)$, for photons with a small transverse momentum. The above equation can be written approximately as

$$E_\gamma^* \frac{dN_\gamma}{d\boldsymbol{p}_\gamma^*} \propto \left(\frac{2E_\gamma^*}{\sqrt{s}}\right)^{A_1^q} \exp^{-2A_2^q E_\gamma^*/\sqrt{s}} \,. \tag{16.76}$$

Therefore, in the center-of-mass system, the photon distribution appears as if it has an effective temperature

$$T_\gamma \sim \frac{\sqrt{s}}{2A_2^q} \,. \tag{16.77}$$

With $A_2^q \sim 4$ as given from Set I of the structure functions of Duke and Owens [15] (Eq. (14.24b)), the effective photon temperature for collisions with $\sqrt{s} = 200$ GeV is of the order of 25 GeV. The large value of the effective photon temperature implies that the photon yield from parton collisions will be greater than the photon yield from the quark-gluon plasma at large photon energies.

§16.8 Experimental Information on Photon Production

In high-energy heavy-ion collisions, a large number of neutral mesons, such as π^o and η, are produced by soft QCD processes. The produced π^o and η decay into two photons with a branching ratio of 98.8% and 38.9% respectively. As a consequence, π^o and η particles are copious sources of photons. Photons can also be produced by heavy-ion bremmstahlung [19]-[21] which is important mainly for soft photon production at low photon energies (in the target rest frame), at forward angles [20,21]. On the other hand, the decay of hadrons leads to photons with a large spread in energy because of the broad rapidity distribution of the produced hadrons.

The number of photons arising from hadron decay is greater than the number of photons arising from other photon sources of interest by 1 to 2 orders of magnitude. It is an experimental challenge to separate out the photon yields into a part arising from the decays of produced hadrons, and the other 'single photon' part arising from other sources, such as due to the interaction between various constituents of the colliding matter or of the created matter.

Experiments to measure the photon yields in high-energy heavy-ion collisions have been carried out at CERN by the WA80/WA93 Collaboration [22]-[24] and by the NA45(CERES) Collaboration [25].

The WA80 Collaboration employs the 'direct' method in which 3,800 lead-glass calorimeter modules covering the pseudorapidity range $2.1 \leq \eta \leq 2.9$ are used to detect photons in nucleus-nucleus collisions. Other quantities such as the total tranverse energy are measured in coincidence to select central and peripheral events. From the detected photons, the differential cross sections of π^o and η mesons can be obtained by reconstructing their yields in the two-photon decay branch. In this method, one evaluates the invariant mass of all photon pairs of the same event to form the two-photon invariant mass distribution. Those photon pairs arising from the decay of the π^o or η will fall into the region of their corresponding meson mass. However, the large photon multiplicity leads to a large number of photon pair combinations, resulting in a large combinatorial background in which many of the photon pairs combine to give a false pair invariant mass in the region of π^o or η masses. This combinatorial background is evaluated by an 'event-mixing' method [26] in which photons are selected from different events having the same global characteristics and combined as in real events to form the purely combinatorial background invariant mass distribution. This event-mixed combinatorial background can then be subtracted from the invariant mass distribution of real events to obtain the actual π^o and η meson yields. Knowing the yield of π^o and η mesons and the efficiencies of the detectors and correcting for small contributions due to the decay of unidentified hadrons other than π^o or η, one can determine that part of the photon yield which does not come from the decay of produced hadrons. Preliminary results from the WA80 Collaboration for the collision of S on Au at 200A GeV indicate that, while there does not appear to be any excess photons above the meson-decay background in peripheral events, an excess of photons above the meson decay background is observed in central collisions for $p_T < 2$ GeV, with an experimental error of 5.8-7.9% [23]. The transverse momentum distribution of the single photons is characterized by a temperature of 180-200 MeV, to be compared with the π^o temperature of about 210 MeV [24]. The results are preliminary and an independent re-analysis will be carried out as a check.

The NA45(CERES) Collaboration [25] measures photon yields by the 'conversion' method. In this method, the target is used as a converter so that a produced photon is converted into an electron and a positron whose momenta are subsequently measured by a Cerenkov ring imaging electron spectrometer. The measurement covers the pseudorapidity range $2.03 < \eta < 2.64$. To subtract the background, the π^o contribution is estimated from the measured distribution of negative pions and the contributions from other mesons, including the relatively large contribution due to the decay of the η mesons, are estimated by m_T scaling. Preliminary results for central collision of S on Au at 200A GeV show that, within the experimental systematic error of 11%, there is no evidence for an excess of photons above the meson decay background [25]. Further work on the data analysis is in progress in order to narrow down the experimental uncertainties.

§**References for Chapter 16**

1. An excellent review on photon production in high-energy heavy-ion collisions can be found in J. Kapusta, P. Lichard, and D. Siebert, Nucl. Phys. A544, 485c (1992) and P. V. Ruuskanen, Nucl. Phys. A544, 169c (1992).

2. E. V. Shuryak, Sov. J. Nucl. Phys. 28 408 (1978).

3. B. Sinha, Phys. Lett. B128, 91 (1983); S. Raha and B. Sinha, Int. J. Mod. Phys. A6, 517 (1991); J. Alam, D. K. Srivastava, B. Sinha and D. N. Basu, Phys. Rev. D48, 1117 (1993).

4. R. C. Hwa and K. Kajantie, Phys. Rev. D32, 1109 (1985).

5. J. Kapusta, P. Lichard, and D. Siebert, Phys. Rev. D44 2774 (1991).

6. C. Itzykson and J.-B. Zuber, *Quantum Field Theory*, McGraw-Hill Book Company, N.Y., 1980.

7. Ta-Pei Cheng and Ling-Fong Li, *Gauge Theory of Elementary Particle Physics*, Clarendon Press, Oxford, 1984.

8. V. B. Berestetskii, E. M. Lifshitz and L. I. Pitaevskii, *Quantum Electrodynamics*, Pergamon Press, Oxford, 1982.

9. J. D. Bjorken and S. D. Drell, *Relativistic Quantum Mechanics*, (McGraw-Hill Book Company, N.Y. 1964) and J. D. Bjorken and S. D. Drell, *Relativistic Quantum Fields*, (McGraw-Hill Book Company, N.Y. 1965).

10. C. Y. Wong, Phys. Rev. C48, 902 (1993).
11. R. Svensson, Astrophys. Jour. 258, 321 (1982).
12. I. S. Gradshteyn and I. M. Ryzhik, *Table of Integrals, Series, and Products*, Acdemic Press, New York 1965.
13. M. Abramowitz and I. A. Stegun, *Handbook of Mathematical Tables*, Dover Publications, New York, 1965.
14. R. D. Pisarski, Nucl. Phys. B309, 476 (1988); R. D. Pisarski, Phys. Rev. Lett. 63, 1129 (1989); E. Braaten and R. D. Pisarski, Nucl. Phys. B337, 569 (1990).
15. D. W. Duke and J. F. Owens, Phys. Rev. D30, 49 (1984).
16. R. Brock *et al.*, *Handbook of Perturbative QCD*, (CTEQ Collaboration), editor George Sterman, Fermilab Report Fermilab-pub-93/094, 1993.
17. J. F. Owens, Rev. Mod. Phys. 59, 465 (1987).
18. J. Alam, D. K. Srivastava, B. Sinha and D. N. Basu, Phys. Rev. D48, 1117 (1993).
19. J. D. Bjorken and L. McLerran, Phys. Rev. D31, 63 (1985).
20. C. M. Ko and C. Y. Wong, Phys. Rev. C33, 153 (1986).
21. L. Xiong and C. M. Ko, Phys. Rev. C37, 880 (1988).
22. K. H. Kampert *et al.*, WA80 Collbaoration, Nucl. Phys. A544, 183 1993.
23. R. Santo *et al.*, WA80 Collbaoration, Nucl. Phys. A566, 61c (1994).
24. K. H. Kampert, Proceedings of the NATO Summer School, Bodrum, September 1993, University of Müster Preprint IKP-MS-94/0101.
25. D. Irmscher *et al.*, NA45/CERES Collbaoration, Nucl. Phys. A566, 347c (1994).
26. T. C. Awes, Nucl. Inst. Meth. A276, 468 (1989).

17. Signatures for the Quark-Gluon Plasma (IV)

§17.1   The Hanbury-Brown-Twiss   Effect   of Intensity Interferometry

Interference is a well-known wave phenomenon associated with the superposition of two or more waves. For example, the appearance of the Young's diffraction pattern in a two-slit experiment is the result of the interference of the amplitudes of two light waves from two slit openings which travel different paths to arrive at the same detection point.

There is another wave interference phenomenon, which is associated with the interference of the intensities when identical particles are detected at different space-time points or energy-momentum points. In a pioneering experiment in 1956, Hanbury-Brown and Twiss measured the angular diameter of a star using the correlation between two photons [1]. When a photon has been detected in one detector, the probability for the detection of a second photon in coincidence is found to exhibit a correlation with respect to the relative transverse separation between the two detectors. The degree of this correlation depends on the angular diameter of the emitting source. Because the correlation pertains to the intensity of one particle at one space-time point and the intensity of another identical particle at another space-time point, measured in coincidence, this type of interferometry is also known as *intensity interferometry*. In many measurements, such as those in high-energy nuclear experiments, the correlation involves the intensity of one particle at one energy-momentum point and the intensity of another identical particle at another energy-momentum point, measured in coincidence. The phenomenon of the space-time correlation or the energy-momentum correlation of detected identical particles emitted from an extended source is known as the *Hanbury-Brown-Twiss (HBT) effect.*

The occurrence of Hanbury-Brown-Twiss intensity interferometry led to the search for theoretical connections between optics and quantum theory which cumulated in the development of the elegant theory of coherent states [2]. From the theory of coherent states, we understand the relationship between the occurrence of correlations and the degree of coherence of the source field: Correlations of the Hanbury-Brown-Twiss type are present for chaotic sources and are absent for coherent sources.

The Hanbury-Brown-Twiss effect has been applied to subatomic physics, using different types of particles [3-20]. The use of pions to measure the space-time dimensions in $p\bar{p}$ annihilation was proposed and successfully applied by Goldhaber, Goldhaber, Lee and Pais [7]. Interference using proton correlations was proposed by Koonin [8] and has been studied experimentally (see for example Reference [5]). In addition to the work of Glauber, there has been a large amount of theoretical and experimental work on the HBT intensity interferometry [4-20]. In this introductory discussion, we shall limit our attention to pion interferometry.

For high-energy heavy-ion collisions, the emission of pions can be treated as due to a partially chaotic source. The momentum correlation of two identical pions detected in coincidence is related to the Fourier transform of a function of the phase-space distribution of the source. Therefore, the Hanbury-Brown-Twiss effect can be used to provide information on the distribution of matter during the late stages of the reaction process, which will be useful to assist in the search for the quark-gluon plasma.

§17.2 Pion Momentum Correlations from Chaotic Sources

Why does a momentum correlation exist in the detection of two identical pions emitted by chaotic sources? To understand the origin of this correlation, we need a description of the state of the pions at the moment of their production at the source points, and of their propagation from the source points to the detection points. Because the two-particle momentum correlation involves the detection of two identical particles in coincidence, we also need to take into account the symmetry with respect to the exchange of these identical particles. The exchange symmetry involving identical particles is the origin of the momentum correlation.

We consider an extended pion source and study first the detection of a single pion (π^-, for example) of four-momentum $k = (\boldsymbol{k}, k^0)$ at the space-time point $x' = (\boldsymbol{x}', t')$ as shown in Fig. 17.1. A point within the extended source is labelled by the space-time coordinate $x = (\boldsymbol{x}, t)$.

For convenience, we shall use the source center-of-mass system as the reference frame to measure all momenta and space-time coordinates. In this coordinate system the center-of-mass of the extended source is at rest. This is a natural coordinate system for the collision of two equal nuclei in a collider. However, in high-energy collisions with fixed targets, the detectors are in motion with respect to the source.

Extended Source

Fig. 17.1 A pion of momentum k is emitted at a typical source point x of an extended source and is detected at the detection point x', where the detector is located. The straight line joining x to x' is the trajectory of the pion from x to x'. The figure is not drawn to scale. The distance between x and x' is many orders of magnitude greater than the linear dimension of the extended source.

The source center-of-mass system may appear to be an awkward coordinate system at first sight, but it turns out to be the most convenient coordinate system because physical quantities of interest such as the single-particle momentum distribution $P(k)$ and the two-particle momentum distribution function $P(k_1, k_2)$ do not depend on the detector coordinates x' but depend on the source coordinates x, as we shall see in Eqs. (17.9), (17.30), and (17.54).

We consider first the pion of four-momentum k to be detected at the detection point x'. The measurements of k and x' do not have fine enough spatial and temporal resolutions to determine the originating point of the pion in the source. It is only known that the pion comes from a point in the extended source. There are many source points x in the extended source where the pion of momentum k can originate and they satisfy the approximate relation which describes the classical trajectory of a pion,

$$x' - x \approx (k/k^0)(t' - t), \tag{17.1}$$

where $k/k^0 = dx'/dt'$ is the velocity of the pion.

Let us focus our attention at a typical source point x as shown in Fig. 17.1. We need to determine, at least approximately, the probability amplitude for the propagation of the pion from x to x'. To obtain this quantum mechanical probability amplitude, we can use the Feynman's path integral method. Accordingly, the quantum mechanical probability amplitude for the pion to start initially at x and to arrive finally at x' is given by summing the phase factor $e^{iS(\text{path})}$ over all paths, as in Eq. (10.2),

$$\text{Amplitude}\,[(x,t) \to (x',t')] \equiv \psi(k : x \to x') = \sum_{all\ paths} e^{iS(\text{path})}, \tag{17.2}$$

where $S(\text{path})$ is the action of the pion particle, whose value depends on the space-time path. All the paths in the sum Eq. (17.1) have the same starting point x and the same end point x', similar to the situation illustrated in Fig. 10.1. The dominant contribution to the sum over paths of Eq. (17.1), comes from the classical trajectory, which is described by the straight line represented approximately by Eq. (17.1). Those contributions coming from paths different from the classical trajectory vary greatly but with opposing signs and tend to cancel out when we perform the sum in Eq. (17.1). It is therefore reasonable to approximate the quantum mechanical probability amplitude by a single term describing the action of the pion with momentum k along the classical trajectory from x to x',

$$\psi(k: \ x \rightarrow x') \cong e^{iS(\text{classical path},k \ : \ x \ \rightarrow \ x')} . \qquad (17.3)$$

We have assumed in the above equation that the pion propagates from the source point x to the detection point x' without absorption or attenuation so that the magnitude of the above probability amplitude is unity.

We shall show in Exercise 17.1 that the action $S(\text{classical path})$ of the pion of momentum k evaluated along the approximate classical trajectory (17.1) from x to x' is simply

$$S(\text{classical path}, k : x \rightarrow x') \approx k \cdot (x - x')$$
$$\approx k^0(t - t') - \boldsymbol{k} \cdot (\boldsymbol{x} - \boldsymbol{x}'). \quad (17.4)$$

From Eqs. (17.3) and (17.4), the probability amplitude for the pion of momentum k to start at x and to arrive at x' is therefore given approximately by

$$\psi(k: \ x \rightarrow x') \approx e^{ik \cdot (x-x')} . \qquad (17.5)$$

For simplicity of notation, we shall henceforth not indicate the approximate nature of the above equality, keeping in mind that this amplitude is obtained from an approximate classical path using the path integral method.

•⟦ Exercise 17.1
Show that the action for a free particle of momentum k evaluated along the classical path (17.1) from x to x' is given by Eq. (17.4).

◇Solution:
We shall first consider the non-relativistic case. The action for the particle is

$$S = \int L \, dt , \qquad (1)$$

where the Lagrangian L for the particle is

$$L = \text{(kinetic energy } T) - \text{(potential energy } V).$$

For the particle with a rest mass m, the kinetic energy is

$$T = \frac{1}{2}m\left(\frac{d\boldsymbol{x}'}{dt'}\right)^2,$$

and the potential energy V is a constant which depends on the chosen energy scale. Because the energy $E = k^0$ is measured by including the rest mass of the particle, the potential energy for the free particle is

$$V = m.$$

Therefore, the Lagrangian of the particle is

$$L = \frac{1}{2}m\left(\frac{d\boldsymbol{x}'}{dt'}\right)^2 - m. \tag{2}$$

Along the classical trajectory, the quantity L is independent of time. Hence, from Eq. (1), the action integral for the particle to start at $(\boldsymbol{x}, t)$ and arrive at $(\boldsymbol{x}', t')$ along the classical trajectory is

$$S(\text{classical path, } k : x \rightarrow x') = \left[\frac{1}{2}m\left(\frac{d\boldsymbol{x}'}{dt'}\right)^2 - m\right](t' - t)$$

$$= m\left(\frac{d\boldsymbol{x}'}{dt'}\right)^2(t' - t) - \left[\frac{1}{2}m\left(\frac{d\boldsymbol{x}'}{dt'}\right)^2 + m\right](t' - t). \tag{3}$$

We note that the classical trajectory Eq. (17.1) relates the time difference $t' - t$ approximately to $\boldsymbol{x}' - \boldsymbol{x}$. Therefore, we have

$$m\left(\frac{d\boldsymbol{x}'}{dt'}\right)^2(t' - t) = m\frac{d\boldsymbol{x}'}{dt'} \cdot \frac{d\boldsymbol{x}'}{dt'}(t' - t)$$

$$\approx \boldsymbol{k} \cdot (\boldsymbol{x}' - \boldsymbol{x}). \tag{4}$$

We also have

$$E = k^0 = \frac{1}{2}m\left(\frac{d\boldsymbol{x}'}{dt'}\right)^2 + m. \tag{5}$$

From Eqs. (3)-(5), we find

$$S(\text{classical path, } k : x \rightarrow x') \approx -k^0(t' - t) + \boldsymbol{k} \cdot (\boldsymbol{x}' - \boldsymbol{x})$$

$$\approx k \cdot (x - x'),$$

as had to be proved.

For the relativistic case, the action S for the free particle is a Lorentz-invariant quantity which has the form [21]

$$S = -m \int \sqrt{ds'^2} = -m \int \sqrt{dt'^2 - d\boldsymbol{x}'^2}$$

$$= -m \int \sqrt{1 - \left(\frac{d\boldsymbol{x}'}{dt'}\right)^2} \, dt' . \tag{6}$$

The validity of Eq. (6) can be easily checked by expanding it in powers of $d\boldsymbol{x}'/dt'$ and comparing the expansion with Eqs. (1) and (2).

We can evaluate the action (6) along the classical trajectory (17.1). We obtain

$$S(\text{classical path}, \ k : x \to x') = -m \sqrt{1 - \left(\frac{d\boldsymbol{x}'}{dt'}\right)^2} \, (t' - t)$$

$$= m \frac{(\frac{d\boldsymbol{x}'}{dt'})^2}{\sqrt{1 - (\frac{d\boldsymbol{x}'}{dt'})^2}}(t' - t) - m \frac{1}{\sqrt{1 - (\frac{d\boldsymbol{x}'}{dt'})^2}}(t' - t)$$

$$\approx \boldsymbol{k} \cdot (\boldsymbol{x}' - \boldsymbol{x}) - k^0(t' - t)$$

$$\approx k \cdot (x - x'),$$

which is the result of Eq. (17.4).

$$]\bullet$$

For any source point, the probability amplitude (17.5) contains the phase $S(\text{classical path}, \ k : x \to x')$. It gives the probability amplitude for the propagation of the pion from the source point x to the detection point x'. This is however not sufficient for our purposes. In order to discuss the probability amplitude for the pion to be *produced* at a source point x, to propagate after production from the source point x, and to arrive at the detection point x', we also need to describe the state of the pion at the moment of its production at the source point x by a production probability amplitude.

We characterize the production probability amplitude for producing a pion of momentum k at x by a magnitude $A(k, x)$ and a phase $\phi(x)$. Without loss of generality, the magnitude $A(k, x)$ can be taken to be real and nonnegative. It depends on the nature of the production process. We shall see later that the production magnitude $A(k, x)$ is related to the phase-space distribution $f(k, x)$ for chaotic sources, and to the spectrum $P(k)$ for coherent sources (see Eqs. (17.14) and (17.47) below). We call $\phi(x)$ the production phase, to distinguish it from the propagation phase $\{ik \cdot (x - x')\}$ in Eq. (17.5) which arises from the propagation of the pion along the classical trajectory. The behavior of the production phase $\phi(x)$ at different source points describes the degree of coherence or chaoticity of the pion production process. For example, a chaotic source can be represented by random production phases at the source points while a coherent source can be represented by a production phase $\phi(x)$ which is a well-behaved (nonrandom) function of the source point coordinate x.

In the remainder of this section, we shall develop a general theoretical framework to study the single-particle momentum distribution and two-particle momentum distribution and shall apply the results to the case of chaotic sources. We shall use the same general framework to study the case of coherent sources and partially coherent sources in the next section.

We examine first the single-particle momentum distribution $P(k)$. Upon including the additional production probability amplitude to describe the state of the pion at the moment of its production at their source points, the complete probability amplitude $\Psi(k : x \to x')$ for a pion of momentum k to be produced from the source point x, to propagate along the classical trajectory, and to arrive at x is

$$\Psi(k : x \to x') = A(k, x) e^{i\phi(x)} \psi(k : x \to x') . \tag{17.6}$$

Besides the production from the source point x, the pion can also be produced from other source points in the extended source. The total amplitude for the detection of a pion at x is the sum of the probability amplitudes from all source points. After taking into account the production probability amplitude $Ae^{i\phi}$ at different source points, the total probability amplitude for a pion with momentum k to be produced from the extended source and to arrive at the detection point x' is given by

$$\Psi(k : \{^{\text{all } x}_{\text{points}}\} \to x') = \sum_x A(k, x) e^{i\phi(x)} \psi(k : x \to x')$$
$$= \sum_x A(k, x) e^{i\phi(x)} e^{ik \cdot (x - x')} . \tag{17.7}$$

The summation over the source points necessitates the specification of the density $\rho(x)$ of the source points per unit space-time volume at the point x. With this specification, the summation $\sum_x$ should be transcribed as an integral over x,

$$\sum_x \cdots \to \int dx \, \rho(x) \cdots \tag{17.8}$$

The single-particle momentum distribution, $P(k)$, which is the probability for a pion of momentum k to be produced from the extended source and arrive at the detection point x', is the absolute square of the total probability amplitude,

$$
\begin{aligned}
P(k) &= |\Psi(k : \{^{\text{all } x}_{\text{points}}\} \rightarrow x')|^2 \\
&= |\sum_x A(k,x)e^{i\phi(x)}\, e^{ik\cdot(x-x')}|^2 \\
&= |\sum_x A(k,x)e^{i\phi(x)}\, e^{ik\cdot x}|^2 , \qquad (17.9)
\end{aligned}
$$

The above expression is for a general extended source. We use the result to examine the single-particle momentum distribution of pions from a chaotic source. A chaotic source is described as a source whose phase $\phi(x)$ associated with the source points x is a random function of the source point coordinate x.

To take advantage of the randomness of the ϕ phases, we expand the righthand side of Eq. (17.9) into terms independent of $\phi(x)$ and terms containing $\phi(x)$. We obtain

$$
P(k) = \sum_x A^2(k,x) + \sum_{\substack{x,y \\ x \neq y}} A(k,x)A(k,y)e^{i\phi(x)}e^{-i\phi(y)}e^{ik\cdot(x-y)} .
$$

$$(17.10)$$

When we take into account the randomness of the phases of the sources at different points, the second term in the above equation (17.10) gives a zero contribution because the large number of terms with slowly varying magnitudes but rapidly fluctuating random phases cancel out one another in the sum (see Supplement 17.1). Therefore, Eq. (17.10) becomes

$$
P(k) = \sum_x A^2(k,x) . \qquad (17.11)
$$

Using the transcription from a summation to an integral, $(\sum_x ...) \rightarrow (\int dx\, \rho(x)...)$, we can rewrite Eq. (17.11) as

$$
P(k) = \int dx\, \rho(x)A^2(k,x) . \qquad (17.12)
$$

We can compare this equation with the property of the phase space distribution function $f(k,x)$ which is the classical analogue of the quantum mechanical Wigner function. As is well known, the integral of the distribution function over the spatial coordinates gives the momentum distribution,

$$
P(k) = \int dx\, f(k,x) . \qquad (17.13)
$$

Therefore, for a chaotic source, the distribution function $f(k, x)$ is related to the production magnitude $A(k, x)$ of the pion source by

$$f(k, x) = A^2(k, x)\rho(x) \, . \tag{17.14a}$$

This relation allows us to express the production magnitude $A(k, x)$ as a function of the phase-space distribution $f(k, x)$,

$$A(k, x) = \sqrt{\frac{f(k, x)}{\rho(x)}} \, , \tag{17.14b}$$

which we shall use later in Eq. (17.31).

$\oplus$[Supplement 17.1
   The second term of Eq. (17.10) is

$$\sum_{\substack{x,y \\ x \neq y}} A(k, x)A(k, y)e^{i\phi(x)}e^{-i\phi(y)}e^{ik\cdot(x-y)} \, . \tag{1}$$

We wish to show that with slowly-varying magnitudes $A(k, x)$ but rapidly fluctuating random phases $\phi(x)$, this term is approximately zero. We have

$$\sum_{\substack{x,y \\ x \neq y}} A(k, x)A(k, y)e^{i\phi(x)}e^{-i\phi(y)}e^{ik\cdot(x-y)}$$

$$= \frac{1}{2}\left[\sum_{\substack{x,y \\ x \neq y}} A(k, x)A(k, y)e^{i\phi(x)}e^{-i\phi(y)}e^{ik\cdot(x-y)} + \sum_{\substack{x,y \\ x \neq y}} A(k, y)A(k, x)e^{i\phi(y)}e^{-i\phi(x)}e^{ik\cdot(y-x)}\right]$$

$$= \sum_{\substack{x,y \\ x \neq y}} A(k, x)A(k, y)\cos[k\cdot x + \phi(x) - (k\cdot y + \phi(y))] \, .$$

The addition of a rapidly fluctuating random phase $\phi(x)$ to a relatively slowly varying phase $k\cdot x$ yields another rapidly fluctuating phase $\Phi(x)$. Therefore, the above equation becomes

$$\sum_{\substack{x,y \\ x \neq y}} A(k, x)A(k, y)e^{i\phi(x)}e^{-i\phi(y)}e^{ik\cdot(x-y)}$$

$$= \sum_{\substack{x,y \\ x \neq y}} A(k, x)A(k, y)\cos[\Phi(x) - \Phi(y)]$$

$$= \sum_{\substack{x,y \\ x \neq y}} A(k, x)A(k, y)\{\cos[\Phi(x)]\cos[\Phi(y)] + \sin[\Phi(x)]\sin[\Phi(y)]\}$$

$$= \sum_x A(k, x)\cos[\Phi(x)]\sum_{\substack{y \\ y \neq x}} A(k, y)\cos[\Phi(y)] + \sum_x A(k, x)\sin[\Phi(x)]\sum_{\substack{y \\ y \neq x}} A(k, y)\sin[\Phi(y)] \, .$$

For a random set of phases $\Phi(y)$, we have

$$\sum_{\substack{y \\ y \neq x}} A(k,y)\cos[\Phi(y)] = \frac{N-1}{N}\sum_{y} A(k,y)\cos[\Phi(y)]$$

$$\sim \sum_{y} A(k,y)\cos[\Phi(y)],$$

where N is the number of source points. Therefore, the expression in Eq. (1) becomes

$$\sum_{\substack{x,y \\ x \neq y}} A(k,x)A(k,y)e^{i\phi(x)}e^{-i\phi(y)}e^{ik\cdot(x-y)} \sim \left|\sum_{x} A(k,x)\cos[\Phi(x)]\right|^2 + \left|\sum_{x} A(k,x)\sin[\Phi(x)]\right|^2 .$$

The coordinate x is a four-vector $(\boldsymbol{x}, t)$. The sum $\sum_x A(k,x)\cos[\Phi(x)]$ can be written as

$$\sum_{x} A(k,x)\cos[\Phi(x)] = \sum_{\boldsymbol{x}}\left\{\sum_{t} A(k,\boldsymbol{x},t)\cos[\Phi(\boldsymbol{x},t)]\right\} .$$

Therefore, for a given spatial point $\boldsymbol{x}$, there are contributions from different time coordinates t in the above summation.

We are interested in situations where the amplitude $A(k,\boldsymbol{x},t)$ varies relatively slowly in time as compared to the fluctuating, random phase $\Phi(\boldsymbol{x},t)$ which varies rapidly in time. In that case, over a period T of the randomly fluctuating phase at the time T_0, we have

$$\sum_{T_0-\frac{T}{2} \leq\, t\, \leq T_0+\frac{T}{2}} A(k,\boldsymbol{x},t)\cos[\Phi(\boldsymbol{x},t)] \sim A(k,\boldsymbol{x},T_0)\sum_{T_0-\frac{T}{2} \leq\, t\, \leq T_0+\frac{T}{2}} \cos[\Phi(\boldsymbol{x},t)]$$

$$\sim 0 .$$

Hence, when the sum over all times is considered, we have

$$\sum_{t} A(k,\boldsymbol{x},t)\cos[\Phi(\boldsymbol{x},t)] \sim 0 ,$$

and similarly

$$\sum_{t} A(k,\boldsymbol{x},t)\sin[\Phi(\boldsymbol{x},t)] \sim 0 .$$

Consequently, from Eq. (1) the term

$$\sum_{\substack{x,y \\ x \neq y}} A(k,x)A(k,y)e^{i\phi(x)}e^{-i\phi(y)}e^{ik\cdot(x-y)}$$

is approximately zero for rapidly fluctuating random phases $\phi(x)$, as had to be proved.

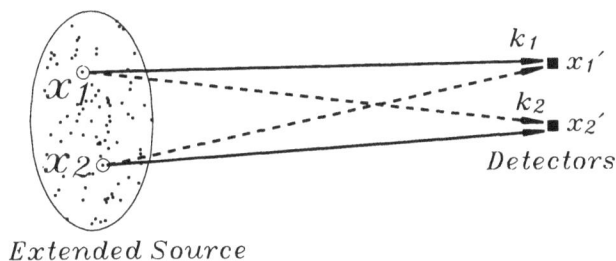

Extended Source

Fig. 17.2 A pion of momentum k_1 is detected at x_1' and another identical pion with momentum k_2 is detected at the space-time point x_2'. They are emitted from the source point x_1 and x_2 of an extended source. The solid lines joining x_1 to x_1' and x_2 to x_2', and the dashed lines joining x_1 to x_2' and x_2 to x_1' are possible trajectories for the pions.

Having completed the analysis for the single-particle momentum distribution $P(k)$ in the detection of a single pion, we turn now to the discussion of the momentum correlations in the detection of two identical pions for a general source. We consider a pion (π^-, for example) with a 4-momentum k_1 detected at the space-time point x_1' and another identical pion with a 4-momentum k_2 detected at the space-time point x_2', x_1' and x_2' being the locations of the detectors (Fig. 17.2). These two pions can originate from any two source points in the extended source. Let us focus our attention on two typical source points of identical pions at x_1 and x_2. We would like to find out first the probability amplitude for the two pions to propagate from these two source points to arrive at their respective detection points, with k_1 at x_1' and k_2 at x_2'.

We consider the case in which the pion of momentum k_1 starts from x_1 and arrives at x_1', and the other identical pion of momentum k_2, starts from x_2 and arrives at x_2', as indicated by the solid lines in Fig. 17.2. The probability amplitude for this occurrence is the product of (i) the probability amplitude $\psi(k_1 : x_1 \to x_1')$ for the pion of momentum k_1 to propagate from x_1 to arrive at x_1', and (ii) the probability amplitude $\psi(k_2 : x_2 \to x_2')$ for pion of momentum k_2 to propagate from x_2 and to arrive at x_2',

$$\psi(k_1 : x_1 \to x_1')\psi(k_2 : x_2 \to x_2'). \qquad (17.15)$$

We can obtain the probability amplitude $\psi(k_i : x_i \to x_i')$ by following Feynman's path integral method again, with the approximation that the dominant contribution to the probability amplitude comes from the classical trajectory. Then, each probability amplitude factor of (17.15) contains only the action evaluated along the classical trajectory, as given by Eq. (17.3),

$$\psi(k : x_i \to x'_i) \cong e^{iS_{cl}(k : x_i \to x'_i)},$$

where $S_{cl}(k_i : x_i \to x'_i)$ is the action of the pion with momentum k along the classical path from the point x_i to x'_i. From Eq. (17.4) and Exercise 17.2, this action is approximately given by

$$S_{cl}(k_i : x_i \to x'_i) = k_i \cdot (x_i - x'_i) \qquad i = 1, 2.$$

The amplitude (17.15) becomes

$$\psi(k_1 : x_1 \to x'_1)\psi(k_2 : x_2 \to x'_2) = e^{iS_{cl}(k_1 : x_1 \to x'_1)} \, e^{iS_{cl}(k_2 : x_2 \to x'_2)},$$

Hence, the probability amplitude for the pion of momentum k_1 to start from x_1 to arrive at x'_1, and the pion of momentum k_2 to start from x_2 to arrive at x'_2 is

$$\psi(k_1 : x_1 \to x'_1)\psi(k_2 : x_2 \to x'_2) = e^{ik_1 \cdot (x_1 - x'_1)}e^{ik_2 \cdot (x_2 - x'_2)}. \qquad (17.16)$$

In order to discuss the probability amplitude for the pion to be *produced* at the source points, to propagate after production from the source points, and to arrive at the detection points, we also need to include the production probability amplitude to describe the states of the pions at the source points. The probability amplitude for the production of the pion with momentum k_i at x_i is given by $A(k_i, x_i)e^{i\phi(x_i)}$. Therefore, the probability amplitude for the two pions to be produced at the source points, to propagate from the source points, and to arrive at the detection points is

$$A(k_1, x_1)e^{i\phi(x_1)}A(k_2, x_2)e^{i\phi(x_2)}\psi(k_1 : x_1 \to x'_1)\psi(k_2 : x_2 \to x'_2),$$
$$(17.17)$$

which can be expressed as

$$A(k_1, x_1)e^{i\phi(x_1)}A(k_2, x_2)e^{i\phi(x_2)}e^{ik_1 \cdot (x_1 - x'_1)}e^{ik_2 \cdot (x_2 - x'_2)}. \qquad (17.18)$$

However, this is not the only probability amplitude contribution for two identical pions produced from x_1 and x_2 to arrive at x'_1 and x'_2. The pion of momentum k_1 detected at x'_1 can also be produced at x_2 and propagate from x_2 to x'_1, and the other identical pion of momentum k_2 detected at x'_2 can be produced at x_1 and propagate from x_1 to x'_2 as indicated by the dashed lines in Fig. 17.2. The probability amplitude for this occurrence is

$$A(k_1, x_2)e^{i\phi(x_2)}A(k_2, x_1)e^{i\phi(x_1)}\psi(k_1 : x_2 \to x'_1)\psi(k_2 : x_1 \to x'_2),$$
$$(17.19)$$

which can be expressed as

$$A(k_1, x_2)e^{i\phi(x_2)}A(k_2, x_1)e^{i\phi(x_1)}e^{ik_1 \cdot (x_2 - x'_1)} \, e^{ik_2 \cdot (x_1 - x'_2)}. \qquad (17.20)$$

Because of the indistinguishability of the pions and the Bose-Einstein statistics of identical bosons, the probability amplitude must be symmetrical with respect to the interchange of the labels of the pions which distinguish them. In this case, the only labels which distinguish the two identical pions are the source point coordinates x_1 and x_2, because one pion of momentum k_1 has been determined to have been detected at x'_1 and the other identical pion of momentum k_2 at x'_2. The probability amplitude must be symmetrical with respect to the interchange of the labels x_1 and x_2. Accordingly, the probability amplitude which satisfies this symmetry is the sum of Eqs. (17.18) and (17.20) divided by $\sqrt{2}$:

$$\frac{1}{\sqrt{2}}\left\{ A(k_1, x_1)e^{i\phi(x_1)}A(k_2, x_2)e^{i\phi(x_2)}e^{ik_1 \cdot (x_1 - x'_1)}e^{ik_2 \cdot (x_2 - x'_2)} \right.$$

$$\left. + A(k_1, x_2)e^{i\phi(x_2)}A(k_2, x_1)e^{i\phi(x_1)}e^{ik_1 \cdot (x_2 - x'_1)}e^{ik_2 \cdot (x_1 - x'_2)} \right\}$$

$$\equiv e^{i\phi(x_1)}e^{i\phi(x_2)}\Phi(k_1 k_2 : x_1 x_2 \to x'_1 x'_2). \qquad (17.21)$$

where $\Phi(k_1 k_2 : x_1 x_2 \to x'_1 x'_2)$ is the part of the probability amplitude in the above equation which does not depend on ϕ. It is defined by

$$\Phi(k_1 k_2 : x_1 x_2 \to x'_1 x'_2)$$

$$= \frac{1}{\sqrt{2}}\left\{ A(k_1, x_1)A(k_2, x_2)e^{ik_1 \cdot (x_1 - x'_1)}e^{ik_2 \cdot (x_2 - x'_2)} \right.$$

$$\left. + A(k_1, x_2)A(k_2, x_1)e^{ik_1 \cdot (x_2 - x'_1)}e^{ik_2 \cdot (x_1 - x'_2)} \right\}. \qquad (17.22)$$

Besides originating from the source points x_1 and x_2, the two pions can also be produced at other source points in the extended source. The total amplitude is the sum of amplitudes from all combinations of two source points. Therefore, the total probability amplitude for two identical pions with momenta k_1 and k_2 to be produced from two source points in the extended source and to arrive at their respective detection points x_1' and x_2' is

$$\Psi(k_1 k_2 : \{^{\text{all } x_1 x_2}_{\text{points}}\} \to x_1' x_2') = \sum_{x_1, x_2} e^{i\phi(x_1)} e^{i\phi(x_2)} \Phi(k_1 k_2 : x_1 x_2 \to x_1' x_2').$$

(17.23)

The two-particle momentum distribution $P(k_1, k_2)$ is defined as the probability distribution for two pions of momenta k_1 and k_2 to be produced from the extended source and to arrive at their respective detection points x_1' and x_2'. From Eq. (17.23), it is given by

$$P(k_1, k_2) = \frac{1}{2!} |\Psi(k_1 k_2 : \{^{\text{all } x_1 x_2}_{\text{points}}\} \to x_1' x_2')|^2.$$

(17.24)

The results obtained in Eqs. (17.22), (17.23), and (17.24) are quite general and are applicable to the study of momentum correlations of two pions from both chaotic and coherent sources. Depending on the properties of the phases $\phi(x)$ associated with the source points, the extended source can be coherent, chaotic, or partially chaotic. The correlation function of two identical pions is dependent on the degree of coherence of the source.

We now apply the results of Eqs. (17.22)-(17.24) to examine the momentum correlations of a chaotic source. A chaotic source is described as a source whose phase $\phi(x)$ associated with the source points x is a random function of the source point coordinate x.

For a chaotic source, we again make use of the random nature of the phases by substituting Eq. (17.23) into Eq. (17.24) and expand the righthand side of Eq. (17.24). We separate out terms which are independent of ϕ and terms which contain ϕ. We obtain

$$\begin{aligned} P(k_1, k_2) = &\frac{1}{2} \sum_{x_1, x_2} \Big\{ \Phi^*(k_1 k_2 : y_1 y_2 \to x_1' x_2')|_{\substack{y_1 = x_1 \\ y_2 = x_2}} \Phi(k_1 k_2 : x_1 x_2 \to x_1' x_2') \\ &+ \Phi^*(k_1 k_2 : y_1 y_2 \to x_1' x_2')|_{\substack{y_2 = x_1 \\ y_1 = x_2}} \Phi(k_1 k_2 : x_1 x_2 \to x_1' x_2') \Big\} \\ +&\frac{1}{2} \sum_{\substack{x_1, x_2, y_1, y_2 \\ \{x_1 x_2\} \neq \{y_1 y_2\}}} \Big\{ e^{i\phi(x_1) + i\phi(x_2) - i\phi(y_1) - i\phi(y_2)} \\ &\times \Phi^*(k_1 k_2 : y_1 y_2 \to x_1' x_2') \Phi(k_1 k_2 : x_1 x_2 \to x_1' x_2') \Big\}. \end{aligned}$$

(17.25)

The two terms in the first summation on the righthand side are equal because of the exchange symmetry. For the chaotic source, the

last term in equation (17.25) gives a zero sum because the contributions of a large number of terms with similar magnitude but random
phases cancel out. Therefore, Eq. (17.25) becomes

$$P(k_1, k_2) = \sum_{x_1, x_2} |\Phi(k_1 k_2 : x_1 x_2 \to x_1' x_2')|^2 . \tag{17.26}$$

Converting the summations in Eq. (17.26) into integrals with the
transcription (17.8), we can rewrite the total probability as a double
integral over the source point coordinates x_1 and x_2,

$$P(k_1, k_2) = \int dx_1 dx_2 \ \rho(x_1)\rho(x_2)|\Phi(k_1 k_2 : x_1 x_2 \to x_1' x_2')|^2. \tag{17.27}$$

Using Eq. (17.22), we find

$$P(k_1, k_2) = \int dx_1 \rho(x_1) A^2(k_1, x_1) \int dx_2 \rho(x_2) A^2(k_2, x_2)$$

$$+ \int dx_1 \rho(x_1) A(k_1, x_1) A(k_2, x_1) e^{i(k_1 - k_2)\cdot x_1}$$

$$\times \int dx_2 \rho(x_2) A(k_2, x_2) A(k_1, x_2) e^{i(k_2 - k_1)\cdot x_2} . \tag{17.28}$$

Using Eq. (17.12), we can rewrite the above equation as

$$P(k_1, k_2) = P(k_1)P(k_2) + \left| \int dx e^{i(k_1 - k_2)\cdot x} \rho(x) A(k_1, x) A(k_2, x) \right|^2 . \tag{17.29}$$

From the above equation, it is convenient to introduce the *effective
density function* $\rho_{\text{eff}}(x; k_1 k_2)$ to rewrite the above equation as

$$P(k_1, k_2) = P(k_1)P(k_2)\left(1 + \left| \int dx \ e^{i(k_1 - k_2)\cdot x} \rho_{\text{eff}}(x; k_1, k_2) \right|^2\right), \tag{17.30}$$

where

$$\rho_{\text{eff}}(x; k_1, k_2) = \frac{\rho(x) A(k_1, x) A(k_2, x)}{\sqrt{P(k_1)P(k_2)}} . \tag{17.31}$$

In terms of the distribution function $f(k, x)$ of Eqs. (17.12) and (17.14)
for chaotic sources, this effective density $\rho_{\text{eff}}(x; k_1 k_2)$ can be expressed
as

$$\rho_{\text{eff}}(x; k_1 k_2) = \frac{\sqrt{f(k_1, x)f(k_2, x)}}{\sqrt{\int dx_1 f(k_1, x_1) \int dx_2 f(k_2, x_2)}} . \tag{17.32}$$

The Fourier transform of $\rho_{\text{eff}}(x; k_1 k_2)$ is defined as

$$\tilde{\rho}_{\text{eff}}(q; k_1 k_2) = \int dx \, e^{iq \cdot x} \rho_{\text{eff}}(x; k_1 k_2) , \qquad (17.33)$$

where $q = k_1 - k_2$. The two-particle momentum distribution function is then a function of $\tilde{\rho}_{\text{eff}}(q; k_1 k_2)$. From Eqs. (17.30)-(17.33), we have

$$P(k_1, k_2) = P(k_1)P(k_2)\left(1 + |\tilde{\rho}_{\text{eff}}(q; k_1 k_2)|^2\right). \qquad (17.34)$$

The *correlation function* $C_2(k_1, k_2)$ is defined as the ratio of the probability for the coincidence of k_1 and k_2 relative to the probability of observing k_1 and k_2 separately,

$$C_2(k_1, k_2) = \frac{P(k_1, k_2)}{P(k_1)P(k_2)} . \qquad (17.35)$$

From Eq. (17.34), we have

$$C_2(k_1, k_2) = 1 + |\tilde{\rho}_{\text{eff}}(q; k_1 k_2)|^2 . \qquad (17.36)$$

Thus, for an extended chaotic source, the two-pion correlation function $C_2(k_1, k_2)$ is directly related to the Fourier transform of the effective density, as given by Eq. (17.32). The two-pion correlation function may be used to probe the phase-space configuration of the source at the time of pion emission.

One can introduce another correlation function $R(k_1, k_2)$ which is related to $C_2(k_1, k_2)$ by

$$R(k_1, k_2) = C_2(k_1, k_2) - 1 . \qquad (17.37)$$

From this definition, $R(k_1, k_2)$ is related to $\tilde{\rho}_{\text{eff}}(q; k_1 k_2)$ by

$$R(k_1, k_2) = |\tilde{\rho}_{\text{eff}}(q; k_1 k_2)|^2 . \qquad (17.38)$$

In many applications, one parametrizes the effective density in the form of a Gaussian distribution,

$$\rho_{\text{eff}}(x; k_1 k_2) = \frac{\mathcal{N}}{4\pi^2 R_x R_y R_z \sigma_t} \exp\left\{ -\frac{x^2}{2R_x^2} - \frac{y^2}{2R_y^2} - \frac{z^2}{2R_z^2} - \frac{t^2}{2\sigma_t^2} \right\},$$
$$(17.39)$$

where the normalization constant $\mathcal{N}(k_1, k_2)$ and the standard deviations $R_x(k_1, k_2)$, $R_y(k_1, k_2)$, $R_z(k_1, k_2)$ and $\sigma_t(k_1, k_2)$ depend on k_1

and k_2. The constant coefficients in Eq. (17.39) are chosen in such a way that the effective density is normalized to

$$\mathcal{N}(k_1, k_2) = \int dx \rho_{\text{eff}}(x; k_1 k_2) \qquad (17.40)$$

Because of the Schwartz inequality [22], we have

$$\int dx \sqrt{f(k_1, x) f(k_2, x)} \leq \sqrt{\int dx_1 f(k_1, x_1) \int dx_2 f(k_2, x_2)}. \qquad (17.41)$$

Therefore, from Eqs. (17.32) and (17.41), we have

$$\int dx \rho_{\text{eff}}(x; k_1 k_2) = \mathcal{N}(k_1, k_2) \leq 1, \qquad (17.42)$$

where the equality holds when $k_1 = k_2$. We therefore have $\mathcal{N}(q = 0) = 1$.

It should be reminded that we have been using the source center-of-mass system to measure all momenta and space-time coordinates. We specify various parameters of the effective density (17.39) in the source center-of-mass system in the following way. We choose the collision axis as the longitudinal z-axis and we place the source center-of-mass at the spatial origin. The parameter R_z measures the longitudinal 'radius' of the source along the beam direction. To measure two-pion correlations, there are two detectors. We choose to place the midpoint between the detectors along the x axis, so that R_x measures the transverse radius of the source along the direction towards the detectors and R_y measures the transverse radius of the source perpendicular to the line joining the source and the detectors.

The Fourier transform of the effective source density (17.39) is

$$\tilde{\rho}_{\text{eff}}(q; k_1, k_2) = \mathcal{N} \exp\left\{-\frac{R_x{}^2 q_x^2}{2} - \frac{R_y{}^2 q_y^2}{2} - \frac{R_z{}^2 q_z^2}{2} - \frac{\sigma_t{}^2 q_t{}^2}{2}\right\}. \qquad (17.43)$$

From Eq. (17.43), the correlation functions become

$$C_2(k_1, k_2) = C_2(q; k_1 k_2) = 1 + \mathcal{N} e^{-R_x{}^2 q_x^2 - R_y{}^2 q_y^2 - R_z{}^2 q_z^2 - \sigma_t{}^2 q_t{}^2}, \qquad (17.44)$$

and

$$R(k_1, k_2) = R(q; k_1 k_2) = \mathcal{N} e^{-R_x{}^2 q_x^2 - R_y{}^2 q_y^2 - R_z{}^2 q_z^2 - \sigma_t{}^2 q_t{}^2}. \qquad (17.44')$$

These results for the Gaussian parametrization can be used to analyze experimental momentum correlation data. It must be emphasized that the parameters $\mathcal{N}(k_1, k_2)$, $R_i(k_1, k_2)$, and $\sigma_t(k_1, k_2)$ in these expressions are, in general, functions of k_1 and k_2. Only in special cases will these parameters be independent of k_1, k_2. For example, we show in Supplement 17.2 that when the distribution function $f(k, x)$ is factorizable as the product of a function of k and a function of x, then the effective density and its parameters $\mathcal{N}$ and R_i are independent of k_1 and k_2 and the effective density is identical to the space-time density distribution of the source points. This occurs when the momentum distribution of the produced pions is independent of the location of the source points. Therefore, a Gaussian parametrization with momentum-independent parameters may be a good description for a static source. However, it is clearly inadequate for a source where the center of the momentum distribution changes from position to position, as in the inside-outside cascade description of particle production or in the case of collective expansion.

In the general cases when the momentum spectrum of the produced particles depend on source coordinates, the parameters which characterize the density function $\rho_{\mathrm{eff}}(x; k_1 k_2)$ will depend on the region of k_1 and k_2 covered in the measurement.

We can compare the effective density $\rho_{\mathrm{eff}}(x; k_1 k_2)$ with the effective density obtained previously by Pratt [13]. If we expand $f(k_i, x)$ with respect to $f(K, x)$ where $K = (k_1 + k_1)/2$, then we have

$$f(k_1, x) = f(K, x) + \left.\frac{\partial f(k, x)}{\partial k^\mu}\right|_{k=K} \frac{q^\mu}{2} + \left.\frac{\partial^2 f(k, x)}{2! \partial k^\mu \partial k^\nu}\right|_{k=K} \frac{q^\mu q^\nu}{4} + \cdots,$$

and

$$f(k_2, x) = f(K, x) - \left.\frac{\partial f(k, x)}{\partial k^\mu}\right|_{k=K} \frac{q^\mu}{2} + \left.\frac{\partial^2 f(k, x)}{2! \partial k^\mu \partial k^\nu}\right|_{k=K} \frac{q^\mu q^\nu}{4} + \cdots,$$

where $q^\mu = k_1^\mu - k_2^\mu$. The quantity $\sqrt{f(k_1, x) f(k_2, x)}$ becomes

$$\sqrt{f(k_1, x) f(k_2, x)} = f(K, x) + O(q^\mu q^\nu), \qquad (17.45)$$

where terms proportional to the first order of q^μ cancel. Therefore, if one neglects terms of second order in q^μ, the effective density is

$$\rho_{\mathrm{eff}}(x; k_1 k_2) \cong \frac{f(K, x)}{\sqrt{\int dx_1 f(k_1, x_1) \int dx_2 f(k_2, x_2)}}. \qquad (17.46)$$

The right-hand side of the above equation (17.46) is the effective density of Pratt [13]. Thus, Eq. (17.46), is an approximation to Eq. (17.32), when terms second order in q^μ are neglected.

In most discussions of the effective density, one can use either Eq. (17.32) or the approximate Eq. (17.46). One should note however that in the case when k_1 and k_2 are quite far apart, so that the overlap of $f(k_1, x)$ and $f(k_2, x)$ are very small, the effective density $\rho(x; k_1, k_2)$ given by Eq. (17.32) is essentially zero as it should be. However, the approximate result of Eq. (17.46) gives the effective density as proportional to $f((k_1 + k_2)/2, x)$ which need not be small, and may therefore be incorrect.

⊕〚 Supplement 17.2
It is useful to discuss a few examples of the effective density ρ_{eff} to show the different types of effective density which may appear in intensity interferometry.

Example I.
We consider the case when the production amplitude at various source points is independent of the source point coordinate,

$$A(k, x) = A(k). \tag{1}$$

In this case, the phase-space distribution $f(k, x)$ as obtained from Eq. (17.14) is factorizable as a product of the momentum distribution $A^2(k)$ and the spatial distribution $\rho(x)$,

$$f(k, x) = A^2(k)\rho(x). \tag{2}$$

The factorization of the phase-space distribution function means that the momentum distribution of the particles produced at one source point is the same as the momentum distribution of the particles produced at another source point. There is no correlation between the momentum distribution and the space-time location.

The momentum distribution $P(k)$ of the particles produced from the entire source is the integral of $f(k, x)$ with respect to x. Therefore, for factorizable $f(k, x)$, $P(k)$ is proportional to $A^2(k)$,

$$P(k) = A^2(k) \int dx \rho(x),$$

and the proportionality constant can be taken to be unity to give the normalization of the density function,

$$\int dx \rho(x) = 1. \tag{3}$$

From Eqs. (2) and (3), the definition of the effective density, Eq. (17.32) , gives

$$\rho_{\text{eff}}(x; k_1 k_2) = \frac{\sqrt{f(k_1, x)f(k_2, x)}}{\sqrt{\int dx_1 f(k_1, x_1) \int dx_2 f(k_2, x_2)}}$$

$$= \rho(x). \tag{4}$$

Therefore, when the phase-space distribution $f(k, x)$ is factorizable as the product of a momentum function and a space-time function, the effective density $\rho_{\text{eff}}(x; k_1 k_2)$ is identical to the source point density $\rho(x)$ and does not depend on k_1 and k_2. Then, the parameters $\mathcal{N}$ and R_i of the effective density do not depend on k_1 and k_2. The momentum correlation function $C_2(k_1, k_2)$ is simply related to the Fourier transform of the space-time distribution of the source.

Example II.

We can consider the case when bosons are produced at the proper time $\tau = \tau_{\text{initial}} = \tau_0$, with an initial distribution in phase space given by $f_0(k_0, x_0; \tau_0)$. They are emitted at the freeze-out proper time $\tau = \tau_{\text{freeze out}}$ after the particles expand according to some equation of motion. The general form of the distribution of pions at the freeze-out time is related to the initial distribution by

$$f(\boldsymbol{k}, \boldsymbol{x}; \tau) = \int d\boldsymbol{k}_0 d\boldsymbol{x}_0 f_0(\boldsymbol{k}_0, \boldsymbol{x}_0; \tau_0) M(\boldsymbol{k}_0, \boldsymbol{x}_0, \tau_0 | \boldsymbol{k}, \boldsymbol{x}, \tau), \tag{5}$$

where the kernel $M(\boldsymbol{k}_0, \boldsymbol{x}_0, \tau_0 | \boldsymbol{k}, \boldsymbol{x}, \tau)$ maps the phase space point $(\boldsymbol{k}_0, \boldsymbol{x}_0)$ at the proper time τ_0 to the point $(\boldsymbol{k}, \boldsymbol{x})$ at the proper time τ.

An example of the mapping function is

$$M(\boldsymbol{k}_0, \boldsymbol{x}_0, \tau_0 | \boldsymbol{k}, \boldsymbol{x}, \tau) = \delta[\boldsymbol{k} - \boldsymbol{K}(\boldsymbol{k}_0, \boldsymbol{x}_0, \tau_0, \tau)]\delta[\boldsymbol{x} - \boldsymbol{X}(\boldsymbol{k}_0, \boldsymbol{x}_0, \tau_0, \tau)], \tag{6}$$

which was used previously to follow the trajectories of phase-space points in the pseudoparticle method of phase-space dynamics [23].

As an illustration of the mapping of the phase-space points, we consider the case of particles produced originally at τ_0 which stream out according to the equation of free motion before they emerge to propagate out of the source at the proper time τ. For free motion, the functions $\boldsymbol{K}$ and $\boldsymbol{X}$ are

$$\boldsymbol{K}(\boldsymbol{k}_0, \boldsymbol{x}_0, \tau_o, \tau) = \boldsymbol{k}_0, \tag{7a}$$

$$\boldsymbol{X}(\boldsymbol{k}_0, \boldsymbol{x}_0, \tau_0, \tau) = \boldsymbol{x}_0 + \frac{\boldsymbol{k}}{k^0}\delta t, \tag{7b}$$

and

$$\delta t = \sqrt{\tau^2 + \boldsymbol{x}^2} - \sqrt{\tau_0^2 + \boldsymbol{x}_0^2}. \tag{7c}$$

The distribution function at τ is then

$$f(\boldsymbol{k}, \boldsymbol{x}; \tau) = \int d\boldsymbol{k}_0 d\boldsymbol{x}_0 f_0(\boldsymbol{k}_0, \boldsymbol{x}_0)\delta(\boldsymbol{k} - \boldsymbol{k}_0)\delta(\boldsymbol{x} - \boldsymbol{x}_0 + \frac{\boldsymbol{k}}{k^0}\delta t)$$
$$= f_0(\boldsymbol{k}, \boldsymbol{x} - \boldsymbol{k}\delta t/k^0). \tag{8}$$

An illustration of the time dependence of the distribution function $f(\boldsymbol{k}, \boldsymbol{x}; \tau)$ is given in Fig. 17.3, with the distribution at different $\boldsymbol{k}$ values shifting to different regions of $\boldsymbol{x}$. The greater the momentum, the greater is the shift due to the expansion.

From Eq. (17.37), the effective density $\rho_{\text{eff}}(\boldsymbol{x}; k_1 k_2)$ depends on the product, $f(\boldsymbol{k}_1, \boldsymbol{x}; \tau) f(\boldsymbol{k}_2, \boldsymbol{x}; \tau)$, of the phase-space distribution at $\boldsymbol{k}_1$ and $\boldsymbol{k}_2$. It is given by

$$P(\boldsymbol{k}_1)P(\boldsymbol{k}_2)\rho_{\text{eff}}(\boldsymbol{x}; k_1 k_2) = \sqrt{f_0(\boldsymbol{k}_1, \boldsymbol{x} - \boldsymbol{k}_1\delta t/k_1^0; \tau_0)f_0(\boldsymbol{k}_2, \boldsymbol{x} - \boldsymbol{k}_2\delta t/k_2^0; \tau_0)}. \tag{9}$$

Phase−Space Distribution f(k, x ; τ)

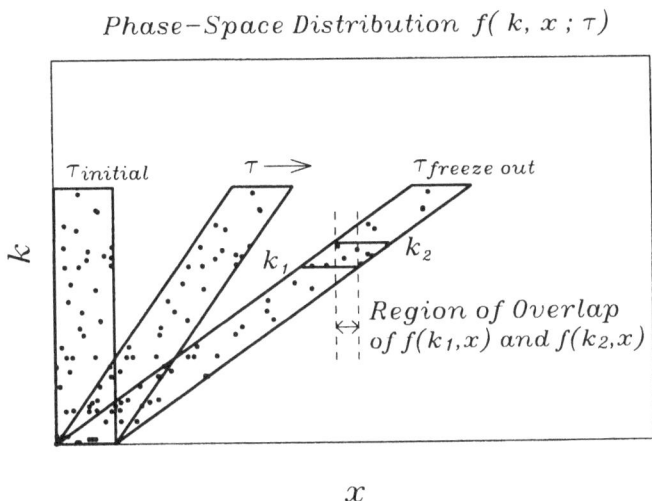

Fig. 17.3 The time-dependence of the phase-space distribution function $f(\mathbf{k}, \mathbf{x}; \tau)$ for the initial proper time τ_{initial}, an intermediate proper time τ, and the freeze-out proper time $\tau_{\text{freeze out}}$. The regions of phase-space density are indicated by the dotted regions. The spatial region of overlap of $f(\mathbf{k}_1, \mathbf{x}; \tau)$ and $f(\mathbf{k}_2, \mathbf{x}; \tau)$ at $\tau_{\text{freeze out}}$ is indicated by the arrow between the dashed lines.

Therefore, at the freeze-out proper time, the product of $f(\mathbf{k}_1, \mathbf{x}; \tau)$ and $f(\mathbf{k}_2, \mathbf{x}; \tau)$ for two different values of $\mathbf{k}_1$ and $\mathbf{k}_2$ will be non-zero essentially only in the region where the nonzero regions of $f(\mathbf{k}_1, \mathbf{x}; \tau)$ and $f(\mathbf{k}_2, \mathbf{x}; \tau)$ overlap, as indicated by the arrow between the dashed lines in Fig. 17.3. We can compare this overlap region with the spatial region spanned by the freeze-out density $\int d\mathbf{k} f(\mathbf{k}, \mathbf{x}; \tau_{\text{freeze out}})$. The spatial region spanned by the effective density $\rho_{\text{eff}}(x; k_1 k_2)$ is much smaller than the spatial region spanned by the freeze-out density $\int d\mathbf{k} f(\mathbf{k}, \mathbf{x}; \tau_{\text{freeze out}})$.   ▌⊕

§17.3 Coherent and Partially Coherent Sources

We find in the last section that for a chaotic source the momentum of one pion is correlated with the momentum of another identical pion. What type of correlation is expected for an extended coherent source?

An extended coherent source is described by a production phase $\phi(x)$ which is a well-behaved function of the source point coordinate x. The phases $\phi(x)$ and $\phi(y)$ at two distinct source points x and y are related. A simple example of a coherent source is one in which the production phase $\phi(x)$ is a constant so that all the production phases at different source points are the same. A second example of a coherent source is the case when the production phase $\phi(x)$ is a common, simple function of time, independent of spatial coordinate. Another example of a coherent source is the production of massive bosons by a separating massless fermion-antifermion pair as described by QED$_2$ which we shall discuss in Eqs. (17.48) and (17.49) below.

We study first the single pion momentum distribution. According to the general result of Eq. (17.7), the total probability amplitude for a pion with momentum k to be produced from an extended source and arrive at the detection point x' is given by

$$\Psi(k : \{^{\text{all } x}_{\text{points}}\} \to x') = \sum_x A(k, x)e^{i\phi(x)}\, e^{ik\cdot(x-x')}\,.$$

For the extended coherent source, because $\phi(x)$ is a well-behaved function of the source point coordinate x, we can perform the summation in the above equation. This summation can be transcribed into an integral. Using the transcription $(\sum_x \ldots) \to (\int dx \rho(x)\ldots)$ according to Eq. (17.8), we write the probability amplitude in the above equation as

$$\Psi(k : \{^{\text{all } x}_{\text{points}}\} \to x') = \int dx\, \rho(x)A(k, x)e^{i\phi(x)}e^{ik\cdot(x-x')}\,.$$

The probability $P(k)$ for a single pion of momentum k to be produced from the extended coherent source and arrive at the detection point x' is the absolute square of the probability amplitude,

$$P(k) = \left|\int dx\, \rho(x)A(k, x)e^{i\phi(x)}e^{ik\cdot(x-x')}\right|^2$$

$$= \left|\int dx\, \rho(x)A(k, x)e^{i\phi(x)}e^{ik\cdot x}\right|^2\,. \tag{17.47}$$

Therefore, for a coherent source, the single-particle momentum distribution is the absolute square of the Fourier transform of the function $\rho(x)A(k, x)e^{i\phi(x)}$.

An example of boson emission from a coherent source is the production of massive bosons by a separating massless fermion and antifermion pair, as described by QED$_2$ in Chapter 6. According to QED$_2$, as can be observed in Eq. (6.53), the momentum distribution of the produced bosons is the absolute square of the Fourier transform of the fermion source current given by Eq. (6.50). One can compare Eq. (17.47) with (6.53) and (6.50). In such a comparison, because QED$_2$ is only two dimensional, we need to limit our attention to the longitudinal coordinate z and the time coordinate t. Particle production according to QED$_2$ is a coherent process whose production phase is characterized by

$$\phi(x) = 0, \tag{17.48}$$

and the product $\rho(z)A(k,z)$ is given by

$$\rho(z)A(k,z) = \sqrt{\frac{1}{4k_0 e^2}} \, j_z(z, t = 0), \tag{17.49}$$

where $j_z(z, t = 0)$ is the initial longitudinal current of the separating fermion-antifermion pair which produces the massive bosons, and e is the QED$_2$ coupling constant (see Chapter 6 for more details about particle production in QED$_2$). If the pion production process in nucleon-nucleon collisions can be described by the formation of two strings, and each string produces pions as described by QED$_2$, then the extended source can be described as containing two separate coherent sources.

We study now the correlation of momentum when two identical pions are produced from an extended coherent source. According to the general result of Eq. (17.25), the probability $P(k_1, k_2)$ for the pions of momentum k_1 and k_2 to be produced from all combinations of two source points in the extended source and to arrive at their respective detection points x_1' and x_2' is

$$P(k_1, k_2) = \frac{1}{2!} |\Psi(k_1 k_2 : \{^{\text{all } x_1 x_2}_{\text{points}}\} \to x_1' x_2')|^2, \tag{17.50}$$

where

$$\Psi(k_1 k_2 : \{^{\text{all } x_1 x_2}_{\text{points}}\} \to x_1' x_2') = \sum_{x_1, x_2} e^{i\phi(x_1)} e^{i\phi(x_2)} \Phi(k_1 k_2 : x_1 x_2 \to x_1' x_2'), \tag{17.51}$$

and

$$\Phi(k_1 k_2 : x_1 x_2 \to x_1' x_2') = \frac{1}{\sqrt{2}} \left\{ A(k_1, x_1) A(k_2, x_2) e^{ik_1 \cdot (x_1 - x_1')} e^{ik_2 \cdot (x_2 - x_2')} \right.$$
$$\left. + A(k_1, x_2) A(k_2, x_1) e^{ik_1 \cdot (x_2 - x_1')} e^{ik_2 \cdot (x_1 - x_2')} \right\}. \tag{17.52}$$

We substitute Eq. (17.52) into (17.51) and we have

$$\Psi(k_1 k_2 : \{^{\text{all } x_1 x_2}_{\text{points}}\} \to x_1' x_2')$$
$$= \frac{1}{\sqrt{2}} \sum_{x_1, x_2} e^{i\phi(x_1)} e^{i\phi(x_2)} \left\{ A(k_1, x_1) A(k_2, x_2) e^{ik_1 \cdot (x_1 - x_1')} e^{ik_2 \cdot (x_2 - x_2')} \right.$$
$$\left. + A(k_1, x_2) A(k_2, x_1) e^{ik_1 \cdot (x_2 - x_1')} e^{ik_2 \cdot (x_1 - x_2')} \right\}. \tag{17.53}$$

The well-behaved property of the phase function $\phi(x)$ for coherent sources makes it meaningful to perform the summation in the above equation. This situation is in contrast to the case of a chaotic source, where because the phases $\phi(x)$ are not well behaved functions of the space-time locations, we must separate out terms with no random phases and terms with random phases, as in Eq. (17.25).

The summation over x_1 and x_2 in Eq. (17.53) can be carried out independently. We get

$$\Psi(k_1 k_2 : \{^{\text{all }x_1 x_2}_{\text{points}}\} \to x_1' x_2')$$

$$= \frac{1}{\sqrt{2}} \sum_{x_1} A(k_1, x_1) e^{i\phi(x_1)} e^{ik_1 \cdot (x_1 - x_1')} \sum_{x_2} A(k_2, x_2) e^{i\phi(x_2)} e^{ik_2 \cdot (x_2 - x_2')}$$

$$+ \frac{1}{\sqrt{2}} \sum_{x_2} A(k_1, x_2) e^{i\phi(x_2)} e^{ik_1 \cdot (x_2 - x_1')} \sum_{x_1} A(k_2, x_1) e^{i\phi(x_1)} e^{ik_2 \cdot (x_1 - x_2')}.$$

which can be simplified to

$$\Psi(k_1 k_2 : \{^{\text{all }x_1 x_2}_{\text{points}}\} \to x_1' x_2')$$

$$= \sqrt{2} \sum_{x_1} A(k_1, x_1) e^{i\phi(x_1)} e^{ik_1 \cdot (x_1 - x_1')} \sum_{x_2} e^{i\phi(x_2)} A(k_2, x_2) e^{i\phi(x_2)} e^{ik_2 \cdot (x_2 - x_2')}$$

$$= \sqrt{2} \int dx_1 \rho(x_1) A(k_1, x_1) e^{i\phi(x_1)} e^{ik_1 \cdot (x_1 - x_1')}$$

$$\times \int dx_2 \rho(x_2) A(k_2, x_2) e^{i\phi(x_2)} e^{ik_2 \cdot (x_2 - x_2')}.$$

From Eq. (17.50) and the above equation, the probability $P(k_1, k_2)$ for the emission of two pions from the extended source arriving at the detection points with k_1 at x_1' and k_2 at x_2' is

$$P(k_1, k_2) = \left| \int dx_1 \, \rho(x_1) A(k_1, x_1) e^{i\phi(x_1)} e^{ik_1 \cdot (x_1 - x_1')} \right.$$

$$\left. \times \int dx_2 \, \rho(x_2) A(k_2, x_2) e^{i\phi(x_2)} e^{ik_2 \cdot (x_2 - x_2')} \right|^2$$

$$= \left| \int dx_1 \rho(x_1) A(k_1, x_1) e^{i\phi(x_1)} e^{ik_1 \cdot x_1} \int dx_2 \rho(x_2) A(k_2, x_2) e^{i\phi(x_2)} e^{ik_2 \cdot x_2} \right|^2.$$

$$\tag{17.54}$$

A comparison with Eq. (17.47) shows that

$$P(k_1, k_2) = P(k_1) \, P(k_2). \tag{17.55}$$

Hence, the two-particle correlation function for a coherent source is

$$C_2(k_1, k_2) = \frac{P(k_1, k_2)}{P(k_1)P(k_2)}$$
$$= 1. \qquad (17.56)$$

From Eq. (17.35), the correlation function $R(k_1, k_2)$ for a coherent source is

$$R(k_1, k_2) = 0. \qquad (17.57)$$

Thus, the probability for the detection of one pion of one momentum is independent of the probability for the detection of another pion of another momentum. There is no momentum correlation in the two-particle momentum distribution function, if the emitting source is a coherent source.

From the detailed derivations of the momentum correlation function, it is possible to trace the origin of the momentum correlation for extended chaotic sources. We observe that in the case of a coherent source, the symmetrization and the factorization of the probability amplitude insure that the two-particle probability function is proportional to the product of the single-particle momentum distributions. Hence, there is no correlation for a coherent source. When one has a chaotic source, the phase function is not well behaved and the summation procedure for the probability function can be carried out only by separating those terms in which the random phases cancel out from terms in which the random phases $\phi(x)$ do not. The randomness of the phases leads to the cancellation of terms which depend on the random phases. One of the terms which is not eliminated arises from the presence of the exchange symmetry involving identical bosons, and this term is proportional to the Fourier transform of a function of the source phase-space distribution function $f(k, x)$. There is no longer the factorization of the two-particle momentum distribution $P(k_1, k_2)$ in terms of the product of single-particle momentum distributions $P(k_1)$ and $P(k_2)$. Because the two-particle momentum distribution $P(k_1, k_2)$ does not equal $P(k_1)P(k_2)$, two-particle momentum correlations exist for chaotic sources and the momentum correlation function contains the Fourier transform of a function of the phase space distribution. The exchange symmetry involving identical particles is the origin of the momentum correlation for chaotic sources.

We have established the relation between the momentum correlation function and the property of coherence or chaoticity. For the analysis of many sets of experimental data, it is convenient to introduce a 'chaoticity' parameter λ to modify the correlation function

$C_2(q)$ of Eq. (17.43) as

$$C_2(q) = 1 + \lambda e^{-R_x{}^2 q_x^2 - R_y{}^2 q_y^2 - R_z{}^2 q_z^2 - \sigma_t{}^2 q_t{}^2}.$$ (17.58)

This correlation function interpolates between the case of a coherent source with $\lambda = 0$ and the case of a completely chaotic source with $\lambda = 1$. It should be realized that Eq. (17.58) is an oversimplified description of states with partial coherence. We examine in Supplement 17.3 the two-particle correlation function for a simple example of a partially coherent source which results in a correlation function more complicated than Eq. (17.58).

⊕〖 Supplement 17.3

The formulation of the last few sections allows one to study here the case of a partially coherent source. As coherence and chaoticity are complementary qualities, a partially coherent source is also a partially chaotic source.

For simplicity, we shall study in this section a simple idealized case in which the production amplitude $A(k, x)$ is independent of coordinates x,

$$A(k, x) = A(k).$$ (1)

We further assume that the density of the chaotic source is proportional to the total density and we define the chaoticity parameter λ to be

$$\rho_\chi(x) = \lambda \rho(x).$$ (2)

The coherent source density is then

$$\rho_c(x) = (1 - \lambda)\rho(x).$$ (3)

We shall normalize the total density according to $\int dx \rho(x) = 1$.

According to Eq. (17.7), the probability amplitude for the emission of a single pion of momentum k from the extended source to arrive at the detection point x' is given by

$$\Psi(k : \{^{\text{all } x}_{\text{points}}\} \rightarrow x') = \sum_x A(k, x) e^{i\phi(x)} e^{ik \cdot (x - x')}.$$ (4)

We divide the summation into the coherent part $\sum^c$ and the chaotic part $\sum^\chi$,

$$\Psi(k : \{^{\text{all } x}_{\text{points}}\} \rightarrow x) = \sum_x{}^c A(k) e^{i\phi_c(x)} e^{ik \cdot (x - x')}$$
$$+ \sum_x{}^\chi A(k) e^{i\phi_\chi(x)} e^{ik \cdot (x - x')},$$ (5)

where $\phi_c(x)$ and $\phi_\chi(x)$ are the production phases of the coherent and the chaotic source points respectively.

The production phases associated with the coherent source can be any well-behaved function of the source coordinates. We study the simple case when the coherent phases are the same and they can be taken to be zero at all source points. Eq. (5) becomes

$$\Psi(k:\{^{\text{all } x}_{\text{points}}\} \to x) = A(k)e^{-ik\cdot x'}\left\{\int dx \rho_c(x)e^{ik\cdot x} + \sum_x^{\chi} e^{i\phi_\chi(x)} e^{ik\cdot x}\right\}$$

$$= A(k)e^{-ik\cdot x'}\left\{\tilde{\rho}_c(k) + \sum_x^{\chi} e^{i\phi_\chi(x)} e^{ik\cdot x}\right\}, \tag{6}$$

where $\tilde{\rho}_c(k)$ is the Fourier transform of $\rho_c(x)$,

$$\tilde{\rho}_c(k) = \int dx\, e^{ik\cdot x}\rho_c(x) \tag{7}$$

$$\equiv (1-\lambda)\, \tilde{\rho}(k).$$

The probability $P(k)$ for the pion of momentum k to be produced from the source and arrive at the detection point x' is the absolute square of the amplitude (6). The chaotic source is described by a random production phase $\phi_\chi(x)$. We expand the absolute square of the probability amplitude into terms which are independent of ϕ_χ and terms which contain ϕ_χ. Terms which depend on the production phase ϕ_χ give zero contribution because of the randomness of the chaotic phases. The probability $P(k)$ becomes

$$P(k) = A^2(k)\left[|\tilde{\rho}_c(k)|^2 + \sum_x^{\chi} 1\right]$$

$$= A^2(k)\left[|\tilde{\rho}_c(k)|^2 + \int dx \rho_\chi(x)\right]$$

$$= A^2(k)\left[(1-\lambda)^2|\tilde{\rho}(k)|^2 + \lambda\right]. \tag{8}$$

Thus, the single-particle momentum distribution depends not only on the amplitude $A(k)$ at the moment of production, but also on the Fourier transform of the coherent source spatial density $\rho_c(x)$ which we have assumed to be $(1-\lambda)\rho(x)$.

We shall now examine the two-particle momentum distribution. According to the general result of Eq. (17.25), the probability $P(k_1, k_2)$ is

$$P(k_1, k_2) = \frac{1}{2!}|\Psi(k_1 k_2 : \{^{\text{all } x_1 x_2}_{\text{points}}\} \to x'_1 x'_2)|^2, \tag{9}$$

where

$$\Psi(k_1 k_2 : \{^{\text{all } x_1 x_2}_{\text{points}}\} \to x'_1 x'_2) = \frac{1}{\sqrt{2}}\sum_{x_1, x_2} e^{i\phi(x_1)}e^{i\phi(x_2)}$$

$$\times \left\{A(k_1, x_1)A(k_2, x_2)e^{ik_1\cdot(x_1-x'_1)}e^{ik_2\cdot(x_2-x'_2)} + A(k_1, x_2)A(k_2, x_1)e^{ik_1\cdot(x_2-x'_1)}e^{ik_2\cdot(x_1-x'_2)}\right\}. \tag{10}$$

To simplify the above equation, we separate out various terms into the coherent part which can be integrated out, and the chaotic part which contains the random phases ϕ_χ. Eq. (10) becomes

$$\Psi(k_1 k_2 : \{^{\text{all } x_1 x_2}_{\text{points}}\} \to x_1' x_2') = \frac{e^{-ik_1 \cdot x_1 - ik_2 \cdot x_2}}{\sqrt{2}} A(k_1) A(k_2)$$

$$\times \left\{ \int dx_1 dx_2 \rho_c(x_1) \rho_c(x_2) \left[e^{ik_1 \cdot x_1 + ik_2 \cdot x_2} + e^{ik_1 \cdot x_2 + ik_2 \cdot x_1} \right] \right.$$

$$+ \sum_{x_1 x_2}^\chi e^{i\phi_\chi(x_1) + i\phi_\chi(x_2)} \left[e^{ik_1 \cdot x_1 + ik_2 \cdot x_2} + e^{ik_1 \cdot x_2 + ik_2 \cdot x_1} \right]$$

$$+ \int dx_1 \rho_c(x_1) e^{ik_1 \cdot x_1} \sum_{x_2}^\chi e^{i\phi_\chi(x_2)} e^{ik_2 \cdot x_2} + \sum_{x_1}^\chi e^{i\phi_\chi(x_1)} e^{ik_1 \cdot x_1} \int dx_2 \rho_c(x_2) e^{ik_2 \cdot x_2}$$

$$+ \left. \int dx_1 \rho_c(x_1) e^{ik_2 \cdot x_1} \sum_{x_2}^\chi e^{i\phi_\chi(x_2)} e^{ik_1 \cdot x_2} + \sum_{x_1}^\chi e^{i\phi_\chi(x_1)} e^{ik_2 \cdot x_1} \int dx_2 \rho_c(x_2) e^{ik_1 \cdot x_2} \right\}.$$

$$(11)$$

This can be simplified to yield

$$\Psi(k_1 k_2 : \{^{\text{all } x_1 x_2}_{\text{points}}\} \to x_1' x_2') = \frac{e^{-ik_1 \cdot x_1 - ik_2 \cdot x_2} A(k_1) A(k_2)}{\sqrt{2}}$$

$$\times \left\{ 2(1-\lambda)^2 \tilde{\rho}(k_1) \tilde{\rho}(k_2) + \sum_{x_1 x_2}^\chi e^{i\phi_\chi(x_1) + i\phi_\chi(x_2)} \left[e^{ik_1 \cdot x_1 + k_2 \cdot x_2} + e^{ik_1 \cdot x_2 + k_2 \cdot x_1} \right] \right.$$

$$+ \left. 2(1-\lambda)\tilde{\rho}(k_1) \sum_{x_2}^\chi e^{i\phi_\chi(x_2)} e^{ik_2 \cdot x_2} + 2(1-\lambda)\tilde{\rho}(k_2) \sum_{x_1}^\chi e^{i\phi_\chi(x_1)} e^{ik_1 \cdot x_1} \right\}. \quad (12)$$

Because of the randomness of the chaotic phases $\phi_\chi(x)$, terms in $P(k_1, k_2)$ containing the random phases do not contribute, and we obtain

$$P(k_1, k_2) = A^2(k_1) A^2(k_2) \left\{ (1-\lambda)^4 |\tilde{\rho}(k_1)\tilde{\rho}(k_2)|^2 \right.$$

$$+ \int dx_1 dx_2 \rho_\chi(x_1) \rho_\chi(x_2) [1 + e^{ik(x_1 - x_2) + ik_2(x_2 - x_1)}] + (1-\lambda)^2 \lambda [|\tilde{\rho}(k_1)|^2 + |\tilde{\rho}(k_2)|^2]$$

$$+ \left. (1-\lambda)^2 \lambda \tilde{\rho}^*(k_1)\tilde{\rho}(k_2)\tilde{\rho}(k_1 - k_2) + (1-\lambda)^2 \lambda \tilde{\rho}^*(k_2)\tilde{\rho}(k_1)\tilde{\rho}(k_2 - k_1) \right\}.$$

This can be simplified as

$$P(k_1, k_2) = A^2(k_1) A^2(k_2) \left\{ (1-\lambda)^4 |\tilde{\rho}(k_1)\tilde{\rho}(k_2)|^2 \right.$$

$$+ \lambda^2 [1 + |\tilde{\rho}(k_1 - k_2)|^2] + (1-\lambda)^2 \lambda [|\tilde{\rho}(k_1)|^2 + |\tilde{\rho}(k_2)|^2]$$

$$+ \left. (1-\lambda)^2 \lambda \tilde{\rho}^*(k_1)\tilde{\rho}(k_2)\tilde{\rho}(k_1 - k_2) + (1-\lambda)^2 \lambda \tilde{\rho}^*(k_2)\tilde{\rho}(k_1)\tilde{\rho}(k_2 - k_1) \right\}. \quad (13)$$

Using the results of Eq. (8), we calculate the quantity

$$P(k_1, k_2) - P(k_1)P(k_2) = A^2(k_1)A^2(k_2)$$
$$\times \left\{ \lambda^2 |\tilde{\rho}(k_1 - k_2)|^2 + \lambda(1 - \lambda)^2 [\tilde{\rho}^*(k_1)\tilde{\rho}(k_2)\tilde{\rho}(k_1 - k_2) + \tilde{\rho}^*(k_2)\tilde{\rho}(k_1)\tilde{\rho}(k_2 - k_1)] \right\}.$$

The two-particle correlation functions $R(k_1, k_2)$ and $C_2(k_1, k_2)$ are

$$R(k_1, k_2) = C_2(k_1, k_2) - 1$$
$$= \left\{ \lambda^2 |\tilde{\rho}(k_1 - k_2)|^2 + \lambda(1 - \lambda)^2 [\tilde{\rho}^*(k_1)\tilde{\rho}(k_2)\tilde{\rho}(k_1 - k_2) + \tilde{\rho}^*(k_2)\tilde{\rho}(k_1)\tilde{\rho}(k_2 - k_1)] \right\}$$
$$\div \left\{ [\lambda + (1 - \lambda)^2 |\tilde{\rho}(k_1)|^2][\lambda + (1 - \lambda)^2 |\tilde{\rho}(k_2)|^2] \right\}, \tag{14}$$

which is a special case of Eq. (4.66) of Ref. [4]. It gives $R(k_1, k_2) = 0$ for $\lambda = 0$ and $R(k_1, k_2) = |\tilde{\rho}(k_1 - k_2)|^2$ for $\lambda = 1$, as it should. We see from the above result that even for a simple idealized case, the correlation function for a partial coherent or a partial chaotic source is a complicated function of λ. The correlation function is both a function of the relative momentum, and also a function of k_1 and k_2. The analysis for a partially chaotic source is not as simple as Eq. (17.58) would indicate.
]⊕

§17.4 Experimental Information on Pion Interferometry

While the above brief discussions of the Hanbury-Brown-Twiss effect provides the basic concepts to relate the two-particle momentum correlation function to the phase-space distribution of the source, there are many corrections which need to be taken into account. The two-particle correlation is influenced by Coulomb interactions, final-state strong interactions, resonance decays. The chaoticity parameter λ is subject to different interpretations. Our example of a partially chaotic source in Supplement 17.3 shows that the two-particle correlation function depends on the degree of chaoticity in a complicated way, much more complicated than Eq. (17.58) would indicate. Furthermore, the chaoticity parameter λ depends not just on the degree of coherence alone. As is reviewed in Ref. [5], it depends also on the non-pion contamination, the number of emitting source points, resonance decays, the multiple scattering process, impact parameter averaging, finite experimental resolutions, and the expansion of the source. The discussion of these topics is beyond the scope of the present introductory book. Interested readers may consult original articles in the literature [4-20] for more thorough discussions.

In the application of pion interferometry to high-energy reactions, we follow the experimental groups of E802/E859 [3,36,37], ISR-SFM

[32,33], UA1 [34], E735 [35], OPAL [24], MARK II [25], CLEO [26], TPC [27], TASSO [28] and AMY [29], and we use the Gaussian parametrization of Eq. (17.58) in the form

$$C_2(q) = 1 + \lambda\, e^{-R_x{}^2 q_x^2 - R_y{}^2 q_y^2 - R_z{}^2 q_z^2 - \sigma_t{}^2 q_t{}^2} \, . \tag{17.58}$$

In the literature, many workers, including those of the NA35 Collaboration [38-42] and the authors of the review article of reference [5], use a different Gaussian parametrization of the following form

$$C_2(q) = 1 + \lambda\, e^{-R'_x{}^2 q_x^2/2 - R'_y{}^2 q_y^2/2 - R'_z{}^2 q_z^2/2 - \sigma_t{}'^2 q_t{}^2/2} \, . \tag{17.59}$$

It is important not to confuse R_i with R'_i when comparing numerical values of radius parameters.

In most analyses, only some of the corrections have been taken into account. Furthermore, as explained earlier, the effective density and size parameters depend in general on the momenta of the two detected pions, k_1 and k_2. Even though not all corrections and effects have been taken into account in these analyses, an approximate and tentative picture may still emerge as we review the experimental data.

We shall first discuss e^+e^- annihilation experiments with a center-of-mass energy $\sqrt{s}$ from 3.095 GeV to 91.3 GeV [24]-[29]. One can extract the radius parameter R_i and the chaoticity parameter λ from two-like-pion momentum correlation data, using Eq. (17.58) or a modification thereof. Within experimental errors, the data are consistent with an isotropic source distribution [26-28], although it may also be described, within experimental errors, as an ellipsoidal distribution with a longitudinal radius twice as large as the transverse radius [27]. As the experimental errors are not small enough to show definitive evidence for an anisotropic density distribution, most analyses of the data make use of an isotropic density so that all radius parameters R_i's are assumed to be equal to a common R.

The systematics of the radius parameter R and the chaoticity parameter λ as a function of the e^+e^- center-of-mass energy are shown in Fig. 17.4. Experimental data indicate that the radius parameter R is of the order of 1 fm and does not appear to increase with energy from $\sqrt{s} =$3.095 GeV to 91.5 GeV. The parameter λ ranges from 0.5 to unity. It reaches a value close to unity when corrections due to the contamination of non-pion tracks are taken into account [24]. One concludes from these values that the emission of pions from the e^+e^- annihilation source is very chaotic.

The radius parameter of about 1 fm is a reasonable description of the transverse dimension of the emission process. However, the

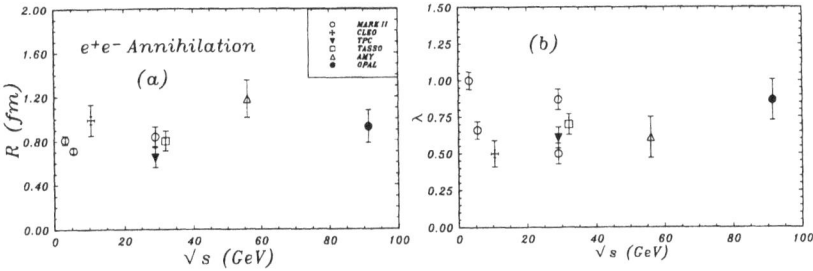

Fig. 17.4 The radius parameter R and the chaoticity parameter λ extracted from e^+e^- annihilations at various center-of-mass energies. The data are from MARK II [25], CLEO [26], TPC [27], TASSO [28], AMY [29], and OPAL [24] and were compiled in Reference [24].

approximate isotropy is a puzzle because the apparent longitudinal radius parameter of 1 fm is much smaller than one would expect for the dimension of the longitudinal hadronization region, which may be as long as many tens of fermi. Clearly, what is measured is not the entire hadronization region but a localized spatial region [24] spanned by the two detected pions having momenta falling within a small range of momentum. How do we understand this result?

In e^+e^- annihilation, the annihilation process leads to the formation of a virtual photon which decays into a quark-antiquark pair. The fragmentation of the separating quark-antiquark pair gives rise to the production of the observed hadrons, which consist mostly of pions. As one would expect from the inside-outside cascade picture of Fig. 6.4 or the string fragmentation picture of Fig. 7.6, the quark q travels in one longitudinal direction while the antiquark $\bar{q}$ travels in the opposite direction. The quark or the antiquark would each carry an energy of $\sqrt{s}/2$. The quark and the antiquark will each travel a distance of $\sqrt{s}/2\kappa$ before they turn back, as in Fig. 7.6. Particles will be produced as the quark and the antiquark pull apart from each other (Fig. 17.5a). Therefore, in the hadronization of the $q\bar{q}$ pair, one expects that the entire hadronization region may be stretched over a distance of about $\sqrt{s}/\kappa$ fermi, which is many tens of fermi for energies of many tens of GeV we consider in Fig. 17.4.

It needs to be pointed out however that the longitudinal extension of the whole hadronization region is not the relevant length measured in interferometry measurements. Pions of different longitudinal momentum come out from different parts of the hadronization region. As pion interferometry measurements are concerned only with pions which are not too far apart in momentum, the relevant length measured by pion interferometry is only the width of the phase space distribution of those pions which fall within a narrow range of momentum.

Experimentally, the rapidity distribution dN/dy of the produced particles has a plateau structure [30]. We also learn from Exercise 7.1 that a rapidity plateau occurs when the $q\bar{q}$-producing vertices leading to the formation of hadrons fall on a curve of a constant proper time $\tau_{\text{freeze out}} \equiv \tau_0$. It is reasonable to assume that the particle production in e^+e^- annihilation is characterized by $q\bar{q}$-production vertices lying on the curve of a constant proper time τ_0. When this happens, we see from Fig. 17.5a and Exercise 7.2 that the longitudinal momentum of the produced particles, as represented by their rapidities, are ordered in space-time.

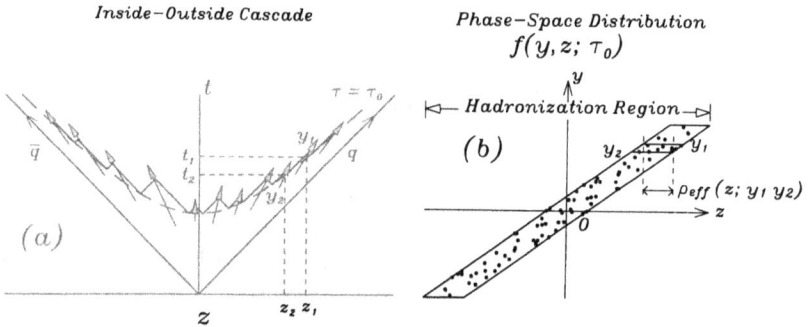

Fig. 17.5 (a) The space-time diagram in the inside-outside cascade picture of particle production. (b) The phase-space distribution function $f(y, z; \tau_0)$ at the freeze-out proper time τ_0 for the rapidity y and the longitudinal coordinate z. The phase-space density is indicated by the dotted region. The spatial region spanned by the effective density $\rho_{\text{eff}}(z; y_1 y_2)$ is indicated by the arrow between the dashed lines.

How do we describe this rapidity-space-time ordering in the source center-of-mass system? We label the space-time coordinates to be (z_i, t_i) at which a particle of rapidity y_i is produced, as is illustrated in Fig. 17.5a. The rapidity-space-time ordering of the produced particles means that

$$\text{if} \quad y_1 > y_2, \quad \text{then} \quad z_1 > z_2, \quad \text{and vice versa}; \qquad (17.60)$$

and

$$\text{if} \quad |y_1| > |y_2|, \quad \text{then} \quad t_1 > t_2, \quad \text{and vice versa}, \qquad (17.61)$$

(see Exercise 7.2). In other words, pions with a greater rapidity y are produced at a greater longitudinal coordinate z, and pions with a greater magnitude of rapidity $|y|$ are produced at a later time t. As time t increases in the center-of-mass system, particles with small rapidities are first produced at small distances from the origin, followed later by particles with greater and greater rapidities at further distances from the origin. The production process occurs from the inside and expands to the outside. This is the inside-outside cascade picture of particle production, first proposed by Bjorken [31] (see Chapters 6 and 7).

The above description of rapidity-space-time ordering is based on a classical description in the limit when the energy is very large, the production proper time is constant, and the neighboring $q\bar{q}$-producing vertices leading to the formation of hadrons are separated only infinitesimally. (See Exercise 7.2 for details). We envisage that the actual production process is a more complicated stochastic process. We expect however that the gross features of the inside-outside cascade picture remain approximately valid: The production vertices should fall approximately on the curve of constant τ, which has a distribution about a mean value τ_0 and the above equations (17.60) and (17.61) should be valid in an average sense. That is, on the average, pions with a rapidity y are produced at a greater longitudinal coordinate z, and pions with a greater magnitude of rapidity $|y|$ are produced at a later time t. In this stochastic description, we can speak of a phase-space distribution $f(k,x)$ involving the longitudinal momentum represented by the rapidity variable y, and the longitudinal coordinate z. The rapidity-space-time ordering of the produced particles leads to the longitudinal phase-space distribution $f(y,z;\tau_0)$ as shown in Fig. 17.5b. For a given value of the rapidity of an ensemble of produced particles, there is an average longitudinal coordinate $\bar{z}(y)$ and an average time $\bar{t}(|y|)$ for the production of the ensemble of particles. The inside-outside cascade picture implies that $\bar{z}(y)$ is a monotonically increasing function of y and $\bar{t}(|y|)$ is a monotonically increasing function of $|y|$. We can assume a Gaussian distribution about these average quantities and an exponential distribution about the transverse mass m_T with a temperature parameter T. The above stochastic description of the rapidity-space-time ordering suggests a phase-space distribution of the form

$$f(k,x) = f(y\boldsymbol{p}_T k_0, z\boldsymbol{x}_T t) = A \int d\bar{y} \frac{dN}{d\bar{y}} \delta(k_0 - \sqrt{\boldsymbol{p}_T^2 + m_T^2 \sinh^2 y + m^2})$$

$$\times \exp\left\{ -\frac{\gamma^2(\bar{y}, \boldsymbol{p}_T)[z - \bar{z}(\bar{y})]^2}{2R_{zo}^2} - \frac{\boldsymbol{x}_T^2}{2R_T^2} - \frac{(y - \bar{y})^2}{2\sigma_y^2} - \frac{m_T - m}{T} - \frac{(t - \bar{t}(\bar{y}))^2}{2\gamma^2(\bar{y}, \boldsymbol{p}_T)\sigma_{to}^2} \right\},$$

$$(17.62)$$

where m is the pion rest mass,

$$\bar{z}(\bar{y}) = \tau_0 \sinh \bar{y}, \tag{17.63a}$$

$$\bar{t}(\bar{y}) = \tau_0 \cosh \bar{y}, \tag{17.63b}$$

$$m_T = \sqrt{p_T^2 + m^2}, \tag{17.63c}$$

and the normalization constant $\mathcal{A}$ is

$$\mathcal{A} = \frac{1}{N(2\pi)^{5/2} R_{zo} R_T^2 \sigma_y \sigma_{to} 2mT(m+T)}, \tag{17.63d}$$

with $N = \int d\bar{y}(dN/d\bar{y})$. Here, the $\gamma(\bar{y}, p_T)$ factor is to take into account the Lorentz contraction and the time dilation for the distribution of pions as they emerge at (z, t) in the neighborhood of $(\bar{z}(\bar{y}), \bar{t}(\bar{y}))$ with an average rapidity $\bar{y}$. For these pions, the average rapidity $\bar{y}$ is $\sinh^{-1}(\bar{z}/\tau_0)$ given by Eq. (17.63a) above. The average velocity of those pions produced around $\bar{z}$ with rapidity around $\bar{y}$ and transverse momentum p_T is

$$\beta = \frac{\sqrt{p_T^2 + m_T^2 \sinh^2 \bar{y}}}{m_T \cosh \bar{y}}.$$

Therefore, the $\gamma(\bar{y}, p_T)$ factor in Eq. (17.62) is

$$\gamma(\bar{y}, p_T) = \frac{1}{\sqrt{1 - \beta^2}}$$

$$= \sqrt{1 + \frac{p_T^2}{m^2}} \cosh \bar{y}. \tag{17.64}$$

The above distribution function is normalized such that

$$\int f(k, x) dx dk = 1. \tag{17.65}$$

With a phase-space distribution in the form of Eq. (17.62) and depicted in Fig. 17.5b, it is easy to understand the small longitudinal size of the emitting source in e^+e^- annihilation. According to Eq. (17.32), the effective density $\rho_{\text{eff}}(x; k_1 k_2)$ is a function of the momentum of the two detected pions. Focussing our attention on the longitudinal phase space, we can consider the distribution function involving the space-time coordinate z and the rapidities y_1 and y_2 of the two detected identical pions. The effective density spans a region which is the overlap of $f(y_1, z)$ and $f(y_2, z)$, as indicated by the narrow region between the dashed curves in Fig. 17.5b. This region is much smaller than the entire hadronization region which extends between the turning points of the quark and the antiquark. We can understand the experimental data as indicating that the effective density in the neighbourhood of k_1 and k_2 is quite narrow, with a longitudinal spatial standard deviation R of only about 1 fm. It is essentially this width of the coordinate spread which is represented by the effective density $\rho_{\text{eff}}(z; y_1 y_2)$ and measured by the Hanbury-Brown-Twiss effect, even though the entire string stretches as long as $\sqrt{s}/\kappa$ fermi. The inside-outside cascade picture also implies that the longitudinal radius decreases with the magnitude of rapidity measured in the center-of-mass system.

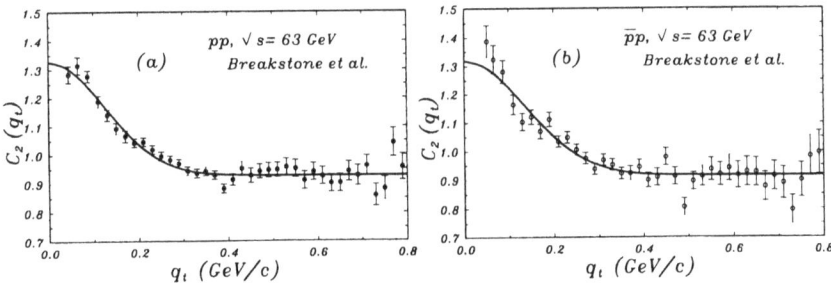

Fig. 17.6 The correlation functions for two identical pions observed for pp reactions at $\sqrt{s} = 63$ GeV are shown in Fig. 17.6a and for $\bar{p}p$ reactions are shown in Fig. 17.6b. The data points are from Refs. [32,33]. The solid curves are fits to the experimental data using equation (17.66).

We examine now the interferometry data for pp and $\bar{p}p$ collisions. As an illustration, we show in Fig. 17.6 the two-pion correlation function observed in the reactions pp and $\bar{p}p$ at $\sqrt{s} = 63$ GeV. The data points are from Breakstone *et al.* [32,33]. Here, q_t is the magnitude of the three-momentum difference q_t, which is selected to be perpendicular to the sum of the two pion three-momenta. The solid curves are the theoretical fits to the data using the parametrization

$$C_2(q_t) = \gamma'(1 + \lambda e^{-R^2 q_t^2}). \qquad (17.66)$$

The parameters of the fits are
  $\gamma'=0.928$, $\lambda=0.432$, and $R = 5.71$ Gev$^{-1} = 1.13$ fm for pp,

γ'=0.916, λ=0.440,   and $R = 5.22$ Gev^{-1} = 1.03 fm for $\bar{p}p$.
These parameters are quite close to each other. The source dimension
for pp and $\bar{p}p$ collisions are of the order of 1 fm and are about the same
as that of e^+e^- annihilation. However, the parameters for pp and $\bar{p}p$
are much smaller than the parameters in e^+e^- annihilations. The
origin of the small value of the λ parameter for pp and $p\bar{p}$ is not fully
understood.

The systematics of the radius parameter and the λ parameter, for
different charged-particle rapidity densities $dN_c/d\eta$ and energies are
shown in Fig. 17.7. As one observes, the radius parameter increases
with the rapidity density, indicating that the size of the emitting
region is apparently greater for a more violent collision with a greater
number of produced particles per unit of rapidity. The parameter λ
decreases as $dN_c/d\eta$ increases.

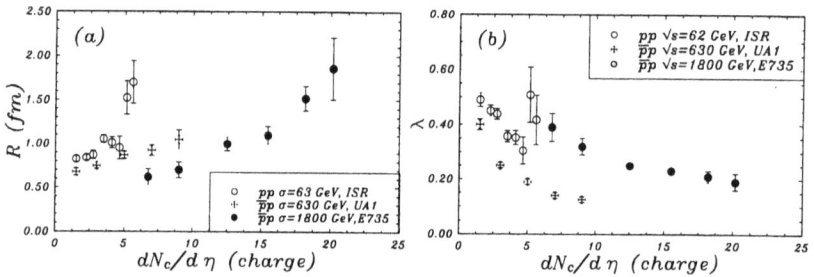

Fig. 17.7 The radius parameter R and the parameter λ extracted from
pp and $\bar{p}p$ collisions at various center-of-mass energies. The data are from
ISR-SFM [33], UA1 [34], and E735 [35].

Since the size parameters for pp and $\bar{p}p$ collisions are about the
same as those for e^+e^- annihilations, we can understand the small
value of the size parameters in the same way as in e^+e^- collisions.
In $\bar{p}p$ or pp collisions, one envisages the formation of one, two or
more strings, as the leading particles spread out from the point of
collision. Particles are produced in the inside-outside cascade picture,
with a rapidity-space-time ordering of the produced particles. Those
pions with small rapidities are produced near the point of collision
while those pions with large values of rapidity are produced later
at larger distances from the point of collision. By the argument of
rapidity-space-time ordering, only those pions within a narrow range
of rapidity are correlated, corresponding to a small section of the
entire hadronization region. The size of this small section is of the
order of 1 fm in the longitudinal direction, as determined by pion
interferometry experiments.

We turn now to the study of pion interferometry in heavy-ion reactions. The first measurement of pion interferometry in heavy-ion reactions was carried out by Fung *et al.* [11]. Since then, a large number of measurements have been made. Ref. [5] gives a review of the present experimental situation. Recent experiments on pion interferometry at very high energies were carried out by the E802/E859 Collaboration at Brookhaven [6,36,37], the NA35 Collaboration [38-42], the NA44 Collaboration [43,44]], the EMU05 Collaboration [45] and the WA80 Collaboration [46] at CERN. We shall briefly review some recent results at high energies.

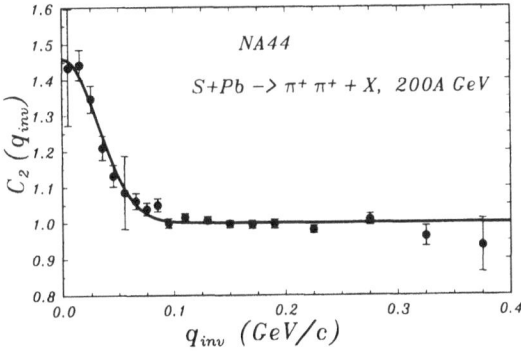

Fig. 17.8 The correlation function measured for central collisions of S on Pb at $E_{lab} = 200A$ GeV [42,43], as a function of the invariant momentum difference q_{inv} of the pion pair. The solid curve is the fit with $\lambda = 0.46$ and $R_{inv} = 4.5$ of Eq. (17.67)

We show in Fig. 17.8 an example of the correlation function observed in high-energy heavy-ion reactions. The data are from the NA44 Collaboration for the collision of S on Pb at 200A GeV. The data are analyzed in terms of a correlation function of the form

$$C_2(q_{inv}) = 1 + \lambda e^{-R_{inv}^2 q_{inv}^2} , \qquad (17.67)$$

where q_{inv}^2 is the invariant quantity defined by

$$q_{inv}^2 = (k_1 - k_2)^2$$
$$= (k_1^0 - k_2^0)^2 - (\boldsymbol{k}_1 - \boldsymbol{k}_2)^2 .$$

The quantity R_{inv} is not a direct measure of the size of the effective density as it contains also the time component. The numerical values

Table 17.1

Radius parameter R_{inv} and parameter λ found in various measurements

System	E_{lab} (GeV)	$R_{\text{inv}}(fm)$	λ	Ref.
^{28}Si on ^{197}Au (Data Set 1)	$14.5A$	4.45 ± 0.44	0.60 ± 0.08	[37]
^{28}Si on ^{197}Au (Data Set 2)	$14.5A$	4.19 ± 0.71	0.44 ± 0.10	[37]
^{28}Si on ^{27}Al (Data Set 1)	$14.5A$	3.44 ± 0.36	0.59 ± 0.08	[37]
^{28}Si on ^{27}Al (Data Set 2)	$14.5A$	4.03 ± 0.71	0.53 ± 0.15	[37]
^{32}S on ^{208}Pb	$200A$	$4.50 \pm .31$	0.46 ± 0.04	[43]

of R_{inv} are greater than the radius parameters R_z and $R_{x,y}$ of Eq. (17.58) by about 30-40% [37]. Nevertheless, the sizes of the emitting region for different collisions can be compared by using the same parametrization.

In Table 17.1, we list the values of R_{inv} extracted from various measurements in central collisions of heavy-ions with various targets from the E802/E859 Collaboration and the NA44 Collaboration. One observes that the value of R_{inv} in the collisions of S on Pb at 200A GeV is not much changed from the values of R_{inv} found in lower energy collisions. This excludes a qualitatively different behavior, such as a large expansion of the pion gas before freezing out as the energy increases from AGS to CERN energies.

NA35 [38-42] made preliminary measurements of the longitudinal and transverse radius parameters. Placing the source center-of-mass at the spatial origin, the collision axis is designated to be the longitudinal z-axis and the detectors are placed along the x-axis. The parameter R_z of Eq. (17.58) measures the longitudinal 'radius' of the source along the beam direction, R_x measures the transverse radius of the source along the direction of the detectors and R_y measures the transverse radius of the source perpendicular to the line joining the source and the detectors. NA35 uses the parameterization of (17.59),

$$C_2(q) = 1 + \lambda e^{-R_{\text{Tout}}^2 q_x^2/2 - R_{\text{Tside}}^2 q_y^2/2 - R_{\text{long}}^2 q_z^2/2 - \sigma_t'^2 q_t^2/2}, \qquad (17.68)$$

where the parameters R_{long}, R_{Tout}, and R_{Tside}, are related to $R_{x,y,z}$

of Eq. (17.58) by

$$R_z = R_{\text{long}}/\sqrt{2},$$

$$R_x = R_{\text{Tout}}/\sqrt{2},$$

and

$$R_y = R_{\text{Tside}}/\sqrt{2}.$$

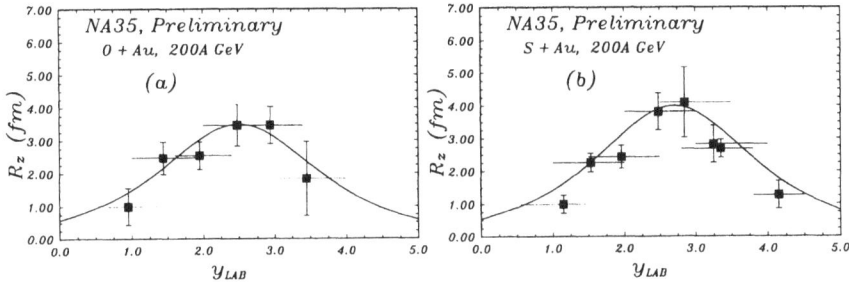

Fig. 17.9 Preliminary NA35 data [38-42] of the longitudinal radius parameter R_z as a function of rapidity, measured in the source center-of-mass system. Fig. 17.9(a) gives the data for O on Au at 200A GeV, and Fig. 17.9(b) gives the data for S on Au at 200A GeV. The curves are obtained by a parameterization of the radius parameter according to Eq. (17.69a)

Fig. 17.10 shows the preliminary NA35 results for R_z which indicates that R_z is a function of the laboratory rapidity y_{LAB}. The solid curves are obtained by a parametrization of the radius parameter as [38]-[40]

$$R_z = \frac{R_{zo}}{\cosh y}. \tag{17.69a}$$

where

$$y = y_{LAB} - y^*, \tag{17.69b}$$

and y^* is the rapidity of the center-of-mass of the source which can be taken to be the rapidity at which the peak of the rapidity distribution dN/dy is located. For collisions of O and S on Au at 200A GeV, the midpoint rapidity between the projectile and target rapidities is 3.025, which is however not the location of the peak of the rapidity distribution dN/dy. Because of the asymmetry of projectile and target masses, the peak of the rapidity distribution is shifted slightly from the midpoint between the beam rapidity and target rapidity toward the lower rapidity region [47]. The curves in Fig. 17.9 are obtained with $R_{zo} = 3.5$ fm, and $y^* = 2.5$ for O on Au and $R_{zo} = 4$ fm and $y^* = 2.7$ for S on Au. The experimental data appear to be described approximately by the relation (17.69). How do we understand this result?

We envisage that whatever the mode of production of particles in nucleus-nucleus collisions, the property of the inside-outside cascade remains approximately valid so that there is a rapidity-space-time ordering of the produced particles. Therefore, we can give a simple description of the source by generalizing the phase-space distribution (17.62) of the inside-outside cascade picture to the case for high-energy nucleus-nucleus collisions. The source of pions is now more extended since it arises from many baryon-baryon collisions and the locations of the baryon-baryon collisions are distributed over a greater space-time region. We can integrate Eq. (17.62) by approximating the gaussian distribution of y with respect to $\bar{y}$ as a delta function $\delta(y-\bar{y})$. Eq. (17.62) then gives

$$f(y\boldsymbol{p}_{T}k_0, z\boldsymbol{x}_{T}t) = A\frac{dN}{dy}\delta(k_0 - \sqrt{p_T^2 + m_T^2\sinh^2 y + m^2})$$

$$\times\exp\left\{-\frac{\gamma^2(y,\boldsymbol{p}_{T})[z-\bar{z}(y)]^2}{2R_{zo}^2} - \frac{x_T^2}{2R_T^2} - \frac{m_T-m}{T} - \frac{(t-\bar{t}(y))^2}{2\gamma^2(y,\boldsymbol{p}_{T})\sigma_t^2}\right\}, \quad (17.70)$$

where

$$\bar{z}(y) = \tau_0\sinh y,$$

and

$$\bar{t}(y) = \tau_0\cosh y.$$

Using this phase-space distribution, the effective density involving the longitudinal phase space coordinates becomes

$$\rho_{\text{eff}}(z, y_1 y_2)\propto\exp\left\{-\frac{\gamma^2(y_1, m_T)[z-\bar{z}(y_1)]^2}{4R_{zo}^2} - \frac{\gamma^2(y_2, m_T)[z-\bar{z}(y_2)]^2}{4R_{zo}^2}\right\}.$$

$$(17.71)$$

The momenta of two pions are correlated only when the magnitude of their momentum difference $|\boldsymbol{q}|$ is of the order of $1/R$. That is, correlations occur only when y_1 and y_2 are close to each other. For y_1 to be close to y_2, we have

$$\rho_{\text{eff}}(z, y_1 y_2) \sim \exp\left\{-\frac{[z-\bar{z}(Y)]^2}{2R_z^2(Y)}\right\}, \quad (17.72a)$$

where

$$Y = (y_1 + y_2)/2, \quad (17.72b)$$

and

$$R_z(Y) = \frac{R_{zo}}{\gamma(Y, \mathbf{p}_T)} . \tag{17.72c}$$

The gamma factor $\gamma(Y, \mathbf{p}_T)$ associated with a given rapidity Y and a given transverse momentum $\mathbf{p}_T$ is given by Eq. (17.64). Using this relation, the radius parameter obtained by sampling an ensemble of two pions in the vicinity of the rapidity Y is therefore

$$R_z(Y) \sim \frac{R_{zo}}{\sqrt{1 + p_T^2/m^2} \cosh Y} . \tag{17.73}$$

The rapidity-dependent part of the above equation is the relation (17.69) used to construct the curves in Fig. 17.9. Therefore, the distribution function of Eq. (17.72) and (17.73) approximately describes the experimental data of the longitudinal radius R_z as a function of rapidity. Eq. (17.73) further suggests that for a given rapidity Y, the longitudinal radius R_z varies with the transverse momentum $\mathbf{p}_T$ as $1/(1 + p_T^2/m^2)^{1/2}$. Preliminary experimental data [42] from NA35 is in agreement with this description, as shown in Fig. 17.10, although an alternative interpretations are also possible [42,20].

We turn now to the topic of the transverse radii. Experimental data from the NA35 Collaboration [38] indicate that the transverse radius parameter increases as the projectile mass number B and target mass number increase (Fig. 17.11a). Fig. 17.11b shows in addition that the two transverse radii R_x and R_y are approximately equal, within the experimental errors. In nucleus-nucleus collisions, the number of pion source points is proportional to the number of baryon-baryon collisions, which varies with the projectile mass number B and the target mass number A as $AB/(A^{2/3} + B^{2/3})$, according to Eqs. (12.15) and (12.26). If one assumes that pions become frozen out when the density of the emitting source is roughly a constant ρ_0, then the radius of the emitting source varies as

$$\rho_0(\text{emitter radius})^3 \propto \frac{AB}{A^{2/3} + B^{2/3}},$$

or

$$(\text{emitter radius}) \propto [AB/(A^{2/3} + B^{2/3})]^{1/3} . \tag{17.74}$$

We plot the transverse radius R_x inferred from interferometry measurements as a function of $r' = [AB/(A^{2/3} + B^{2/3})]^{1/3}$. The solid curve in Fig. (17.11a) is the parametrization

$$R_x = 0.52 + 0.58r' , \tag{17.75}$$

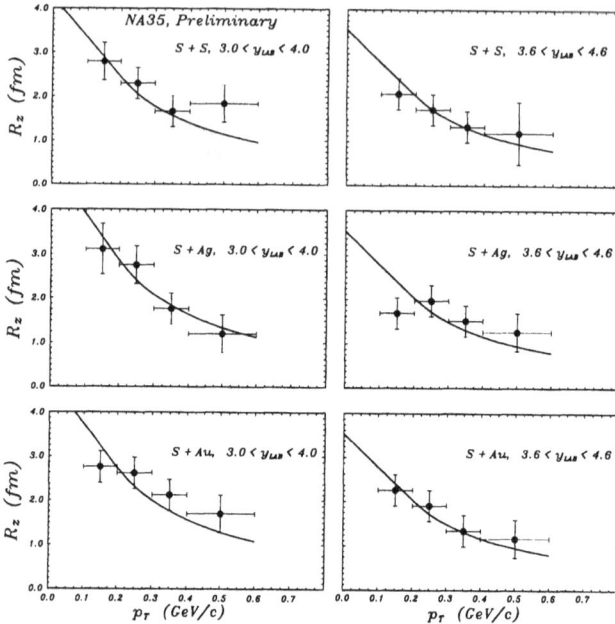

Fig. 17.10 The longitudinal radius R_z as a function of the transverse momentum p_T for various projectile target combinations and rapidity intervals, for a bombarding energy of 200 GeV per projectile nucleon. The data points are from Ref. [42] of the NA35 Collaboration and the solid curves are obtained from the parametrization of $(\text{constant})/(1 + p_T^2/m^2)^{1/2}$, in accordance with Eq. (17.73).

indicating an approximately linear correlation, and the possibility of a constant density of emitters at freeze out. More data are needed to allow one to construct a better description of the transverse part of the space-time picture of the pion emitting source at the late stages of the reaction.

In future experiments, intensity interferometry will be one of the many tools used to search for the quark-gluon plasma. Research into the specific signals to indicate the presence of a phase transition remains a subject of current research. The pertinent quantities of interest are the radius parameters $R_{x,y,z}$ of the emitting source and how these size parameters and their momentum dependence are affected by the occurrence of a phase transition.

It has been suggested that [14,16,19] for a first order phase transition, when the energy density of the quark-gluon plasma is only slightly higher than the threshold value, the system will remain in

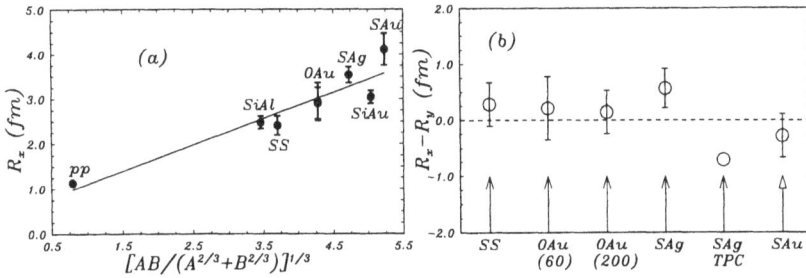

Fig. 17.11 (a) Transverse radius R_x in central collisions with different projectile and target combinations plotted as a function of the quantity $r' = [AB/(A^{2/3} + B^{2/3})]^{1/3}$. The solid curve is the line $R_x = 0.52 + 0.58r'$. (b) The difference of $R_x - R_y$ for various projectile and target combinations. The data point for pp collisions is from Ref. [32] and the other data are from the NA35 Collaboration [38]-[42].

the mixed phase during the transition from the plasma phase to the hadronic phase for a long lifetime. This arises because for a system not far from the phase transition threshold, the magnitude of its velocity field is small when the system evolves from the plasma phase to the mixed phase at the transition temperature. Furthermore, the pressure of the system in the mixed phase is a constant over a large range of the combined mixed density (see the Maxwell construction in Ref. [48]). Due to the constancy of the pressure in the mixed phase and the small initial velocity field, there is no large scale hydrodynamical motion in the mixed phase. The mixed-phase system will remain in that configuration for a long time, until the phase transition is completed, with the quark-gluon plasma system changed into the hadron system.

What is the consequence of a long lifetime of the emitting source? For a given pair of pions to be detected by detectors placed along the x-axis, the p_x momentum of these pions along the x-axis must be greater than the p_y momentum along the y-axis, which is perpendicular to the beam axis and the detector x-axis. The pions which come out from the source and detected by the detectors in the x-direction will span a larger region of space in the x-direction as compared to space spanned in the y-direction. Consequently, the radius parameter along the detector direction R_x will be larger then the radius parameter R_y perpendicular to the detector direction. The measurement of R_x and R_y will indicate whether the system goes through this stage of a long lifetime or not [14,19,16]. In this scenario, the long lifetime arises when the initial velocity field of the mixed phase is not large, so that the system can sustain itself for a long time. On the other hand, if one starts with an highly excited quark-gluon plasma system for which the hydrodynamical motion can be substantial by the time it reaches the mixed phase, the velocity field will be large and the the lifetime of the mixed-phase system will be shortened. The effect of an increase in R_x as compared to R_y will not be as pronounced.

§References for Chapter 17

1. R. Hanbury-Brown and R. Q. Twiss, Phil. Mag. 45, 633 (1954); R. Hanbury-Brown and R. Q. Twiss, Phil. Mag. Nature 177, 27 (1956); R. Hanbury-Brown and R. Q. Twiss, Phil. Mag. Nature 178, 1046, (1956).
2. R. J. Glauber, Phys. Rev. Lett. 10, 84 (1963); R. J. Glauber, Phys. Rev. 130, 2529 (1963); R. J. Glauber, Phys. Rev. 130, 2766 (1963).
3. Excellent theoretical and experimental reviews of the Hanbury-Brown-Twiss interferometry and its application to subatomic physics can be found in References 4, 5 and 6.
4. M. Gyulassy, S. K. Kauffman, and L. W. Wilson, Phys. Rev. C20, 2267 (1979).
5. D. Boal, C.-K. Gelbke, and B. K. Jennings, Rev. Mod. Phys. 62, 553 (1990).
6. W. A. Zajc, in *Particle Production in Highly Excited Matter*, Edited by H. H. Gutbrod and J. Rafelski, Plenum Press, New York, 1993, page 435.
7. G. Goldhaber, S. Goldhaber, W. Lee, and A. Pais, Phys. Rev. 120, 300 (1960).
8. S. E. Koonin, Phys. Lett. B70, 43 (1977); F. B. Yano and S. E. Koonin, Phys. Lett. B78, 556 (1978).
9. G. I. Kopylov and M. J. Podgoretsky, Yad. Fiz. 18, 656 (1973) [Sov. J. Nucl. Phys. 18, 336 (1974)].
10. G. N. Fowler and R. M. Weiner, Phys. Lett. 70B, 201 (1977).
11. S. Y. Fung, W. Gorn, G. P. Kiernan, J. J. Lu, Y. T. Oh, and R. T. Poe, Phys. Rev. Lett. 41, 1592 (1978).
12. M. Biyajima, Phys. Lett. B92, 193 (1980); M. Biyajima, Prog. Theo. Phys. 66, 1378 (1981); M. Biyajima, Prog. Theo. Phys. 68, 1273 (1982).
13. S. Pratt, Phys. Rev. Lett. 53, 1219 (1984).
14. S. Pratt, Phys. Rev. D33, 72 (1986); S. Pratt, Phys. Rev. D33, 1314 (1986).
15. J. Bolz, U. Ornik, M. Plümer, B. R. Schlei, and R. M. Weiner, Phys. Lett. B300, 404 (1993); I. V. Andreev, M. Plümer, and R. M. Weiner, University of Marbury Preprint, Print-92-0518, September 1992.
16. Y. Hama and S. S. Padula, Phys. Rev. D37, 3237 (1988).
17. C. C. Shih and P. Carruthers, Phys. Rev. D38, 56 (1988).
18. M. Gyulassy, and S. S. Padula, Phys. Lett. B217, 181 (1988).
19. G. F. Bertsch, Nucl. Phys. A498, 173c (1989).
20. Yu. M. Sinyukov, Nucl. Phys. A498, 151c (1989)
21. John D. Jackson, *Classical Electrodynmaics*, John Wiley and Sons, N. Y., 1962, page 573.
22. P. M. Morse and H. Feshbach, *Mathods of Theoretical Physics*, McGraw-Hill Book Company, New York, 1953, Vol I, page 82.
23. C. Y. Wong, Phys. Rev. C25, 1460 (1982).
24. P. D. Acton *et al.*, OPAL Collaboration, Phys. Lett. B267, 143 (1991).
25. I. Juricic *et al.*, MARK II Collaboration, Phys. Rev. D39, 1 (1989).
26. P. Avery *et al.*, CLEO Collaboration, Phys. Rev. D32, 2294 (1985).
27. H. Aihara *et al*, TPC Collaboration, Phys. Rev. D31, 996 (1985).

28. M. Althoff *et al*, TASSO Collaboration, Zeit. Phys. C30, 355 (1985).
29. R. Walker *et al*, AMY Collaboration, B-E Correlations in pion production, at TRISTAN, KEK preprint 90-60-AMY 90-5.
30. H. Aihara *et al.*, Lawrence Berkeley Laboratory Report LBL-23737 (1988), as reported by W. Hofmann, Ann. Rev. Nucl. Sci. 38, 279 (1988)
31. J. D. Bjorken, Lectures presented in the 1973 Proceedings of the Summer Institute on Particle Physics, edited by Zipt, SLAC-167 (1973).
32. A. Breakstone *et al.*, Phys. Lett. 162B, 400 (1985).
33. A. Breakstone *et al.*, Zeit. Phys. C33, 333 (1987).
34. C. Albajar *et al* UA1 Collaboration, Phys. Lett. B226, 410 (1989).
35. T. Alexopoulos *et al* E735 Collaboration, Phys. Rev. D48, 984 (1993).
36. W. A. Zajc *et al*, E802/E859 Collaborations, Nucl. Phys. A544, 237c (1992).
37. T. Abbott *et al*, E802 Collaborations, Phys. Rev. Lett. 69, 1030 (1992).
38. P. Seyboth *et al*, NA35 Collaboration, Nucl. Phys. A544, 293c (1992).
39. P. Seyboth *et al*, NA35 and NA44 Collaboration, Multiparticle Dynamics, 1992, Edited by C. Pajares, World Scientific, 1993, p. 307.
40. D. Ferenc *et al*, NA35 Collaboration, Nucl. Phys. A544, 531c (1992).
41. H. Ströbele, NA35 Collaboration, in Proceedings of the 2nd International Workshop on Relativistic Aspect of Nulcear Physics, Editors T. Kodama *et al.*, World Scientific, 1992, p. 65.
42. H. Ströbele, NA35 Collaboration, in Proceedings of the 3nd International Workshop on Relativistic Aspect of Nulcear Physics, Editors T. Kodama *et al.*, World Scientific, to be published; G. Roland *et al.*, NA35 Collaboration, A566, 527c (1994).
43. T. Sugitate, NA44 Collaboration, in Proceedings of the International Symposium, Kyoto, Japan, Editors M. Biyajima *et al.*, World Scientific, 1992, p. 264.
44. H. Boggild *et al*, NA44 Collaboration, Phys. Lett. B302, 510 (1993).
45. Y. Takahashi, EMU05 Collaboration, in Proceedings of the International Symposium, Kyoto, Japan, Editors M. Biyajima *et al.*, World Scientific, 1992, p. 276.
46. K. H. Kampert *et al*, WA80 Collaboration, Nucl. Phys. A544, 183c (1992); R. Albrecht *et al*, Zeit. Phys. C53, 225 (1992).
47. J.-Y. Zhang, X. He, C. C. Shih, S. Sorensen, and C. Y. Wong, Phys. Rev. C46, 748 (1992).
48. K. Huang, *Statistical Mechanics*, John Wiley and Sons, New York, 1963.

18. Signatures for the Quark-Gluon Plasma (V):

§18.1 Strangeness Content in Matter at Thermal and Chemical Equilibrium

The strangeness content in hadron matter and in a quark-gluon plasma are different [1]-[14]. In nuclear matter, the valence quarks consist of up and down quarks. The content of strange quarks and antiquarks is small. In nucleon-nucleon collisions, $u\bar{u}$, $d\bar{d}$ and $s\bar{s}$ pairs are produced. The strange quark and antiquark subsequently combine with neighboring quarks and antiquarks to form strange particles. In the Schwinger model of particle production, the production mechanism can be approximately thought of as arising from the spontaneous creation of a $q\bar{q}$ pair in a strong linear Abelian electric field, as described in Chapter 4. The production probability is proportional to the Schwinger factor, $\exp\{\pi m_q^2/\kappa\}$, in Eq. (5.21),

$$\frac{\Delta N}{\Delta t \Delta x \Delta y \Delta z} = \frac{\kappa^2}{8\pi^3} \exp\left\{-\frac{\pi m_q^2}{\kappa}\right\}.$$

where m_q is the mass of the quark q and κ is the string tension. In a nucleon-nucleon collision, to produce a strange quark which is a constituent of a strange hadron, the analysis of Eq. (5.24) gives an estimate of about 0.1 for the ratio of $s\bar{s}$ pairs to nonstrange ($u\bar{u}$ or $d\bar{d}$) pairs [Eq. (5.42b)] (with a *constituent mass* of about 450 MeV for strange quark and 325 MeV for up and down quarks).

If we count the valence quarks of a system of hadrons consisting of pions and kaons, the ratio K^+/π^+ can be related to the ratio $s\bar{s}/(u\bar{u}\,d\bar{d})$ by

$$\frac{s+\bar{s}}{u+\bar{u}+d+\bar{d}} = \frac{K^+/\pi^+}{1.5 + K^+/\pi^+}, \tag{18.1}$$

as discussed in Supplement 5.1. For p-Be collisions at 14.6 GeV, the experimental ratio of K^+/π^+ is about 0.08 [9]. Therefore, in p-Be collisions at 14.6 GeV, which should be quite similar to pp collisions, the ratio $s\bar{s}$ to ($u\bar{u}\,d\bar{d}$) is of the order of 0.05. The probability for the production of strange quarks and antiquarks is small in nucleon-nucleon collisions.

In nucleus-nucleus collisions, a large number of hadrons are produced. What is the ratio of strange to nonstrange particles in the hadron gas, if it is allowed to reach thermal and chemical equilibrium?

The produced hadrons consist mainly of pions, and kaons. The state of the hadron matter is said to be in chemical equilibrium when the interaction of the hadrons do not alter the densities of different types of hadrons in the system. We can obtain an estimate of the strangeness content fraction by treating the system of pions and kaons as an electrically neutral boson gas in thermal and chemical equilibrium. The occupation probabilities are given by the Bose-Einstein distribution characterized by a temperature T and the chemical potential of the particle. Because the hadron gas is electrically neutral, we have $n_{\pi^+} = n_{\pi^-}$ and $n_{K^+} = n_{K^-}$. As a consequence, the chemical potentials of the charged mesons are zero. The chemical potential is also zero for neutral mesons, which has no conserved charge [14]. The density n_i of particle i with rest mass m_i at temperature T is given by

$$n_i = \frac{1}{(2\pi)^3} \int_0^\infty \frac{4\pi p^2 d|\mathbf{p}|}{e^{\sqrt{\mathbf{p}^2 + m_i^2}/T} - 1}. \tag{18.2}$$

We shall show in Supplement 18.1 that the above integral leads to the result

$$n_i = \frac{Tm_i^2}{2\pi^2} \sum_{k=1}^\infty \frac{1}{k} K_2\left(\frac{km_i}{T}\right), \tag{18.3}$$

where K_2 is the modified Bessel function of order 2. Therefore, for a hadron gas in thermal and chemical equilibrium, the ratio of n_{K^+} to n_{π^+} is

$$\frac{n_{K^+}}{n_{\pi^+}} = \left(\frac{m_{K^+}}{m_{\pi^+}}\right)^2 \frac{K_2(m_{K^+}/T) + K_2(2m_{K^+}/T)/2 + K_2(3m_{K^+}/T)/3 + \cdots}{K_2(m_{\pi^+}/T) + K_2(2m_{\pi^+}/T)/2 + K_2(3m_{\pi^+}/T)/3 + \cdots}. \tag{18.4}$$

At a temperature of 200 MeV, which is the order of magnitude of the temperature one encounters in high energy hadron collision processes [9-11], we have $m_{K^+}/T \sim 2.5$ and $m_{\pi^+}/T \sim 0.7$. The tabulated values of the modified Bessel functions on page 417 of Ref. [15] gives

$$\frac{n_{K^+}}{n_{\pi^+}} = \left(\frac{493.6}{139.6}\right)^2 \frac{0.12144 + .002654 + 1.066 \times 10^{-4} + \cdots}{3.6614 + 0.35099 + 0.072562 + \cdots} = 0.3792.$$

The above ratio and Eq. (18.1) lead to the ratio of the strange quark density to the nonstrange quark density given by

$$\frac{n_s + n_{\bar{s}}}{n_u + n_{\bar{u}} + n_d + n_{\bar{d}}} = \frac{n_{K^+}/n_{\pi^+}}{1.5 + n_{K^+}/n_{\pi^+}}$$
$$= 0.2018.$$

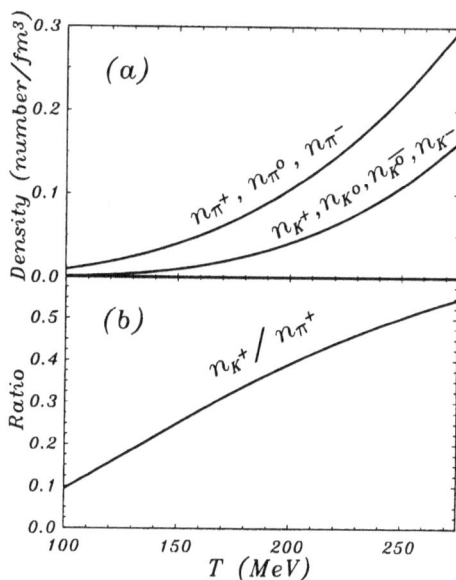

Fig. 18.1 (a) Pion and kaon densities as a function of hadron temperature T. (b) The ratio of K^+ density to π^+ density, as a function of temperature.

Therefore, for a hadron gas in thermal and chemical equilibrium at T=200 MeV, the K^+/π^+ ratio is about 0.38 and the the strangeness content is about 0.2. The K^+/π^+ ratio and $s\bar{s}/(u\bar{u}\,d\bar{d})$ ratio are considerably enhanced over the respective ratios of about 0.08 and 0.05 observed in p-Be collisions [9].

The pion and kaon density as a function of temperature for a hadron gas in thermal and chemical equilibrium, calculated numerically with Eq. (18.2), are shown in Fig. 18.1(a). The ratio n_{K^+}/n_{π^+} is shown in Fig. 18.1(b). As the temperature increases, the pion density and kaon density increase. However, the kaon density increases at a more rapid rate as compared to the pion density. Therefore, the ratio n_{K^+}/n_{π^+} increases as T increases.

As yet, it is not determined whether hadron matter produced in nucleus-nucleus collisions can react sufficiently frequently to reach chemical equilibrium within the time available during the collision process [8-13]. An early detailed analysis by Koch, Müller, and Rafelski [4] indicated that if one starts with hadron matter with non-strange particles at $T \sim 200$ MeV, the reaction rate is not fast enough for the hadron gas to approach chemical equilibrium, because of the large threshold energy for strange hadron pair production as compared to the temperature of the hadron gas. However, many models have been put forth to describe additional strangeness enhancement due to the lowering of the of strange particles masses as the temperature approaches the phase transition temperature [8], Δ-Δ interactions [12], and (higher meson resonance)-nucleon interactions [13]. There is much current research activity in this area and much work remains to be done to clarify the situation.

$\bigoplus$[Supplement 18.1

To show the result of Eq. (18.2), we have from Eq. (18.1),

$$
\begin{aligned}
n_i &= \frac{T^3}{2\pi^2} \int_0^\infty \frac{z^2 dz}{e^{\sqrt{z^2+(m_i/T)^2}} - 1} \\
&= \frac{T^3}{2\pi^2} \int_0^\infty z^2 dz e^{-\sqrt{z^2+(m_i/T)^2}} \sum_{k=0}^\infty e^{-k\sqrt{z^2+(m_i/T)^2}} \\
&= \frac{T^3}{2\pi^2} \sum_{k=0}^\infty \int_0^\infty z^2 dz\, e^{-(k+1)\sqrt{z^2+(m_q/T)^2}}.
\end{aligned}
\tag{1}
$$

By using Eq. (9.6.23) of Ref. [15], we have

$$
\int_0^\infty z^2 dz\, e^{-(k+1)\sqrt{z^2+(m_i/T)^2}} = \frac{1}{k+1}\left(\frac{m_i}{T}\right)^2 K_2\left(\frac{(k+1)m_i}{T}\right).
$$

The density n_i is therefore,

$$
n_i = \frac{T m_i^2}{2\pi^2} \sum_{k=1}^\infty \frac{1}{k} K_2\left(\frac{km_i}{T}\right).
\tag{2}
$$

As a special case for large values of m_i such that $m_i/T \gg 1$, Eq. (9.7.2) of Ref. [15] gives

$$
K_2\left(\frac{km_i}{T}\right) \cong \sqrt{\frac{\pi T}{2km_i}} e^{-km_i/T}.
$$

In this limit, the density n_i is

$$
n_i = \left(\frac{m_i T}{2\pi}\right)^{3/2} \sum_{k=1}^\infty \frac{1}{k^{3/2}} e^{-km_i/T}.
\tag{3}
$$

The ratio of two different densities of two types of particles at the same temperature is

$$
\frac{n_i}{n_j} \sim \left(\frac{m_i}{m_j}\right)^{3/2} e^{-(m_i-m_j)/T}.
\tag{4}
$$

For $T = 200$ MeV, because the pion rest mass is less than the temperature, Eq. (4) cannot be a good approximation to estimate the ratio of kaon density to pion density. It is necessary to use the modified Bessel function directly, as in Eqs. (18.3) and (18.4).

If a quark-gluon plasma is produced, the strangeness content is governed by the dynamical state of the plasma. In the plasma, quarks, antiquarks and gluons interact to change their momentum states and to transform from one type of particle into another. The situation is much like a chemical mixture in which chemical compounds interact to change to other momentum states and to transform to other chemicals. In this mixture, thermal equilibrium is reached when the momentum distributions of the particles do not change, even though momentum exchanges continue through the interaction between particles. The gain in the momentum distribution from one reaction is balanced by the loss in the momentum distribution from the inverse reaction or other reactions. The momentum distributions of the particles at thermal equilibrium are then governed by the temperature T. In a similar way, chemical equilibrium is reached when the densities of different particles reach a steady state even though the particles continue to interact and transform from one kind to another. The gain in the density of one kind of particle from one reaction is counterbalanced by the loss in the density from the inverse reaction or other reactions. Equilibrium of this kind is called a chemical equilibrium. In other words, the state of the quark-gluon plasma is said to be in chemical equilibrium when the interactions of the constituents do not alter the densities of different types of particles in the plasma. The state of the plasma is then characterized by the temperature T and various chemical potentials μ_i for the different particles i.

We shall examine the approach to chemical equilibrium in a quark-gluon plasma in the next section. Here, we wish to ask the following question: What are the densities of the different kinds of quarks if the plasma has a lifetime long enough to establish thermal and chemical equilibrium? Because deconfined quarks, antiquarks and gluons follow the dynamics of perturbative QCD, we can obtain an idea of these densities by using an approximate description of free quarks and gluons with quark masses given by the quark *current masses* of Table 9.1.

At thermal and chemical equilibrium, the occupation probabilities of the quarks are given by the Fermi-Dirac distribution. The quark density n_q at chemical potential μ_q and temperature T is given by

$$n_q(\mu_q) = \frac{N_c N_s}{(2\pi)^3} \int_0^\infty \frac{4\pi p^2 d|p|}{1 + e^{(\sqrt{p^2+m_q^2}-\mu_q)/T}}, \tag{18.5a}$$

where $N_c (= 3)$ is the number of colors and $N_s (= 2)$ is the number of spins.

Given the chemical potential μ_q, we can also obtain the density of antiquarks $n_{\bar{q}}$. The presence of an antiquark $\bar{q}$ corresponds to the absence of a quark q in a negative energy state. Following the arguments leading to Eq. (7) of Supplement 15.1, the number density of antiquarks is

$$n_{\bar{q}}(\mu_q) = \frac{N_c N_s}{(2\pi)^3} \int_0^\infty \frac{4\pi p^2 d|p|}{1 + e^{(\sqrt{p^2+m_q^2}+\mu_q)/T}}. \tag{18.5b}$$

We consider a quark-gluon plasma with three quark flavors: u quarks, d quarks, and s quarks. The density of various quarks at temperatures of 200 MeV and 400 MeV are shown in Fig. 18.2.

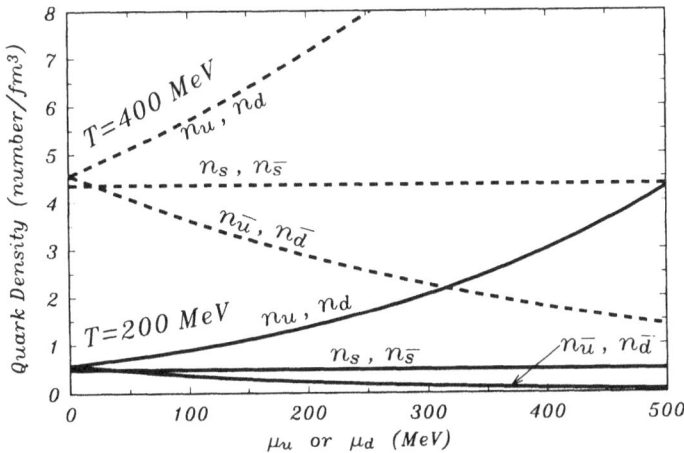

Fig. 18.2 Densities of various types of quarks and antiquarks as a function of the chemical potentials of the up or down quarks, μ_u or μ_d, at $T = 200$ and 400 MeV. The chemical potential μ_s of the strange quarks is constrained to be zero to give $n_s = n_{\bar{s}}$.

To study the results of Fig. 18.2, we consider first the case $\mu_u = \mu_d = 0 = \mu_s$ corresponding to a quark-gluon plasma with no net baryon content, as might be produced in the central rapidity region with heavy-ion collisions at very high energies (see Chapter 13). The results of Supplement 18.2 show that for the case when $\mu_u = \mu_d = \mu_s = 0$ and

when the temperature T is of the same order as the strange quark
mass m_s, the density of all quarks and antiquarks are nearly the same
(see Fig. 18.2). In such a plasma, the content of strange quarks and
strange antiquarks is much greater than what one would expect either
in colliding nuclear matter or in an equilibrated hadron gas without
a phase transition. Hence, an enhancement of the number of strange
quarks and antiquarks is suggested as a signal for the presence of a
quark-gluon plasma [1]. The enhancement of the number of strange
quarks and antiquarks leads to an enhancement of the production of
mesons with an s or an $\bar{s}$ as a constituent. From the Particle Table
of Ref. [16], some mesons with an s or an $\bar{s}$ as a constituent are

$$K^+ = u\bar{s}, \quad K^0 = d\bar{s}, \quad \overline{K}^0 = \bar{d}s, \quad K^- = \bar{u}s \ \text{ and } \ \phi = s\bar{s},$$

where the right-hand side of each equality sign lists the dominant
component of each meson. According to the convention that the
strangeness quantum number S of a quark follows the same sign as its
electric charge [16], S is 1 for K^+ and K^o, S is -1 for $\overline{K}^o$ and K^-, and
S is 0 for ϕ. It has been suggested that the enhancement of the number
of strange quarks and antiquarks will lead to an enhancement of the
production of K mesons [1] and ϕ mesons [17] and may be used as a
signal for the quark-gluon plasma. Furthermore, the enhancement of
the number of $\bar{s}$ is accompanied by the enhancement of the number of
$\bar{u}$ and $\bar{d}$ when $\mu = 0$ (Fig. 18.2). With nearly equal numbers of quarks
and antiquarks, the probability of forming antihyperons by combining
$\bar{u}, \bar{d}$ and $\bar{s}$ quarks in the plasma is nearly the same as the probability
of forming nonstrange and strange baryons by combining u, d and s
quarks. Hyperons and antihyperons are classified according to their
strangeness S and isospin I as in Table 18.1 [16]. The enhancement of
the production of $\overline{\Lambda}, \overline{\Sigma}, \overline{\Xi}$ and $\overline{\Omega}$ antihyperons can be used as signals
for the production of the quark-gluon plasma with $\mu_u = \mu_d = 0$ [1].
In contrast, in nucleon-nucleon collisions, the number of antihyperon
produced is greatly suppressed by the Schwinger factor where it is
necessary to tunnel the mass of the heavy diquark and the strange
quark in the strong color electric field against the string tension κ, in
order to produce a hyperon-antihyperon pair.

In the heavy-ion stopping regime where one hopes to produce a
quark-gluon plasma with a large net baryon density, the baryons
which participate in the collision contain valence u and d quarks.
Because of the conservation of baryon number, the baryon content
of the colliding nuclei give rise to the resultant baryon-rich quark
matter, with nonzero quark chemical potentials μ_u and μ_d. On the
other hand, with no valence strange quarks in the colliding nuclei the

strange quark chemical potential μ_s is zero and the densities of s and $\bar{s}$ are the same.

Table 18.1

Hyperon				Anti-hyperon		
	Quarks	S	I		Quarks	S
Λ	uds	-1	0	$\bar{\Lambda}$	$\bar{u}\bar{d}\bar{s}$	1
Σ^+	uus	-1	1	$\bar{\Sigma}^+$	$\bar{u}\bar{u}\bar{s}$	1
Σ^0	uds	-1	1	$\bar{\Sigma}^0$	$\bar{u}\bar{d}\bar{s}$	1
Σ^-	dds	-1	1	$\bar{\Sigma}^-$	$\bar{d}\bar{d}\bar{s}$	1
Ξ^0	uss	-2	$\frac{1}{2}$	$\bar{\Xi}^0$	$\bar{u}\bar{s}\bar{s}$	2
Ξ^-	dss	-2	$\frac{1}{2}$	$\bar{\Xi}^-$	$\bar{d}\bar{s}\bar{s}$	2
Ω^-	sss	-3	0	$\bar{\Omega}^-$	$\bar{s}\bar{s}\bar{s}$	3

At a temperature of $T = 0$, the results of Eq. (9.16b) give $\mu_u = \mu_d = 434$ MeV as the value of the up quark and the down quark phase transition chemical potential, above which the transition from the hadron phase to the quark-gluon plasma phase occurs. As the temperature T increases, there is an additional contribution to the pressure, one expects therefore that the value of the phase transition chemical potential μ_u or μ_d decreases. Since the quark matter created in the heavy-ion reactions in the stopping regime has a temperature of the order of a hundred MeV, the phase transition chemical potential μ_u or μ_d should be lower than the value of $\mu_{u,d} = 434$ MeV at $T = 0$. In this case with nonzero chemical potentials $\mu_{u,d}$, the densities of u and d are greater than the density of s and $\bar{s}$ which in turn are greater than the densities of $\bar{u}$ and $\bar{d}$ (Fig. 18.2).

Consider the fate of an s quark and an $\bar{s}$ antiquark in a medium with nonzero chemical potential $\mu_{u,d}$. Because the densities of u and

d are greater than the densities of $\bar{u}$ and $\bar{d}$, it is much more likely for the $\bar{s}$ antiquark to combine with a u or a d quark to form $K^+(u\bar{s})$ or $K^o(d\bar{s})$ than it is for the strange quark s to combine with a $\bar{u}$ or a $\bar{d}$ to form $\overline{K}^o(\bar{u}s)$ and $K^-(\bar{d}s)$. For the strange quark s, a more likely outcome is for it to combine with u and d quarks to form $\Lambda(uds)$, $\Sigma^+(uus)$, $\Sigma^o(uds)$, or $\Sigma^-(dds)$, instead of combining with $\bar{u}$ and $\bar{d}$ to produce $\overline{K}^o$ and K^-. Experimental measurements which can probe the numbers of s and $\bar{s}$ relative to the numbers of u, d, $\bar{u}$ and $\bar{d}$ can be used to find out the thermodynamical state of the quark-gluon plasma in the stopping region.

$\oplus\mathbb{[}$ Supplement 18.2

The results of Fig. 18.2 were obtained by a direct numerical integration of Eqs. (18.5a) and (18.5b). For certain regions of chemical potential and mass, one can also write out the density of quarks and antiquarks as an analytical series as in Supplement 18.1.

We first examine the case with a small value of the chemical potential μ_q such that $\mu_q \ll T$. We have from Eq. (18.5a),

$$
\begin{aligned}
n_q(\mu_q) &= \frac{N_c N_s T^3}{2\pi^2} \int_0^\infty \frac{z^2 dz}{1 + e^{\sqrt{z^2+(m_q/T)^2}-\mu_q/T}} \\
&= \frac{N_c N_s T^3}{2\pi^2} \int_0^\infty z^2 dz\, e^{-\sqrt{z^2+(m_q/T)^2}-\mu_q/T} \sum_{k=0}^\infty (-1)^k\, e^{-k(\sqrt{z^2+(m_q/T)^2}-\mu_q/T)} \\
&= \frac{N_c N_s T^3}{2\pi^2} \sum_{k=0}^\infty (-1)^k \int_0^\infty z^2 dz\, e^{-(k+1)(\sqrt{z^2+(m_q/T)^2}-\mu_q/T)}.
\end{aligned} \tag{1}
$$

As in Supplement 18.1, we again use (9.6.23) of Ref. [15],

$$
\int_0^\infty z^2 dz\, e^{-(k+1)\sqrt{z^2+(m_q/T)^2}} = \frac{1}{k+1}\left(\frac{m_q}{T}\right)^2 K_2\left(\frac{(k+1)m_q}{T}\right),
$$

where K_2 is the modified Bessel Function of order 2. The quark density n_q is therefore,

$$
n_q(\mu_q) = \frac{N_c N_s T m_q^2}{2\pi^2} \sum_{k=1}^\infty \frac{(-1)^{k-1}}{k} K_2\left(\frac{k m_q}{T}\right) e^{k\mu_q/T}. \tag{2}
$$

The antiquark density $n_{\bar{q}}$ is obtained from Eq. (1) by replacing μ_q by $-\mu_q$. From the above equation, we have the antiquark density

$$
n_{\bar{q}}(\mu_q) = \frac{N_c N_s T m_q^2}{2\pi^2} \sum_{k=1}^\infty \frac{(-1)^{k-1}}{k} K_2\left(\frac{k m_q}{T}\right) e^{-k\mu_q/T}. \tag{3}
$$

We can consider different limits of the above equations. For large values of m_q such that $m_q/T \gg 1$, Eq. (9.7.2) of Ref. [15] gives

$$K_2\left(\frac{km_q}{T}\right) \cong \sqrt{\frac{\pi T}{2km_q}} e^{-km_q/T} \, .$$

In this limit, the quark density is

$$n_q(\mu_q) = N_c N_s \left(\frac{m_q T}{2\pi}\right)^{3/2} \sum_{k=1}^{\infty} \frac{(-1)^{k-1}}{k^{3/2}} e^{-k(m_q - \mu_q)/T} \, , \tag{4}$$

and the antiquark density is

$$n_{\bar{q}}(\mu_q) = N_c N_s \left(\frac{m_q T}{2\pi}\right)^{3/2} \sum_{k=1}^{\infty} \frac{(-1)^{k-1}}{k^{3/2}} e^{-k(m_q + \mu_q)/T} \, . \tag{5}$$

In the other limit, as the quark mass approaching zero, we have from Eq. (9.6.9) of Ref. [15]

$$m_q^2 K_2\left(\frac{km_q}{T}\right) \cong m_q^2 \frac{1}{2}\Gamma(2)\left(\frac{1}{2}\frac{km_q}{T}\right)^{-2}$$
$$= \frac{2T^2}{k^2}$$

The quark density Eq. (2) is then

$$n_q(\mu_q) = \frac{N_c N_s T^3}{\pi^2} \sum_{k=1}^{\infty} \frac{(-1)^{k-1}}{k^3} e^{k\mu_q/T} \, . \tag{6}$$

When $\mu_q = 0$, this gives the result of Eq. (3) of Supplement 9.2,

$$n_q(\mu_q = 0) = \frac{N_c N_s T^3}{2\pi^2} \frac{3}{2} \, 1.202 \, . \tag{7}$$

Using Eqs. (2) and (7), we can compare the densities for the u, d and s quarks at $T = m_s$ and at $T = 2m_s$ for $\mu_q = 0$. At a temperature of $T = m_s$, the sum of the first three terms of the expansion (2) yields the strange quark and strange antiquark density

$$n_s = n_{\bar{s}} = \frac{N_c N_s T^3}{2\pi^2} 1.518 \, . \tag{8}$$

On the other hand, Eq. (7) gives the up quark and down quark density at $\mu_{u,d} = 0$,

$$n_u = n_d = \frac{N_c N_s T^3}{2\pi^2} 1.803 \, . \tag{9}$$

By comparing Eqs. (8) and (9), we observe that the densities of $u, d, s, \bar{u}, \bar{d}$, and $\bar{s}$ are nearly the same, for $\mu = 0$ and $T = m_s$.

At a temperature of $T = 2m_s$, we have $m_s/T = 0.5$. From Ref. [15], the sum of the first three terms of the expansion (2) gives

$$n_s = n_{\bar{s}} = \frac{N_c N_s T^3}{2\pi^2} 1.733 \, . \tag{10}$$

By comparing Eqs. (9) and (10), we observe again that the densities of $u, d, s, \bar{u}, \bar{d}$, and $\bar{s}$ are even more similar for $\mu = 0$ and $T = 2m_s$. We conclude from the above that at $\mu = 0$ and $T = m_s$ (or $T = 2m_s$), the densities of $u, d, s, \bar{u}, \bar{d}$, and $\bar{s}$ are about the same, as shown in Fig. 18.2.

We consider next the case of a large value of μ_q such that μ_q is much greater than m_q and T. The integral of Eq. (18.5a) can be rewritten in a different form by introducing the kinetic energy variable z as

$$\boldsymbol{p}^2 + m_q^2 = (z + m_q)^2 \, .$$

In terms of an integral in z, Eq. (18.5a) becomes

$$
\begin{aligned}
n_q(\mu_q) &= \frac{N_c N_s}{2\pi^2} \int_0^\infty \frac{|\boldsymbol{p}|(z + m_q)dz}{1 + e^{(z+m_q-\mu_q)/T}} \\
&= \frac{N_c N_s}{2\pi^2} \int_0^\infty \frac{\sqrt{(z + m_q)^2 - m_q^2}\,(z + m_q)\,dz}{1 + e^{[z-(\mu_q-m_q)]/T}} \, .
\end{aligned} \tag{11}
$$

We can compare this equation with the result of Eq. (58.1) of Landau and Lifshitz [5]

$$\int_0^\infty \frac{f(z)dz}{1 + e^{(z-\mu)/T}} = \int_0^\mu f(z)dz + \frac{\pi^2}{6}T^2 \frac{df}{dz}\bigg|_{z=\mu} + \frac{7\pi^4}{360}T^4 \frac{d^3 f}{dz^3}\bigg|_{z=\mu} + \dots, \tag{12}$$

which is valid for $\mu \gg T$. Upon using the result of Eq. (12) to evaluate Eq. (11), the quark density is

$$n_q(\mu_q) = \frac{N_c N_s}{2\pi^2} \left[\int_0^{\mu_q - m_q} f(z)dz + \frac{\pi^2}{6}T^2 \frac{df}{dz}\bigg|_{z=\mu_q-m_q} + \frac{7\pi^4}{360}T^4 \frac{d^3 f}{dz^3}\bigg|_{z=\mu_q-m_q} + \dots \right], \tag{13}$$

where

$$f(z) = \sqrt{(z + m_q)^2 - m_q^2}\,(z + m_q) \, .$$

The righthand side of Eq. (13) can be evaluated to give n_q as a function of μ_q and T.

As a special example, we consider the density of u and d quarks at large values of the chemical potential $\mu_{u,d}$. For the u and d quarks, it is reasonable to take the quarks to be massless and

$$f(z) = z^2 \, .$$

Eq. (13) gives

$$n_{u,d}(\mu_{u,d}) = \frac{N_c N_s}{2\pi^2}\left[\frac{\mu_{u,d}^3}{3} + \frac{\pi^2}{6}T^2\, 2\mu_{u,d}\right].\tag{14}$$

As a numerical example, we can consider the case of $T = 200$ MeV and $\mu_{u,d} = 2T = 400$ MeV, to obtain get

$$n_{u,d}(\mu_{u,d}) = \frac{N_c N_s T^3}{2\pi^2}\left[2.6667 + 6.5797\right].\tag{15}$$

A comparison of Eq. (15) with Eq. (9) shows that the density of u or d quarks at $\mu_{u,d} = 400$ MeV is about 6 times the density of u or d quarks at $\mu_{u,d} = 0$, as exhibited in Fig. 18.2. $\rrbracket\oplus$

§18.2 Rate of Approach to Chemical Equilibrium in a Quark-Gluon Plasma

The discussions in the last Section dealt with the state of the plasma at thermal and chemical equilibrium. We shall examine here the rate of approach to chemical equilibrium in order to obtain the time scale for the strangeness content to reach chemical equilibrium. For simplicity, we consider a thermalized plasma to consist initially of u, d, $\bar{u}$, $\bar{d}$ and g, with negligible strangeness content. The plasma will evolve toward chemical equilibrium, by reactions among its constituents. The strangeness content of the plasma will change through strangeness-producing reactions, approaching the equilibrium density characterized by the temperature T and the strangeness chemical potential μ_s. As mentioned before, the production of a strange quark is accompanied by the production of a strange antiquark. The density of s and the density of $\bar{s}$ are the same and the strange quark chemical potential μ_s is constrained to be zero.

In the plasma, strange quarks and antiquarks can be produced by collisions among the constituents of the plasma. There are two processes by which $s\bar{s}$ pairs can be produced. Strange quarks and antiquarks can be produced by light quark and antiquark collisions through the reactions

$$u + \bar{u} \to s + \bar{s},\tag{18.6a}$$

and

$$d + \bar{d} \to s + \bar{s}.\tag{18.6b}$$

The above reactions are special cases of the more general reaction

$$q + \bar{q} \to Q + \overline{Q},\tag{18.7}$$

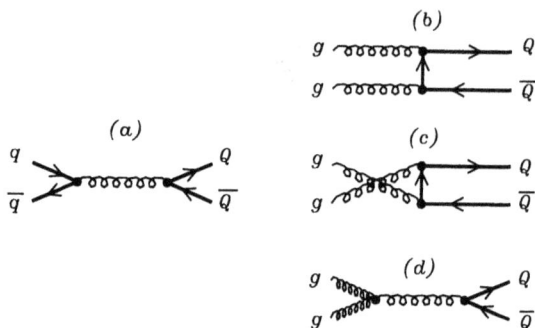

Fig. 18.3 Strangeness-producing processes in a quark-gluon plasma.

where q is a light u or d quark and Q is a moderately-heavy or a heavy quark such as an s quark (or a c quark).

The reaction (18.7) proceeds by the conversion of the $q\bar{q}$ pair into a gluon, with the subsequent emission of the $Q\bar{Q}$ pair, as represented by the Feynman diagram 18.3a. The integrated cross section for this reaction (18.7), averaged over initial colors and spins and summed over final colors and spins is [19,20,4,5]

$$\sigma_{q\bar{q}}(M) = \frac{8\pi\alpha_s^2}{27M^2}(1 + \frac{\eta}{2})\sqrt{1 - \eta}, \qquad (18.8)$$

where $M(= \sqrt{(q+\bar{q})^2})$ is the center of mass energy of the colliding $q\bar{q}$ pair, the quantity η is

$$\eta = 4m_Q^2/M^2,$$

and m_Q is the mass of the quark Q. Obviously, the reaction (18.7) cannot proceed unless the energy M exceeds the threshold energy $2m_Q$, which is the rest mass of the final particles Q and $\bar{Q}$. For low energies when M is close to the threshold energy $2m_Q$, the cross section is

$$\sigma_{q\bar{q}}(M) \sim \frac{\pi\alpha_s^2}{3M^2}\frac{4}{3}\sqrt{1 - \eta}, \qquad (18.9a)$$

which shows a rapid rise of the cross section just above the threshold. In the limit of very high energies such that $M \gg m_Q$, the cross section is

$$\sigma_{q\bar{q}}(M) \sim \frac{\pi\alpha_s^2}{3M^2}\frac{8}{9}. \qquad (18.9b)$$

Reaction (18.7) is not the only one which produces strangeness in the plasma. Strange quarks and antiquarks can also be produced by the collision of gluons in the plasma through the reaction

$$g + g \to s + \bar{s}, \tag{18.10}$$

which is a special case of the more general reaction

$$g + g \to Q + \bar{Q}. \tag{18.11}$$

The processes by which a pair of gluons turn into a strange quark-antiquark pair are represented by the Feynman diagrams Figs. 18.3b, 18.3c and 18.3d. The integrated cross section for the reaction (18.11), averaged over the initial gluon types and gluon polarizations and summed over the final colors and spins, is [19]

$$\sigma_{gg}(M) = \frac{\pi\alpha_s^2}{3M^2}\left[(1+\eta+\frac{1}{16}\eta^2)\ln\left(\frac{1+\sqrt{1-\eta}}{1-\sqrt{1-\eta}}\right) - \left(\frac{7}{4}+\frac{31}{16}\eta\right)\sqrt{1-\eta}\right]. \tag{18.12}$$

The threshold energy for the reaction (18.11) is $2m_Q$. For energies M just above the threshold $2m_Q$, the cross section is

$$\sigma_{q\bar{q}}(M) \sim \frac{\pi\alpha_s^2}{3M^2}\frac{7}{16}\sqrt{1-\eta}, \tag{18.13a}$$

while for very high energies when $M \gg m_Q$, the cross section is

$$\sigma_{q\bar{q}}(M) \sim \frac{\pi\alpha_s^2}{3M^2}\left[(1+\eta)\ln\left(\frac{M^2}{m_Q^2}\right) - \left(\frac{7}{4}+\frac{17\eta}{16}\right)\right]. \tag{18.13b}$$

We plot in Fig. 18.4 the two cross sections as a function of the center-of-mass energy M of the colliding $q\bar{q}$ or gg system, for a typical value of $\alpha_s = 0.4$ and $m_s = 200$ MeV. As one observes, the cross sections rise rapidly above the threshold and the magnitudes of the cross sections for the two processes are comparable and are of the order of $(\alpha_s^2/M^2)(1 - 4m_q^2/M^2)^{1/2}$. They are about 0.1 mb, at energies in the range 0.4 GeV $\leq M \leq 1$ GeV.

Knowing the strangeness-producing cross sections (18.8) and (18.12) involving the constituents of the plasma, we can determine the rate of $s\bar{s}$ production in the plasma. This can be carried out in the same way as in Exercise 14.2 of Section 14.

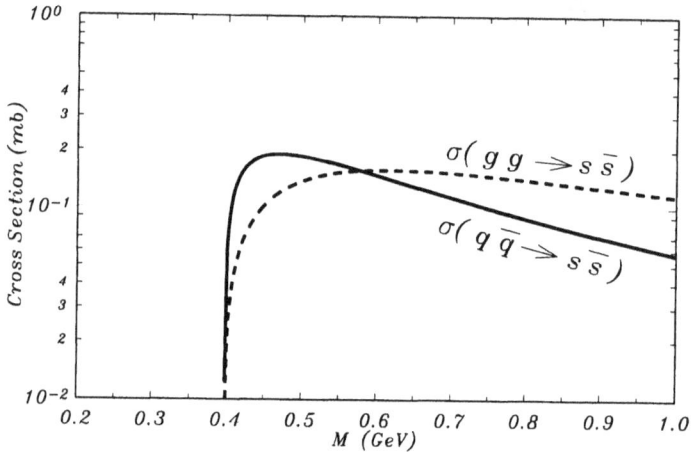

Fig. 18.4 Cross sections for the process $q\bar{q} \to s\bar{s}$ (solid curve) and $gg \to s\bar{s}$ (dashed curve), assuming $m_s = 200$ MeV.

We first consider the process $q + \bar{q} \to s + \bar{s}$. In the quark-gluon plasma, the number of quarks with one flavor and momentum $\boldsymbol{p}_1$, in the volume element d^3x and the momentum element d^3p_1, is

$$dN_q = N_c N_s \frac{d^3x\, d^3p_1}{(2\pi)^3}\, f_q(E_1) \tag{18.14}$$

where $f_q(E_1)$ is the quark energy distribution in the form of the Fermi-Dirac distribution with a temperature T and a chemical potential μ,

$$f_q(E_1) = \frac{1}{e^{(E_1-\mu)/T} + 1}. \tag{18.15}$$

Not all the reactions between a quark and an antiquark lead to the production of $s\bar{s}$ pairs. To produce an $s\bar{s}$ pair, the quark and the antiquark must have the same flavor.

We denote the relative velocity between the quark and the anti-quark by v_{12}. The cross section for a quark q and an antiquark $\bar{q}$ of the same flavor to produce an $s\bar{s}$ pair of invariant mass M, averaged over the initial colors and spins, is given by $\sigma_{q\bar{q}}(M)$ of Eq. (18.8). The volume swept by the cross section $\sigma_{q\bar{q}}(M)$ per unit time due to the relative motion of the quark and the antiquark is $\sigma_{q\bar{q}}(M)v_{12}$. Consequently, the number of antiquarks of the same flavor having momentum $\boldsymbol{p}_2$ and energy $E_2 = (\boldsymbol{p}_2^2 + m_q^2)^{1/2}$ in the momentum element d^3p_2, with which a quark can interact per unit time (to produce $s\bar{s}$ pairs), is

$$\sigma_{q\bar{q}}(M)v_{12} \, N_c N_s \, f_{\bar{q}}(E_2)d^3p_2/(2\pi)^3 \, , \qquad (18.16)$$

where $f_{\bar{q}}(E_2)$ is the antiquark energy distribution

$$f_{\bar{q}}(E_2) = \frac{1}{e^{(E_2+\mu)/T} + 1} \, . \qquad (18.17)$$

Each interaction between a quark and an antiquark produces one $s\bar{s}$ pair. The number of $s\bar{s}$ pairs produced per unit time due to the $q\bar{q} \rightarrow s\bar{s}$ process is the integral of the product of Eqs. (18.14) and (18.16), summed over the flavors f which can be u and d,

$$\frac{dN_{s\bar{s}}}{dt}(q\bar{q} \rightarrow s\bar{s}) = N_c^2 N_s^2 \sum_{f=1}^{N_f=2} \int \frac{d^3x \, d^3p_1 \, d^3p_2}{(2\pi)^6} f_q(E_1)f_{\bar{q}}(E_2)\sigma_{q\bar{q}}(M)v_{12} \, .$$

$$(18.18)$$

Therefore, from Eq. (18.18), the rate of production of $s\bar{s}$ pairs per unit time per unit spatial volume is

$$\frac{dN_{s\bar{s}}}{dt \, d^3x}(q\bar{q} \rightarrow s\bar{s}) = N_c^2 N_s^2 N_f \int \frac{d^3p_1 d^3p_2}{(2\pi)^6} f_q(E_1)f_{\bar{q}}(E_2)\sigma_{q\bar{q}}(M)v_{12} \, .$$

$$(18.19)$$

We also need to evaluate the rate of production of $s\bar{s}$ pairs by the interaction of gluons in the plasma. In the plasma, the number of gluons with momentum p_1, in the volume element d^3x and the momentum element d^3p_1, is

$$dN_g = N_g N_\epsilon \frac{d^3x \, d^3p_1}{(2\pi)^3} f_g(E_1) \qquad (18.20)$$

where $N_g(=8)$ is the number of different gluons, $N_\epsilon(=2)$ is the number of polarizations, and $f_g(E_1)$ is the Bose-Einstein distribution

$$f_g(E_1) = \frac{1}{e^{E_1/T} - 1} \, . \qquad (18.21)$$

We again denote the relative velocity between the two interacting gluons by v_{12}. The cross section for two gluons to interact to produce an $s\bar{s}$ pair of invariant mass M, averaged over the initial gluon type and gluon polarizations is given by $\sigma_{gg}(M)$ of Eq. (18.12). The volume swept by the cross section $\sigma_{gg}(M)$ per unit time due to the relative motion of the two gluons is $\sigma_{gg}(M)v_{12}$. Consequently, the number of gluons having momentum p_2 and energy $E_2 = |p_2|$ in the momentum element d^3p_2, with which the first gluon can interact per unit time (to produce $s\bar{s}$ pairs), is

$$\sigma_{gg}(M)v_{12}\, N_g N_\epsilon\, f_g(E_2) d^3p_2/(2\pi)^3 \,. \tag{18.22}$$

The interaction of one gluon with another gluon produces one $s\bar{s}$ pair. The number of $s\bar{s}$ pairs produced per unit time due to the $gg \to s\bar{s}$ process is the integral of the product of Eqs. (18.20) and (18.22),

$$\frac{dN_{s\bar{s}}}{dt}(gg \to s\bar{s}) = N_g^2 N_\epsilon^2 \int \frac{d^3x\, d^3p_1\, d^3p_2}{(2\pi)^6} f_g(E_1) f_g(E_2) \sigma_{gg}(M) v_{12} \,. \tag{18.23}$$

Consequently, the rate of $s\bar{s}$ pair production per unit space-time volume by the reaction $gg \to s\bar{s}$ is given by

$$\frac{dN_{s\bar{s}}}{dt\, d^3x}(gg \to s\bar{s}) = N_g^2 N_\epsilon^2 \int \frac{d^3p_1\, d^3p_2}{(2\pi)^6} f_g(E_1) f_g(E_2) \sigma_{gg}(M) v_{12} \,. \tag{18.24}$$

In arriving at Eqs. (18.19) and (18.24), we have assumed that the initial densities of the strange quarks and antiquarks are so small that the reduction of the rate from the reversed reaction and from the Pauli exclusion principle of occupied strange quark states can be neglected.

An examination of Eqs. (18.19) and (18.24) indicates that they have the same mathematical structure. We can treat them in a parallel way as special cases of the more general expression

$$\frac{dN_{s\bar{s}}}{d^4x} = N \int \frac{d^3p_1\, d^3p_2}{(2\pi)^6} f_1(E_1) f_2(E_2) \sigma(M) v_{12} \,, \tag{18.25}$$

where N, $\sigma(M)$, $f_1(E_1)$ and $f_2(E_2)$ depend on the process. For the process $q\bar{q} \to s\bar{s}$, N from Eq. (18.19) is

$$N = N_{q\bar{q}} = N_c^2 N_s^2 N_f = 72, \tag{18.26a}$$

$f_1(E_1)$ is the quark distribution $f_q(E_1)$ of Eq. (18.15), $f_2(E_2)$ is the antiquark distribution $f_{\bar{q}}(E_2)$ of Eq. (18.17), and $\sigma(M)$ is $\sigma_{q\bar{q}}(M)$ of

Eq. (18.8). For the process $gg \to s\bar{s}$, the quantity N from Eq. (18.24) is

$$N = N_{gg} = N_g^2 N_\epsilon^2 = 256 \,, \qquad (18.26b)$$

$f_1(E)$ and $f_2(E)$ are the gluon distribution $f_g(E)$ of Eq. (18.21), and $\sigma(M)$ is $\sigma_{gg}(M)$ of Eq. (18.12).

It is clear that because the basic cross sections $\sigma_{q\bar{q}}$ and σ_{gg} are of the same order of magnitude, the difference in the multiplicative numbers $N_{q\bar{q}}$ and N_{gg} from Eqs. (18.26a) and (18.26b), makes the rate for the reaction $gg \to s\bar{s}$ much greater than that for the reaction $q\bar{q} \to s\bar{s}$.

A further comparison of Eq. (18.24) with Eq. (14.1) of Section 14.1 shows they are of the same form as Eq. (14.1). As the latter equation can be integrated out analytically by the saddle-point method, with results given by Eq. (14.2), we can use the results there to write down the number of $s\bar{s}$ pairs produced per unit space-time 4-volume. The mass m_q in Eq. (14.2) is the mass of the initial particles in the $q\bar{q} \to l^+l^-$ reaction. For our present problem with the reactions $q\bar{q} \to s\bar{s}$ and $gg \to s\bar{s}$, instead of $q\bar{q} \to l^+l^-$ treated in Eq. (14.1), the initial particles are massless gluons and light quarks, which can be taken to be massless. When the masses of the initial particles are set equal to zero, Eq. (14.2) can be used to write Eq. (18.25) as

$$\frac{dN_{s\bar{s}}}{dM^2 d^4 x} = N \frac{\sigma(M)M^2}{2(2\pi)^4} f_1(\epsilon(M)) F_2\left(\frac{M^2}{4\epsilon(M)}\right) \sqrt{\frac{2\pi}{w(\epsilon)}} \,, \qquad (18.27)$$

where $F_2(E)$ is the indefinite integral of $[-f_2(E)]$,

$$F_2(E) = - \int_\infty^E f_2(E') dE' \,, \qquad (18.28)$$

ϵ is the root of the extremum condition

$$\left\{ \frac{d}{dE} \left[\ln f_1(E) + \ln F_2\left(\frac{M^2}{4E}\right) \right] \right\}_{E=\epsilon} = 0 \,, \qquad (18.29)$$

and the quantity $w(\epsilon)$ is related to the second derivatives with respect to E at the extremum ϵ given by

$$w(\epsilon) = - \left\{ \frac{d^2}{dE^2} \left[\ln f_1(E) + \ln F_2\left(\frac{M^2}{4E}\right) \right] \right\}_{E=\epsilon} \,. \qquad (18.30)$$

We shall show in Supplement 18.3 that for the distributions of Eqs. (18.15), (18.17) and (18.21), we find

$$F_2(E) = \pm T \ln(1 \pm e^{-(E+\mu)/T}), \qquad (18.31)$$

where the upper sign is for $q\bar{q} \to s\bar{s}$ and the lower sign is for $gg \to s\bar{s}$. The quantity ϵ in Eq. (18.27) obtained from Eq. (18.29) is

$$\epsilon = \frac{M}{2}\left[\frac{1 \pm e^{-(\epsilon-\mu)/T}}{\ln(1 \pm e^{-(M^2/4\epsilon+\mu)/T})(1 \pm e^{(M^2/4\epsilon+\mu)/T})}\right]^{1/2} \qquad (18.32)$$

When $\mu = 0$ and $M \gg T$, an approximate solution of ϵ is

$$\epsilon \cong M/2, \qquad (18.33)$$

and the quantity in the square bracket is approximately unity. A more accurate solution of ϵ can be obtained from Eq. (18.32) by iteration, starting with the approximate solution of (18.33). The quantity $w(\epsilon)$ in Eq. (18.27) is

$$w = \frac{4}{MT}\left[\left(\frac{M}{2\epsilon}\right)^3 \frac{1 \pm [1+(\epsilon/2T)]e^{-(\epsilon-\mu)/T}}{(1 \pm e^{-(\epsilon-\mu)/T})\ln(1 \pm e^{-(M^2/4\epsilon+\mu)/T})(1 \pm e^{(M^2/4\epsilon+\mu)/T})}\right]. \qquad (18.34)$$

For the case of $\mu = 0$ and $M \gg T$, an approximate value of w is

$$w \cong \frac{4}{MT}. \qquad (18.35)$$

What is the rate of change of the strange quark density n_s? Because the number of strange quarks N_s is the same as the number of $s\bar{s}$ pairs, the density n_s of produced strange quarks, dN_s/d^3x, is equal to $dN_{s\bar{s}}/d^3x$. Therefore, Eq. (18.27) can be rewritten to represent the rate of strange quark density change as given by

$$\frac{dn_s}{dM^2 dt} = N\frac{\sigma(M)M^2}{2(2\pi)^4}f_1(\epsilon(M))F_2\left(\frac{M^2}{4\epsilon(M)}\right)\sqrt{\frac{2\pi}{w(\epsilon)}}, \qquad (18.36)$$

The rate of change of the strange quark density can be obtained by integrating the above equation with respect to M^2,

$$\frac{dn_s}{dt} = N\int dM^2 \frac{\sigma(M)M^2}{2(2\pi)^4}f_1(\epsilon(M))F_2\left(\frac{M^2}{4\epsilon(M)}\right)\sqrt{\frac{2\pi}{w(\epsilon)}}. \qquad (18.37)$$

With the cross sections given explicitly as a function of M as in Eqs. (18.8) and (18.12), we can carry out the integration over M^2 numerically.

There is however a simplifying approximate expression for the case of $\mu = 0$ and $T \ll 2m_Q$. Because the cross section $\sigma(M)$ is zero below the threshold energy $2m_Q$ (see Fig. 18.4), the contribution to the integral comes from energies $M > 2m_Q$. We are justified to consider $T \ll M$ when $T \ll 2m_Q$. Then, for the case $\mu = 0$, we have $\epsilon \cong M/2$ and $w \cong 4/MT$ and Eq. (18.36) becomes

$$\frac{dn_s}{dM^2 dt} \cong N \frac{\sigma(M)M^2}{2(2\pi)^4} T^2 \frac{[\pm \ln(1 \pm e^{-M/2T})]}{e^{M/2T} \pm 1} \sqrt{\frac{\pi M}{2T}}. \qquad (18.38)$$

The rate of strange quark density change is

$$\frac{dn_s}{dt} \cong N_i \int dM^2 \frac{\sigma_i(M)M^2}{2(2\pi)^4} T^2 \frac{[\pm \ln(1 \pm e^{-M/2T})]}{e^{M/2T} \pm 1} \sqrt{\frac{\pi M}{2T}}. \qquad (18.39)$$

The result is not as simple when $\mu \neq 0$ and T is not small. It becomes necessary to evaluate ϵ and w numerically by Eqs. (18.32) and (18.34) to obtain the rate of change of the strangeness density.

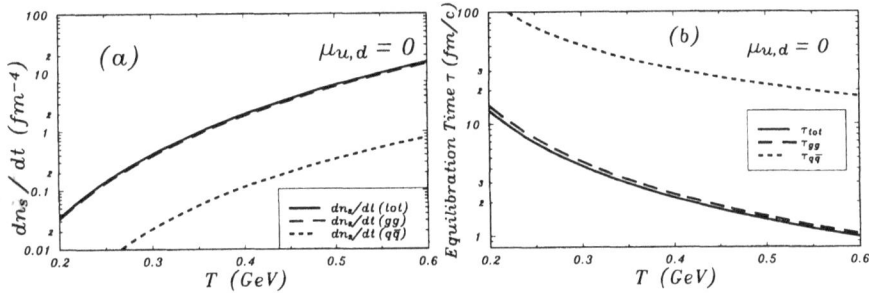

Fig. 18.5 (a) The rate of strangeness density change dn_s/dt as a function of the temperature T. The solid curve gives the total rate, while the long-dashed and the short-dashed curves give the rate from the $gg \to s\bar{s}$ process and the $q\bar{q} \to s\bar{s}$ process, respectively. (b) The equilibration time τ as a function of the temperature T. The long-dashed and the short-dashed curves give the equilibration time for the $gg \to s\bar{s}$ process and the $q\bar{q} \to s\bar{s}$ process, respectively. The solid curve shows the equilibration time τ_{tot} when both processes are considered.

With chemical potentials $\mu_{u,d} = 0$, we find the rate of strangeness density change as shown in Fig. 18.5a obtained by using Eqs. (18.28)-(18.34) and (18.37). As expected, the rate for the $gg \to s\bar{s}$ production process is much greater than the rate for the $q\bar{q} \to s\bar{s}$ process. The total rate of change of the strangeness density comes predominantly from the $gg \to s\bar{s}$ process. Therefore, the dynamics of the change of the strangeness content is dominated by the collision of gluons [2]-[7].

The strange quark density n_s at chemical equilibrium as obtained from Eq. (18.5) is exhibited in Fig. 18.2. If one starts from a zero strange quark density, then in order to reach the equilibrium density n_s(chemical equilibrium) with a rate of change of dn_s/dt, the equilibration time constant τ is given by

$$(\text{equilibration time } \tau) = \frac{n_s(\text{chemical equilibrium})}{dn_s/dt}. \qquad (18.40)$$

This gives the order of magnitude of the time scale for a quark-gluon plasma, initially void of strange quarks and antiquarks, to approach a chemical equilibrium involving strange quarks. The effects of the reverse reaction and Pauli exclusion, which occur when the strange quark density become substantial, will modify the magnitude of the equilibration time. The use of a different value of the strong coupling constant or a running coupling constant will also modify the magnitude of the equilibration time. Nevertheless, the estimate (18.40) gives the order of magnitude of the time scale for the approach to strangeness equilibrium.

The knowledge of the equilibrium strange quark density and the rate of change of the strange quark density in Fig. 18.5a allows one to calculate the equilibration time τ from Eq. (18.40). The result is shown in Fig. 18.5b. As one observes, τ is of the order of 10 fm/c for a temperature of about 200 MeV, it decreases to a few fm/c for $T = 300$ MeV. Because the collision process in a heavy nucleus-nucleus collision takes place within a time scale of 5-10 fm/c, we can conclude that the equilibration of strangeness may not be complete for temperatures of about 200 MeV, but may be close to completion if the temperature of the plasma reaches about 400 MeV.

The rate of strangeness production arising from gluon collisions is independent of the chemical potential $\mu_{u,d}$, while the rate from the $q\bar{q} \to s\bar{s}$ decreases as $\mu_{u,d}$ increases. Because of the dominance of the strange production from the gluon collision process, the total production rate is insensitive to the chemical potential $\mu_{u,d}$. The relation between temperature and strangeness equilibration given above for $\mu_{u,d} = 0$ is therefore also applicable to the case for $\mu_{u,d} \neq 0$. That is, even for a baryon-rich quark-gluon plasma, the equilibration of strangeness may not be complete for temperatures of about 200 MeV, but may be close to completion if the temperature of the plasma reaches about 400 MeV.

⊕〖 Supplement 18.3

We shall show here how Eqs. (18.31)-(18.35) are obtained. We have $f_i(E)$ given by

$$f_1(E) = \frac{1}{e^{(E-\mu)/T} \pm 1}, \tag{1}$$

and

$$f_2(E) = \frac{1}{e^{(E+\mu)/T} \pm 1}, \tag{2}$$

where the upper sign is for the process $q\bar{q} \to s\bar{s}$ while the lower sign is for the process $gg \to s\bar{s}$ for which μ is zero. Consequently, the function $F_2(E) = -\int_\infty^E f_2(E')dE'$ is

$$F_2(E) = -\int^E \frac{dE'}{e^{(E+\mu)/T} \pm 1}$$
$$= \pm T \ln(1 \pm e^{-(E+\mu)/T}). \tag{3}$$

We need to find the location ϵ which is the root of the extremum condition (18.29),

$$\left\{ \frac{d}{dE} \left[\ln f_1(E) + \ln F_2(\frac{M^2}{4E}) \right] \right\}_{E=\epsilon} = 0. \tag{4}$$

We can evaluate the first derivatives and find

$$\frac{d}{dE} \ln f_1(E) = \frac{-1}{T(1 \pm e^{-(E-\mu)/T})}, \tag{5}$$

and

$$\frac{d}{dE} \ln F_2\left(\frac{M^2}{4E}\right) = \frac{1}{\pm \ln(1 \pm e^{-(M^2/4E+\mu)/T})} \frac{1}{(\pm 1 + e^{(M^2/4E+\mu)/T})} \frac{M^2}{4E^2 T}. \tag{6}$$

The equation for ϵ is then

$$\frac{1}{T(1 \pm e^{-(E-\mu)/T})} = \frac{1}{\pm \ln(1 \pm e^{-(M^2/4E+\mu)/T})} \frac{1}{(\pm 1 + e^{(M^2/4E+\mu)/T})} \frac{M^2}{4E^2 T}\bigg|_{E=\epsilon}. \tag{7}$$

The above equation can be written as

$$\epsilon = \frac{M}{2} \left[\frac{1 \pm e^{-(\epsilon-\mu)/T}}{\ln(1 \pm e^{-(M^2/4\epsilon+\mu)/T})(1 \pm e^{(M^2/4\epsilon+\mu)/T})} \right]^{1/2}. \tag{8}$$

For the case when $\mu = 0$ and $M \gg T$, the above equation is satisfied when

$$\epsilon = \frac{M}{2},$$

and the quantity in the square bracket of Eq. (8) is approximately unity. Eq. (8) has been written in such a form that it can be used to obtain a more accurate value of ϵ by iteration, using an approximate solution on the righthand side to give a new solution for the quantity ϵ on the lefthand side. The iteration can be started by substituting $\epsilon = M/2$ on the righthand side, and can be continued by taking an improved solution for ϵ as the average of the old and the new solution for ϵ.

Having shown how we can find the quantity ϵ, we need to obtain $w(\epsilon)$ which involves the second derivative of $[\ln f_1(E)]$ and $[\ln F_2(M^2/4E)]$ with respect to E at the extremum,

$$w(\epsilon) = -\left\{\frac{d^2}{dE^2}\left[\ln f(E) + \ln F\left(\frac{M^2}{4E}\right)\right]\right\}_{E=\epsilon}. \tag{9}$$

The second derivative of $[\ln f_1(E)]$ is

$$\frac{d^2}{dE^2}\ln f_1(E) = \frac{\mp e^{-(E-\mu)/T})^2}{T^2(1 \pm e^{-(E-\mu)/T}}, \tag{10}$$

and the second derivative of $[\ln F_2(M^2/4E)]$ with respect to E is

$$\frac{d^2}{dE^2}\ln F_2\left(\frac{M^2}{4E}\right)$$

$$= \frac{-1}{\pm[\ln(1 \pm e^{-(M^2/4E+\mu)/T})]^2}\frac{\pm e^{-(M^2/4E+\mu)/T}}{(1 \pm e^{-(M^2/4E+\mu)/T})}\frac{1}{(\pm 1 + e^{(M^2/4E+\mu)/T})}\left(\frac{M^2}{4E^2T}\right)^2$$

$$+ \frac{1}{\pm\ln(1 \pm e^{-(M^2/4E+\mu)/T})}\frac{e^{(M^2/4E+\mu)/T}}{(\pm 1 + e^{(M^2/4E+\mu)/T})^2}\left(\frac{M^2}{4E^2T}\right)^2$$

$$+ \frac{1}{\pm\ln(1 \pm e^{-(M^2/4E+\mu)/T})}\frac{1}{(\pm 1 + e^{(M^2/4E+\mu)/T})}\left(\frac{-2M^2}{4E^3T}\right)$$

$$= \frac{-1 \pm \ln(1 \pm e^{-(M^2/4E+\mu)/T})e^{(M^2/4E+\mu)/T}}{[\ln(1 \pm e^{-(M^2/4E+\mu)/T})]^2(1 \pm e^{(M^2/4E+\mu)/T})^2}\left(\frac{M^2}{4E^2T}\right)^2$$

$$+ \frac{1}{T\ln(1 \pm e^{-(M^2/4E+\mu)/T})}\frac{1}{(1 \pm e^{(M^2/4E+\mu)/T})}\left(\frac{-M^2}{2E^3}\right).$$

The first term is approximately zero since $\ln(1 \pm e^{-}x)e^x \sim \pm 1$ for large values of x. Eq. (9) becomes

$$-w = \frac{\mp e^{-(\epsilon-\mu)/T}}{T^2(1 \pm e^{-(\epsilon-\mu)/T})^2} + \frac{1}{T\ln(1 \pm e^{-(M^2/4\epsilon+\mu)/T})}\frac{1}{(1 \pm e^{(M^2/4\epsilon+\mu)/T})}\left(\frac{-M^2}{2\epsilon^3}\right).$$

Using Eq. (8), we can write the above equation as

$$-w = \left[\frac{\mp e^{-(\epsilon-\mu)/T}}{1 \pm e^{-(\epsilon-\mu)/T}}\frac{M^2}{4\epsilon^2T} - \frac{M^2}{2\epsilon^3}\right]\frac{1}{T\ln(1 \pm e^{-(M^2/4\epsilon+\mu)/T})(1 \pm e^{(M^2/4\epsilon+\mu)/T})}.$$

The quantity w is therefore equal to

$$w = \frac{M^2}{2\epsilon^3T}\left[1 \pm \frac{\epsilon}{2T}\frac{e^{-(\epsilon-\mu)/T}}{1 \pm e^{-(\epsilon-\mu)/T}}\right]\frac{1}{\ln(1 \pm e^{-(M^2/4\epsilon+\mu)/T})(1 \pm e^{(M^2/4\epsilon+\mu)/T})}$$

$$= \frac{4}{MT}\left[\left(\frac{M}{2\epsilon}\right)^3\frac{1 \pm (1 + \epsilon/2T)e^{-(\epsilon-\mu)/T}}{(1 \pm e^{-(\epsilon-\mu)/T})\ln(1 \pm e^{-(M^2/4\epsilon+\mu)/T})(1 \pm e^{(M^2/4\epsilon+\mu)/T})}\right]. \tag{11}$$

After the value of ϵ has been obtained by using Eq. (8), the above equation can be used to evaluate the value of w. The quantity in the square bracket is of the order of unity and an approximate value of w is $4/MT$. The expressions for $F_2(E)$, $\epsilon(M)$ and $w(\epsilon)$ in Eqs. (3), (8), and (11) are the equations which had to be proved.]⊕

§18.3 Experimental Information on Strangeness Production

Experiments to study strangeness production in high-energy heavy-ion collisions have been carried out by many groups: E802 [9,21,22], and E810 [23] at BNL, and NA34 [24,25], NA35 [26,27], NA38 [28] and WA85 [29] at CERN. We shall discuss some of the most notable results from these investigations.

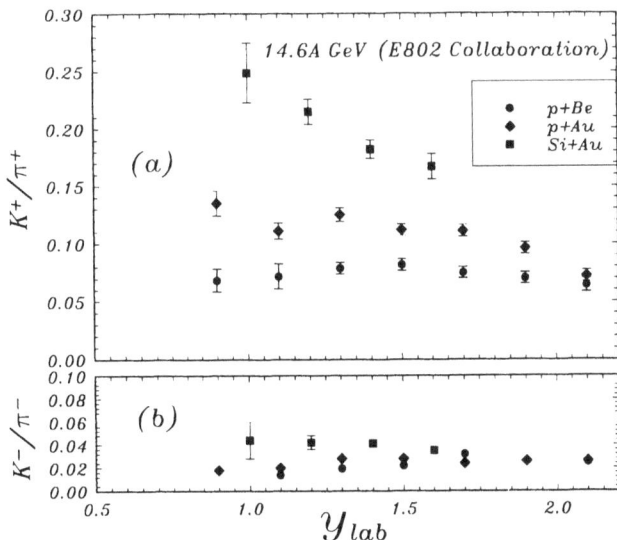

Fig. 18.6 The ratios (a) K^+/π^- and (b) K^-/π^- as a function of rapidity in the laboratory system for various projectile and target combinations at $14.6A$ GeV. The data are from the E802 Collaboration [9,21,22].

As the mass number increases, the number of produced pions and kaons increase. However, the number of K^+ increases more rapidly than the number of π^+ while the number of K^- increases only very slightly faster than the number of π^-. Figs. 18.6a and 18.6b show the K^+/π^+ and K^-/π^- ratios for p+Be, p+Au and Si+Au collisions at

14.6A GeV from the E802 Collaboration. The data indicate that the K^+/π^+ ratio for $p+^9$Be is about 0.08 over a large range of rapidity. The K^+/π^+ ratio is much enhanced for p+Au and for Si+Au. The enhancement is greater in the lower rapidity region as compared to the higher rapidity region. For the same target nucleus, the enhancement increases as the mass of the projectile increases. In the collision of Si on Au measured at $y_{lab} = 1$, the ratio K^+/π^+ increases by a factor of about 3 over that of $p+^9$Be.

Fig. 18.6b shows the K^-/π^- ratio for various projectile and target combinations as a function of rapidity. While the K^-/π^- ratio ranges from 0.02 to 0.04 in p+Be collisions, the K^-/π^- ratio is not much enhanced for p-Au collisions, and is only slightly increased for Si+Au collisions. Because the yield of π^- is only slightly different from the yield of π^+, the data in Fig. 18.6b show that the yield of K^+ is quite different from the yield of K^- in nucleus-nucleus collisions.

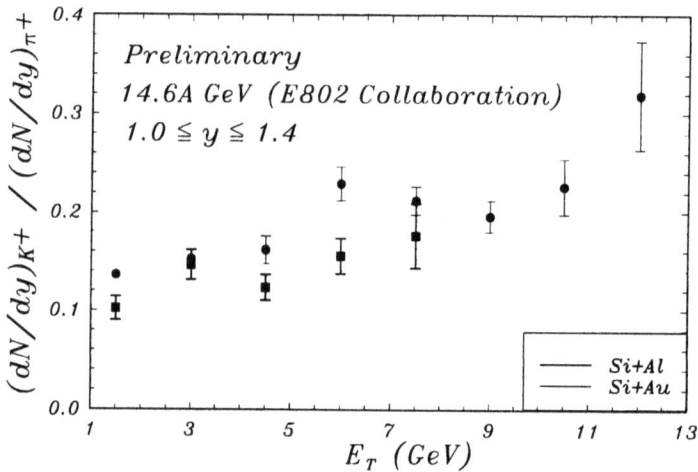

Fig. 18.7 The ratio of $(dN/dy)_{K^+}/(dN/dy)_{\pi^+}$ for collisions of Si on Al and Au at 14.6A GeV, in the interval $1.0 \leq y \leq 1.4$. The preliminary data are from the E802 Collaboration [9].

In Fig. 18.7, we show the ratio $(dN/dy)_{K^+}/(dN/dy)_{\pi^+}$ for Si+Au and Si+Al in the rapidity region $1 \leq y_{lab} \leq 1.4$, as a function of the transverse energy E_T. One observes that the K^+/π^+ ratio increase as a function of transverse energy, indicating that the K^+/π^+ ratio is correlated with the degree of nuclear overlap. The greater the degree of overlap of the two nuclei, the greater is the ratio K^+/π^+.

The results of the above figures may at first be interpreted as the possible production of the quark gluon plasma with enhanced content of strange quarks and antiquarks. In a quark-gluon plasma with a net baryon content, the plasma is rich in u and d quarks, and depleted in $\bar{u}$ and $\bar{d}$, with the density of s and $\bar{s}$ intermediate in between, as shown in Fig. 18.2 for $\mu_{u,d} \neq 0$. In this environment of a dense baryon-rich quark-gluon plasma, which may occur in collisions in the stopping regime, it is likely for the $\bar{s}$ quark to find a u quark or a d quark to form a K^+ meson or a K^o meson. It is however difficult for the $\bar{s}$ quark to find a $\bar{u}$ or a $\bar{d}$ quark to form a K^- or a $\overline{K}^o$ meson. Hence, the number of K^+ is much enhanced compared to numbers of K^-. There are however alternative explanations in terms of hadron scenarios [8,12,13] or double strings [30]. We can expect some difference in the K^+ and K^- yields from hadron interactions in the baryon-rich region, without the assumption of a baryon-rich quark-gluon plasma. Kaons and Λ particles can be produced by nucleon-nucleon and pion-nucleon collisions via the reactions

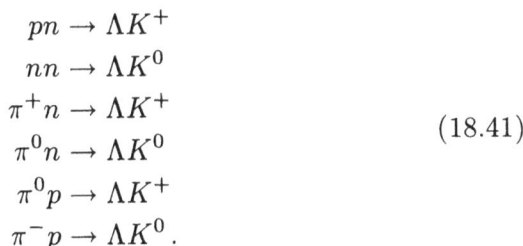

$$pn \to \Lambda K^+$$
$$nn \to \Lambda K^0$$
$$\pi^+ n \to \Lambda K^+$$
$$\pi^0 n \to \Lambda K^0 \tag{18.41}$$
$$\pi^0 p \to \Lambda K^+$$
$$\pi^- p \to \Lambda K^0 \, .$$

For Si projectiles on heavy targets, the peak of the pion rapidity distribution occurs not at midrapidity but at a rapidity shifted from the midrapidity region towards the baryon-rich target rapidity region. In this region, the presence of pions and nucleons near the target rapidity region enhances the production of K^+, Λ and K^o, by the nucleon-nucleon and pion-nucleon reactions (18.41). Thus, the occurrence of secondary collisions in the baryon-rich environment after the primary collisions will lead to an enhanced production of K^+, Λ, and K^o. Kaons can also be produced by the collision of energetic pions via the reactions

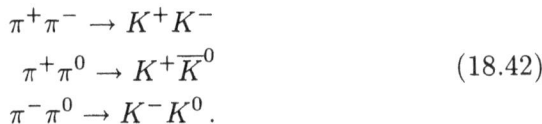

$$\pi^+ \pi^- \to K^+ K^-$$
$$\pi^+ \pi^0 \to K^+ \overline{K}^0 \tag{18.42}$$
$$\pi^- \pi^0 \to K^- K^0 \, .$$

On the other hand, K^- mesons and $\overline{K}^o$ are readily absorbed. Their densities can be depleted by the reactions

$$K^- p \to \Lambda \pi^0$$
$$\overline{K}^0 n \to \Lambda \pi^0, \qquad (18.43)$$

which release energy. Because these reactions are exothermic reactions which occur without an energy threshold, they occur readily. The K^- and $\overline{K}^o$ particles which are produced in primary collisions or in secondary reactions (18.42) may be absorbed by interacting with nucleons to be converted into Λ hyperons. These reactions deplete the numbers of K^- and $\overline{K}^o$ particles. The result of these reactions is to lead to an enhancement of K^+, Λ_0 and K^o, but a depletion of K^- and $\overline{K}^o$. Quantitative calculations based on these pictures is however not sufficient to describe the experimental K^+ enhancement and additional ingredients are needed [8,12,13]. It has been proposed that the dominant factor for the enhancement of K^+ production arises from the lowering of the meson masses as the temperature approaches the phase transition temperature [8], from the inclusion of Δ-Δ interactions [12], or from (higher meson resonance)-nucleon interactions [13]. The subject matter remains under debate and is the subject of current research. Interested readers may consult Refs. [8], [12] and [13] for further details.

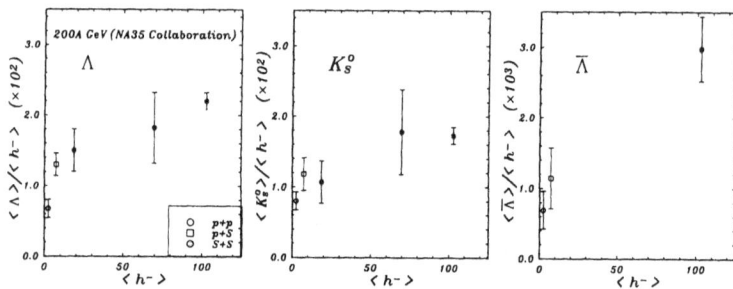

Fig. 18.8 The ratio $< \Lambda > / h^- >$, $< K_s^o > / h^- >$, $< \overline{\Lambda} > / < h^- >$, for various nucleus-nucleus collisions at 200A GeV as a function of $< h^- >$. The data are from the NA35 Collaboration [26].

The NA35 Collaboration studied the production of Λ, K_s^o, and $\overline{\Lambda}$ for the reaction S+S at 200 GeV per projectile nucleon at CERN [27]. The ratios $< \Lambda > / < h^- >$, $< K_s^o > / < h^- >$ and $< \overline{\Lambda} > / < h^- >$ are exhibited in Fig. 18.8. These ratios show that as the number of negative particles (mostly pions) increases, the production of Λ, K_s^o, and $\overline{\Lambda}$ increases. They reach values which are 2-3 times that expected from nucleon-nucleon collisions. The origin of the enhancement of Λ and $\bar{\Lambda}$ is not completely understood. There is a recent proposal to explain the enhancement of $\bar{\Lambda}$ in terms of the lowering of the baryon mass in dense matter [31].

The production of Ξ and $\bar{\Xi}^-$ has also been observed by the WA85 collaboration in $p+$W and S$+$W collisions at 200A GeV. The transverse momentum distribution can be represented by $dN/dm_T^2 \propto e^{-m_T/T}$, with slope parameters T in the range of 200-250 MeV. Over the rapidity range of $2.3 < y < 3.0$ and $m_T > 1.72$ GeV for the collision of S on W, the measurements yield the ratios given in Table 18.2.

Table 18.2

Ratio	S+W data [29]
$\bar{\Lambda}/\Lambda$	0.13±0.03
$\bar{\Xi}^-/\Xi^-$	0.39±0.07
Ξ^-/Λ	0.20±0.04
$\bar{\Xi}^-/\bar{\Lambda}$	0.6±0.2

A preliminary study of the production of Ξ particles gives the ratio of $\bar{\Xi}^-/\Xi = 0.27 \pm 0.06$ for p+W collisions, not corrected for acceptance or reconstructed efficiency [32]. The ratio $\bar{\Xi}^-/\Xi = 0.39 \pm 0.07$ for S+W collisions appear to be much higher than in $p+$W reactions. The ratio of $(\Xi^- + \bar{\Xi}^-)/(\Lambda + \bar{\Lambda})$ for e^+e^- and $\bar{p}p$ collisions and the ratio $\bar{\Xi}^-/\bar{\Lambda}$ for pp collisions are all about 0.1 [32,33]. Thus, the ratio $\bar{\Xi}^-/\bar{\Lambda} = 0.60 \pm 0.20$ for S+W collisions is also very much higher. There appears to be some degree of flavor equilibration. How these observations can fit into the picture of the dynamics of a hadron system or alternatively the quark-gluon plasma remains to be seen.

§**References for Chapter 18**

1. J. Rafelski, Phys. Rep. 88, 331 (1982); excellent reviews on the subject of strangeness production can be found in Ref. 2, 3 and 4.

2. J. Rafelski, Nucl. Phys. A544, 279c (1992).
3. H. C. Eggers and J. Rafelski, Int. J. Mod. Phys. A6, 1067 (1991).
4. P. Koch, B. Müller and J. Rafelski, Phys. Rep. 142, 167 (1986).
5. T. Matsui, B. Svetitsky and L. D. McLerran, Phys. Rev. D34, 783 (1986).
6. T. Matsui, B. Svetitsky and L. D. McLerran, Phys. Rev. D34, 2047 (1986).
7. C. Greiner, P. Koch, and H. Stöcker, Phys. Rev. Lett. 58, 1825 (1987), T. S. Biro, P. Lévai, and B. Müller, Phys. Rev. D42, 3078 (1990).
8. C. M. Ko and L. H. Xia, Nucl. Phys. A498 561C-566C (1989); G. E. Brown, C. M. Ko, Z. G. Wu, and L. H. Xia, Phys. Rev. C43, 1881 (1991).
9. S. Nagamiya, Nucl. Phys. A544, 5c (1992).
10. G. Odyniec, in Proceedings of International Workshop on Quark Gluon Signatures, Strassburg, France, 1990.
11. J. Satchel and G. R. Young, Ann. Rev. Nucl. Part. Sci. 42, 537 (1992).
12. Y. Pang, T. J. Schlagel, and S. H. Kahana, Phys. Rev. Lett. 18, 2743 (1992).
13. R. Matiello et al, Nucl. Phys. B24, 221 (1991); H. Sorge, A. von Keitz, R. Mattiello, H. Stöcker, and W. Greiner, Phys. Lett. B243, 7 (1990).
14. J. I. Kapusta, *Finite-tempertaure Field Theory*, Cambridge University Press, 1989, page 7.
15. M. Abramowitz and I. A. Stegun, *Handbook of Mathematical Tables*, Dover Publications, New York, 1965.
16. Particle Data Group, Review of Particle Properties, Phys. Lett. 204, 1 (1988).
17. A. Shor, Phys. Rev. Lett. 54, 1122 (1985).
18. L. D. Landau and E. M. Lifshitz, *Statistical Physics*, Pergamon Press, Oxford, Third Edition, 1980.
19. M. Glück and E. Reya, Phys. Lett. 79B, 453 (1978).
20. B. L. Combridge, Nucl. Phys. B151, 429 (1979).
21. T. Abbott et al, E802 Collaboration, Phys. Rev. Lett. 66, 1567 (1991).
22. W. A. Zajc et al, E802 Collaboration, Nucl. Phys. A544, 237c (1992).
23. K. J. F. Foley, et al, E810 Collaboration, Nucl. Phys. A544, 335c (1992).
24. H. van Hecke et al, HELIOS Collaboration, Nucl. Phys. A525, 227c (1991). (Menton)
25. M. A. Mazzoni et al, HELIOS Collaboration, Nucl. Phys. A544, 623c (1992). (Menton)
26. J. Bartke et al, NA35 Collaboration, Zeit. Phys. C48, 191 (1990).
27. P. Seyboth et al, NA35 Collaboration, Nucl. Phys. A544, 293c (1992); M. Kowalski et al, NA35 Collaboration, Nucl. Phys. A544, 609c (1992).
28. M. C. Abreu et al, NA38 Collaboration, Nucl. Phys. A544, 209c (1992); R. Ferreira, NA38 Collaboration, Nucl. Phys. A544, 497c (1992).
29. S. Abatzia et al, WA85 Collaboration, Nucl. Phys. A544, 321c (1992).
30. K. Werner and J. Aichelin, Phys. Lett. B308, 372 (1993).
31. C. M. Ko, M. Asakawa, and P. Lévai, Phys. Rev. C46, 1072 (1992).
32. S. Abatzia et al, WA85 Collaboration, Nucl. Phys. A525, 441c (1992).
33. T. Akesson et al., AFS Collaboration, Nucl. Phys. B246, 1 (1984); R. E. Ansorge et al., UA5 Collaboration, CERN-EP/89-41; M. Althhoff et al., TASSO Collaboration, Phys. Lett. 130B, 340 (1983); S. R. Klein et al., MARK II Collaboration, Phys. Rev. Lett. 58, 664 (1987); H. Yamamoto et al., TPC Collaboration, Proc. Recontres de Moriond (1986); M. S. Alam et al., CLEO Collaboration, Phys. Rev. Lett. 53, 24 (1984); A. Albrecht et al., ARGUS Collaboration, Phys. Lett. 183B, 419 (1987); S. Abachi et al., HRS Collaboration, Phys. Rev. Lett. 58, 2627 (1987).

19. Summary

The physics of high-energy heavy-ion collisions is an emerging field of research. It encompasses the study of strongly interacting systems in thermal and chemical equilibrium, and the dynamics of these systems in energetic collisions or in nonequilibrium evolution. This book has been written to provide the background material for the major topics in this field, at an elementary level. These topics are summarized in this chapter. The summary also serves as an introduction to the contents of the book.

The first chapter explains the reasons why there is much optimism for the possibility of creating regions of very high energy density using high-energy heavy-ion collisions. It is pointed out that an important characteristic of nuclear collision is the large amount of longitudinal energy which may be converted into the production of particles, with the energy deposited in a small region of space within a short time. The mechanism for such an occurrence renders itself useful to explore unknown regions of matter at very high energy densities and temperatures. On the other hand, theoretical studies of strongly interacting matter at high temperatures and densities suggest that as the temperature increases or as the baryon density increases, there may be a confinement and deconfinement (or symmetry-breaking and symmetry-restoring) phase transition, It is therefore of interest to use high-energy heavy-ion collisions to study the physics of symmetry breaking, and the deconfinement transition of quarks.

In nuclear collisions at high energies, constituents of the colliding particles actively participate in the collision process and a large number of particles are produced. In Chapter 2, we introduce kinematic variables which are used to describe the constituents and the produced particles. The light-cone variable, the Feynman scaling variable, the rapidity variable, and the pseudorapidity variable are studied, as well as the mutual relations between them. These variables allows us to describe the dynamics of nuclear collisions and the intrinsic structure of hadrons.

Nucleon-nucleon collisions provide valuable information for the discussion of nucleus-nucleus collisions. Making use of the kinematic variables introduced in Chapter 2, we study the characteristics of nucleon-nucleon collisions in Chapter 3. We note that nucleon-nucleon collisions are highly inelastic and that the cross section for losing a large fraction of the longitudinal energy in a nucleon-nucleon collision is large. The loss in longitudinal energy is converted into the energy of the produced particles. We discuss the rapidity distribution and the transverse momentum distribution of the produced particles and the concept of the transverse mass and the m_T scaling of the produced particles.

An important part of the hadron collision dynamics occurs in hard processes at the constituent level. They are specially significant for the production of particles with transverse momentum much greater than 1 GeV/c, or with a momentum fraction x close to unity. To discuss the dynamics of hard processes, we introduce in Chapter 4 the infinite momentum frame and use it to examine parton-parton dynamics in the relativistic hard-scattering model for hadron-hadron collisions. The light-cone variable which was introduced earlier in Chapter 2 plays a natural role in the infinite momentum frame. The counting rules for the structure function of hadrons at very high momenta are derived to provide a simple explanation of the form of the structure function often encountered in the literature. We discuss some simple applications of the hard-scattering model and the direct fragmentation model.

In inelastic hadron collisions, the majority of particles produced are in the middle rapidity region and have an average transverse momentum of about 0.3 GeV. These particles are soft particles. The process of soft particle production is associated with a length scale that is large in the context of quantum chromodynamics. The dynamics of soft particle production belongs to the realm of nonperturbative QCD. The lack of nonperturbative solutions of quantum chromodynamics for the soft particle production process leads to a proliferation of models. In Chapters 5 through 8, we discuss different theoretical models which have been proposed to examine the process of soft particle production in e^+e^- annihilation and nuclear collisions.

In Chapter 5, we examine the Schwinger particle production mechanism in which particle-antiparticle pairs are produced spontaneously in the presence of a strong linear potential, which is assumed to represent the state of the system as a quark and an antiquark pull apart from each other. The rate of particle production is written in terms of the mass of the produced particle and the strength of the field, which is related to the string tension coefficient.

In Chapter 6, we study the process of particle production in massless two-dimensional quantum electrodynamics (QED_2), based on the model first proposed by Schwinger and studied extensively by Casher, Kogut, and Suskind and by Bjorken. The fermions in QED_2 play the role of quarks in QCD_4 and the stable bosons in QED_2 play the role of pions in QCD_4. We study the nature and the origin of the bound bosons in QED_2 and examine the dynamics of the produced boson field when a fermion-antifermion pair is allow to pull apart from each other. The dynamics of the produced bosons in QED_2 is best described by the inside-outside cascade description of particle production. The rapidity distribution of the produced bosons have features which mimic those observed in high-energy collision processes.

In Chapter 7, we discuss another model of particle production based on the concept of classical strings. In this model, massless quarks reside at the end points of strings. A stable particle is depicted as a basic unit of a yo-yo state in which the string undergoes stable contraction and expansion. Particle production occurs when the end points of a string are stretched with new quark-antiquark pairs being produced at points on the string, resulting in the breaking of the string into many shorter segments. The phenomenological Lund model of string fragmentation provides a quantitative description of the production mechanism.

In Chapter 8, we examine the dual parton model of particle production. In this model, a hadron is described as consisting of dual partons which carry the momentum of the parent hadron. The nonperturbative property of the scattering of hadrons, which comes from the dual resonance model of hadrons, are exploited to provide information on the momentum distribution of the partons. In addition, by using the optical theorem, those diagrams which describe the hadron-hadron elastic amplitude in the dual resonance model are used to infer the nature of particle production.

In subsequent Chapters 9, 10, and 11, we turn our attention to the equilibrium states of a system of quarks, antiquarks, and gluons at high temperatures and/or high baryon densities. To provide intuitive understanding of the possible states of a quark-gluon plasma, we examine in Chapter 9 the bag model of hadrons and the dependence of the pressure of a quark-gluon system on temperature and quark chemical potential. By comparing the quark-bag pressure and the pressure of the degenerate quark-gluon matter, one finds that there is a temperature or a baryon density beyond which the bag pressure can be overwhelmed by the pressure of the quark-gluon matter. A transition from one form of matter to another form is therefore made plausible when the temperature or the baryon density increases.

Although qualitative and intuitive considerations in Chapter 9 suggest the possible transition to a new form of matter consisting of quarks, antiquarks, and gluons – the quark-gluon plasma, a better theoretical model is needed to provide a more reliable way to assess the equilibrium states of a quark-gluon system. As the gauge theory of QCD is generally considered to be the correct theory for the interaction of quarks and gluons, it is necessary to study the states of the quark-gluon matter using the theory of QCD. A concrete nonperturbative implementation of the theory of QCD is the lattice gauge theory. We introduce the formulation of the lattice gauge theory in Chapter 10. The basic concept of the path integral is used for a single particle in a field and generalized to the partition function in quantum chromodynamics. In this generalization from the path integral to the partition function, it is more convenient to work with the Euclidean action in Euclidean space than the Minkowski action in Minkowski space. We compare and contrast the basic quantities in Euclidean space and in Minkowski space so as to provide the basic tools to understand the lattice gauge theory.

Some of the results of lattice gauge theory are discussed in Chapter 11. How a linear confining potential can arise between a quark and an antiquark in the lattice gauge theory is studied by examining the Wilson loop parameter. Since the order of a phase transition is an important concept, we illustrate the origin of different orders of phase transitions with the analogous spin system or gauge system in the mean-field approximation. Numerical results in lattice QCD reveal the possible transition from the confined hadron phase to the quark-gluon plasma phase as the temperature increases.

After studying the static properties of a system of quarks, antiquarks and gluons at high temperatures and densities, we seek ways to produce quark-gluon matter in the deconfined quark-gluon plasma phase. In the laboratory, nucleus-nucleus collisions at very high energies provide a promising way to produce high temperature or high density matter. They can be utilized to explore the existence of the quark-gluon plasma. In Chapter 12, we examine the geometrical aspect in nucleus-nucleus collisions based on the Glauber model of mean-free path and multiply collisions. The occurrence of multiple collision allows the conversion of a large fraction of the energy of a nucleus from the longitudinal motion to the energy deposited in the vicinity of the center of mass of the colliding system.

The occurrence of multiple collisions is accompanied by a high degree of stopping of nuclear matter in nuclear collision. In Chapter 13, we discuss the evidence of nuclear stopping and the deposition of energy in the central rapidity region in nucleus-nucleus collisions. Bjorken's estimate of energy density in nucleus-nucleus collisions is derived and the energy density is found to be quite large in energetic nucleus-nucleus collisions. The magnitude of the energy density is of the same order as the energy density one expects for the confinement-deconfinement phase transition to occur. We examine the dynamics of the quark-gluon plasma following the simple hydrodynamical model of Bjorken.

However favorable theoretical predictions and optimistic experimental assessments from initial findings may turn out to be, the search for the existence of the quark-gluon plasma is ultimately an experimental undertaking. It is important to sharpen the experimental tools for its discovery and exploration. The next five chapters in the book introduce the reader to the experimental probes which have been suggested to study the possible production of the quark-gluon plasma.

Since leptons and photons interact with the particles in the collision region only through the electromagnetic interaction, the interaction is not strong. Consequently, the mean-free paths of the produced leptons and photons are quite large and they may not suffer further collisions after they are produced. Thus, they carry information on the thermodynamical state of the medium at the moment of their production. In Chapter 14, we discuss dilepton production to investigate the thermodynamical state of matter created in the collision process. The rates of dilepton production from the quark-gluon plasma and from other processes are worked out to provide the necessary tools to investigate the states of the dilepton sources.

In Chapter 15, we discuss the occurrence of J/ψ suppression. The physics of Debye screening is examined in the Abelian approximation for a simple model of the quark-gluon plasma, so as to bring out the salient features of the physics to the screening phenomenon. How this effect may lead to the dissociation of a J/ψ particle is studied. We subsequently examine other possible modes of J/ψ suppression based on J/ψ-hadron interactions.

In Chapter 16, we study the electromagnetic probe of photons to investigate the state of the system during the collision process. The rate of photon production in the quark-gluon plasma and in a hadron gas is worked out in detail to provide the formalism for its use in the search effort.

In Chapter 17, we discuss the use of the Hanbury-Brown-Twiss effect to study the space-time extent of the emission source at the late stages of the reaction process. The origin of the Hanbury-Brown-Twiss effect is discussed starting from first principles. The dependence of the two-body intensity correlations on the chaoticity or the coherence of the emitting sources are derived. The relationship between the correlation function and the phase-space distribution of the emitting source is worked out to provide a way to study the space-time dynamics of the evolving system. Recent experimental date on pion interferometry are reviewed.

In Chapter 18, we study the strangeness production in high-energy collisions. We first examine the densities of pions and kaons for a hadron gas in thermal and chemical equilibrium and note that there is an enhancement of kaons relative to pions if thermal and chemical equilibrium is reached. It is as yet not known whether chemical equilibrium can be reached for a hadron gas within the short time available during the collision. We next study a quark-gluon plasma in thermal and chemical equilibrium. The densities of quarks of various flavors as a function of temperature are then examined to provide a possible signature of quark-gluon plasma formation. There is a substantial enhancement of strangeness content in a quark-gluon plasma. Our investigation of the rate of strangeness production indicates that the equilibration of strangeness in a quark-gluon plasma may not be complete for temperatures of about 200 MeV, but may be close to completion if the temperature of the plasma reaches about 400 MeV. Recent experimental data on strangeness production are reviewed.

The present book has been written at a time when the research in the field of high-energy heavy-ion collisions is making rapid progress. With the availability of heavier beams and higher energies, man's aspiration to fathom the mystery of the early universe may be realized with the creation of matter which kindles the early moments of the universe. The advances in this field will push man's knowledge to new and unknown frontiers.

Index

ERRATA

INTRODUCTION
TO HIGH-ENERGY HEAVY-ION COLLISIONS

C. Y. WONG

p. 6, Ref. 2, Bubbia should read Rubbia.

p. 498, line 8 from bottom, $e^- x$ should read e^{-x}.

p. 500, Fig. 18.7, solid square points are data points for Si+Al and solid circle points are data points for Si+Au.

p. 501, Eq. (18.4), $pn \to \Lambda K^+$ should read $pn \to \Lambda K^+ n$, and $nn \to \Lambda K^0$ should read $nn \to \Lambda K^0 n$.